Vassil Sgurev, Mincho Hadjiski, and Janusz Kacprzyk (Eds.)

Intelligent Systems: From Theory to Practice

# Studies in Computational Intelligence, Volume 299


**Editor-in-Chief**

Prof. Janusz Kacprzyk
Systems Research Institute
Polish Academy of Sciences
ul. Newelska 6
01-447 Warsaw
Poland
*E-mail:* kacprzyk@ibspan.waw.pl


---

Further volumes of this series can be found on our homepage: springer.com

Vol. 278. Radomir S. Stankovic and Jaakko Astola
*From Boolean Logic to Switching Circuits and Automata,* 2010
ISBN 978-3-642-11681-0

Vol. 279. Manolis Wallace, Ioannis E. Anagnostopoulos, Phivos Mylonas, and Maria Bielikova (Eds.)
*Semantics in Adaptive and Personalized Services,* 2010
ISBN 978-3-642-11683-4

Vol. 280. Chang Wen Chen, Zhu Li, and Shiguo Lian (Eds.)
*Intelligent Multimedia Communication: Techniques and Applications,* 2010
ISBN 978-3-642-11685-8

Vol. 281. Robert Babuska and Frans C.A. Groen (Eds.)
*Interactive Collaborative Information Systems,* 2010
ISBN 978-3-642-11687-2

Vol. 282. Husrev Taha Sencar, Sergio Velastin, Nikolaos Nikolaidis, and Shiguo Lian (Eds.)
Intelligent Multimedia Analysis for Security Applications, 2010
ISBN 978-3-642-11754-1

Vol. 283. Ngoc Thanh Nguyen, Radoslaw Katarzyniak, and Shyi-Ming Chen (Eds.)
*Advances in Intelligent Information and Database Systems,* 2010
ISBN 978-3-642-12089-3

Vol. 284. Juan R. González, David Alejandro Pelta, Carlos Cruz, Germán Terrazas, and Natalio Krasnogor (Eds.)
*Nature Inspired Cooperative Strategies for Optimization (NICSO 2010),* 2010
ISBN 978-3-642-12537-9

Vol. 285. Roberto Cipolla, Sebastiano Battiato, and Giovanni Maria Farinella (Eds.)
*Computer Vision,* 2010
ISBN 978-3-642-12847-9

Vol. 286. Zeev Volkovich, Alexander Bolshoy, Valery Kirzhner, and Zeev Barzily
*Genome Clustering,* 2010
ISBN 978-3-642-12951-3

Vol. 287. Dan Schonfeld, Caifeng Shan, Dacheng Tao, and Liang Wang (Eds.)
*Video Search and Mining,* 2010
ISBN 978-3-642-12899-8

Vol. 288. I-Hsien Ting, Hui-Ju Wu, Tien-Hwa Ho (Eds.)
*Mining and Analyzing Social Networks,* 2010
ISBN 978-3-642-13421-0

Vol. 289. Anne Håkansson, Ronald Hartung, and Ngoc Thanh Nguyen (Eds.)
*Agent and Multi-agent Technology for Internet and Enterprise Systems,* 2010
ISBN 978-3-642-13525-5

Vol. 290. Weiliang Xu and John Bronlund
*Mastication Robots,* 2010
ISBN 978-3-540-93902-3

Vol. 291. Shimon Whiteson
*Adaptive Representations for Reinforcement Learning,* 2010
ISBN 978-3-642-13931-4

Vol. 292. Fabrice Guillet, Gilbert Ritschard, Henri Briand, Djamel A. Zighed (Eds.)
*Advances in Knowledge Discovery and Management,* 2010
ISBN 978-3-642-00579-4

Vol. 293. Anthony Brabazon, Michael O'Neill, and Dietmar Maringer (Eds.)
*Natural Computing in Computational Finance,* 2010
ISBN 978-3-642-13949-9

Vol. 294. Manuel F.M. Barros, Jorge M.C. Guilherme, and Nuno C.G. Horta
*Analog Circuits and Systems Optimization based on Evolutionary Computation Techniques,* 2010
ISBN 978-3-642-12345-0

Vol. 295. Roger Lee (Ed.)
*Software Engineering, Artificial Intelligence, Networking and Parallel/Distributed Computing,* 2010
ISBN 978-3-642-13264-3

Vol. 296. Roger Lee (Ed.)
*Software Engineering Research, Management and Applications,* 2010
ISBN 978-3-642-13272-8

Vol. 297. Tania Tronco (Ed.)
*New Network Architectures,* 2010
ISBN 978-3-642-13246-9

Vol. 298. Adam Wierzbicki
*Trust and Fairness in Open, Distributed Systems,* 2010
ISBN 978-3-642-13450-0

Vol. 299. Vassil Sgurev, Mincho Hadjiski, and Janusz Kacprzyk (Eds.)
*Intelligent Systems: From Theory to Practice,* 2010
ISBN 978-3-642-13427-2

Vassil Sgurev, Mincho Hadjiski, and
Janusz Kacprzyk (Eds.)

# Intelligent Systems: From Theory to Practice

Springer

Academician Vassil Sgurev,
Professor, Ph.D., D.Sc.
Institute of Information Techologies
Bulgarian Academy of Sciences
2, Acad. G. Bonchev Str.
P.O. Box 161
Sofia 1113
Bulgaria
E-mail: sgurev@bas.bg

Academician Janusz Kacprzyk,
Professor, Ph.D., D.Sc.
Systems Research Institute
Polish Academy of Sciences
ul. Newelska 6
01-447 Warsaw
Poland
E-mail: kacprzyk@ibspan.waw.pl

Academician Mincho Hadjiski,
Professor, PhD., D.Sc.
Institute of Information Techologies
Bulgarian Academy of Sciences
2, Acad. G. Bonchev Str.
P.O. Box 161
Sofia 1113
Bulgaria
E-mails: bgimcssmc@yahoo.com

ISBN 978-3-642-13427-2 e-ISBN 978-3-642-13428-9

DOI 10.1007/978-3-642-13428-9

Studies in Computational Intelligence ISSN 1860-949X

Library of Congress Control Number: 2010928589



*Typeset & Cover Design:* Scientific Publishing Services Pvt. Ltd., Chennai, India.

Printed on acid-free paper

9 8 7 6 5 4 3 2 1

springer.com

# Foreword

In the modern science and technology there are some research directions and challenges which are at the forefront of world wide research activities because of their relevance. This relevance may be related to different aspects. First, from a point of view of researchers it can be implied by just an analytic or algorithmic difficulty in the solution of problems within an area. From a broader perspective, this relevance can be related to how important problems and challenges in a particular area are to society, corporate or national competitiveness, etc. Needless to say that the latter, more global challenges are probably more decisive a driving force for science seen from a global perspective.

One of such "meta-challenges" in the present world is that of *intelligent systems*. For a long time it has been obvious that the complexity of our world and the speed of changes we face in virtually all processes that have impact on our life imply a need to automate many tasks and processes that have been so far limited to human beings because they require some sort of intelligence. Examples may be found everywhere, from a need to support decision making through a need to cope with a rapidly growing amount of all kinds of data that clearly exceed human comprehension and cognitive capacity, a need to cope with aging societies who need some intelligent systems to support the elderly, to a growing need from the military for tools and techniques, and then systems and devices, that would make it possible to eliminate (or limit) the human presence in the battlefield. All these challenges call for advances systems which will exhibit some intelligence and will therefore be useful to their human users.

The area of broadly perceived intelligent systems has emerged, in its present form, just after World War II, and was initially limited to some theoretical attempts to emulate human reasoning, notably by using tool from formal logic. The advent of digital computers has clearly played a decisive role by making it possible to solve difficult problems. In the mid-1950 the term *artificial intelligence* was coined. The early research efforts in this area, heavily based on symbolic computations alone, though have had some successes, have not been able to solve many problems in which numerical calculations have been needed, and new, more constructive approaches have emerged, notably *computational intelligence* which have been based on various tools and techniques, both related to symbolic and numerical calculations. This modern direction has produced many relevant theoretical results and practical applications in what may be termed *intelligent systems*.

It is quite natural that a field, like that of intelligent systems, which is both scientifically challenging and has such a tremendous impact on so many areas of human activity at the level of an individual, small social groups and entire societies, has triggered attempts to discuss basic topics and challenges involved at scientific gatherings of various kinds, from small and informal seminars, through specialized workshops and conferences to large world congresses. The first gatherings have been mainly concerned with a presentation of more specific technical issues and solutions, and then people have tried more and more to present the area in a multifaceted way by presenting both recent development and challenges, and by finding time and space to discuss more general issues relevant for the area and beyond.

This volume has been triggered by vivid discussions on various aspects related to intelligent systems at IEEE-IS'2008 – The 4$^{th}$ International IEEE Conference on Intelligent Systems: Methodology, Models and Applications in Emergent Technologies held in Varna, Bulgaria, on September 6-8, 2008. The Conference gather more than 150 participants – both senior, well known researchers, and young scientists just starting their careers – from all corners of the globe, included more than 150 papers, including seven plenary presentations by distinguished speakers: R. Yager, G. Klir, J. Kacprzyk, J. Zurada, K. Hirota, Y. Popkov and K. Atanassov. The time and venue of the Conference, in a Black Sea resort of an international reputation, have contributed to an atmosphere that has just naturally stimulated many formal and informal discussions. As a result, by a general consent, we have decided to edit this volume to present a broad coverage of various novel approaches that – in view of the conference participants, modern trends, and our opinion – play a crucial role in the present and future development of a broadly perceived area of intelligent systems.

A remark which is due here is that though the volume is related to the recent IEEE-IS'2008 conference, one has to take into account that this conference is the forth one in the row of IEEE-IS conferences which were launched in Bulgaria some years ago to respond to a need of the international research community for a forum for the presentation of results and an open discussion of various approaches, sometimes controversial, that could guarantee open mindedness. Luckily enough, this has been achieved and the consecutive IEEE-IS conferences, held in Bulgaria and the UK, have become such unique places.

This volume contains a selection of peer reviewed most interesting extended versions of papers presented at IEEE-IS'2008 complemented with some relevant works of top people who have not attended the conference. The topics covered include virtually all areas that are considered to be relevant for the development of broadly perceived intelligent systems. They start from logical foundations, including works on classical and non-classical logics, notably fuzzy and intuitionistic fuzzy logic, and – more generally – foundations of computational intelligence and soft computing. A significant part of about 30 contributions included in this volume is concerned with intelligent autonomous agents, multi-agent systems, and ontologies. Issues related to intelligent control, intelligent knowledge discovery and data mining, and neural/fuzzy-neural networks are discussed in many papers. Intelligent decision support systems, sensor systems, group decision making and negotiations, etc. are also discussed. Much attention has been paid to a very promising direction in the area of intelligent systems, namely that of hybrid systems that through a synergistic combination of best features and strengths of particular approaches help attain a new functionality, effectiveness and efficiency.

We hope that this volume will be a significant part of the scientific literature devoted to intelligent systems, will provide a much needed comprehensive coverage of recent developments, and will help clarify some difficult problems, indicate future directions and challenges, and even initiate some new research efforts.

We wish to thank all the contributors for their excellent work. We hope that the volume will be interesting and useful to the entire intelligent systems research community as well as other communities in which people may find the presented tools and techniques useful to formulate and solve their specific problems.

We also wish to thank Dr. Tom Ditzinger and Ms. Heather King from Springer for their multifaceted support and encouragement.

January, 2010
Sofia, Bulgaria

Vassil Sgurev
Mincho Hadjiski
Janusz Kacprzyk

# Contents

# Tagging and Fuzzy Sets

Ronald R. Yager and Marek Reformat

**Abstract.** The Introduction of web 2.0 and social software make significant changes in users' utilization of the web. User involvement in processes restricted so far to system designers and developers is more and more evident. One of the examples of such involvement is tagging. Tagging is a process of labeling (annotating) digital items – called resources – by users. The labels – called tags – assigned to those resources reflect users' ways of seeing, categorizing, and perceiving particular items. As the result a network of interconnected resources and tags is created. Connections between resources and tags are weighted with numbers reflecting how many times a given tag has been used to label a resource.

A network of resources and tags constitutes an environment suitable for building fuzzy representations of those resources, as well as tags. This simple concept is investigated here. The paper describes principles of the concept and shows some examples of its utilization. A short discussion dedicated to interrelations between tagging and fuzziness is included.

**Keywords:** fuzzy sets, tagging, membership degree value, resources, tags, tag-clouds, linguistic labels, users, web 2.0, social software, search, classification, mapping.

## 1 Introduction

Internet becomes an immense inventory of digital items of different types. Those items could be songs, photos, documents, or any entities that can be stored on the Internet. All web resources require annotation for classification and searching purposes. So far, the tasks of annotation and categorization of items are performed

Ronald R. Yager
Machine Intelligence Institute
Iona College, New Rochelle, NY 10801

Marek Reformat
thinkS$^2$: thinking software and system laboratory
Electrical and Computer Engineering
University of Alberta, Edmonton, Canada, T6G 2V4

V. Sgurev et al. (Eds.): Intelligent Systems: From Theory to Practice, SCI 299, pp. 1–17.
springerlink.com 

in a top-down manner by designers and developers of systems. Those people are experts in the domain, and their expertise is used to construct annotations, and to divide items into different categories. A good example of that is a process of organizing manuscripts and books [Rowley92]. Users should "understand" the principles used to categorize resources and follow them in order to find things they are looking for.

One of the ways of describing items is utilization of metadata [Moen06]. Metadata, according to the National Information Standards Organization (NISO), is "structured information that describes, explains, locates, or otherwise makes it easier to retrieve, use or manage an information resource" [NISO04]. In other words, metadata is a data that describe other data.

It becomes popular, especially in the framework of the Semantic Web [Berners01], to think of an ontology [Gruber95] as a metadata structure for semantic annotation of items. However, establishing an ontology as source of metadata for annotating a large number of web resources is not easy [Shirky05]. There are multiple issues related to differences between contents of an ontology and items/resources stored on the web.

In the dawn of social software and web 2.0, users become more involved in all aspects related to building, controlling and maintaining the web. Popular social services on the web, like Delicious[1], Furl[2], Flickr[3], CiteULike[4], and Yahoo My Web 2.0[5], allow users to annotate and categorize web resources. In those cases, users have introduced an important type of metadata – *tags*. The annotation processes performed by users is nothing else but labeling resources with tags. Users annotate resources easily and freely without knowing any taxonomies or ontologies. Tags represent any strings that user consider appropriate as descriptions of resources on the web. Resource, on the other hand, could be any items that have been posted on the Internet, and are accessible by users. The process of labeling – annotating – resources performed by users is called *tagging* [Mathes04] [Hammond05].

The fact that a resource is described by a number of tags, and a single tag is associated with a number of resources creates an opportunity to look at tags and resources from the perspective of fuzzy sets [Klir97][Pedrycz07]. The concept of fuzzy representation of resources and tags is developed here, and simple examples of it utilization are shown and explained.

Section 2 is a short introduction to tagging. It contains an explanation of main components of tagging systems and an illustrative example. It also introduces a concept of tag-clouds. Formal definitions of a tagging system and a network of interconnected resources and tags are presented in Section 3. Section 4 shows how fuzzy sets can be derived from the network of interconnected resources, and how their mappings to linguistic label can be preformed. The discussion about importance of tagging for applications of fuzziness to real-world systems is the subject of Section 5.

---

1 del.isio.us
2 furl.net
3 flickr.com
4 citeulike.org
5 myweb2.search.yahoo.com

# 2 Tag Clouds as User-Driven Resource Descriptors

## 2.1 Concepts and Example

A simple need to find relevant items among all resources stored on the web leads to a growing demand for suitable ways of organizing, categorizing and classification of web resources. Usefulness of the web resources depends on users' abilities to find things they are looking for, and this depends on a proper identification – annotation – of resources. It is difficult to find a resource that is related to such keywords as "vacation", "joy", or "sun" without a schema that allows resources to be annotated with such keywords.

Up to recently, a process of finding digital items on the web has been supported with building hierarchical structures of categories. Such structures allow for "sorting" all items into categories and organizing them into well defined collections. This means that finding an item requires some knowledge about possible categories to which the item could belong. A lack of that knowledge makes the task quite difficult.

A growing involvement of users in building repositories of digital items and being responsible for their maintenance brings a new approach to a process of describing resources for the identification processes. This new approach is called *tagging*. All items are described by anyone who "sees" them and wants to provide his/her description and/or comment. Experts are not required to provide category structures and their descriptions. This bottom-up approach becomes quite popular due to its simplicity, effectiveness, and enjoyableness. The principle of this approach is very simple – every user assigns to any item a number of different labels that he/she sees as appropriate ones [Methes04][Smith08].

Tagging becomes a source of information that is used for a number of research topics: for discovering changes in behavioral patterns of users [Fu09], for studying semiotic dynamics – how populations of humans establish and share semiotic systems [Cattuto07], for inferring about global semantics fitting a bottom-up approach to semantic annotation of web resources [Zhang06]. There is also an interesting work targeting the issues of tag similarity and relatedness [Cattuto08], as well as discovering regularities in user activities, tag frequencies, and kinds of tags used [Golder06].

A simple real-world example illustrates the results of a tagging process. The LibraryThing[6] is a web site with descriptions of more than 15 millions of books. It allows users to attach labels to books. More than 200,000 users are involved in that process, and they have used more than 20 millions tags.

Let us search for a book entitled *Catch-22*. As the result, we obtain not only info about the book, but also a list of related tags – keywords "attached" to this book by different users. Such tags are "American literature", "fiction", "humor", and "satire", Figure 1.

---

[6] LibraryThing.com

**Fig. 1** Tags for *Catch-22* (TBR stands for To Be Read)

As it can be seen, the list represents a "user-based" description of the book. The tags express users' perception of the book, as well as the meanings and associations that the book brings to them. Some tags represent users' interests, for example "American literature", "historical fiction", "humor", "military", while some other ones represent users' ways of describing the book – "own", "favorite", "to read", "unread".

Multiple tags are assigned to a single book so the book is annotated with variety of keywords. Different combinations of those keywords may lead us to the book. Additionally, the same tags can be used to describe other books. This represents a new way of connecting *Catch-22* to other books. Each tag can be associated with a link that leads to other books that have been "associated" with the same tag.

The tags play the role of "hooks" for pulling information from other sites. In this case they can represent links to other resources, located anywhere on the web, that are "associated" with those tags [Smith08].

## 2.2 *Web Resources, Tags, and Users*

The simple example presented above provides us with an intuitive understanding what tagging is about. In a nutshell, tagging describes a process of labeling resources performed by people. Tagging can be represented with a very simple model, Figure 2. The model contains three main elements: users, resources, and tags.

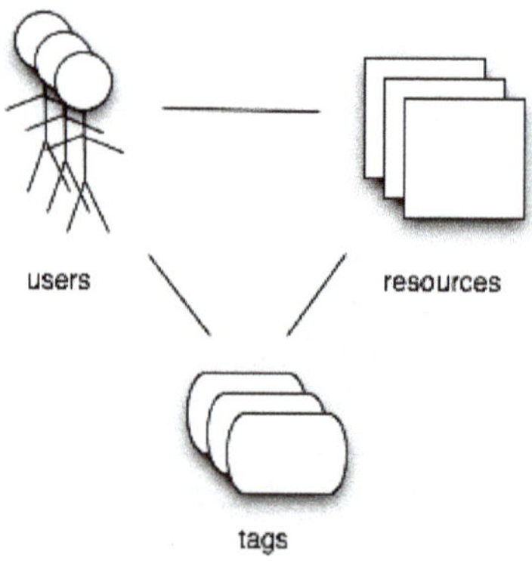


**Fig. 2** Tagging: *users* add *tags* to *resources*

*Users* are the people who are involved in a tagging process – use a tagging system. They construct keywords, assign them to resources, and add resources (in some cases). Users have a variety of interests, needs and goals. It is assumed that they tag to achieve one goal – sharing and labeling a document so it is easy to find it later.

*Resources* are items that users label with keywords. A resource can be any entity representing any real or imaginary thing that is uniquely identified on the web. It is very often that in a single system all resources are of the same type.

Users label resources using keywords called *tags*. Usually, tagging is open-ended, so tags can be of any kind. Tags can be: *descriptive* providing details about a resource, for example, its title/name, its author/creator, its location, its intended use, as well as representing users' individual emotions and feelings towards a resource; *administrative* used to manage a collection of resources, for example, the data a resource was created/acquired, the person who owns the rights to the resource; *structural* used to associate the resource with other resources. Tags are treated as metadata that describe a given resource. Among all three types of tags the administrative and structural ones are the most unambiguous, while the descriptive ones are the most subjective – they could require personal interpretation.

## 2.3 Tag-Clouds and Their Importance

A tagging process performed by multiple users means that many tags are used to annotate a single resource, and multiplicity of those tags can vary. A graphical representation of such scenario is called a tag-cloud. It is a way of presenting tags where more frequently assigned tags, to a given resource, are emphasized – usually in size or color. Tag clouds tell at a glance which tags are more popular.

The previously presented set of tags for *Catch-22* (Figure 1) is an example of such a tag-cloud. Figure 3 shows the same cloud, but this time there are numbers besides tags. Those numbers represent how many times a given tag was used by different users to label the resource. In our example, the book *Catch-22* has been labeled with the tag "fiction" by 2,545 users, with the tag "WWII" by 799 users, the "war" by 725, the "satire" by 610, the "classic" by 479, the "novel" by 374, the "humor" by 345, and the "literature" by 332 users.

**Fig. 3** Tags for *Catch-22* – numbers in brackets represent number of occurrence of each tag

### 2.4 Tagging Systems

Tagging takes place in the context of so called a tagging system. Such a system provides an infrastructure supporting users in their resource labeling activities. Architecture of such a system identifies what kind of tagging can be performed. Some systems allow users to add their own tags, some systems allow only to choose from a set of tags, yet other systems allow for adding resources, or tagging all or only specific resources. In other words, a system contains rules about who can tag, what can be tagged, and what kind of tags can be used.

Tagging systems, called also collaborative tagging systems, constitute social bookmarking sites – a form of a social software. Here, users reach an agreement about a description of the resource that is being tagged. Individuals describe the resource in their own way. A "common" consensus regarding that description emerges via the most popular tags.

Currently, there are a number of systems that use the tagging approach for organizing items and simplifying a searching process (Section 1). Those systems have different rules regarding what users can do. For example, www.del.icio.us.com is a social bookmarking system that allows users to submit and label different URLs. This system imposes very minimal restrictions regarding what users can submit and what kind of tags they can use. On the other hand, www.amazon.com provides more limited tagging capabilities. The only resources users can tag are system (amazon) resources, and the tags that can be used are controlled by amazon.com.

## 3 Tagging Definitions and Structure

### 3.1 General Overview

Following [Hotho06], a tagging system can be represented as a tuple. The formal definition is as follows:

**Definition 1.** A tagging system is a tuple TS=(U, T, R, Y) where U, T and R are finite sets with elements *users*, *tags* and *resources* respectively, while Y is a relation between the sets, i.e., Y is subset of U x T x R. A *post* is a triple (u, $T_{ur}$, r) with u $\in$ U, and r $\in$ R, and a non-empty set $T_{ur} = \{t \in T | (u, t, r) \text{ from } Y\}$.

The relation ($u_k$, $t_i$, $r_j$) means that the user $u_k$ assigned the tag $t_i$ to the resource $r_j$. Such definition allows us to define other quantities, for example, an occurrence.

**Definition 2.** An occurrence of a given tag $t_i$ as a label for a resource $r_j$ is given by the number of triples ($u_k$, $t_i$, $r_j$) where $u_k$ is any user that belongs to U, i.e.,

$$occur_{i,j}(t_i, r_j) = card\{(u_k, t_i, r_j) \in Y \mid u_k \in U\} \tag{1}$$

## 3.2 Tags and Resources

Given a tagging system (U, T, R, Y), we define the Resource-Tag Graph – RTG – as a weighted undirected graph whose set of vertices is a union of sets T and R. A tag $t_i$ and a resource $r_j$ are connected by an edge, iff there is at least one post ($u_k$, $T_{ur}$, $r_j$) where $t_i$ is $\in T_{ur}$, and $u_k \in U$. The weight of this edge is given by the occurrence $occur_{i,j}(t_i, r_j)$.

The previously described concept of a tag-cloud is a description of a single resource. At the same time, it is a snippet of a whole network of interconnected tags and resources – a part of RTG. The tag-cloud in Figure 4 represents all tags that have been used to annotate a single resource, let us called it a resource **r1**. Different sizes of fonts representing tags reflect "popularity" of those tags. The *tag01* (**t1**) is the most popular, while the *tag03* (**t3**) is the least popular.

**Fig. 4** An example of a tag-cloud

The sizes of fonts of different tags that constitute a tag-cloud are calculated based on occurrences of tags associated with a given resource. There are two mappings used for that purpose – one of them is a linear mapping (Eq. 2) where a scaling factor for font sizes is calculated based on the values of tag occurrences; and a logarithmic mapping (Eq. 3) where calculations are performed based on the log values of tag occurrences. The calculated scaling factor is used to determine the font size based on the provided minimum and maximum font sizes (Eq. 4). The formulas below are used for calculating the scaling factor for a tag $t_i$ associated with a resource $r_j$.

$$scale_{i,j}^{Lin} = \frac{occur_{i,j}(t_i, r_j) - \min_k(occr_{k,j}(t_k, r_j))}{\max_k(occr_{k,j}(t_k, r_j)) - \min_k(occr_{k,j}(t_k, r_j))} \quad (2)$$

$$scale_{i,j}^{Log} = \frac{\log(occur_{i,j}(t_i, r_j)) - \log(\min_k(occr_{k,j}(t_k, r_j)))}{\log(\max_k(occr_{k,j}(t_k, r_j))) - \log(\min_k(occr_{k,j}(t_k, r_j)))} \quad (3)$$

$$font\_size_{i,j} = min\_font\_size + scale_{i,j} * (max\_font\_size - min\_font\_size) \quad (4)$$

A different representation of the tag-cloud, Figure 4, together with a bigger fragment of RTG containing this tag-cloud is presented in Figure 5. It contains three resources **r1**, **r2**, and **r3**, and a number of tags – from **t1** to **t10**. Each connection of this network is associated with a number representing how many times a given tag has been used in describing a given resource – $occur_{i,j}(t_i, r_j)$.

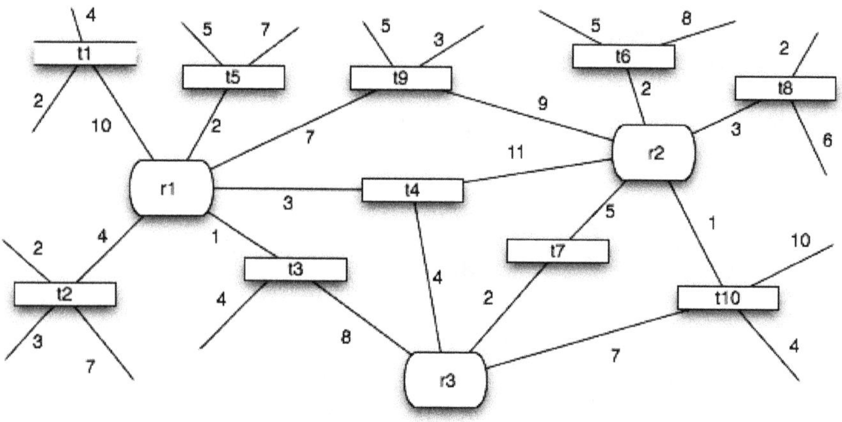


**Fig. 5** A snippet of RTG with resources (**r1-3**) and tags (**t1-10**). The tags also label other resources not shown here. Numbers indicate how many times a tag was used to label a resource – occurrences.

## 4 Tagging-Based Fuzzy Sets

An interesting conclusion can be drawn based on the fragment of RTG presented in Figure 5. The network provides us with a two "types" of interconnections:

- a single resource is linked with a number of tags, and those tags describe this resource;
- a single tag is linked with a number of resources, and those resources describe this tag.

Such observation leads us to the concept of defining two fuzzy sets based on RTG interconnections. Those two types of sets are described below.

### *4.1 Resource as Fuzzy Set over Tags*

The RTG presented in Figure 6 shows a setting where a single resource (**r1**) is labeled with a number of tags. Such a look illustrates a scenario where a resource can be represented by a fuzzy set defined based on tags. A formal definition of such case is included below.

**Definition 3. Fuzzy Representation of Resource.** A resource $r_j$ in the RTG can be represented by a fuzzy set $\Phi_r(r_j)$ as $\Phi_r(r_j)=\{\mu^r_{j,1}(t_1)/t_1, \mu^r_{j,2}(t_2)/t_2,..., \mu^r_{j,m}(t_m)/t_m\}$, where $\{t_1, t_2,..., t_m\}$ is the set of tags in RTG and $\mu^r_{j,i}$ is the membership of $r_j$ with a tag $t_i$ in RTG. $\Phi_r(r_j)$ is called the fuzzy representation of $r_j$.

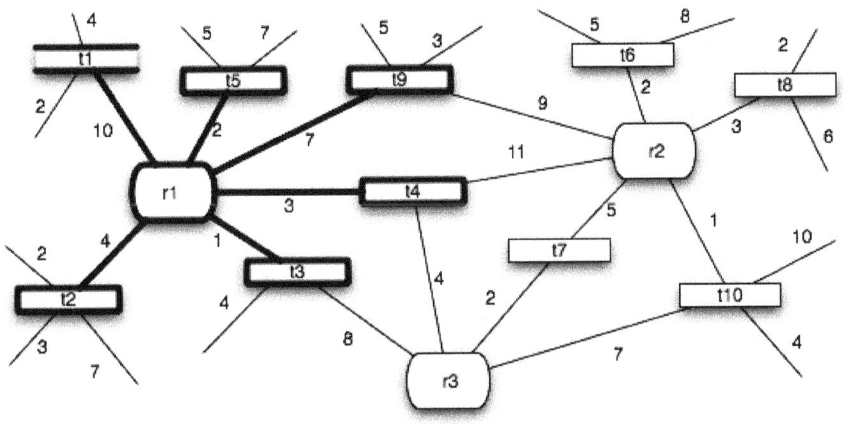


**Fig. 6** A snippet of RTG illustrating the resource **r1** described by tags

The weights of connections between a given resource and the tags describing it are used to derive the values of membership degrees. There are many different ways how the weights can be used to determine the membership values. One of them is shown below. It is a straightforward method that applies the linear and logarithmic mappings used to determine the scaling factor for font sizes (Section 3.2).

The membership degree values are calculated based on the occurrences of tags used to label a single resource. For **r1**, the occurrences used for calculations are the ones marked in gray in Table 1.

**Table 1** Occurrences of tags describing resources

| | t1 | t2 | t3 | t4 | t5 | t6 | t7 | t8 | t9 | t10 |
|---|---|---|---|---|---|---|---|---|---|---|
| **r1** | 10 | 4 | 1 | 3 | 2 | | | | 7 | |
| **r2** | | | | 11 | | 2 | 5 | 3 | 9 | 1 |
| **r3** | | | 8 | 4 | | | 2 | 2 | | 7 |
| **r…** | 2 | 2 | 4 | | 5 | 5 | | 6 | 3 | 4 |
| **r…** | 4 | 3 | | | 7 | 8 | | | 5 | 10 |
| **r…** | | 7 | | | | | | | | |
| | | | | | | | | | | |

The membership values can be calculated using the linear and logarithmic mapping equations:

$$\mu_{j,i}^{r,Lin} = \frac{occur_{i,j}(t_i, r_j) - \min_k(occr_{k,j}(t_k, r_j))}{\max_k(occr_{k,j}(t_k, r_j)) - \min_k(occr_{k,j}(t_k, r_j))} \tag{5}$$

$$\mu_{j,i}^{r,Log} = \frac{\log(occur_{i,j}(t_i, r_j)) - \log(\min_k(occr_{k,j}(t_k, r_j)))}{\log(\max_k(occr_{k,j}(t_k, r_j))) - \log(\min_k(occr_{k,j}(t_k, r_j)))} \tag{6}$$

The calculated values are presented in Table 2.

**Table 2** Membership degree values for the fuzzy representation of **r1**

| | **t1** | **t2** | **t3** | **t4** | **t5** | **t6** | **t7** | **t8** | **t9** | **t10** |
|---|---|---|---|---|---|---|---|---|---|---|
| **occurrences** | 10 | 4 | 1 | 3 | 2 | | | | 7 | |
| $\mu_{1,i}^{r,Lin}$ **(linear mapping)** | 1 | .3333 | 0 | .2222 | .1111 | | | | .6667 | |
| $\mu_{1,i}^{r,Log}$ **(logarithmic mapping)** | 1 | .6021 | 0 | .4771 | .3010 | | | | .8451 | |
| | | | | | | | | | | |

So, the fuzzy sets representing **r1** are:

$$\Phi_r^{Lin}(r_1) = \{\frac{1.0}{t1}, \frac{0.33}{t2}, \frac{0.0}{t3}, \frac{0.22}{t4}, \frac{0.11}{t5}, \frac{0.67}{t9}\} \tag{7}$$

using the linear mapping

$$\Phi_r^{Log}(r_1) = \{\frac{1.0}{t1}, \frac{0.60}{t2}, \frac{0.0}{t3}, \frac{0.48}{t4}, \frac{0.30}{t5}, \frac{0.85}{t9}\} \tag{8}$$

and using the logarithmic mapping.

It can be noticed that the membership values of $\Phi_r^{Log}(r_1)$ are higher that the one of $\Phi_r^{Lin}(r_1)$. This means that $\Phi_r^{Log}(r_1)$ is "more liberal" in the case of estimating levels of significance of tags in describing the resource **r1**. The $\Phi_r^{Lin}(r_1)$ is much more striker. It requires much higher values of occurrences to assign higher values of membership degree.

## 4.2 *Tag as Fuzzy Set over Resources*

The same RTG can be also looked at from the perspective of tags, and this is shown in Figure 7. Right now, the tag **t4** is annotated with the resources **r1**, **r2**, and **r3**. It is opposite to the scenario in Figure 6 where the resource **r1** is annotated with tags **t1**-**t5** and **t9**. This leads to the definition of a tag as a fuzzy set of resources. A formal definition of such a set is as follows.

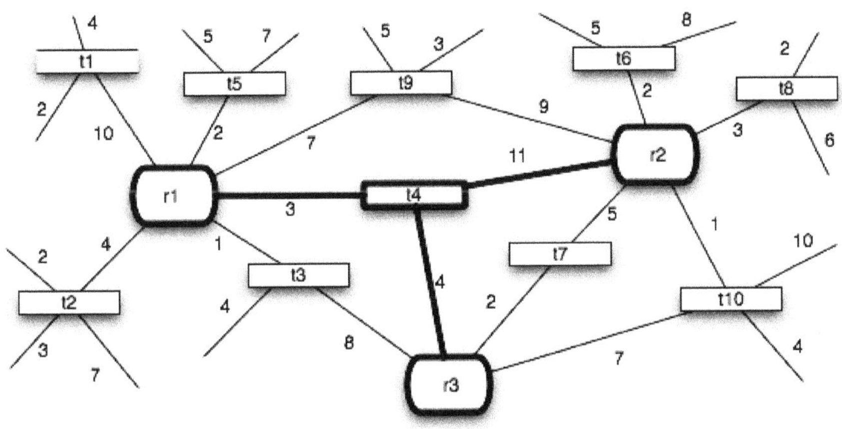


**Fig. 7** A snippet of RTG illustrating the tag **t4** described by resources

**Definition 4. Fuzzy Representation of Tag.** A tag $t_i$ in the RTG can be represented by a fuzzy set $\Phi_t(t_i)$ as $\Phi_t(t_i)=\{\mu^t_{i,1}(r_1)/r_1, \mu^t_{i,2}(r_2)/r_2,..., \mu^t_{i,n}(r_n)/r_n\}$, where $\{r_1, r_2,..., r_n\}$ is the set of tags in RTG and $\mu^t_{i,j}$ is the membership of $t_i$ with a resource $r_j$ in RTG. $\Phi_t(t_i)$ is called the fuzzy representation of $t_i$.

The values of membership degrees can be calculated in the similar way as it has been done for fuzzy representation of a resource. This time, Eq. 2 and Eq. 3 are modified to accommodate the fact that a tag is "a central point" and multiple resources are used for its annotation. The modified equations are presented below:

$$\mu^{t,Lin}_{i,j} = \frac{occur_{i,j}(t_i,r_j) - \min\limits_n(occr_{i,n}(t_i,r_n))}{\min\limits_n(occr_{i,n}(t_i,r_n)) - \min\limits_n(occr_{i,n}(t_i,r_n))} \tag{9}$$

$$\mu^{t,Log}_{i,j} = \frac{\log(occur_{i,j}(t_i,r_j)) - \log(\min\limits_n(occr_{i,n}(t_i,r_n)))}{\log(\max\limits_n(occr_{i,n}(t_i,r_n))) - \log(\min\limits_n(occr_{i,n}(t_i,r_n)))} \tag{10}$$

The following table is created based on Figure 7, this time the occurrences associated with the tag **t4** are marked with gray.

**Table 3** Occurrences of tags describing resources

| | **t1** | **t2** | **t3** | **t4** | **t5** | **t6** | **t7** | **t8** | **t9** | **t10** |
|---|---|---|---|---|---|---|---|---|---|---|
| **r1** | 10 | 4 | 1 | 3 | 2 | | | | 7 | |
| **r2** | | | | 11 | | 2 | 5 | 3 | 9 | 1 |
| **r3** | | | 8 | 4 | | | 2 | 2 | | 7 |
| **r...** | 2 | 2 | 4 | | 5 | 5 | | 6 | 3 | 4 |
| **r...** | 4 | 3 | | | 7 | 8 | | | 5 | 10 |
| **r...** | | 7 | | | | | | | | |
| | | | | | | | | | | |

Using Eqs. 9 and 10 the membership degree values are obtained, Table 4.

**Table 4** Membership values for **t4**

| | **r1** | **r2** | **r3** |
|---|---|---|---|
| **occurrences** | 3 | 11 | 4 |
| $\mu_{i,4}^{t,Lin}$ **(linear mapping)** | 0 | 1 | 0.1250 |
| $\mu_{i,4}^{t,Log}$ **(logarithmic mapping)** | 0 | 1 | 0.2214 |
| | | | |

So, for **t4** the fuzzy sets are as follow for the linear mapping:

$$\Phi_t^{Lin}(t_4) = \{\frac{0.0}{r1}, \frac{1.0}{r2}, \frac{0.13}{r3}\} \tag{11}$$

and for the logarithmic mapping:

$$\Phi_t^{Log}(t_4) = \{\frac{0.0}{r1}, \frac{1.0}{r2}, \frac{0.22}{r3}\} \tag{12}$$

### *4.3 Mapping to Linguistic Labels*

The values of membership degree can be further mapped into linguistic labels. It means that a single membership value calculated using one of the equations 5, 6, 8 or 10 is transformed into activation levels of a number of linguistic labels. Once again such transformation can be performed in a number of ways. General steps of a mapping schema are presented in Figure 8. A value enters the schema, and it passes through a transformation function. The transformed value is fuzzified via fuzzy membership functions representing different linguistic labels. A number of membership functions depends on a number of used/required linguistic labels. In Figure 8, there are three functions associated with the labels *high*, *medium*, and *low*. The output represents the levels of "activation" of used membership functions.

The solid line in Figure 8 represents a linear transformation function. However, different functions, such as (A) or (B), can be applied to make the transformation more "pessimistic" (leading to higher activation of linguistic labels representing lower degrees of significance), or more "optimistic" (favoring linguistic labels representing higher degrees of significance).

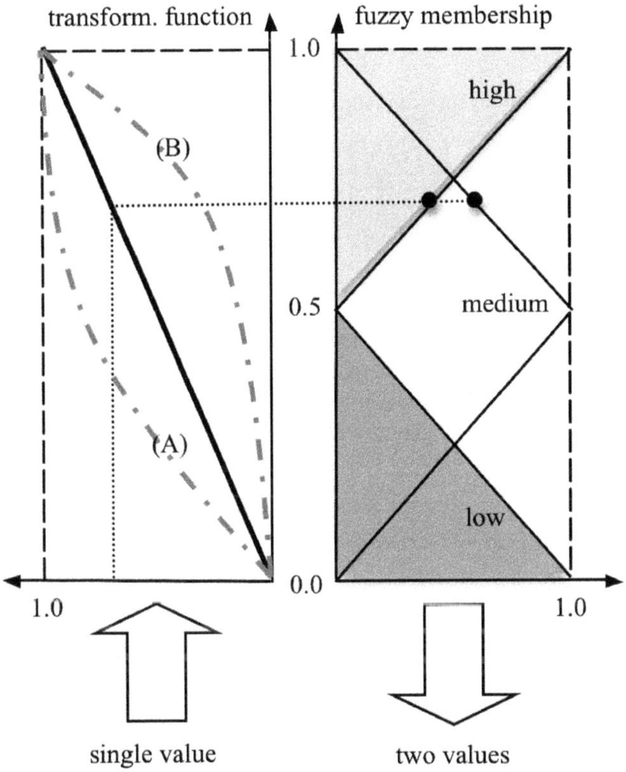


**Fig. 8** An illustrative mapping schema (different transformation functions can be used, here a linear one is shown in solid black)

A simple mapping schema with a linear transformation function is used to "translate" the values calculated for tags describing the resource **r1** (Section 4.1), and resources describing the tag **t4** (Section 4.2).

For the resource **r1**, the process of mapping the value $\mu^r_{1,2}(t_2)$ for the tag **t2** is presented. Table 5 shows the values (gray column) used for calculations. The transformation uses a linear function and is performed into three linguistic labels: *low, medium*, and *high*. The results of the process are included in Table 6.

**Table 5** Values of $\mu^r_{1,i}(t_i)$ using linear and logarithmic mappings

| | t1 | t2 | t3 | t4 | t5 | t6 | t7 | t8 | t9 | t10 |
|---|---|---|---|---|---|---|---|---|---|---|
| $\mu^{r,Lin}_{1,i}$ **(linear mapping)** | 1 | .3333 | 0 | .2222 | .1111 | | | | .6667 | |
| $\mu^{r,Log}_{1,i}$ **(logarithmic mapping)** | 1 | .6021 | 0 | .4771 | .3010 | | | | .8451 | |
| | | | | | | | | | | |

**Table 6** Membership values for linguistic labels *low*, *medium* and *high*

| | | t2 | | *low* | *medium* | *high* | |
|---|---|---|---|---|---|---|---|
| **linear mapping** | | .3333 | | 0.3333 | 0.6667 | | |
| **logarithmic mapping** | | .6021 | | | 0.7582 | 0.2418 | |
| | | | | | | | |

The transformation means that right now we have three forms of the tag **t2**: **t2-low**, **t2-medium**, and **t2-high**. For example, assuming that **t2** stands for a keyword "funny" the transformation would imply "little funny", "funny" and "very funny". Each of those new tags is associated with the resource to a different degree, Figure 9.

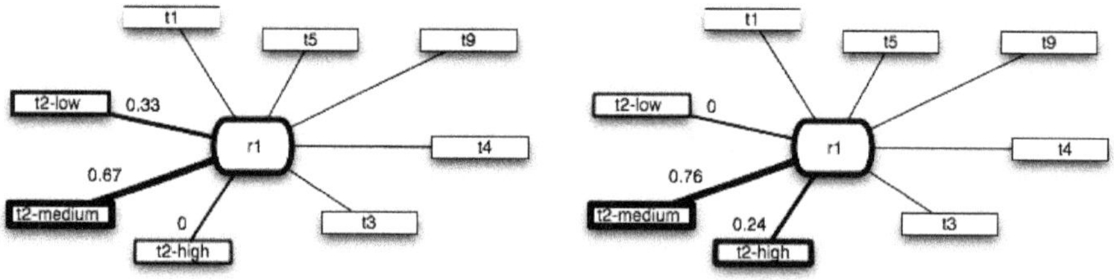


(a) (b)

**Fig. 9** Linguistic labels for the tag **t2** describing the resource **r1**, for linear mapping (a), and logarithmic mapping (b)

The same process of transforming a membership value of a fuzzy set $\Phi$ into linguistic labels is performed for the value of $\mu^t_{4,3}(r_3)$ of the fuzzy set $\Phi_r$ for the tag **t4**. Table 7 shows the values (gray column) used for calculations. As above, the transformation is linear, and the conversion is performed into three linguistic labels: *low*, *medium*, and *high*. The results are in Table 8.

**Table 7** Values of $\mu^{t}_{4,j}(r_j)$ using linear and logarithmic mappings

| | **r1** | **r2** | **r3** |
|---|---|---|---|
| $\mu_{4,j}^{t,Lin}$ **(linear mapping)** | 0 | 1 | 0.1250 |
| $\mu_{4,j}^{t,Log}$ **(logarithmic mapping)** | 0 | 1 | 0.2214 |
| | | | |

**Table 8** Membership values for: *low*, *medium* and *high*

| | | **r3** | | ***low*** | ***medium*** | ***high*** | |
|---|---|---|---|---|---|---|---|
| **linear mapping** | | .1250 | | 0.7500 | 0.2500 | | |
| **logarithmic mapping** | | .2214 | | 0.5572 | 0.4428 | | |
| | | | | | | | |

The interpretation of that transformation goes back to the concept of importance of connections between resources and tags. However, this time the importance is estimated in the context of all resources annotated with the same tag. The illustration of that is shown in Figure 10. We can see that connection between **t4** and **r3** is split into three connections – each of them representing different strength of connection. In Figure 10 a) the connection "low" is dominating; while in Figure 10 b) the strengths of the connections "low" and "medium" are comparable. It can be observed that the differences in calculating membership degrees of the original $\Phi_t$ fuzzy set (Eq. 9 or Eq 10) influences the interpretation of the importance of the connection.

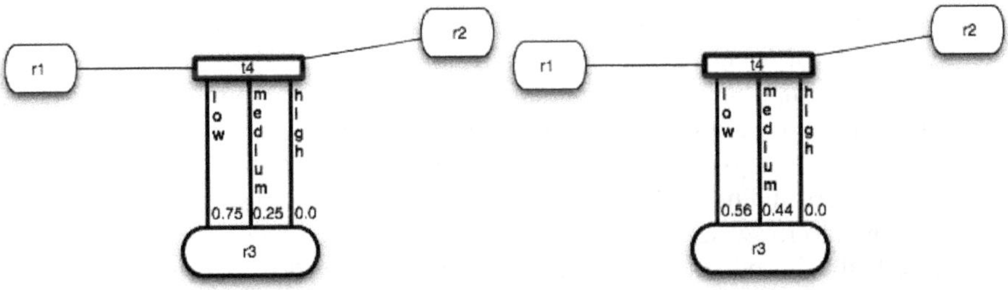


(a) (b)

**Fig. 10** Linguistic labels for the resource **r3** describing the tag **t4**, for linear mapping (a), and logarithmic mapping (b)

## 5 Discussion and Conclusions

The paper addresses an issue of interrelation between tagging systems and fuzziness. It seems quite intuitive to see resemblance between the tag-cloud concept and the concept of a fuzzy set. As it can be seen in Sections 4.1 and 4.2 it is relatively easy to "transform" a fragment of RTG (a different representation of a tag-cloud) into different fuzzy sets.

Representation of a single resource as a fuzzy set based on tags (Section 4.1) provides a very interesting aspect related to the annotation of a resource by tags. The mapping of those tags into linguistic labels introduces different levels (degrees) of meaning of those tags. Depending on a number of linguistic labels (and associated with them membership functions – Section 4.3) we can introduce tags that are able to express degrees of significance of the original tag. For example, for a single tag – "**like**" we can obtain degrees of this tag: "**like** a little", "**like** so-so" and "**like** a lot", all by transforming a value of membership degree into fuzzy linguistic labels. This can easily bring a new dimension to a tag-cloud – a dimension of degree.

Representation of a tag as a fuzzy set of resources is a different look at RTG. This look can enhance search activities in a tagging system. This "opposite" concept – a tag is annotated using resources – allows us to look at a single connection between a tag and a resource in the context of other resources labeled with this tag. Selection of linguistic labels such as "**low**", "**medium**" and "**high**" leads to a more human-like "ranking" of resources according to their degree of relatedness to the tag, Section 4.3. Using that approach, we can easily indicate to what degree a given tag should be associated with a resource.

Introduction of fuzziness to a tagging process provides a means to formalize imprecision. The tagging by its nature is imprecise – the concepts of occurrences and co-occurrences of tags create an environment where descriptions of resources are not always uniform and consistent. Such a situation could be quite common for distributed tagging systems, or interconnected collection of different tagging systems.

Once the relations between tags and resources are expressed using fuzziness, a number of techniques based on fuzzy sets and logic can be used to reason about similarities between resources or tags, to build decision schemas supporting search activities, or to better maintain local and distributed tagging systems.

Application of a tagging concept for front-end interfaces to fuzzy based systems brings an important method of gathering details about membership functions. Dynamic nature of tagging – constant changes in occurrences, as well as possible changes in sets of tags and resources – provide unparallel abilities to "track" human-like understanding and perception of resources and tags. Incorporating that information into fuzzy sets provides a chance to built truly human-centric systems.

Overall, the introduction of tagging approach as a way of describing items posted on the Internet introduces an opportunity for application of fuzzy sets and systems techniques in real-world web-based systems supporting users in their search for relevant items. All this can be done in the framework of truly human-conscious systems where aspects of human imprecision and perception are addressed by tagging preformed by users of those systems.

## References

[Berners01] Berners-Lee, T., Hendler, J., Lassila, O.: The Semantic Web. Scientific American 284, 34–43 (2001)

[Catutto08] Catutto, C., Benz, D., Hotho, A., Stumme, G.: Semantic Grounding of Tag Relatedness in Social Bookmarking Systems. In: Proceedings of the 7th International Conference on the Semantic Web, pp. 615–631 (2008)

[Cattuto07] Cattuto, C., Loreto, V., Pietronero, L.: Semiotic Dynamics and Collaborative Tagging. Proceedings of National Academy of Sciences PNAS 104(5), 1461–1464 (2007)

[Fu09] Fu, W.-T., Kannampallil, T., Kang, R.: A Semantic Imitation Model of Social Tag Choices. In: Proceedings of the 2009 International Conference on Computational Science and Engineering, pp. 66–73 (2009)

[Golder06] Golder, S., Huberman, B.: The Structure of Collaborative Tagging Systems. Journal of Information Sciences 32, 198–208 (2006)

[Gruber95] Gruber, T.: Toward Principles for the Design of Ontologies used for Knowledge Sharing. International Journal of Human-Computer Studies 43(4-5), 907–928 (1995)

[Hamm05] Hammond, T., Hannay, T., Lund, B., Scott, J.: Social Bookmarking Tools (I): A General Review. D-Lib. Magazine 11 (2005), `http://www.dlib.org/dlib/april05/hammond/04hammond.html`

[Hotho06] Hotho, A., Jaschke, R., Schmitz, C., Stumme, G.: Information Retrieval in Folksonomies: Search and Ranking. In: Sure, Y., Domingue, J. (eds.) ESWC 2006. LNCS (LNAI), vol. 4011, pp. 411–426. Springer, Heidelberg (2006)

[Klir97] Klir, G., Clair, U., Yuan, B.: Fuzzy Set Theory: Foundations and Applications. Prentice-Hall, Englewood Cliffs (1997)

[Methes04] Mathes, A.: Folksonomies – Cooperative Classification and Communication Through Shared Metadata, `http://www.adammathes.com/academic/computer-mediated-communication/folksonomies.html`

[Moen06] Moen, W., Miksa, S., Eklund, A., Polyakov, S., Snyder, G.: Learning from Artifacts: Metadata Utilization Analysis. In: Proceedings of the 6th ACM/IEEE-CS Joint Conference on Digital Libraries, pp. 270–271 (2006)

[Pedrycz07] Pedrycz, W., Gomide, F.: Fuzzy Systems Engineering: Toward Human-Centric Computing. Wiley/IEEE Press (2007)

[Rowley92] Rowley, J.: Organizing Knowledge: An Introduction to Information Retrieval, 2nd edn. Gower Publishing, Aldershot (1992)

[Shirky05] Shirky, C.: Ontology is Overrated: Categories, Links and Tags, `http://www.shirky.com/writings/ontology_overrated.html`

[Smith08] Smith, G.: Tagging: People-Powered Metadata for the Social Web. New Riders (2008)

[Zhang06] Zhang, L., Wu, X., Yu, Y.: Emergent Semantics from Folksonomies: A Quantitative Study. Journal on Data Semantics VI 4090, 168–186 (2006)

[NISO04] National Information Standards Organization (NISO), Understanding Metadata. NISO Press, Bethesda, USA (2004)

# Intelligent Control of Uncertain Complex Systems by Adaptation of Fuzzy Ontologies

Mincho Hadjiski, Vassil Sgurev, and Venelina Boishina

**Abstract.** A new approach for intelligent control is proposed for complex uncertain plants using synergism between multi-agent and ontology based frameworks. A multi stage procedure is developed for situation recognition, strategy selection and control algorithm parameterization following coordinated objective function. Fuzzy logic based extension of conventional ontology is implemented to meet uncertainties in the plant, its environment and sensor information. Ant colony optimization is applied to realize trade-off between requirements and control resources as well as for significant reduction of the communication rate among the intelligente agents. To react on unexpected changes in operational conditions certain adaptation functionality of the fuzzy ontology is foreseen. A multi-dimensional cascade system is considered and some simulation results are presented for variety of strategies implemented.

**Index Terms:** Adaptation, control system, fuzzy logic, multi - agent system, ontology.

## 1 Introduction

Industrial control systems are under strong pressure from business conditions connected with globalization, competition, environmental requirements, and

Mincho B. Hadjiski
Sofia, Bulgaria, Institute of Information Technologies - Bulgarian Academy of Sciences (BAS) and University of Chemical Technologies and Metallurgy –Sofia, 8 Kliment Ohridski Blvd., 1756 Sofia, Tel.: (+3592)8163329
e-mail: hadjiski@uctm.edu

Vassil S. Sgurev
Sofia, Bulgaria, Institute of Information Technologies - Bulgarian Academy of Sciences (BAS) "Acad. G. Bonchev" str., Bl. 29A 1113, Tel.: (+359.2) 708-087
e-mail: sgurev@bas.bg

Venelina G. Boishina
Sofia, Bulgaria,University of Chemical Technologies and Metallurgy –Sofia, 8 Kliment Ohridski Blvd., 1756 Sofia, Tel.: (+3592)8163329
e-mail: venelina_b@yahoo.com

V. Sgurev et al. (Eds.): Intelligent Systems: From Theory to Practice, SCI 299, pp. 19–40.
springerlink.com 

societal movements. Additionally, they in the last decade become increasingly complex in order to meet the challenges of the insistent requirements for higher and higher efficiency. This is result of the ambition to subject in a general control strategy a lot of plant's properties which have been ignored or considered very approximately up to day. Control systems in industry tend to be large–scale, distributed and network-based. It becomes necessary to take into account the real world problems like inaccuracy, vagueness, fuzziness. In many cases formulation of current settings, constraints and estimation criteria consider significant subjective part. Additional sensor information for softsensing and inference control often is a source of external and/or internal inconsistency. In many cases the plant's behaviour is so complicated and poorly determined that conventional control methods based on crisp algorithms only become inadequate and are impossible or non effective for real application. In the present contribution the hybrid approach in process control systems via combination of conventional control methods with some information technologies, proposed in (Hadjiski and Boishina 2007, Hadjiski, Sgurev and Biosina 2006, Hadjiski and Bioshina 2005), is developed further in a case of multidimensional cascade control. The main using components are:

- Conventional PID controllers at a basic level.
- Multi-agent systems (MAS), considering intelligent agents with both autonomous functionality and ontologies service in dynamic conditions (Gonzales et al. 2006, JADE 2009,Wooldridge 2002, W3C 2009).
- Standard static ontologies (O) aimed to represent the existing (and new acquired) knowledge about the plant as well as to structure a new information (FIPA 2009, Hadjiski and Boishina 2007, W3C 2009).
- Fuzzy Ontologies (FO) in order to treat classes (concepts) with unsharp or fuzzy boundaries (Calegary and Sanchez 2007, Lin 2007, Stoilos and al 2005, Stracia 2001, Teong 2007).
- Ant Colony Optimization (ACO) (Dorigo and al. 2006, Chang 2005) for dynamic optimization of the control system based on fuzzy ontology destined for radical reduction of the volume of inter agent communication in the conditions of uncertainty (Hadjiski and Boishina 2007, 2005).
- On–line optimization of the hybrid Agent – Ontology control system as reactive behaviour against the variations of the plant, its environment, control purposes, possible failures.

The main problem in this work is to gain a full integration of the above functionalities of the separate elements included in the control system.

## 2 Problem Definition

*A. Plant properties*

Below a generalized class of industrial and related plants is under consideration. A typical cases are:

- In power industry – e. g. combustion control in steam generator, where via flow rate of inlet air the boiler efficiency and ($NO_x$, $SO_x$) – emissions must be controlled in cascade control system under changeable operation conditions – load of the block boiler–turbine, fossil caloricity and composition, mill–fan capacity etc (Hadjiski, Sgurev and Biosina 2006, Hadjiski and Bioshina 2005).
- In metallurgical industry, e. g. control of agglomeration process in a sinter plant, where optimal sintering conditions must be maintained using a very limited number of manipulated variables, mainly via ignition control in multi cascade control systems (Mitra and Gangedaran 2005, Terpak and al. 2005).
- In ecological industry, e.g. anaerobic wastewater treatment plants control where the multistage process (mechanical, biological, chemical and sludge treatment) represent a typical case of multi cascade control with limitations in the number and capacity of the stage's manipulated variables. At the same time a significant changes in the operation conditions could decrease the plant efficiency: changes in the quantity or quality of the wastewater to be treated can lead to destabilization of the process, faults based on the signals obtained from different points, meteorological changes etc. (Manesis , Sardis and King 1998, Smirnov and Genkin 1989, Yang and al. 2006).

The listed instances as well as a great number of relevant plants possess some common characteristics:

1. A presence of slow and fast dynamic channels which are suitable for cascade/multi-cascade control (Hadjiski and Boishina 2007, Mitra and Gangedaran 2005, Smirnov and Genkin 1989)
2. A lack of enough manipulated variables according the desirable control variables, i. e. the plants (or stages) are with nonsquare matrix (Hadjiski, Sgurev and Biosina 2006, Mitra and Gangedaran 2005, Yang and al. 2006).
3. Variability of the plant and operational conditions (load, disturbances, faults) (Hadjiski and Bioshina 2007.Hadjiski, Sgurev and Biosina 2006, Hadjiski and Bioshina 2005, Manesis,Sardis and King 1998).
4. Constraints could be both hard and soft and often which are time–variant.
5. The monitoring of a number of completementary variables allows inference cascade control.

*B. Situation Control Definition*

For such kind of plans with possibility for different way of control according to the current system situation caused by changes in the control environment (PiTNavigator 2009, Valero at al. 2004) the following strategy of control is proposed (Fig. 1).In this structure of control each of the time control period is given as a unit of the all sequence of actions:

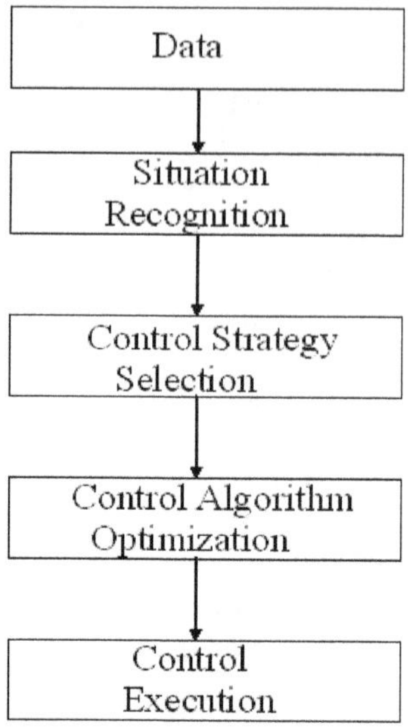


**Fig. 1** Intelligent Strategy Selection

- Data- presented from current system measurements, predicted signals, system disturbances, and inference parameters.

-Situation Recognition – based on the expert knowledge about the system behavior, data about system parameters.

-Control Strategy Selection – based on system objective functions, different procedures for strategy selection, system priorities, control weights. The possible strategy selection is inspired from different control structures, algorithms, different ways of coordination control and current situations. The given research is based on the changes of the system environment. A new way of strategy selection depended on current system situations is proposed.

-Control Algorithm Optimization – after chooses of strategy of control there is a need to optimize the selected strategy. There are different algorithms to do that based of: common for the current situation multi-criteria objective function, some kind of coordination of various partial optimization tasks.

-Control Execution – after optimization of the selected strategy the chosen control is been executed in a way to do an optimal system control.

### *C. Structure of the control system*

For the described above class of plants a generalized cascade control system is accepted in this investigation (Fig. 2).

The presented cascade control system consists of two plants: Single Input Two Outputs (SITO) plant and Two Inputs Two Outputs (TITO) plant. The cascade system is represented from the main control loop for TITO plant and internal control loops for the SITO plant. The TITO plant is strongly influenced by external ($v_2$) and internal ($v_3$) disturbances. The changes in the parameters of

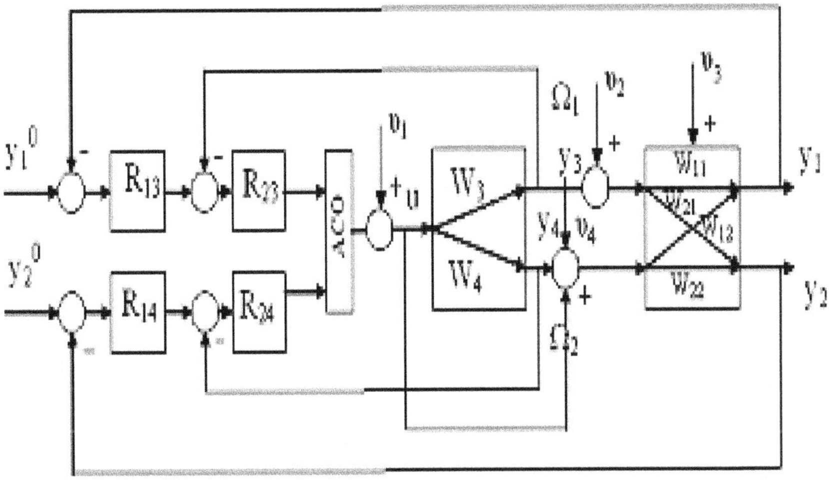


**Fig. 2** Structure of generalized cascade control system

transfer functions $W_{ij}(s)$ due to generalized disturbances $v_3$ make necessary to adapt the knowledge in the plant ontology.

Four possible situations for cascade control system are considered:

- *Case 1: $\Omega_1 = 0$ and $\Omega_2 = 0$* – the system is not cascade.
- *Case 2: $\Omega_1 = 1$ and $\Omega_1 = 0$* – the system is one-sided cascade.
- *Case 3: $\Omega_1 = 0$ and $\Omega_1 = 1$* – the system is one-sided cascade.
- *Case 4: $\Omega_1 = 1$ and $\Omega_1 = 1$* – the system is multi-cascade.

### *D. Conceptual model of the intelligent control system*

The conceptual model of the considered intelligent cascade control systems is presented in Fig. 3.

The developed control system is composed from two Multi-Agent Systems which work in cascade. The MAS1 is used for tracking the SITO plant outputs. The MAS2 is been used for the TITO plant control. The basic agents in Cascade Multi-Agent System (CMAS) are: Ant Colony Optimization (ACO) agents, control agents, constraint agents, model agents.

Both Multi–Agent Systems (MAS) are developed in order to work in cascade for keeping the system outputs among given constraints. The present agents' systems work autonomously with usage of separate ontologies for knowledge sharing in MAS.

Ant Colony Optimization (ACO) approach (Dorigo at al. 2006, Chang 2005) is used for on-line optimization in both MAS1 and MAS2.

Based on generalized plants properties "soft" constraints are accepted in the form:

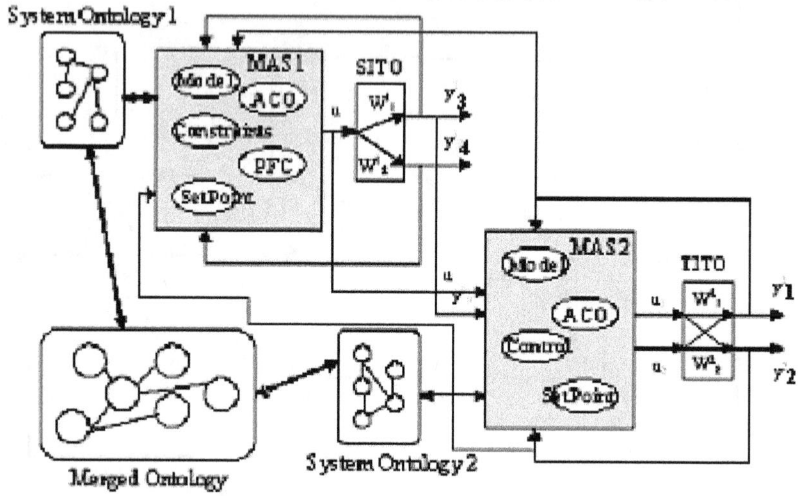


**Fig. 3** Conceptual model of the intelligent cascade control system

$$
\begin{aligned}
u_{\min} &\le u_{j+1/t} \le u_{\max,} & t \le j \\
y_{\min} &\le y_{j+1/t} \le y_{\max,} & t \le j
\end{aligned}
\qquad (1)
$$

where: $u_{min}$ and $u_{max}$ are defined constraints over the manipulated variables; $y_{min}$ and $y_{max}$ – defined constraints over the controlled variables

*E. Control problem statement*

To estimate the dynamic behaviour of the cascade multi-agent system (CMAS) the optimization criteria is accepted which include "penalized" function in order to take into account the "soft" constraints (1):

$$
\begin{aligned}
J^{i} = J + \| R_1^{l}(y_1(k+l/k) - y_1^{l}(k+1+l/k))^2 + R_1^{u}(y_1(k+l/k) - \\
- y_1^{u}(k+l/k))^2 \| + \| R_2^{l}(y_2(k+l/k) - y_2^{l}(k+l/k))^2 + \\
+ R_2^{u}(y_2(k+l/k) - y_2^{u}(k+l/k))^2 \| \rightarrow \min
\end{aligned}
\qquad (2)
$$

where: $R_{1/2}^{l/u}$ – weight matrix which depends on the constraints violations: $R_{1/2}^{l/u} \rightarrow 0$ - without constraints, $R_{1/2}^{l/u} \rightarrow \infty$ - with hard constraints. These "penalized" matrixes are defined dynamically by Ant Colony Optimization (ACO) algorithm and shared knowledge from the ontology. The optimal control is determined in the ACO agents by minimization of the common functional $J^1$ presented in eq. (2).

The two MAS work in cascade aiming to rich an optimal control for the plant outputs $y_1$, $y_2$ (MAS2) and intermediate variables $y_3$, $y_4$ (MAS1). Each of the MAS is using Ant Colony Optimization (ACO) in order to track the outputs into corresponding constraints and to minimize the value of $J^1$ for both plants. The MAS1 is developed in order to control the SITO plant with nonsquare matrix and the MAS2 is designed to control a plant with changeable structure (changes in the values of structural parameters $\Omega_i$).

The considered multi-cascade system is investigated as subjected on strong disturbances (up to 20% of the set points). The optimization procedure must be fast in order to avoid instability from one side and to provide trade-off between specifications addressed to all outputs $y_i$ (i=1, 4) from another. Multi-criteria problem here is solved via scalarization of the partial criteria

$$J = \sum_{i=1}^{4} \alpha_i J_i \tag{3}$$

where $J_i$ – is a partial criterion for i-th output, $\alpha_i$ – weight coefficient.

## 3 Intelligent Control System Components

*A. Multi- agent system (MAS)*

As it was considered above two MAS have been created for CMAS.

*B. Ontologies(O)*

The ontology can be describe with a formal model like a triple (Calagary and Sanchez 2007, Hadjiski and Boishina 2007, Straccia 2001):

$$O = \langle X, R, \Sigma \rangle \tag{4}$$

where:

- $X$ – the finit set of concepts. For the cascade control system those concepts are: the plant's parameters, the kind of constraints, control signals and etc.
- $R$ – describes the relationships and dependences among the different concepts in the system. For the present system $R$ describes the system constraints ($u^l$, $u^u$, $y_1^l$, $y_1^u$, $y_2^l$, $y_2^u$), relations among the plant parameters, disturbances and etc.
- $\Sigma$ – represent the interpretations for instance in the form of axioms. In our system they are: fuzzy rules, rules for making optimal decision and etc.

The developed CMAS uses separate ontologies for knowledge sharing among subsystems. The MASs are running on one host into different containers (JADE 2009). This leads to big communication rate and slow control. Usage of ontologies to share the knowledge improves the cascade control system (Hadjiski and Boishina 2007) and decrease the communication rate.

### C. *Ant Colony Optimization (ACO)*

The convincing results received in previous works (Hadjiski and Boishina 2007, Hadjiski,Sgurev and Boishina 2006, Hadjiski and Boishina 2005)have stimulated further implementation of ACO in given contribution.

According the Fig. 2 the agents and ontologies are interacting in order to retrieve the knowledge, update the pheromones in MASs and to adapt the system environment (Chang 2005). The ants communicate each other trough the system environment represented by ontology. The optimal control is chosen according the probability the control to be “bad” or “good” (Dorigo and al. 2006):

$$P_u(m) = \frac{(u_m + g)^b}{(u_m + g)^b + (l_m + g)^b} \tag{5}$$

$$P_l(m) = 1 - P_u(m) \tag{6}$$

where: $u_m$ is the number of ants accepted decision for good; $l_m$- number of ants accepted decision for bad; $P_u(m)$ - the probability of decision to be good; $P_l(m)$ - the probability of decision to be bad; $g$ and $b$ – parameters; m – number of all ants in the colony;

When the knowledge in the systems is updated, the ants spread the information about their believes for a "good" or “bad” decision and the quality of approved decisions in the current time $k$ can be describe by the relation:

$$Q(k) = \frac{TP(k)}{TP(k) + FN(k)} \frac{TN(k)}{FP(k) + TN(k)} \tag{7}$$

where: $TP(k)$ is the number of ants accepted $P_u(m)$ for good decision; $TN(k)$ - number of ants accepted $P_u(m)$ for bad decision; $FP(k)$ - number of ants accepted $P_l(m)$ for good decision; $TN(k)$ - number of ants accepted $P_l(m)$ for bad decision.

The equation (7) gives the possibilities to define the qualities of pheromones which are necessary to update the information in the system $\tau_{ij}(k+1)$. That could be represented in the form:

$$\tau_{ij}(k+1) = \tau_{ij}(k) + \tau_{ij}(k)Q(k) \tag{8}$$

where $\tau_{ij} = \sum_{i=1}^{a} b_i^{-1}$ is the quality of dispersed pheromones; a- number of attributes includes in decision making; b – the possible attribute values.

From equations (7) and (8), the change of knowledge can be given as follows:

$$SK(k+1) = SK(k) + SK(k)Q(k) \tag{9}$$

where: SK(k), SK(k+1) – are current and foregoing system knowledge.

## 4 Adaptive Fuzzy Ontology

### A. *Fuzzy Ontologies*

The formal ontology model (FOM) (4) is presented in this work via W3C specified ontology language OWL in the version of OWL-DL (W3C 2009) This kind "crisp" ontology becomes less suitable in our case of cascade control, where some of the concepts have not a precise definition or posses uncaps or fuzzy boundaries. The problem to deal with imprecise concepts has been considered more than 40 years ago in (Zadeh 1965) on the base of Fuzzy Sets and Fuzzy Logic. Resently in a number of works a variety of extensions of description logic (DL) have been proposed in order to define Fuzzy Ontology (Calagary and Sanchez 2007, Lin 2007, Stoilos and al. 2005, Straccia 2001, Toeng 2007, Widyantorn 2001). Introduction of formal Fuzzy Ontology (FO) strongly depends on the application domain. In this paper we use fuzzy extension of the FOM in the form:

$$O_F = <X_F, R_F, \Sigma_F> \tag{10}$$

where: $X_F$ is a set of fuzzy concepts, $R_F$ is a set of fuzzy relations, and $\Sigma_F$ is a set of fuzzy interpretations. As "everything is a meter of degree" (Zadeh 1965) the degree of fuzziness in the $O_F$ components could be very different, depending on the application domain. In this investigation the largest degree of fuzziness is concentrated in the component $\Sigma_F$.

### B. *Ontology merging*

In the two cascade Multi-Agents Systems each of the MASs is serviced by separate ontology. The merge of ontologies is needed to assure a stable ontology. In order to create a stable fuzzy ontology certain checks for accuracy and fulfillness must be done. A part of the knowledge in the first MAS is equal to the knowledge in the second MAS. The goal of ontology merge is to assure a stable agent's work, to avoid the conflicts among the agents, and to prevent a possibility to work with wrong data.

The main steps of ontology merging are: to define equal concepts, relations and interpretations; to recognize the common indications as well as all differences in ontologies structure and semantics. The knowledge change must be admitted and process of ontology merging becomes a cyclic one.

For indication of stable ontology merge the Kolmogorov algorithm is used (Vitorino and Abraham 2009). The mechanism of knowledge merge can be described in the next steps:

1) To detect the common knowledge domains for both ontologies.

2) To define the system merge knowledge ($SK_{12}$) retrieved from the first and second MAS in case when for the fuzzy ontology knowledge from the first one ($SK_1$) has a priority. The merge knowledge ($SK_{21}$) shows that the knowledge retrieved from the second system ($SK_2$) is with higher priority.

3) To determine the destination among the present knowledge in two MAS and the common knowledge:

$$D(k+1) = \frac{(SK_{12}(k) - SK_1(k)) + (SK_{21}(k) - SK_2(k))}{SK_1(k) + SK_2(k)} \tag{11}$$

In case when the couple $\{SK_{12}, SK_{21}\} > \{SK_1, SK_2\}$ the D(k) is always positive. That assures the knowledge covering and stable new merge ontology.

*C. Fuzzy Ontology merging*

The mechanism of merge the ontology structure and semantic is shown in Fig. 4. In the process of merge, each knowledge is checked whether it belongs to some existing cluster or to individual cluster. When D(k) is positive, the knowledge from two ontologies could be merged. When D is negative new cluster for each separate knowledge is formed in order to assure stable information processing in the system.

The value of D(k) is computed for each new ontology cluster. After that the "rank of relation" represents the relation among the terms and concepts in the ontologies. According the value of the "rank of relation" the merge knowledge is been fuzzificated (Fig. 5).

The "rank of fuzzification" for each element in ontology cluster can be defined as:

$$K_{fuzzy}(i) = \mu_{SK_{12}}(i) D(i) \tag{12}$$

where: $i$ - number of Ontology element; $K_{fuzzy}$ – merge fuzzy knowledge rank; $\mu_{SK_{12}}(i) = \frac{S_{12}(i)}{S_{total}}$ - "rank of relation", where $S_{12}$ is the number of related terms and concepts from the Ontology 1 and Ontology 2 with current elements; $S_{total}$ – the total number of relations in the system.

The ants in the system search the optimal control in new fuzzy area of the knowledge represented from Fuzzy Ontology. Each of the MAS works with corresponding cluster for current situation (values of $\Omega_1 = 0, 1$ and $\Omega_1 = 0, 1$ and $\alpha_i(k) = 0{,}1$). In case of Ontologies merge with fuzzification of knowledge the two MAS work together like common system, which is using the classical ACO (Dorigo and al. 2006).

To illustrate the generation of Fuzzy Ontology a part of corresponding source code is shown in Fig. 5. The "rank of fuzzification" is noted as a parameter - fuzzyRank. The differences among ontologies are denoted with the property – diferentFrom. The equivalent parts of the knowledge is denoted as sameAs. The terms, denoted with concepts, and common regions for the two ontologies are (intersectionOf).

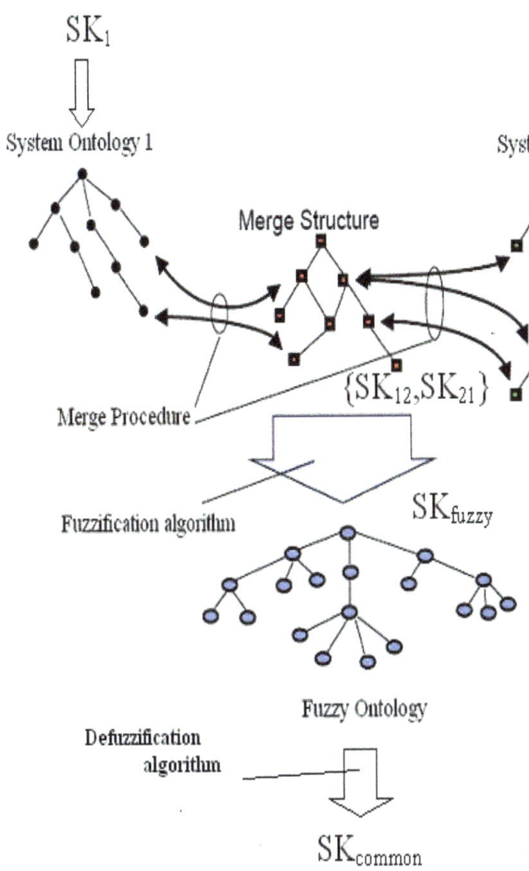


**Fig. 4** Ontology merge and fuzzification procedure

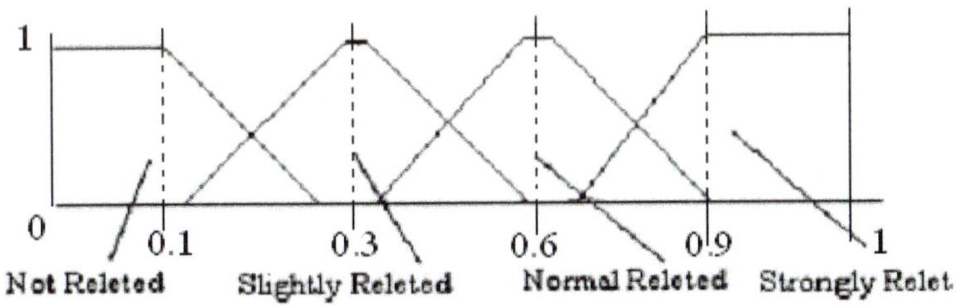


**Fig. 5** "Rank of relation" the merge knowledge

### *D. Fuzzy Ontology adaptation*

At it was mentioned above in this investigation a big disturbance $v_3$ (Fig. 7) influencing over the gain coefficients in TITO plant transfer functions have been accepted. The developed CMAS reacts adaptively in order to prevent the unstable behaviour. The agents estimate the size of this disturbance using fuzzy logic (Zadeh 1965) and compute the rank of possible gain variances Agents define the fuzzy gain variance constraints. The gain's fuzzyfication is done in Agent Knowledge Layer (Fig. 6) in dependence on the given system of rules and information represented in

```
<owl:Class rdf:ID="HVACControl">
        <rdfs:subClassOf>
                <owl:Restriction>
                        <owl:onProperty rdf:resource="#hasAirControl"/>
                        <owl:cardinality rdf:datatype="&xsd;nonNegativeInteger">1
   </owl:cardinality>
                </owl:Restriction>
                <owl:ObjectProperty rdf:ID="forAirControl">
                        <rdfs:domain rdf:resource="#Control"/>
                        <rdfs:range rdf:resource="#AirConotrol"/>
                        <rdfs:parameter rdf:resource="#fuzzyRank"/>
                </owl:ObjectProperty>
        </rdfs:subClassOf>
        <rdf:HVACControl rdf:ID="computeVentilationControl">
                <owl:differentFrom rdf:resource="#control1"/>
                <owl:differentFrom rdf:resource="#control2"/>
                <rdfs:parameter rdf:resource="#fuzzyRank"/>
        </rdf:HVACControl>
        <rdf:AirControl rdf:ID="setpoint1">
                <owl:sameAs rdf:resource="#control1"/>
                <owl:diferentFrom rdf:resource="#control1second"/>
                <rdfs:parameter rdf:resource="#fuzzyRank"/>
        </rdf:AirControl>
        <rdf:AirControl rdf:ID="setpoint2">
                <owl:sameAs rdf:resource="#output1"/>
                <owl:diferentFrom rdf:resource="#output1second"/>
                <rdfs:parameter rdf:resource="#fuzzyRank"/>
        </rdf:AirControl>
        <owl:intersectionOf rdf:parseType="Control">
                <owl:Class rdf:about="#HVACControl"/>
                <owl:Restriction>
                        <owl:onProperty rdf:resource="#hasAirControl"/>
                        <owl:hasValue rdf:resource="#control1"/>
                </owl:Restriction>
        </owl:intersectionOf>
</owl:Class>
```

**Fig. 6** Merge Fuzzy Ontology representation

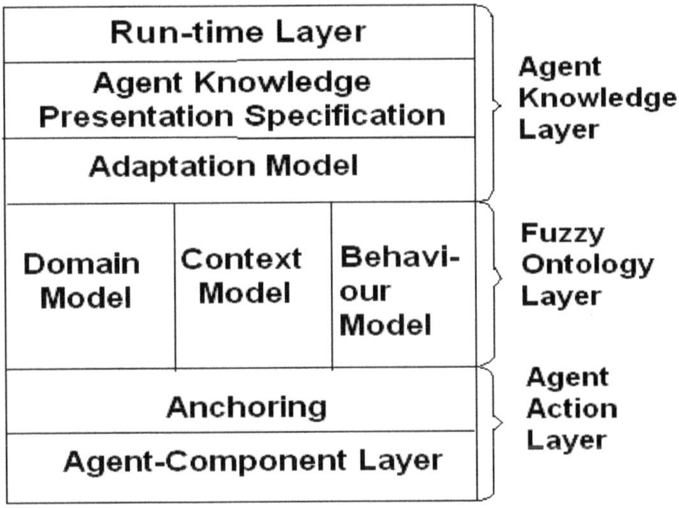


**Fig. 7** Layers in CMAS using adaptive fuzzy ontology

Run-time Layer. This rules are a part of the Knowledge Presentation Layer. In that layer the adaptive system model is been defined (Adaptation Model Layer). To provide a better control action the pheromone trace is been fuzzificated (Dorigo and al. 2006) and that leads to fuzzification of the current knowledge system. The knowledge is updated via fuzzy SK bringing the ontologies for adaptation the clusters. The CMAS chooses a stabile control behaviour based on the knowledge defined in the Behaviour Model Layer (which is a part of the Fuzzy Ontology Layer) and the corresponding cluster. After that the received retrieve control is been applied by Agent Component Layer. The Agent Component Layer is a physical layer that forms the control actions in the system.

## 5 Situation Control

To obtain the proper strategy for control, first we need to estimate the proper system situation. In this kind of systems, a lot of system parameters can be changed which will lead to different strategy selection. After identification of the proper system parameters the following scheme is been used to obtain the current system situation and strategy selection (Fig. 8).

Following the proposed structure for the system adaptation according the new situation (Correas and al. 1999, haase and al. 2008) there is a need first to detect it and after that to chose the proper control strategy. When the system received the new data and some knowledge about the controlled plant the agents and ontologies formed the model of the current situation. The situation model is formed from different fuzzy sets describing the system parameters and their possible variations. According the model and the system knowledge about the situation some system

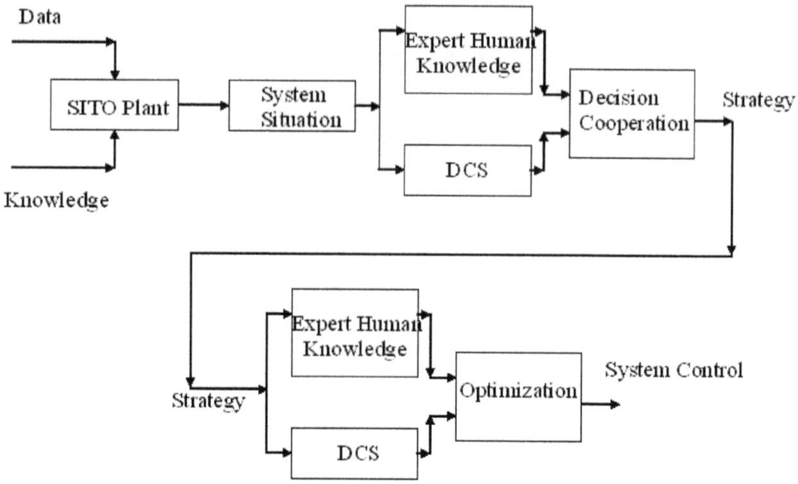


**Fig. 8** Algorithm for Strategy Selection

behavior is been defined. The system behavior is given below. In many of the system situation the control system, which is present from multi-agent control system and composed Decentralized Control System (DCS) can take decision and estimate the current situation. There are a lot off situation with the present of uncertainness in which the agent couldn't found out the situation by selves (harrera and al. 2008). So, in our system the expert knowledge is been used. The expert knowledge is composed as a fuzzy ontology which supports the agents system. The decision cooperation implements the following structure (Fig. 9):

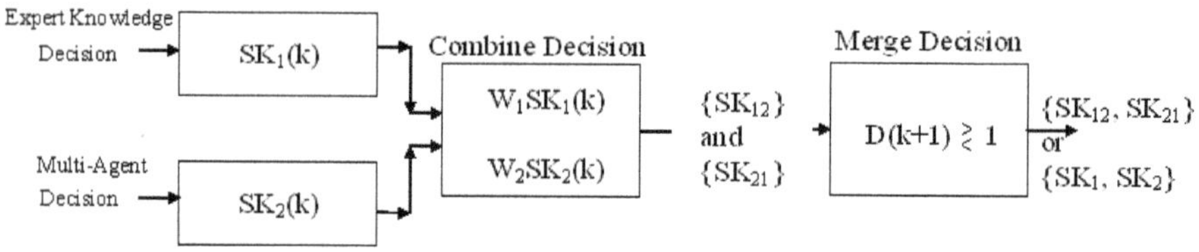


**Fig. 9** Decision Coperation

In the presence of uncertainness into the system there are possibility of different expert knowledge decision and agent decision. So, in our system that decisions are been weighting. When the weight of the expert decision is bigger the weight of agent decision the $\{SK_{12}\}$ is formed. Into the other case the $\{SK_{21}\}$ is obtained. In a way to assure a stable decision taken about the current situation the Kolmogorov algorithm is been executed (Vitorino and Abraham 2009). The obtained strategy must be again estimated and optimized in a way to be defined the optimal system control.

## 6 Strategy Selection

According to above statements, the strategy selection algorithm must be pre-defined for multi-dimensional control systems. The algorithm couldn't choose only one proper strategy for control according to the current system situation. The strategy selection in a way to adapt the system, become a complex problem with determination of a combinatorial strategy selection.

The present system of SITO plant control using the cascade way of control has a possibility to choose different way of control based of the decision making of agent system. The given system can make a reconfiguration of the control loops, change the parameters of control in try to adapt to the current system environment.

Some of the possible strategies of control are given into table 1.

**Table 1**

| № | Situation | MVs | SVs | Transfer Functions | Controler(s) |
|---|---|---|---|---|---|
| 1 | SITO-1 | $u_1$ | $y_1, y_2$ | $W_{11}, W_{12}$ | $R_1, R_2$ |
| 2 | SITO-2 | $u_2$ | $y_1, y_2$ | $W_{21}, W_{22}$ | $R_1, R_2$ |
| 3 | SISO-1 | $u_1$ | $y_1$ | $W_{11}$ | $R_1$ |
| 4 | SISO-2 | $u_2$ | $y_2$ | $W_{22}$ | $R_2$ |

The mechanism for merging of ontology knowledge can be used into represented of the strategy combination:

$$D(k+1) = \frac{(S_{12}(k) - S_1(k)) + (S_{21}(k) - S_2(k))}{S_1(k) + S_2(k)} \tag{13}$$

where: D(k+1) – difference between the strategies, $S_{12}/S_{21}$ - combination between strategies, $S_1/S_2$ – stand alone strategies.

In the cases when the following combination can assure the positive value of D(k+1), its means that the chosen combination of strategies is good. In other way the implemented Ant Colony Optimization algorithm searches for better control strategies.

$$\langle S_{12}, S_{21}\rangle > \langle S_1, S_2\rangle \tag{14}$$

$$D(k+1) > 1 \tag{15}$$

This kind of searching of optimal control strategy can be represented as a combinatory task of:

$$\langle \eta^0, NO_x^{\ 0}, \rho^0\rangle = S_i! \tag{16}$$

This task can be illustrated as (Fig. 10):

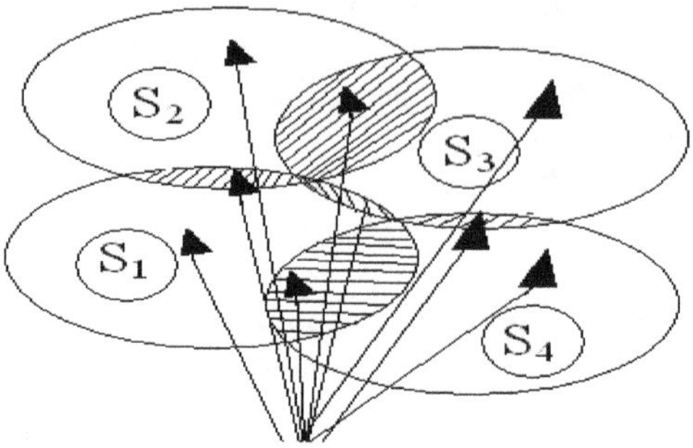


**Fig. 10** Strategy combination selection

### *A. Situation Ontology*

There is designed a Situation Ontology, which describes the different situation which are possible for the cascade system for SITO plant control. The Situation Ontology (SitOnto) is designed by system experts, and contains a various scenarios of system behavior. The experts designed the SitOnto from number of sets, which contains the information about system parameters. When the system parameters are into some kind of possible range, based on expert knowledge about the system, they defined the sets of :

- excellent behavior: all system parameters are in range, the system error of control is minimal;
- normal behavior: some of the system parameters are into the "soft" system constrains, the priority system output is into constrains, may be there is some noxious emissions out of range, but the SITO system is under optimal control;
- alert behavior: some of the system parameters are not good – big quality of $NO_x$, low power, the system constrains start to be punished;
- dangerous behavior: a lot of the system parameters are not good –$NO_x$, are out of the control, the system seems to become unstable;
- emergency behavior: everything is out of range. May be there is system fault, but all of the constraints are been punished. The SITO system is unstable.

This SitOnto integer the values, ranges, constraints about the most of the system perimeters and tried to described the all possible behaviors of the system.

### *B. Strategy Selection and Situation Ontology*

The agents in MAS run asynchronously and choose a different strategy for control in dependence of their knowledge about the current system situation. Each of the agents of the system reacts according its knowledge and the knowledge shared into the system. The chosen combination of strategy of control can lead the SITO system to different behavior (Oyrazabal 2005,Volero, Correas and Serra 1998) (Fig. 11).

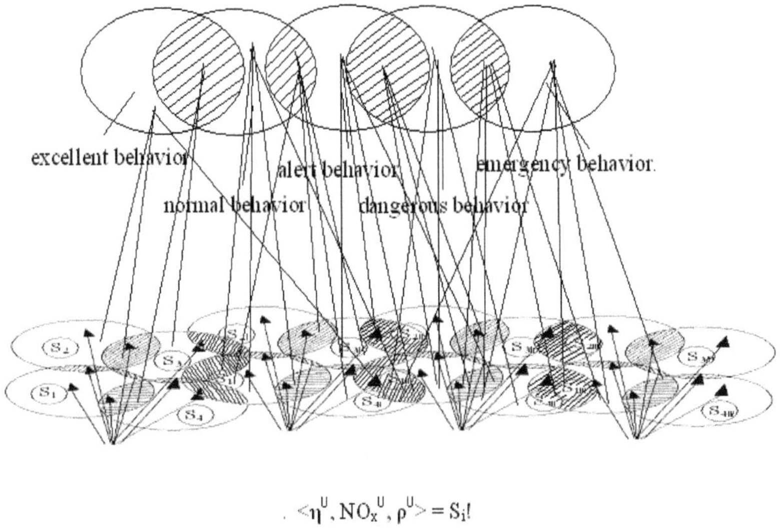


**Fig. 11** Strategy Selection according the system behavior

Assuming that the experts can't predict the whole behavior of the multiagent system and how that going to reflect to the system behavior. In this way into the mapping is include the current value of the agents decision according to the Ant Colony Optimization the control behavior to be good (eq. 5).

The experts supposed that there are some possible levels of control according to the value of the $P_u(m)$, which is represent in fig. 12.

To be more accurate into creating of the mapping between the Strategy Ontology and the Situation Ontology, the system uses the fuzzyficated value of the possibility for good control - $K_{fuzzy}$. To obtain this value it is been include like approximated weight parameter $Wsk_{ij}$ (11). This weighting parameter is been used to include the decision of the expert knowledge which reflect to the ants decisions.

$$P_u^*(m) = P_u W_{skij} \tag{17}$$

The choice of strategy reacts to the system behavior, which can be described with non-linear objective function:

$$System_{behavior} = f\{O_{SitOnto}, O_{StragOnto}\} P_u^*(m) \tag{18}$$

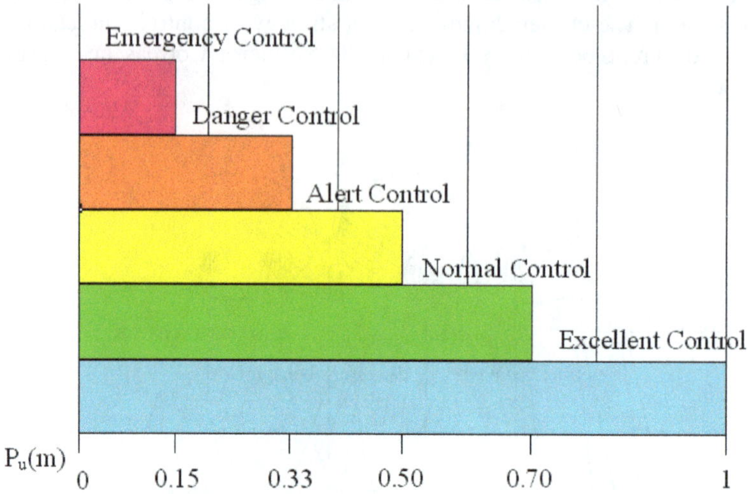


**Fig. 12** Possible system behaviors

## 7 Software Realization

All software realization is accomplished on the base of standardized specifications. For the intelligent agents in MAS the JADE (Java Agent Development Environment) (FIPA 2009) platform is used, which corresponds to FIPA (Foundation of Intelligent Physical Agents) (FIPA 2009) requirements. The FIPA Ontology Service Reference Model (FIPA 2009) have been used for development of ontology – agent co–operation. RDF (Resource Description Framework) and OWL (Ontology Web Language) as a model and syntax for a basic ontology creation have been used according W3C (World Wide Web language)(W3C 2009) specifications. Because the lack of clear specifications for fuzzy extending of the standard OWL the most established methods based on description logic formalism (Calagary and Sanchez 2007, Lin 2007, Straccia 2001, Toeng 2007, Widyantron and al. 2001) have been adopted in this work.

All together the developed CMAS considers 100 agents (58 in MAS1 and 42 in MAS2), 60 classes in both ontologies as well as corresponding number of relations and axioms.

## 8 Simulation Results

*A. Control System without ontology merge*

When the system is controlled without coordination among the system ontologies the agents loose coordination and possibility to consolidate knowledge and communication among the agents. The result of is that agents die. They loose the control functions (Fig. 13) and the cascade control system become invalid.

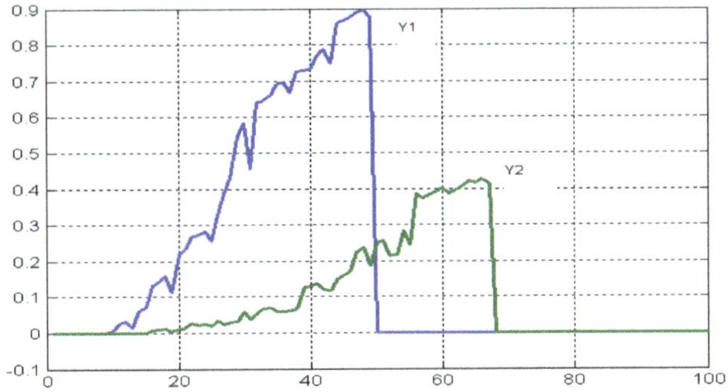


**Fig. 13** Reaction of the system without ontology merging

*B. Control system using Fuzzy Ontology with knowledge merge*

In the case when agents' system uses merged fuzzy ontology, the knowledge in the two system ontologies is consolidated. That results in avoiding the conflicts among agents and the CMAS becomes stable. The agents communicate each other using clusters from the fuzzy ontology. When some changes appeare in the plant the ontology merge mechanism is applied and cascade control system remains stable.

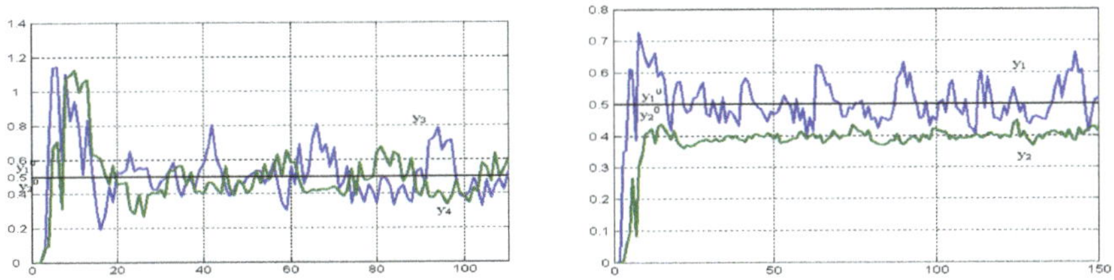

a) First Plant ($y_2$ and $y_3$) b) Second Plan t($y_1$ and $y_2$)

**Fig. 14** Case 1: $\Omega_1 = 0$ and $\Omega_2 = 0$

In Fig. 14 is presented the case when the CMAS is trying to keep the four outputs in the range only with one manipulated variable u. The CMAS uses the

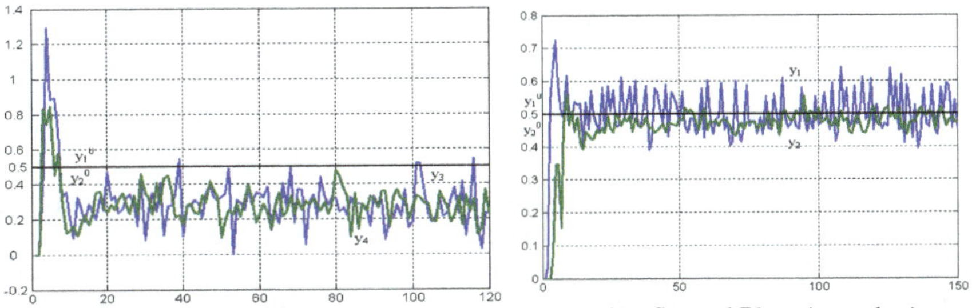

a) First Plant ($y_2$ and $y_3$) b) Second Plan t($y_1$ and $y_2$)

**Fig. 15** Case 2: $\Omega_1 = 1$ and $\Omega_1 = 0$

information shared in Adaptive Fuzzy Ontology and according the value of $K_{fuzzy}$ decides how to use the information. That situation is very complicated, because these kind of plants belong to the type Single Input Two Outputs (SITO) which are very hard to control in requested borders [5, 6, 7].

In Fig. 15 is given the reaction of the system in case with two possible manipulated variables – $y_3$ and u are available. In the showed situation $y_1$ and $y_2$ are in the specified ranges.

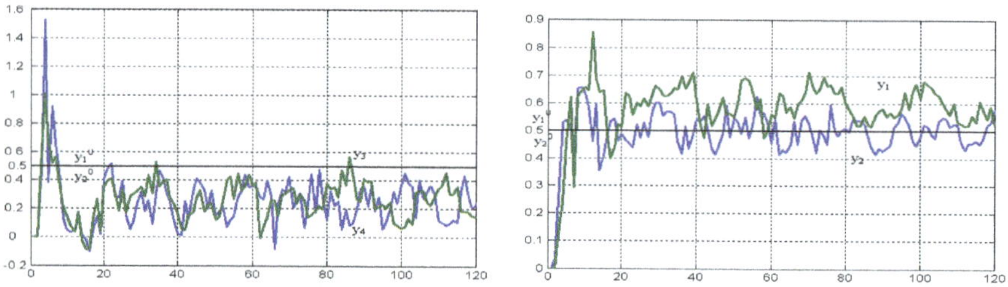


a) First Plant ($y_2$ and $y_3$) b) Second Plan t($y_1$ and $y_2$)

**Fig. 16** Case 3: $\Omega_1 = 0$ and $\Omega_1 = 1$

Fig. 16 illustrates the case when again two possible control variables – $y_4$ and u are available

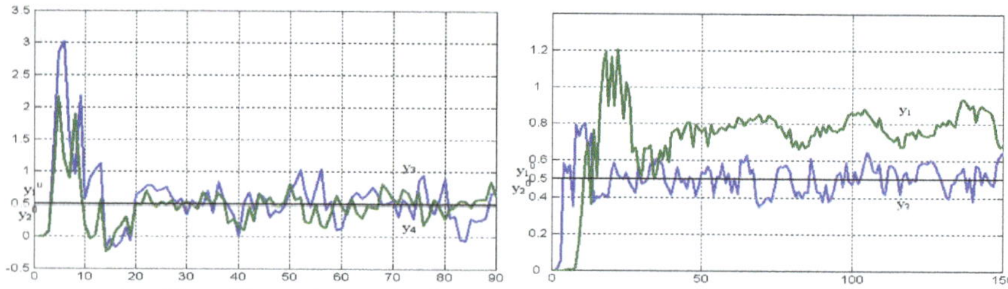


a) First Plant ($y_2$ and $y_3$) b) Second Plan t($y_1$ and $y_2$)

**Fig. 17** Case 4: $\Omega_1 = 1$ and $\Omega_1 = 1$

Fig. 17 shows the simulation results for a case with tree manipulated variables – $y_3$,$y_4$ and u.

## 9 Conclusions

Multi-agent and ontology collaboration is promising approach to control of complex industrial plants with large uncertainty. The Ant Colony Optimization is relevant method for integration, coordination and communication rate reduction in hybrid agent/ontology systems. The Fuzzy Ontology with adaptation is suitable functionality in order to overcome unforeseen variations in the plant behavior, constraints, disturbances, control settings. The obtained results validate adopted approach for hybrid agent/ontology control with adaptation. The developed control structures could be incorporated successfully into the established industrial control systems.

## References

[1] Calegary, S., Sanchez, E.: A Fuzzy Ontology – Approach to Improve Semantic Information Retrieval. In: Proc. of 6th Int. Semantic Web Conference, Korea (2007)

[2] Correas, L., Martinez, A., Volero, A.: Operation Diagnosis of a Combined Cycle based on Structural Theory of Thermoiconomics. In: ASME Int. Mechanical Engineering Congress and Exposition, Nashvill, USA (1999)

[3] Dorigo, M., Birattari, M., Stützle, T.: Ant Colony Optimization. IEEE Computational Intelligence Magazine 1(4) (2006)

[4] FIPA Specification (2006), `http://www.fipa.org`

[5] Gonzalez, E.J., Hamilton, A., Moreno, L., Marichal, R.L., Toledo, J.: A MAS Implementation for System Identification and Process Control. Asian Journal of Control 8(4) (2006)

[6] Haase, T., Weber, H., Gottelt, F., Nocke, J., Hassel, E.: Intelligent Control Solutions for Steam Power Plants to Balance the Fluctuation of Wind Energy. In: Proc. of the $17^{th}$ World IFAC Congress, Seoul, Korea (2008)

[7] Hadjiski, V.B.: Dynamic Ontology–based Approach for HVAC Control via Ant Colony Optimization. In: DECOM 2007, Izmir, Turkey (2007)

[8] Hadjiski, M., Sgurev, V., Boishina, V.: Intelligent Agent-Based Non-Square Plants Control. In: Proc. of the 3-d IEEE Conference on Intelligent Systems, IS 2006, London (2006)

[9] Hadjisk, M., Boishina, V.: Agent Based Control System for SITO Plant Using Stigmergy. In: Intern. Conf. Automatics and Informatics 2005, Sofia, Bulgaria (2005)

[10] Herrera, S.I., Won, P.S., Reinaldo, S.J.: Multi-Agent Control System of a Kraft Recovery Boiler. In: Proc. of the $17^{th}$ World IFAC Congress, Seoul, Korea (2008)

[11] JADE 2007 (2007), `http://jade.tilab.com`

[12] Lin, J.N.K.: Fuzzy Ontology-Based System for Product Management and Recommendation. International Journal of Computers 1(3) (2007)

[13] Manesis, A., Sardis, D.J., King, R.E.: Intelligent Control of Wastewater Treatment Plants. Artifical Intelligence in Engineering 12(3) (1998)

[14] Mitra, S., Gangadaran, M., Rajn, M., et al.: A Process Model for Uniform Transverse Temperature Distribution in a Sinter Plant. Steel Times International (4) (2005)

[15] PiT Navigator, Advanced Combustion Control for Permanent Optimized ail/fuel Distribution, `http://www.powitec.de`

[16] Valero, A., Correas, L., Lazzsreto, A., et al.: Thermoeconomic Philosophy Applied to the Operating Analysis and Diagnosis of Energy Systems. Int. J. of Thermodynamics 7(N2) (2004)

[17] Ramos, V., Abraham, A.: ANTDIS: Self-organized Ant based Clustering Model for Intrustion Detection System, `http://www.arxiv.org/pdf/cs/0412068.pdf`

[18] Volero, A., Correas, L., Serra, L.: Online Thermoeconomic Diagnosis of Thermal Power Plants. In: NATO ASI, Constantza, Rumania (1998)

[19] Lee, C.-S.: Introduction to the Applications of Domain Ontology (2005), http://www.mail.nutn.edu.tw/~leecs/pdf/Leecs-SMC_Feature_Corner.pdf
[20] Smirnov, D.N., Genkin, B.E.: Wastewater Treatment in Metal Processing, Metallurgy, Moskow (1989) (in Russian)
[21] Stoilos, G., Stamon, G., Tzonvaras, V., Pan, J.Z., Horrocks, I.: Fuzzy OWL: Uncertainty and the Semantic Web. In: Proc. Int. Workshop OWL: Experience and Directions (2005)
[22] Straccia, U.: Reasoning with Fuzzy Description Logics. Journal of Artificial Intelligence Research 14(2) (2001)
[23] Oyarzabal, J.: Advanced Power Plant Scheduling. Economic and Emission Dispatch, Dispower (19) (2005)
[24] Terpak, J., Dorcak, L., Kostial, I., Pivka, L.: Control of Burn – Through Point for Aglomeration Belt. Metallurgia 44(4) (2005)
[25] Toeng, H.C.: Internet Application with Fuzzy Logic and Neural Network: A Survey. Journal of Engineering, Computing and Architecture 1(2) (2007)
[26] Yang, Z., Ma, C., Feng, J.Q., Wu, O.H., Mann, S., Fitch, J.: A Multi – Agent Framework for Power System Automation. Int. Journal of Innovations in Energy System and Power 1(1) (2006)
[27] Widyantorn, D.H., Yenn, J.: Using Fuzzy Ontology for Query Refinement in a Personalized Abstract Search Engine. In: Proc. of 9th IFSA World Congress, Vancouver, Canada (2001)
[28] Wooldridge, M.: An Introduction to Multi–Agent Systems. John Wiley, Chichester (2002)
[29] W3C, http://www.w3.org
[30] Zadeh, L.: Fuzzy sets. Information and Control 8(3) (1965)

# NEtwork Digest Analysis Driven by Association Rule Discoverers

Daniele Apiletti, Tania Cerquitelli, and Vincenzo D'Elia

**Abstract.** An important issue in network traffic analysis is to profile communications, detect anomalies or security threats, and identify recurrent patterns. To these aims, the analysis could be performed on: (i) Packet payloads, (ii) traffic metrics, and (iii) statistical features computed on traffic flows. Data mining techniques play an important role in network traffic domain, where association rules are successfully exploited for anomaly identification and network traffic characterization. However, to discover (potentially relevant) knowledge a very low support threshold needs to be enforced, thus generating a large number of unmanageable rules. To address this issue, efficient techniques to reduce traffic volume and to efficiently discover relevant knowledge are needed. This paper presents a NEtwork Digest framework, named NED, to efficiently support network traffic analysis. NED exploits continuous queries to perform real-time aggregation of captured network data and supports filtering operations to further reduce traffic volume focusing on relevant data. Furthermore, NED exploits two efficient algorithms to discover both traditional and generalized association rules. Extracted knowledge provides a high level abstraction of the network traffic by highlighting unexpected and interesting traffic rules. Experimental results performed on different network dumps showed the efficiency and effectiveness of the NED framework to characterize traffic data and detect anomalies.

## 1 Introduction

Nowadays computer networks have reached a very large diffusion and their pervasiveness is still growing, passing from local, departmental, and company networks to more complex interconnected infrastructures. The rapid expansion of the Internet is a typical example introducing also the problems of such global

---

Daniele Apiletti · Tania Cerquitelli · Vincenzo D'Elia
Politecnico di Torino, Dipartimento di Automatica e Informatica
Corso Duca degli Abruzzi 24, 10129 Torino, Italy
e-mail: {daniele.apiletti,
tania.cerquitelli,vincenzo.delia}@polito.it

V. Sgurev et al. (Eds.): Intelligent Systems: From Theory to Practice, SCI 299, pp. 41–71.
springerlink.com 

networks. The evolution of applications and services based on Internet, e.g., peer-to-peer file sharing, instant messaging, web-services, etc., allows the users to exchange and share almost every type of information. In this complex scenario, efficient tools for network traffic monitoring and analysis are needed. Network traffic analysis can be summarized as the extraction of relevant knowledge from the captured traffic to keep it under the control of an administrator. However, due to the continuous growth in network speed, terabytes of data may be transferred through a network every day. Thus, it is hard to identify correlations and detect anomalies in real-time on such large network traffic traces. Hence, novel and efficient techniques, able to deal with huge network traffic data, need to be devised.

A significant effort has been devoted to the application of data mining techniques to network traffic analysis [6]. The application domains include studying correlations among data (e.g., association rule extraction for network traffic characterization [4], [11] or for router misconfiguration detection [14]), extracting information for prediction (e.g., multilevel traffic classification [13], Naive Bayes classification [16]), grouping network data with similar properties (e.g., clustering algorithms for intrusion detection [18], or for classification [7], [9], [15], [21]). While classification algorithms require previous knowledge of the application domain (e.g., a labeled traffic trace), association rule extraction does not. Hence, the latter is a widely used exploratory technique to highlight hidden knowledge in network flows. The extraction process is driven by enforcing a minimum frequency (i.e., support) constraint on the mined correlations. However, to discover (potentially relevant) knowledge a very low support constraint has to be enforced, thus generating a huge number of unmanageable rules [4]. To address this issue, a network digest representation of traffic data and a high-level abstraction of the network traffic are needed.

Since continuous queries [3] are an efficient technique to perform real-time aggregation and filtering, they can be exploited to effectively reduce traffic volume. Instead, association rule extraction is un unsupervised technique to efficiently represent correlation among data. This paper presents a novel approach jointly taking advantage of both continuous queries and association rules to efficiently perform network traffic analysis.

We propose the NEtwork Digest framework, named NED, which performs network traffic analysis by means of data mining techniques to characterize traffic data and detect anomalies. NED performs (i) on-line stream analysis to aggregate and filter network traffic, and (ii) refinement analysis to discover relationships among captured data.

NED allows on-line stream analysis concurrently with data capture by means of user-defined continuous queries. This step reduces the amount of network data, thus obtaining meaningful network digests for pattern discovery. Furthermore, NED provides a refinement analysis exploiting two mining algorithms to efficiently extract both association rules and generalized association rules. NED's final output could be either a set of association rules [12] or a set of generalized association rules [32], which are able to characterize network traffic and to show correlation and recurrence of patterns among data. Generalized association rules provide a powerful tool to efficiently extract hidden knowledge, discarded by previous approaches. Taxonomies, user-provided or automatically inferred from data, drive the pruning phase of the extraction process. Experiments performed on different network dumps showed the efficiency and effectiveness of the NED framework in characterizing traffic data and highlighting meaningful features.

## 2 NED's Architecture

NED (NEtwork Digest) is a framework to efficiently perform network traffic analysis. NED addresses three main issues: (i) *Data stream processing* to reduce the amount of traffic data and allow a more effective use, both in time and space, of data analysis techniques, (ii) *Taxonomy generation* to automatically extract a hierarchy of values for each traffic flow attribute, and (iii) *Hidden knowledge extraction* from traffic data to characterize network traffic, detect anomalies, and identify recurrent patterns. The last step can be performed by means of two different data mining techniques (i.e., association rules and generalized association rules).

Fig. 1 shows NED's main blocks: data stream processing, taxonomy generation, and refinement analysis. Traffic packets can be collected by an external network capture tool (i.e. Tstat [41], Analyzer [40], Tcpdump [42]) or directly sniffed at runtime from an interface of the machine running the framework. These data are the input of the stream processing block, which summarizes the traffic while preserving structural similarities among temporally contiguous packets. Furthermore, it discards irrelevant data to reduce traffic volume. Data stream processing is performed concurrently with data capture by means of *continuous queries*, whereas hidden knowledge is extracted from the stored continuous query results in a refinement analysis step, which currently implements two efficient *association rule mining* algorithms. Other data mining techniques [12] may be easily integrated in this step. Continuous queries perform *aggregation* (i.e., similar records can be summarized by a proper digest) and *filtering* (i.e., meaningless data for the current analysis is discarded) of network traffic. The output *flows* (i.e., filtered and aggregated packet digests) can be saved into a permanent data store. The storage is required only when different refinement analysis sessions need to be performed.

The taxonomy generation block extracts a taxonomy for each traffic flow attribute. A taxonomy is a hierarchy of aggregations over values of one attribute in the corresponding value domain. It is usually represented as a tree. Taxonomies drive the pruning phase of the knowledge extraction process to efficiently discover unexpected and more interesting traffic rules. Furthermore, the extracted knowledge provides a high level abstraction of the network traffic. In the NED framework taxonomies could also be provided directly by the user.

The aim of the refinement analysis step is to discover interesting correlations, recurrent patterns and anomalies on traffic data. Currently, interesting patterns can be extracted in the form of either association rules or generalized association rules. While association rules represent correlations and implications among network traffic data, generalized association rules provide a high level abstraction of the network traffic and allows the discovery of unexpected and more interesting

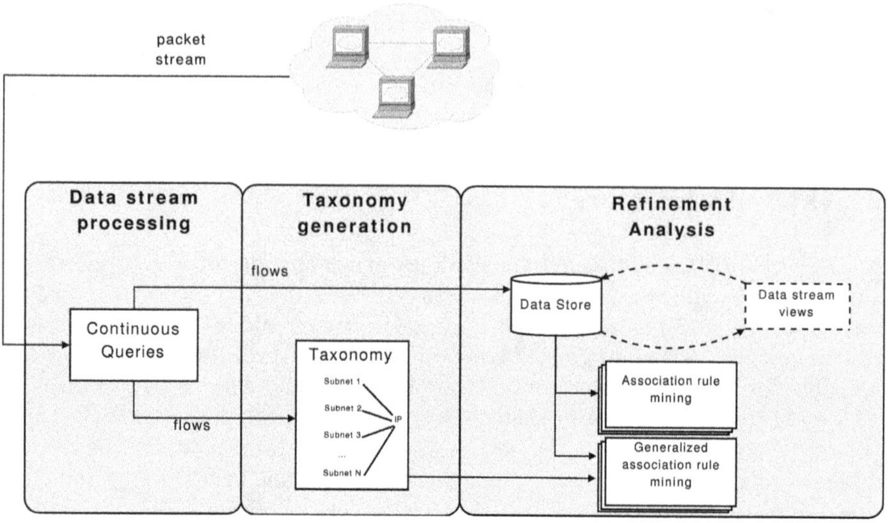


**Fig. 1** NeD's architecture

traffic rules. Furthermore, the framework allows other data mining techniques to be easily integrated.

The refinement analysis is a two step process: (i) An optional data stream view block selects a suitable user-defined subset of flows to focus the following analysis on. (ii) Rule extraction is performed either on the data stream view, which contains the selected flows, or on all the flows in the permanent data store. Taxonomies drive the mining process when generalized association rules are extracted. To describe NED we will use a running example, which will be validated on real datasets in the experimental validation section.

## 3 Data Stream Processing

The data stream processing block of NED reduces the volume of traffic data by grouping similar packets and discarding irrelevant ones. The network data traffic could be collected trough dedicated tool called *sniffer*. Those tools are able to reconstruct the data packets starting from the bit flows transmitted on channels to which tools are connected via hardware interfaces. A sniffer can reconstruct the traffic, for example, by following the general standard ISO/OSI. Thus, network traffic can be considered as a stream of structured data. Each packet is a record whose attributes (i.e., tags) are defined by network protocols. Each record is characterized by at most one value for each tag. In our running example, we focus on source and destination IP addresses, source and destination TCP ports, the level-4 protocol (e.g., TCP, UDP), and the size of the packet.

The NED framework is able to analyze either a traffic trace previously saved with a network capture tool (e.g., Analyzer tool [40]), or a live capture from the network interface of the machine on which the analysis is performed. To this aim an ad-hoc implementation of a sniffer using the libpcap libraries [43] for the language ANSI C has been developed.

Since traffic packets are captured as an unbounded stream, a conventional aggregation process would never terminate. To overcome this issue, continuous queries [3] are exploited and CQL (Continuous Query Language [2]) is used. Queries are issued once and then logically run continuously over a sliding window of the original stream. Hence, the following parameters need to be defined: (i) Aggregation and filtering rules expressed in a subset of SQL instructions, (ii) a sliding window, whose duration $l_{length}$ is expressed in seconds, which identifies the current set of data on which rules are applied, (iii) the step $s_{step} \leq l_{length}$, which defines how often the window moves and consequently the output is produced. In NED a record produced as output by the continuous query is a flow, which summarizes a group of similar and temporally contiguous packets, as shown in the following examples.

| Timestamp | Source | Size |
|---|---|---|
| 0 | A | $size_{A,0}$ |
| 1 | A | $size_{A,1}$ |
| 2 | B | $size_{B,2}$ |
| 3 | A | $size_{A,3}$ |
| 4 | B | $size_{B,4}$ |
| 5 | B | $size_{B,5}$ |
| 6 | A | $size_{A,6}$ |
| 7 | B | $size_{B,7}$ |
| 8 | C | $size_{C,8}$ |

(a) A toy packet capture

| Time | Current flows: Source | Size | Output |
|---|---|---|---|
| 0 | A | $size_{A,0}$ | - |
| 1 | A | $\sum_{i\in\{0,1\}} size_{A,i}$ | - |
| 2 | A | $\sum_{i\in\{0,1\}} size_{A,i}$ | - |
| | B | $size_{B,2}$ | |
| . | . | . | . |
| 6 | A | $\sum_{i\in\{0,1,3,6\}} size_{A,i}$ | $\{A, B\}$ |
| | B | $\sum_{i\in\{2,4,5\}} size_{B,i}$ | |
| 7 | A | $\sum_{i\in\{1,3,6\}} size_{A,i}$ | - |
| | B | $\sum_{i\in\{2,4,5,7\}} size_{B,i}$ | |
| 8 | A | $\sum_{i\in\{3,6\}} size_{A,i}$ | $\{A, B, C\}$ |
| | B | $\sum_{i\in\{2,4,5,7\}} size_{B,i}$ | |
| | C | $size_{C,8}$ | |

(b) Flows in window

**Fig. 2** Packet aggregation. $l_{length} = 6$, $s_{step} = 2$

*Example 1*: Fig. 2(a) reports a toy packet capture to describe how the sample query works. The length of the window is 6 UOT (Units Of Time) and the output is produced every 2 UOT (step = 2 UOT). Fig. 2(b) shows the output produced by the continuous query and how the window evolves. Some trivial steps have been omitted.

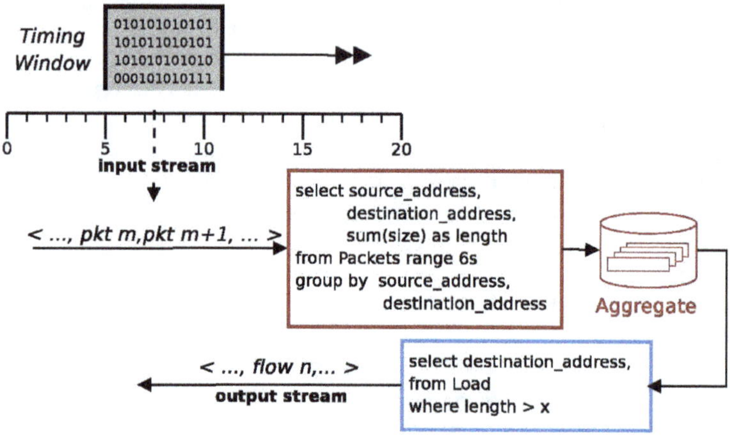


**Fig. 3** Pipeline of continuous queries

To improve CQL query readability, aggregation queries and filtering queries are decoupled (see Fig. 3). Aggregation is performed concurrently with data capturing, while filtering can be executed both on line and off line. The packet filtering is performed in the stream analysis block and discards meaningless packets from the aggregation, whereas flow filtering is performed in the data stream view block and discards undesired flows for the specific analysis purpose.

Three types of continuous queries are implemented in NED as described in the following sections.

## Query 1

The purpose of this query is to reduce the volume of traffic data by preserving information about TCP flows, their participants, their size and their fragmentation. Since this query targets all TCP flows in network traffic, it does not perform any data filtering, but simply aggregates by IP source address, TCP source port, IP destination address, and TCP destination and port. It also computes the total size and the number of packets of the flow.

**Aggregate**

| | |
|---|---|
| Select | source-IP, source-Port, destination-IP, destination-Port, Sum(size) as flow-size, Count(*) as packets |
| From | Packets [Range by 60 seconds] |
| Where | level4 = 'TCP' |
| Group by | source-IP, source-Port, destination-IP, destination-Port |

**Filter**

| | |
|---|---|
| Select | source-IP, source-Port, destination-IP, destination-Port, flow-size |
| From | Aggregate |

## Query 2

This query targets the extraction of the longest IP traffic flows. Once packets have been aggregated by source and destination addresses, and source and destination ports, flows whose length is lower than a given threshold are discarded. The threshold is expressed as a percentage of the total traffic of the current window. Both filtering and aggregation considerably reduce the dataset size.

**Aggregate**

| | |
|---|---|
| Select | source-IP, source-Port, destination-IP, destination-Port, Sum(size) as flow-size, Count(*) as packets |
| From | Packets [Range by 60 seconds] |
| Where | Level3 = 'IP' |
| Group by | source-IP, source-Port, destination-IP, destination-Port |

**Filter**

| | |
|---|---|
| Select | source-IP, destination-IP, flow-size |
| From | Aggregate |
| Where | flow-size > ratio * (Select Sum(flow-size) From Aggregate) |

## Query 3

This query targets the recognition of unconventional TCP traffic, which is usually exchanged on ports different from the well-known ones (i.e., port number > 1024).

Query 3 has two filtering stages. Firstly, only flows which do not have well-known ports as source and destination are kept. Secondly, the longest flows are selected. If these two filtering stages are both performed in the continuous query, the output flows are significantly reduced, but different analysis types become unfeasible. To avoid this limitation, filters maybe applied in the data stream view block.

| | **Aggregate** |
|---|---|
| Select | source-IP, source-Port, destination-IP, destination-Port, Sum(size) as flow-size, Count(*) as packets |
| From | Packets [Range by 60 seconds] |
| Where | level4 = 'TCP' |
| Group by | source-IP, source-Port, destination-IP, destination-Port |
| | **Port-Filtering** |
| Select | * |
| From | Aggregate |
| Where | source-Port > 1024 and destination-Port> 1024 |
| | **Size-Filtering** |
| Select | * |
| From | Port-Filtering |
| Where | flow-size > ratio * (Select Sum(flow-size) From Port-Filtering) |

## 4 Taxonomy Generation

Given a network trace stored in the data store, NED extracts relevant patterns (i.e., rules), which provide an abstract representation of interesting correlations in network data. To drive the mining process for extracting more abstract and interesting knowledge, taxonomies need to be defined. A taxonomy is a hierarchy of aggregations over values of one attribute (e.g., IP address, TCP port) and it is usually represented as a tree. NED allows to automatically infer interesting taxonomies directly from the data. To this aim three different algorithms have been devised and implemented to automatically extract taxonomies for the considered attributes (i.e., IP address, TCP port, packet number, and flow size).

While the port and the IP addresses are actually hierarchical attributes, the packet and byte size are numerical ones. Thus, different algorithms have been devised.

## *4.1 Generation of the Taxonomy Over IP Addresses*

The taxonomy generator for IP addresses (Fig. 4) takes as input the *flows* produced by the continuous query block and a support threshold, i.e. the minimum number of flows that a network prefix is supposed to aggregate. It returns as output a taxonomy where i) the root of the tree is an empty node aggregating all addresses, ii) the leaves of the tree are specific IP addresses, and iii) the internal nodes are network prefixes which aggregates the nodes below. The procedure focuses on the automatic creation of the internal nodes of the taxonomy.

```
1:  loop
2:     flow := read a new aggregated flow
3:     convert source and destination addresses of the flow in binary form
4:     insert addresses into the BST
5:  end loop
6:  prune branches with lower frequency than the support threshold
7:  loop
8:     branch := walk trough a remaining branch of the BST
9:     reconstruct the bynary address
10:    convert the address in decimal form
11:    print address on output file
12: end loop
```

**Fig. 4** Taxonomy generator for IP addresses

To aggregate network prefixes, the procedure exploits Binary Search Tree (BST), which is populated with all the available addresses in the data store. For each flow in the data store (lines 1-5), the source and destination addresses of each flow are converted in the binary notation over 32 bit (line 3). Then, addresses are inserted in BST with 32 levels of generalization (line 4). Each node of the BST structure consists in a record which holds the information about the value of the $i^{th}$ bit, its support and the level in the hierarchy. The insertion is performed starting from a root node and adding the address bits, starting from the most significant one. If the branch already exists, the support counter is updated, otherwise a new branch with support counter initialized to 1 is built on the left in case of a bit equal to 0, and on the right in case of 1.

Then, the algorithm walks through all the nodes of the tree (lines 6), removing nodes whose support is below the support threshold. The objective of this step is to prune prefixes which aggregate a number of flows below the given threshold.

Finally, the algorithm traverses the remaining branches in the tree (lines 7-12). For each node, it extracts the related subnet address, and builds the taxonomy according to the father-child relationships of addresses in the prefix tree.

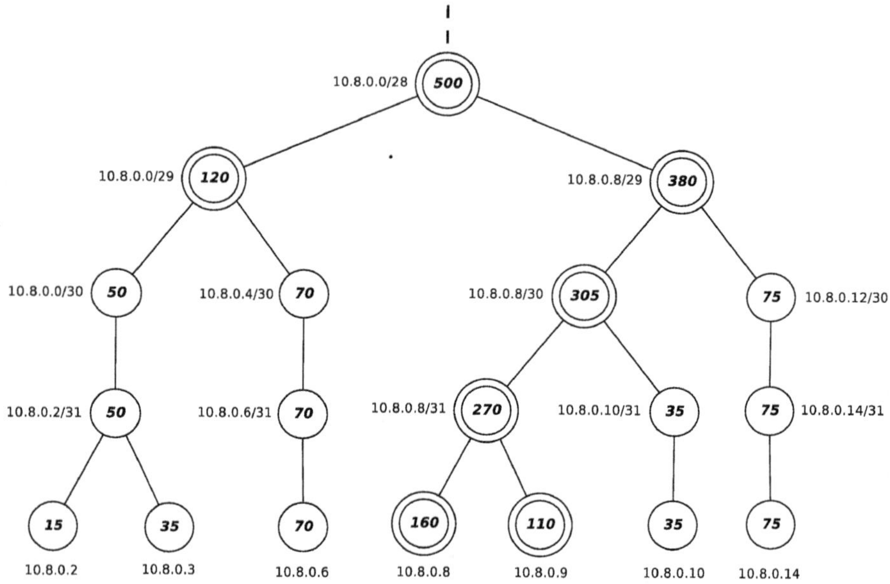


**Fig. 5** An example of Tree used for address aggregation

Fig. 5 shows an example of the BST after the analysis of 500 flows. Seven distinct IP addresses have been read, and constitute the leaves of the tree.

For each node, Fig. 5 shows the corresponding IP address/netmask pair, and its support, i.e., the number of flows aggregated by that node.

If the support threshold is set to 100, only the double-circled nodes are preserved by the pruning phase and then used to build the IP addresses taxonomy.

## 4.2 *Generation of the Taxonomy over Port Numbers*

For the creation of the taxonomy over TCP ports, the framework exploits the IANA classification [38]. The TCP ports which are read from the flows in the data store constitute the leaves of the taxonomy. Then, TCP ports from 0 to 1023 are aggregated into the *well known* category, ports from 1024 to 49151 are aggregated into the *registered* category, and ports from 49152 to 65535 are aggregated into the *dynamic* category.

## 4.3 *Generation of the Taxonomy over Flow Size and Number of Packets*

The taxonomies over the flow size and number of packet variables are created by using a vector for each attribute to store all the different values in the aggregated flows. These values constitute the leaves of the taxonomy. Then, the framework exploits the Equal Frequency Discretization [39] to create the upper levels of the taxonomy. For a given taxonomy level, each node in that level aggregates the same number of flows.

## 4.4 *Taxonomy Conversion*

In this phase, the taxonomy is also exported in the eXtensible Markup Language (XML) [30] format and validated against an XML Schema Definition (XSD). The use of this language allows a precise definition of the hierarchy of generalization for our tags and gives the possibility of taxonomy visualization in a tree view using a common browser. Fig. 6 shows an example of rules encoded in XML language.

```
<?xml version="1.0" encoding="UTF-8"?>
<rules xmlns:xsi="http://www.w3.org/2001/XMLSchema-instance">
  <ip_address default="external-ip">
    <net octet_1="130" octet_2="192" octet_3="24" octet_4="0" netmask="21">subnet_1</net>
    <net octet_1="130" octet_2="192" octet_3="16" octet_4="0" netmask="20">subnet_2</net>
    <net octet_1="130" octet_2="192" octet_3="0" octet_4="0" netmask="19">subnet_3</net>
    <net octet_1="130" octet_2="192" octet_3="0" octet_4="0" netmask="18">subnet_4</net>
    <net octet_1="130" octet_2="192" octet_3="0" octet_4="0" netmask="17">subnet_5</net>
    <net octet_1="130" octet_2="192" octet_3="0" octet_4="0" netmask="16">subnet_6</net>
    <net octet_1="130" octet_2="0" octet_3="0" octet_4="0" netmask="7">subnet_7</net>
    <net octet_1="128" octet_2="0" octet_3="0" octet_4="0" netmask="6">subnet_8</net>
    <net octet_1="128" octet_2="0" octet_3="0" octet_4="0" netmask="4">subnet_9</net>
    <net octet_1="64" octet_2="0" octet_3="0" octet_4="0" netmask="3">subnet_10</net>
    <net octet_1="128" octet_2="0" octet_3="0" octet_4="0" netmask="3">subnet_11</net>
    <net octet_1="192" octet_2="0" octet_3="0" octet_4="0" netmask="3">subnet_12</net>
    <net octet_1="128" octet_2="0" octet_3="0" octet_4="0" netmask="2">subnet_13</net>
    <net octet_1="0" octet_2="0" octet_3="0" octet_4="0" netmask="1">subnet_14</net>
    <net octet_1="0" octet_2="0" octet_3="0" octet_4="0" netmask="0">subnet_15</net>
  </ip_address>
  <port_number default="default-port_number">
    <range minport_number="1" maxport_number="1023">well_known</range>
    <range minport_number="1024" maxport_number="49151">registered</range>
    <range minport_number="49152" maxport_number="65535">dynamically_allocated</range>
  </port_number>
  <flow_size default="default-flow_size">
    <range minval="142665" maxval="524317">flow_size_1</range>
    <range minval="524321" maxval="861383">flow_size_2</range>
    <range minval="862506" maxval="1417728">flow_size_3</range>
    <range minval="1438216" maxval="2665500">flow_size_4</range>
  </flow_size>
  <number_of_packets default="default-number_of_packets">
    <range minval="99" maxval="503">number_of_packets_1</range>
    <range minval="509" maxval="917">number_of_packets_2</range>
    <range minval="921" maxval="1777">number_of_packets_3</range>
  </number_of_packets>
</rules>
```

**Fig. 6** An example of XML describing taxonomies

# 5 Refinement Analysis

NED discovers interesting correlations and recurrent patterns in network traffic data by means of association rule mining, which is performed in the refinement analysis phase.

## 5.1 Association Rules

Let $T_{Traffic}$ be a network traffic dataset whose generic record $F_{Flow}$ is a set of $F_{Feature}$. Each $F_{Feature}$, also called item, is a couple (attribute, value). An attribute models a characteristic of the flow (e.g., source address, destination port). Such a $T_{Traffic}$ dataset is available in the NED data store, i.e., the input of the refinement analysis block.

Association rules identify collections of itemsets (i.e., sets of $F_{Feature}$) that are statistically related (i.e., frequent) in the underlying dataset. An association rule is represented in the form X → Y where X and Y are disjoint conjunctions of $F_{Feature}$. Rule quality is usually measured by support and confidence. Support is the percentage of items containing both X and Y. It describes the statistical relevance of a rule. Confidence is the conditional probability of finding Y given X. It describes the strength of the implication. Association rule mining is a two-step process: (i) Frequent itemset extraction and (ii) association rule generation from frequent itemsets. Given a support threshold s%, an itemset (i.e., a set of $F_{Feature}$) is said frequent if it appears in at least s% of flows.

*Example 2*: Consider the toy dataset in Fig. 2(a) for the itemset mining process. With a support threshold greater than 25%, the 2-itemsets

{<*Dest-addr* : DA1> , <*Dest-port* : DP1>}, s = 50%

{<*Dest-addr* : DA2 >, <*Dest-port* : DP1>}, s = 50%

are frequent. Hence, the flows directed to DA2 or DA1 at port DP1 are frequent. Once mined frequent itemset, association rules [1], [12] are used to analyze their correlations. Given these, the following association rule

{<*Dest-addr* : DA1>} → { <*Dest-port* : DP1>}, *s*% support, *c*% confidence

states that <Dest-port :DP1> appears in c% of $F_{Flow}$ which contains also <Dest-addr :DA1>.

## 5.2 Generalized Association Rules

Association rule extraction, driven by support and confidence constraints as described in the previous sections, sometimes involves (i) generation of a huge number of unmanageable rules [4] and (ii) pruning rare itemsets even if their hidden knowledge might be relevant. Since rare correlations are pruned by the mining process, the (potentially relevant) knowledge hidden in this type of patterns may be lost.

To address the above issue, generalized association rule extraction [31] can be exploited. The concept of generalized (also called multi-level) association rules has been first proposed in [12]. This data mining technique automatically extracts higher level, more abstract correlations from data, by preventing the pruning of hidden knowledge discarded by previous approaches. The extraction process is performed in two steps: (i) Generalized itemsets extraction and (ii) generalized association rule generation. Itemset generalization is based on a set of predefined taxonomies which drive the pruning phase of the extraction process.

The following example highlights the need of a more powerful abstraction of association rules. Consider a web server on port 80 having IP address 130.192.5.7. To describe the activity of a client connecting to this server, a rule with the form

{<*Source-addr*: 11.12.13.14>} → {<*Dest-add* : 10.0.0.1>, <*Dest-port*: 80>}, *s%*, *c%*

should be extracted. Since a single source IP address is a 1-itemset which may be unfrequent in a very large traffic network trace, extracting such rule would require enforcing a very low support threshold which makes the task unfeasible. However, a higher level view of the network may be provided by the following generalized association rule

{<*Source-addr*: 11.12.13.0/24>} → {<*Dest-addr*: 10.0.0.1>, <*Dest-port*: 80>}, *s%*, *c%*

which shows a subnet generating most of the traffic and provides knowledge that could be even more valuable for network monitoring. The number of different tagged items in network traffic may be very large (e.g., different value ranges for packet size) and information related to single tagged items does not provide useful knowledge. Generalized rules are a powerful tool to address this challenge.

The algorithm, which extracts generalized itemsets, takes in input a set of taxonomies, the dataset, and a minimum support threshold. It is an *Apriori* [12] variant. Apriori is a level-wise algorithm, which, at each step, generates all frequent itemsets of a given length. At arbitrary iteration *l*, two steps are performed: (i) Candidate generation, the most computationally and memory intensive step, in which all possible *l*-itemsets are generated from *l-1*-itemsets, (ii) candidate pruning, which is based on the property that all subsets of frequent itemsets must also be frequent, to discard candidate itemsets which cannot be frequent. Finally, actual candidate support is counted by reading the database. The generalized association rule algorithm follows the same level-wise pattern. Furthermore, it manages rare itemsets by means of taxonomies. Further candidate pruning is based on uniqueness of attributes in a given transaction by optimizing the candidate generation step with respect to the Apriori algorithm. Finally, only generalized itemsets derived from rare itemsets are kept.

When a generalized itemset is generated only by frequent itemsets, it is discarded because its knowledge is already provided by the corresponding (frequent) non-generalized itemsets.

## *5.3 Data Stream View*

The data stream view block allows to select a subset of the flows obtained as continuous query outputs. The following example, focusing on the SYN flooding attack, shows its usefulness. The SYN flooding attack occurs when a victim host receives more incomplete connection requests that it can handle. To make this attack more difficult to detect, the source host randomizes the source IP address of the packets used in the attack. An attempt of SYN flooding [10] can be recognized by mining rules in the form

{ victim-IP, victim-port}$\rightarrow$ size, s% support, c% confidence

Suppose that, to reduce the amount of stored data, the network traffic has been aggregated with respect to address and port of both source and destination. For each flow the size is computed (e.g., packets differing in one of these features belong to different flows). This step is performed by running the following continuous query on the data stream.

**Aggregate**

| | |
|---|---|
| Select | source-IP, source-Port, destination-IP, destination-Port, Sum(size) as flow-size |
| From | Packets [Range by 60 seconds] |
| Where | level4 = 'TCP' |
| Group by | source-IP, source-Port, destination-IP, destination-Port |

Since the complete dataset contains hundreds of flows, the support of the SYN-flooding rule may be too low to be relevant. To overcome this issue, the output of the previous continuous query may be appropriately filtered. Since we are interested in flows whose size is lower than a threshold *x* expressed in bytes, the following query may be exploited to create a data stream view.

**Filter**

| | |
|---|---|
| Select | source-IP, source-Port, destination-IP, destination-Port, flow-size |
| From | Aggregate |
| Where | Flow-size • x |

The refinement analysis, performed on the results of the described data stream view, extracts a small number of association rules characterized by high support. These rules highlight more effectively any specific traffic behavior.

# 6 Experimental Validation

To validate our approach we have performed a large set of experiments addressing the following issues: (i) stream analysis, (ii) taxonomy generation, and (iii) refinement analysis.

Refinement analysis is based on association rule mining and generalized association rules. For each mining technique two algorithms have been run. Association rule mining is performed by means of the frequent itemset extraction based on the LCM v.2 algorithm [20] (FIMI'04 best implementation algorithm), and association rule generation is performed using Goethal's implementation of the Apriori algorithm [5]. Generalized association rule mining is based on the Genio algorithm [31] to extract generalized frequent itemsets, and Goethal's implementation [5] to extract generalized association rules.

## *6.1 Experimental Settings*

Three real datasets have been exploited to perform the NED validation. Network datasets have been obtained by performing different capture stages using the Analyzer traffic tool [17] on a backbone link of our campus network. We will refer to each dataset using the ID shown in Table 1, where the number of packets and the size of each dataset are also reported.

Experiments have been performed by considering three window lengths (60s, 120s, and 180s) and a link speeds of 100 Mbps. The value of the step $s_{step}$ has been set to half of the window length.

**Table 1** Network traffic datasets

| ID | Number of packets | Size [MByte] |
|---|---|---|
| A | 25969389 | 2621 |
| B | 24763699 | 2500 |
| C | 26023835 | 2625 |

To avoid discarding packets, a proper buffer size has to be determined. The buffer must be able to store all possible flows in a time window, whose worst case value is the maximum number of captured packets (i.e., each packet belongs to a different flow). Thus, the buffer size has been set to the following number of flows

$$\text{size} = \frac{\text{link speed} * \text{window length}}{\text{minimum size of a packet}}$$

Experiments have been performed on a 2800 MHz Pentium IV PC with 2Gb main memory running Linux (kernel 2.7.81). All reported execution times are real times, including both system and user times. They have been obtained using the Linux time command as in [8]. The values of the memory occupation are taken from the */proc/PID/status* file, which collects many information about the running processes. We consider the value of the *VmPeak*, which contains the maximum size of the virtual memory allocated to the process during its execution.

## 6.2 Stream Analysis

To validate the performance of the stream analysis block the proposed continuous queries have been run (see Section "Data stream processing"). Performance, in terms of execution time and required main memory, have been analyzed for each query. However, due to lack of space, only query 2 is reported as representative. The aim of query 2 is to monitor the traffic on a backbone link and select the flows which generate an amount of traffic greater than a certain percentage of the total traffic in a given window. Thus, the query receives two input parameters: (i) the traffic percentage (i.e., traffic ratio) and (ii) the windows length. Different values for each parameter have been analyzed. Traffic ratio has been set to 10%, 20%, and 50%, while window length has been set to 60s, 120s, and 180s. Table 2 and Table 3 report the CPU time and the main memory required to perform the stream analysis process, and the number of extracted flows. Results are reported in Fig. 7(a) for dataset A and in Fig. 7(b) for dataset C, which have been selected as representative datasets.

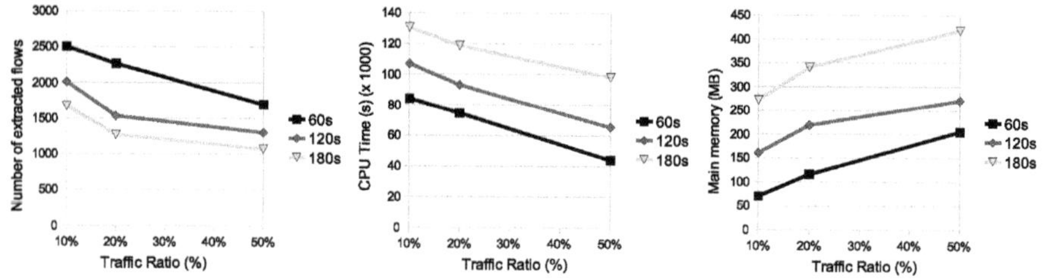


**(a) Number of extracted flows (b) CPU Time** **(c) Memory usage**

**Fig. 7(a)** Dataset A: Experimental results of the data stream processing

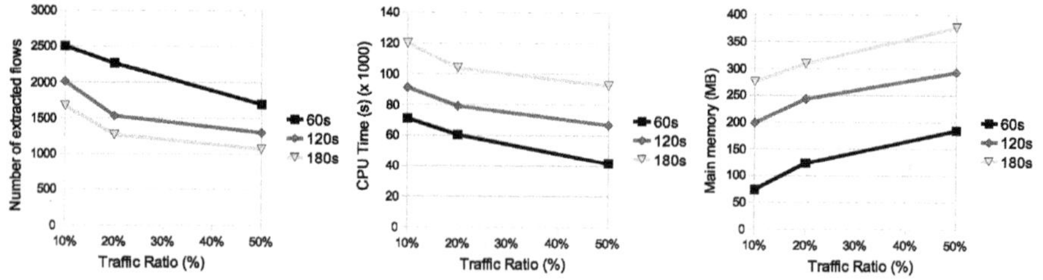


**(a) Number of extracted flows** **(b) CPU Time** **(c) Memory usage**

**Fig. 7(b)** Dataset C: Experimental results of the data stream processing

The experimental results show that the CPU time needed for the stream analysis process increases for higher values of the window length. A lower traffic ratio leads the program to store an increasing number of flows, affecting both the insertion and the extraction operations from the data structures, thus requiring more memory. On the contrary, when the traffic ratio increases, the CPU time decreases, because there will be less flows that satisfy the size constraint.

The analysis of the main memory highlights that the required memory grows when window length is increased, because more flows need to be stored. The traffic ratio parameter slightly affects the amount of required memory, because the percentage value is enforced in the printing phase.

Furthermore, we have analyzed the maximum number of aggregated flows that is extracted in a time window. As shown in Table 2 and Table 3, the number of aggregated flows decreases when the traffic percentage value decreases, because less flows satisfy the traffic ratio constraint. Furthermore, the flow number decreases when the window length increases. Since each flow represents traffic data received in a longer time interval, only the most relevant flows (in terms of transferred data in the observing window) are extracted.

By comparing results reported in Fig. 7(a) with that reported in Fig. 7(b), the general trend of the stream processing analysis is the same in both considered datasets A and C. The main difference is in the number of flows, which is slightly less in the second considered trace.

## *6.3 Taxonomy Generation*

The taxonomy generation block aims at automatically generating taxonomies for each flow attribute. Performance, in terms of execution time and memory consumption, depends on the considered attribute.

In the case of the *port number*, the taxonomy is predetermined according to the IANA classification. Thus, the taxonomy generation involves a single scan of the set of flows, with constant and limited memory consumption.

In the case of a numerical attribute, such as the *number of packets* and *flow size*, the process performs an Equal Frequency Discretization. If the set of flows contains $n$ different values for the considered attribute, the process requires the allocation of two arrays of $n$ integers, where the first contains the different values of the attribute, and the second contains the frequency of each value. Thus, the memory allocation is limited to $2 * n *$ the size of an integer. The discretization process requires a single scan of the set of flows, to collect the values, then a sort of the vector which contains the values, then a single scan of the vector containing the frequency to determine the ranges.

The creation of the taxonomy for IP addresses is more demanding in terms of time and memory resources, due to the required data structure. In particular, performances are affected by the support threshold, and the distribution of input data (i.e., the input dataset and the continuous query used to aggregate network traffic data). Thus, experiments have been performed by considering different sets of flows on input, and different support thresholds. The sets of flows have been obtained by running query 2 (see Section "Stream analysis") on different datasets.

Results are reported for dataset A and C, selected as representative datasets. Different values for each parameter have been considered. The window length of query 2 has been set to 30s, 120s and 180s, while the traffic ratio has been set to 10% (other values have been omitted due to lack of space). The minimum support threshold has been set to 20%, 40%, 60% and 80%.

Fig. 8(a) and 9(a) show, for both datasets, the number of extracted subnets. Results are reported for different settings of window length and minimum support threshold.

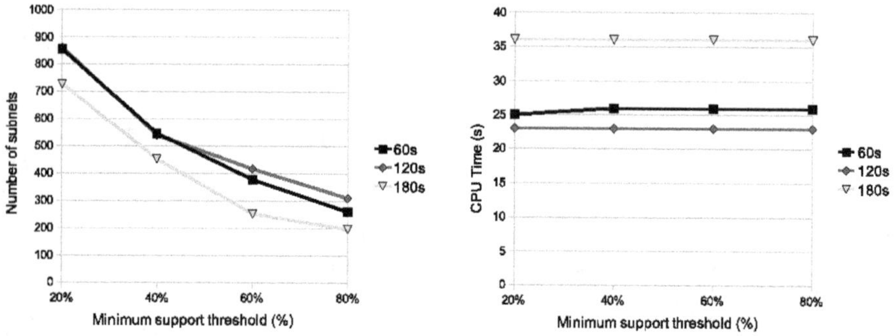


(a) Number of generated subnets (b) Execution Time

**Fig. 8** Dataset A: Experimental results for taxonomy generation

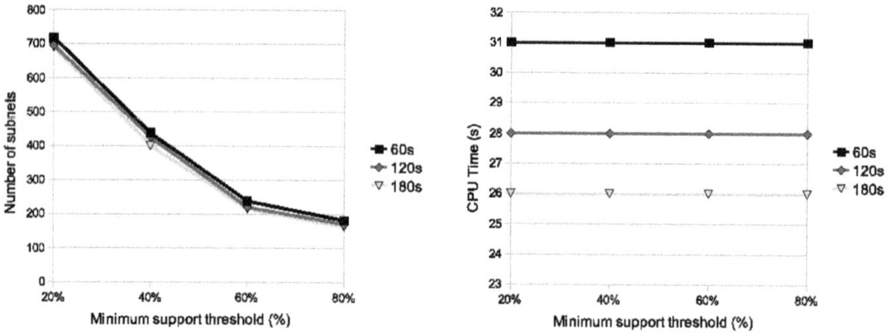


(a) Number of subnets (b) Execution Time

**Fig. 9** Dataset C: Experimental results for taxonomy generation

The choice of the window length parameter affects the number of extracted flows, as well as their distribution. Hence, a larger window length does not necessarily lead to an increasing number of extracted subnets. When the minimum support threshold increases, the taxonomy generator creates subnets which aggregate more specific IP addresses. Thus, for a specific window length, the trend of the number of extracted subnets is decreasing.

Fig. 8(b) and 9(b) show, for the same settings above, the CPU time required. The time required for the taxonomy generation process is mainly affected by the scan of the set of flows, and by the time required to insert a new IP address in the prefix tree. Hence, the minimum support threshold has a little impact and, for a specific window length, the trend of the CPU time is constant.

Fig. 10 shows the main memory usage obtained by varying the window length parameter for both datasets. The memory usage is not affected by the minimum support threshold, since the size of the prefix tree is independent from the threshold used in the pruning phase. Thus, the size of the allocated memory for a

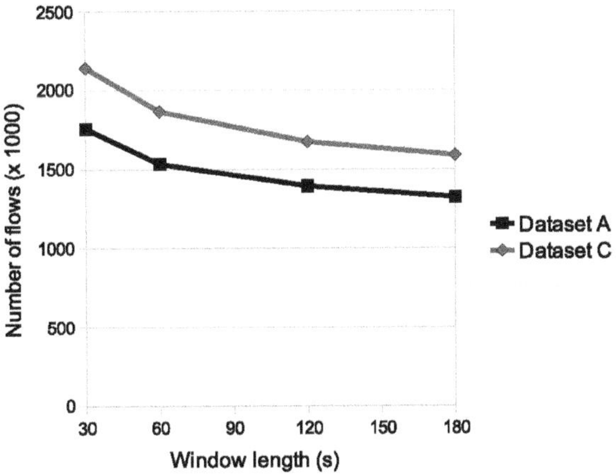


**Fig. 10** Memory usage for taxonomy generation

specific window length is the same for every value of minimum support threshold. Instead, the size of the prefix tree is related to the number of different IP addresses read from the set of flows and, in particular, to the number of different prefixes found. For the considered datasets, this value depends on the number of flows generated by the continuous query. Thus, the memory usage decreases when the window length increases, because fewer flows are generated.

## 7 Refinement Analysis

To validate the performance of the refinement analysis block of the NED framework, different analysis sessions have been performed. We analyzed the effect of the support and confidence thresholds on both classes of rule mining (i.e., association rules, generalized association rules) and the effectiveness of the proposed approach in extracting hidden knowledge. Furthermore, the analysis of a subset of interesting rules in the network context and some interesting analysis scenarios has been discussed.

## *7.1 Association Rule Extraction*

Association rule extraction is driven by two quality indices: Support and confidence. The support evaluates the observed frequency of itemsets in the dataset, while the confidence characterizes the "strength" of a rule. Different minimum support and confidence thresholds significantly affect the cardinality of the extracted rule set and the nature of the rules. We analyzed the number of extracted association rules for different combinations of the threshold values. Analysis has been performed for each proposed query and results are discussed in the following of this section.

**Query 1**

Query 1 aggregates packets with respect to source address, source port, destination address and destination port. Thus, it significantly reduces the data cardinality, while preserving general traffic features.

Fig. 6 report the number of extracted rules for each dataset considering different support and confidence thresholds. Since the three datasets have similar behavior, we focus on dataset A, where we observe that some 1-itemsets are highly frequent, such as <source address: 130.192.a.b> and <destination address: 130.192.a.b>. To further investigate the meaning of the rules, we consider the following examples.

*Example4.* Considering minimum support s ≥ 0.1% and minimum confidence c ≥ 50% leads to the extraction of a large amount of rules in the following form.

{<*-*port* : x>} → { <*-address : 130.192.a.b >}

Since port x is frequent (regardless of the port number), these rules state that the address 130.192.a.b (i) generates remarkable traffic both as receiver and as transmitter, (ii) it is likely to be a server which provides many services, because it uses a wide range of ports. We can conclude that 130.192.a.b is probably the public IP address of a router implementing NAT. An inspection of the network topology confirms such result.

Fig. 11 reports the number of extracted rules for query 1 when varying support and confidence thresholds. By enforcing high support and confidence thresholds, the number of extracted patterns decreases. The decreasing trend is particularly evident for high support thresholds, whereas most of the rules have high confidence values for any support threshold. Thus, in this scenario, support is more selective in filtering patterns.

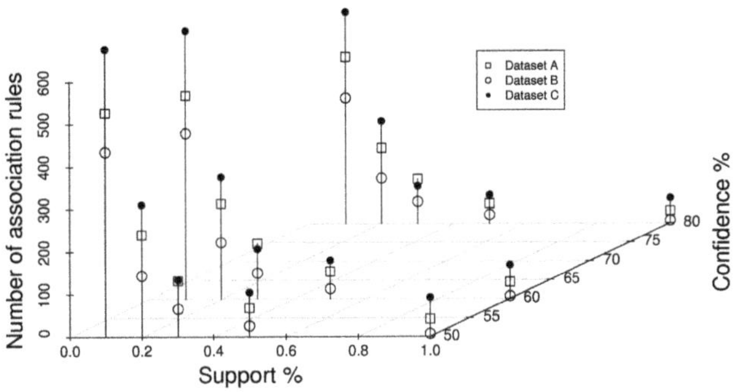


**Fig. 11** Query 1: Number of extracted association rules

*Example 5.* By setting the minimum support to 0.3% and the minimum confidence to 0.5%, some interesting patterns are extracted. NAT rules are still present, and other rules become more evident. For example

{< *source-address* : 130.192.c.d >} → {< *source-port* :443 >}, s = 0.3%, c = 99%

identifies 130.192.c.d as an https server. It was confirmed to be the student webmail server.

{<*destination-port* : 6101>, <*flow-size* : 96>} → {<*source-address*: *x.y.z.w*>}, s = 0.3%, c = 98%

highlights that Synchronet-rtc service is frequently and mostly used by *x.y.z.w*.

Analyses performed on rules extracted from datasets B and C confirm the results obtained on dataset A. The traffic network features, inferred from the rules, highlight the same NAT routing and servers.

To identify patterns arising from long flows, another step of filtering is required. This issue, addressed by the second query, has been analyzed in the next section.

## Query 2

The second query selects the flows which generate an amount of traffic greater than a certain percentage of the total traffic in a window. The aim is to describe more accurately rules extracted by Query 1.

Fig. 12 shows the number of association rules extracted from the results of Query 2 applied to the datasets A, B, and C. Rules discovered in this case predominantly have the following form.

{<*source-address*: SA>, <*source-port* : SP>} → {<*destination-address*: DA>, <*destination-port* : DP>}

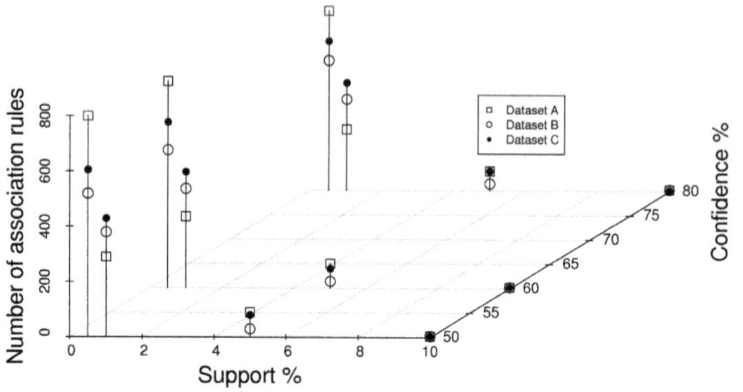


**Fig. 12** Query 2: Number of extracted association rules

Many extracted rules describe general network features such as NAT routing or main servers. Furthermore, this analysis assesses the pervasiveness of different services. Mined rules highlight the importance of several protocols like netrjs, systat and QMTP in the examined network. Some rules are worth further investigation. Many flows have source and destination ports greater than 1024. This fact may highlight unconventional traffic, such as peer-to-peer communications. Another filtering step is necessary to clearly identify involved hosts. This issue has been addressed by Query 3.

### Query 3

The third query extracts long flows whose source and destination ports are beyond 1024. Fig. 13 shows the number of association rules extracted from the result of Query 3 applied to datasets A, B and C. Because of the additional filtering step, the number of rules is significantly lower than the ones extracted by Query 1 and Query 2. Furthermore, these rules are even more specific than previous ones, as shown by the following example.

*Example 6.* Consider the following rule.

{<*source-address* : 130.192.e.f >} → {<*destination-port* : 4662 >},
s = 1.98%, c=77%

The address 130.192.e.f is identified as having a remarkable amount of traffic toward remote hosts on port 4662. Since this is the default port for eDonkey2000 servers [19] of the ED2K peer to peer network, we can conclude that (i) the source host is exchanging data with the ED2K servers, and (ii) its amount of traffic on not well-known ports is mainly related to such *peer-to-peer* network.

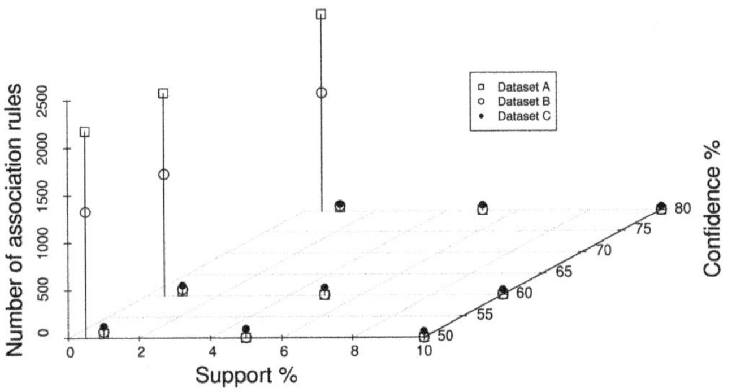


**Fig. 13** Query 3: Number of extracted association rules

## *7.2 Generalized Association Rules*

Generalized rule mining exploits taxonomies to drive the pruning phase of the extraction process. To allow comparison among different datasets, a fixed user-provided taxonomy has been used. The taxonomy used in these experiments aggregate infrequent items according to the following hierarchies. TCP ports are aggregated into three ranges: Well known ports (between 1 and 1023), registered ports (between 1023 and 49151) and dynamic ports (otherwise). IP addresses which are local to the campus network are aggregated by subnet. IP addresses which do not belong to the campus network are aggregated in a general external address group. The flow size attribute is aggregated over 4 bins, whose intervals are [1,1000), [1000, 2000), [2000, 3000) and equal or grater than 3000 bytes.

To perform a full mining session, Query 2 has been run by setting the window size of 60s and the ratio parameter to 10%. The number of extracted generalized association rules has been analyzed for different combinations of the support and confidence threshold values. The absolute number of extracted rules is different among the datasets, because of the different number of flows (see Table 1) and data distribution. However, the overall trend is similar.

Fig. 14, Fig. 15, Fig. 16, and Fig. 17 report, for the datasets A and C, the number of extracted rules for different combinations of the support and confidence threshold values. Furthermore, the number of generalized association rules and the number of specific (non-generalized) rules, built using non-aggregated items, are also reported. This result gives a measure of the number of rules which would have been discarded if a traditional approach had been used with the same support threshold.

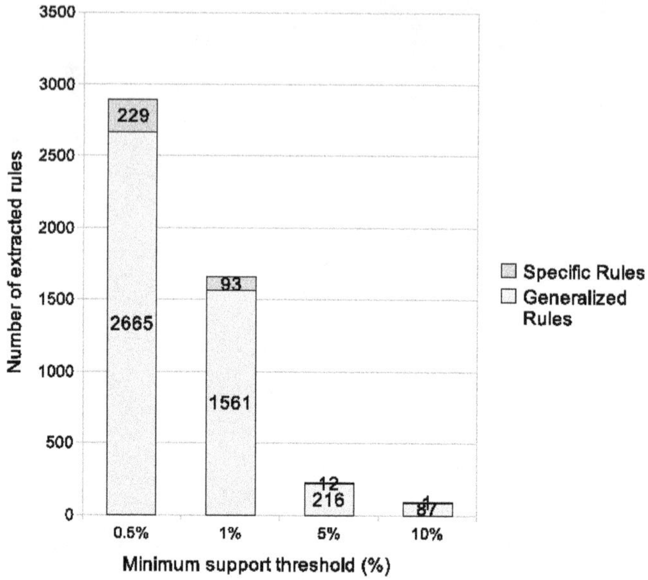


**Fig. 14** Dataset A: Number of extracted rules for different values of minimum support with minimum confidence=20%

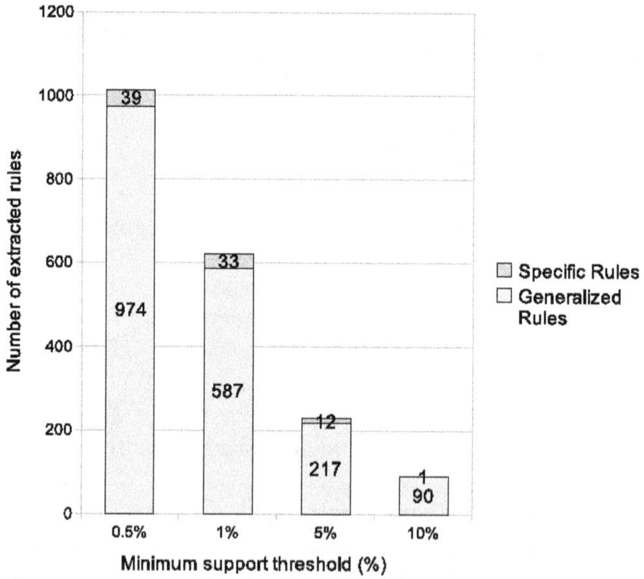


**Fig. 15** Dataset C: Number of extracted rules for different values of minimum support with minimum confidence=20%

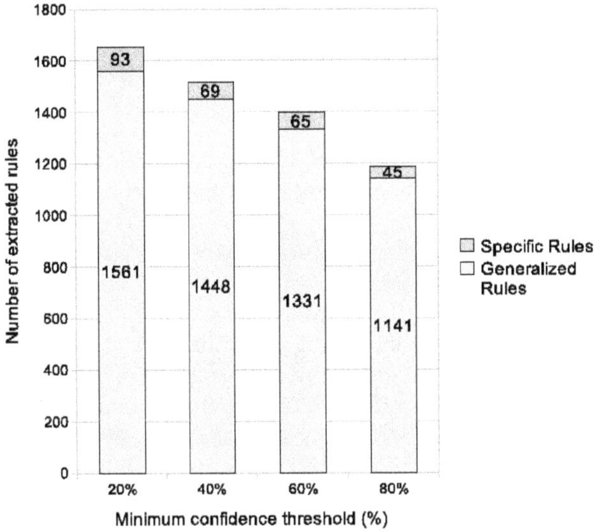


**Fig. 16** Dataset A: Number of extracted rules for different values of minimum confidence with minimum support=1%

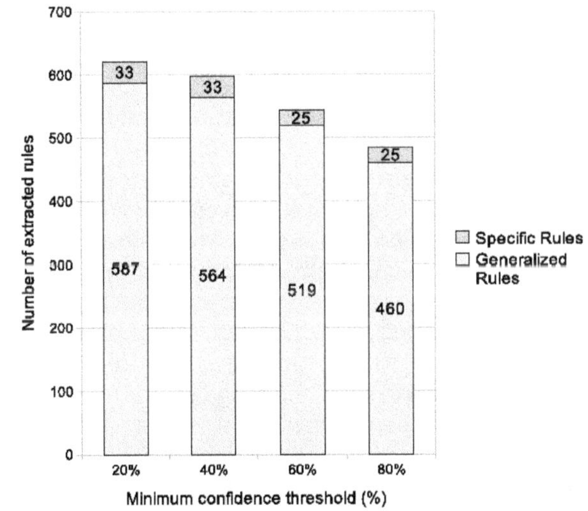


**Fig. 17** Dataset C: Number of extracted rules for different values of minimum confidence with minimum support=1%

The number of generalized association rules decreases when both the minimum support and the minimum confidence thresholds are increased. The support threshold is rather selective. For high support values, only a small number of rules is extracted (see Fig. 14 and Fig. 15). However, it is important to note that frequent is not necessarily a synonym of interesting. A rather high number of strong correlations is instead extracted also for high confidence values (see Fig. 16 and Fig. 17). Furthermore, other values of minimum confidence yield analogous results, as rules with high confidence are rather uniformly distributed over a wide support range.

Generalized association rules may include many different combinations of attributes. Thus, for low support thresholds, a large variety of combinations may satisfy the support constraint. These rules are suitable for capturing unexpected peculiar knowledge. Many examples of generalized rules highlight correlations relating basic attributes *destination-address, source-address, destination-port* with the *size* attribute. This effect is referable to the particular taxonomy of the *size* tag. Its values are discretized into 4 bins only, leading to a very dense aggregation. Hence, each single aggregation value becomes rather frequent. Diverse discretization techniques or intervals may lead to a different behavior. A similar behavior is shown by the *source-port* attribute, which is often aggregated as registered or dynamically-assigned. This reveals the allocation policy for the client source port, which is typically dynamically assigned on the client host, always excluding the well-known ports.

The support and confidence thresholds also affect the kind of extracted rules. By setting high support thresholds, only very frequent patterns are extracted. However, their interest may be marginal. To satisfy the high selectivity of the minimum support threshold, the generalization process has led to a rule which is too general to provide interesting knowledge. Instead, the use of a low support threshold coupled with different quality indices (e.g., confidence) leads to the extraction of a higher number of rules where peculiar patterns arise.

Fig. 18 shows the number of the generalized 2-itemsets obtained from dataset C. For better visualization, results have been restricted to addresses of the campus network. Thus, no external IP address has been considered. The 2-itemsets are in the form (destination-address, destination-port), where the address is automatically aggregated to the subnet when single IP support is under the minimum support threshold (set to 1%). Fig. 13 provides a characterization of the traffic on the campus network. Many extracted itemsets describe general network features. For example, the first top support itemset identifies the VPN concentrator of the campus network. Larger itemsets allow focusing on specific traffic behaviors. For example, the itemsets (destination-address=130.192.e.e, destination-port=57403, source-address=x.x.x.x, source-port=registered-port) with support equal to 2.3% highlights an unconventional high-volume traffic toward a specific host of the campus network, whereas the itemsets (source-address=y.y.y.y, destination-address=130.192.a.a, destination-port=registered-port, source-port=well-known) with support equal to 2% identifies connections to the VPN concentrator by means of a client using well-known source ports.

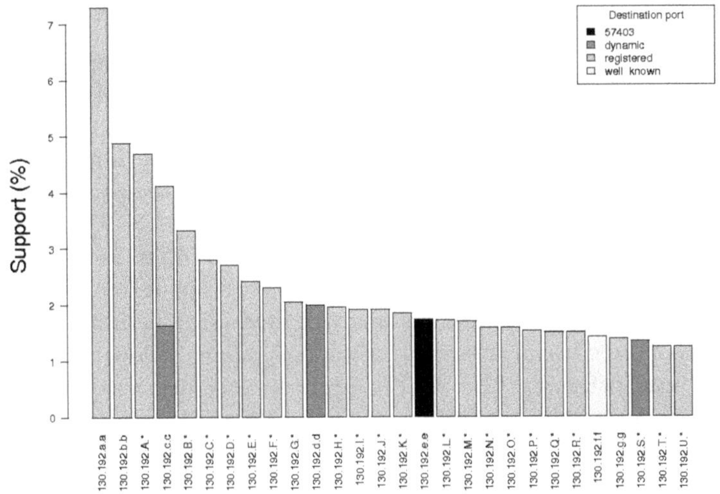


**Fig. 18** Number of extracted itemsets for different destination IP addresses and ports

## 8 Related work

A significant effort has been devoted to the application of data mining techniques to the problem of network traffic analysis. The work described in [6] presents some theoretical considerations on the application of data mining techniques to network monitoring. Since the network traffic analysis domain is rather broad, research activities have addressed many different application areas, e.g., web log analysis [22] and enterprise-wide management [23]. To the best of the authors' knowledge, less results are available in the network traffic characterization domain. Data mining techniques have been used to identify correlations among data (e.g. association rules extraction for network traffic characterization [3] or for router misconfiguration detection [14]), to build prediction models (e.g., multilevel traffic classification [13], Naive Bayes classification [16]), or to characterize web usage [24]. Knowledge discovery systems have also been used to learn models of relevant statistics for traffic analysis. In this context, [26] is an example of how neural networks can be used to determine the parameters which most influence packet loss rate.

Traffic data categorization, addressed by means of classification techniques, is an effective tool to support network management [33]. In general, classification techniques can be divided in supervised and unsupervised. While the first group requires previous knowledge of the application domain, i.e., new unlabeled traffic flows are assigned a class label by exploiting a model built from traffic network data with known class labels, the second does not. Furthermore, network traffic classification can be performed by analyzing different features: (i) packet payloads

[33][34], (ii) traffic metrics [36], [37], and (iii) statistical features computed on traffic flows [16], [35].

Traditional traffic classification techniques perform a deeper inspection of packet payloads [34] to identify application signatures. To apply these approaches, the payload must be visible and readable. Both assumptions limit the feasibility of these approaches [33]. First of all, payloads could be encrypted, making the deep packet inspection impossible. Furthermore, the classifier has to know the syntax of each application payload, to be able to interpret it.

A parallel effort has been devoted to the application of the continuous queries in different domains. Continuous queries have been applied in the context of network traffic management to real-time monitoring of network behavior. In particular, they have been exploited to detect congestions and their causes [2] and to support load balancing [25]. In [2], [25], network data analysis is directly performed by means of continuous queries, without data materialization and further data exploration by means of data mining techniques.

Data mining techniques have also played a central role in studying correlations in intrusion detection systems, also called IDSs. The first application of data mining to IDS required the use of *labeled data* to train the system [15], [27], [28], [29]. A trace where the traffic has already been marked as "normal" or "intrusion" is the input to the algorithm in the learning phase. Once this phase is concluded, the system is able to classify new incoming traffic. This approach is obviously effective and efficient in identifying known problems. While it is, in general, not effective against novel attacks which are unknown. Another widely used technique in this context is clustering, as proposed in [9], [18], [21], where it is used to detect "normal" traffic and separate it from outlier traffic, which represents anomalies. Finally, the work in [11] targets intrusion detection by means of frequent itemset mining, which characterize standard (i.e., frequent) traffic behavior.

## 9 Conclusions

NED is a framework to efficiently perform network traffic analysis. It provides efficient and effective techniques to perform data stream analysis and refinement analysis for network traffic data. The former reduces the amount of traffic data while the latter automatically extracts interesting and useful correlations and recurrence of patterns among traffic data. Experimental results performed on real traffic traces show the effectiveness of the NED framework in characterizing traffic data and performing different kind of analyses.

**Acknowledgments.** We are grateful to Fulvio Risso for providing the real traffic datasets captured from the campus network, and to Claudio Testa and Felice Iasevoli for developing parts of the NED framework.

## References


[1] Agrawal, R., Srikant, R.: Fast Algorithms for Mining Association Rules in Large Databases. In: Proceedings of the 20th International Conference on Very Large Data Bases, pp. 487–499 (1994)

[2] Arasu, A., Babu, S., Widom, J.: The CQL continuous query language: semantic foundations and query execution. The VLDB Journal, The International Journal on Very Large DataBases 15(2), 121–142 (2006)

[3] Babu, S., Widom, J.: Continuous queries over data streams. ACM SIGMOD Record 30(3), 109–120 (2001)

[4] Baldi, M., Baralis, E., Risso, F.: Dipt. di Autom. e Inf. Data mining techniques for effective and scalable traffic analysis. In: 2005 9th IFIP/IEEE International Symposium on Integrated Network Management, IM 2005, pp. 105–118 (2005)

[5] Goethals, B.: Frequent Pattern Mining Implementations, `http://www.adrem.ua.ac.be/~goethals/software`

[6] Burn-Thornton, K., Garibaldi, J., Mahdi, A.: Pro-active network management using data mining. In: Global Telecommunications Conference, GLOBECOM 1998, vol. 2 (1998)

[7] Erman, J., Arlitt, M., Mahanti, A.: Traffic classification using clustering algorithms. In: MineNet 2006, pp. 281–286. ACM Press, New York (2006)

[8] FIMI, `http://fimi.cs.helsinki.fi/`

[9] Guan, Y., Ghorbani, A., Belacel, N.: Y-Means: A clustering method for intrusion detection. In: Proceedings of Canadian Conference on Electrical and Computer Engineering, pp. 4–7 (2003)

[10] Harris, B., Hunt, R.: TCP/IP security threats and attack methods. Computer Communications 22(10), 885–897 (1999)

[11] Hossain, M., Bridges, S., Vaughn Jr., R.: Adaptive intrusion detection with data mining. In: IEEE International Conference on Systems, Man and Cybernetics, vol. 4 (2003)

[12] Han, J., Kamber, M.: Data Mining: Concepts and Techniques. In: Gray, J. (ed.) The Morgan Kaufmann Series in Data Management Systems. Morgan Kaufmann Publishers, San Francisco (August 2000)

[13] Karagiannis, T., Papagiannaki, K., Faloutsos, M.: Blinc: multilevel traffic classification in the dark. In: SIGCOMM, pp. 229–240 (2005)

[14] Le, F., Lee, S., Wong, T., Kim, H.S., Newcomb, D.: Minerals: using data mining to detect router misconfigurations. In: MineNet 2006, pp. 293–298. ACM Press, New York (2006)

[15] Lee, W., Stolfo, S.: A framework for construction features and models for intrusion detection systems. ACM Transactions on Information and System Security (TISSEC) 3(4), 227–261 (2000)

[16] Moore, A.W., Zuev, D.: Internet traffic classification using Bayesian analysis techniques. In: SIGMETRICS 2005, pp. 50–60. ACM Press, New York (2005)

[17] NetGroup, Politecnico di Torino. Analyzer 3.0, `http://analyzer.polito.it/30alpha/`

[18] Portnoy, L., Eskin, E., Stolfo, S.: Intrusion detection with unlabeled data using clustering. In: Proceedings of ACM CSS Workshop on Data Mining Applied to Security, PA (November 2001)

[19] The SANS Institute. Port 4662 details, http://isc.sans.org/port.html?port=4662
[20] Uno, T., Kiyomi, M., Arimura, H.: LCM ver. 2: Efficient mining algorithms for frequent/closed/maximal itemsets. In: FIMI (2004)
[21] Wang, Q., Megalooikonomu, V.: A clustering algorithm for intrusion detection. Proc. SPIE 5812, 31–38 (2005)
[22] Yang, Q., Zhang, H.: Web-log mining for predictive Web caching. IEEE Transactions on Knowledge and Data Engineering 15(4), 1050–1053 (2003)
[23] Knobbe, A., Van der Wallen, D., Lewis, L.: Experiments with data mining in enterprise management. In: Proceedings of the Sixth IFIP/IEEE International Symposium on Distributed Management for the Networked Millennium, Integrated Network Management, pp. 353–366 (1999)
[24] Bianco, A., Mardente, G., Mellia, M., Munafo, M., Muscariello, L.: Web User Session Characterization via Clustering techniques. In: GLOBECOM, New York, vol. 2, p. 1102 (2005)
[25] Duffield, N.G., Grossglauser, M.: Trajectory sampling for direct traffic observation. IEEE/ACM Trans. Netw. 9(3), 280–292 (2001)
[26] Lee, I., Fapojuwo, A.: Data Mining Network Traffic. In: Canadian Conference on Electrical and Computer Engineering (2006)
[27] Roesch, M.: Snort–Lightweight intrusion detection for networks. In: Proceeding of the 13th Systems Administration Conference, LISA 1999, pp. 299–238 (1999)
[28] Lee, W., Stolfo, S., Mok, K.: A data mining framework for building intrusion detection models. In: IEEE Symposium on Security and Privacy, vol. 132 (1999)
[29] Lee, W., Stolfo, S.: Data mining approaches for intrusion detection. In: Proceedings of the 7th USENIX Security Symposium, vol. 1, pp. 26–29 (1998)
[30] World Wide Web Consortium. eXtensible Markup Language, http://www.w3.org/XML
[31] Baralis, E., Cerquitelli, T., D'Elia, V.: Generalized itemset discovery by means of opportunistic aggregation. Technical report, Politecnico di Torino (2008), https://dbdmg.polito.it/twiki/bin/view/Public/NetworkTrafficAnalysis
[32] Han, J., Fu, Y.: Mining multiple-level association rules in large databases. IEEE Trans. Knowl. Data Eng. 11(5), 798–804 (1999)
[33] Naguyen, T., Armitage, G.: A Survey of Techniques for Internet Traffic Classification Using Machine Learning. In: IEEE Communications Surveys and Tutorials 2008 (October 2008)
[34] Haffner, P., Sen, S., Spatscheck, O., Wang, D.: ACAS: automated construction of application signatures. In: Proceedings of the 2005 ACM SIGCOMM workshop on Mining network data, pp. 197–202. ACM, New York (2005)
[35] Auld, T., Moore, A., Gull, S.: Bayesian Neural Networks for Internet Traffic Classification. IEEE Trans. on Neural Networks 18(1), 223 (2007)
[36] Bernaille, L., Akodkenou, I., Soule, A., Salamatian, K.: Traffic classification on the fly. ACM SIGCOMM Computer Communication Review 36(2), 23–26 (2006)
[37] McGregor, A., Hall, M., Lorier, P., Brunskill, J.: Flow Clustering Using Machine Learning Techniques. LNCS, pp. 205–214. Springer, Heidelberg (2004)
[38] Internet Assigned Numbers Authority (IANA). Port Numbers, http://www.iana.org/assignments/port-numbers

[39] Tan, P., Steinbach, M., Kumar, V.: Introduction to Data Mining. Addison-Wesley, Reading (2005)
[40] NetGroup, Politecnico di Torino. Analyzer 3.0, http://analyzer.polito.it
[41] Telecommunication Network Group, Politecnico di Torino. Tstat 1.01, http://tstat.polito.it
[42] Network Research Group, Lawrence Berkeley National Laboratory. Tcpdump 4.0.0, http://www.tcpdump.org
[43] Network Research Group, Lawrence Berkeley National Laboratory. Libpcap 1.0.0, http://www.tcpdump.org

# Group Classification of Objects with Qualitative Attributes: Multiset Approach

Alexey B. Petrovsky

**Abstract.** The paper considers a new approach to group classification of objects, which are described with many qualitative attributes and may exist in several versions. Clustering and sorting techniques are based on the theory of multiset metric spaces. New options of the multi-attribute objects' aggregation and features of the classes generated are discussed.

## 1 Introduction

Aggregation of objects (alternatives, variants, options, actions, situations, persons, items, and so on) in several classes (clusters, categories, groups) is one of the most popular tools for discovering, extracting, formalizing and fixing knowledge. Properties of objects are specified usually with a set of characteristics or attributes, values of which can be quantitative (numerical) and qualitative (symbolic or verbal). A class includes objects having common peculiarities. Classes may be predefined or appeared and built during a process of classification.

Classification problems are considered in data analysis, decision making, pattern recognition and image analysis, artificial intelligence, biology, sociology, psychology, and other areas. Constructed classification of objects allows us to reveal interrelations of different kinds and investigate possible ties between objects based on their features. A result of classification is new concepts and rules of their creation.

To solve the problems of classification, there are developed and used various techniques, which may be divided on such groups: classification without a teacher or clustering and classification with a teacher, nominal classification and ordinal classification. In the clustering methods, objects are aggregated together in groups (clusters) based on degree of them closeness, which formally established with

Alexey B.Petrovsky
Institute for Systems Analysis, Russian Academy of Sciences, Moscow 117312, Russia
e-mail: pab@isa.ru

V. Sgurev et al. (Eds.): Intelligent Systems: From Theory to Practice, SCI 299, pp. 73–97.
springerlink.com 

some distance between objects in an attribute space. The number of generated clusters can be arbitrary or fixed. In the methods of classification with a teacher, it is searched a general rule to assign an object to one of the given classes, that is built on basis of preliminary information about membership of some part of objects to certain classes. The presence or absence of ordering classes by some property or quality distinguishes methods for ordinal and nominal classification. In these methods, it is necessary to find the values of object attributes or their combinations, which are typical for each class.

Most of the known classification methods operate with objects described by many quantitative characteristics [1, 3, 5, 8, 11, 14, 19, 20, 21]. In these cases, as a rule, each object is associated with a vector, consisting of the numerical values of attributes, or a row of data table "Objects-Attributes". If the object is described by qualitative features, such symbolic or verbal variables are transformed usually by one or another way in the numerical ones, for example, using the lexicographic scale or fuzzy membership functions [9, 22, 25]. In this case, unfortunately, attention to the admissibility and validity of similar transformations of qualitative data into quantitative ones is not always given. Significantly fewer methods of classifying objects described by qualitative features, where these attributes are not transformed into numerical [6, 7, 12, 13, 16, 18]. In such cases, each object is represented by a tuple (cortege) consisting of symbolic or verbal values of attributes.

The situation becomes more complicated when one and the same object can exist in multiple versions or copies. For example, object's characteristics were measured in different conditions or using different instruments, either an object was independently evaluated by several experts upon many criteria. Then, not one vector or tuple will correspond to everyone object, but a group of vectors or tuples, which should be considered and treated as a whole. In this case, obviously, the values of similar components of different vectors/tuples can vary and even be contradictory. It is clear that a collection of such multi-attribute objects can have very complicated structure, which is very difficult to analyze.

Typically, a group of objects, represented by several vectors, is replaced by a single vector. For example, this vector has the components derived by averaging or weighting the values of attributes of all members of the group, is the center of a group or a vector, the closest to all the vectors within the group. Note, however, that the properties of all objects in a group may be lost after such replacement. For qualitative variables the operations of averaging, weighing, mixing and similar data transformations are mathematically incorrect and unacceptable. Thus, a group of objects, represented by several tuples, can not be connected with a single tuple. So, we need new ways of aggregating similar objects and work with them.

In this paper, objects, which are described by many qualitative and/or quantitative values of attributes, are represented as multisets or sets with repeating elements. New techniques for operating with groups of such objects are discussed. New methods of group clustering and sorting multi-attribute objects in multiset metric space are suggested.

## 2 Multisets and Multiset Metric Spaces

A multiset (also called a bag) is a known notion that is used in combinatorial mathematics, computer science and other fields [3, 10, 17, 24]. A multiset $\boldsymbol{A}$ drawn from an ordinary (crisp) set $X=\{x_1, x_2,..., x_j,...\}$ with different elements is defined as the following collection of element groups

$$\boldsymbol{A}=\{k_A(x_1)\circ x_1,..., k_A(x_j)\circ x_j,...\}=\{k_A(x)\circ x|x\in X, k_A\in \mathbf{Z}_+\}. \quad (1)$$

Here $k_A$: $X\to\mathbf{Z}_+=\{0,1,2,...\}$ is called a counting or multiplicity function of multiset, which defines the number of times the element $x_j\in X$ occurs in the multiset $\boldsymbol{A}$, and this is indicated by the symbol $\circ$. The multiset generalizes the notion of ordinary set.

The theoretical model of multiset is very suitable for structuring and analyzing a collection of objects that are described by many qualitative and/or quantitative attributes and also may exist in several versions or copies. Give some examples of such objects that can be represented as multisets.

Let $\mathcal{A}=\{A_1,...,A_n\}$ be a collection of recognized graphic objects (printed or handwritten symbols, lines, images, pages) [2]. The set $X=\{x_1,...,x_h\}$ is a base of standard samples that consists of whole symbols or separate structural fragments of symbols. In the process of recognition, each graphic object $A_i$ is compared with the sample set $X$ and is related to any standard symbol $x_j$. The result of recognition of the graphic object $A_i$ can be represented in the form $\boldsymbol{A}_i=\{k_{Ai}(x_1)\circ x_1,..., k_{Ai}(x_h)\circ x_h\}$, where $k_{Ai}(x_j)$ is equal to a valuation of recognized object $A_i$ computed with respect to the standard symbol $x_j$.

Let $\mathcal{A}=\{A_1,...,A_n\}$ be a file of textual documents related to any problem field, for instance, reports, references, projects, patents, reviews, articles, and so on [15]. The set of lexical units (descriptors, keywords, terms, etc) $X=\{x_1,...,x_h\}$ is a problem-oriented terminological dictionary or thesaurus. The content of document $A_i$ can be written as the collection of lexical units in the form $\boldsymbol{A}_i=\{k_{Ai}(x_1)\circ x_1,..., k_{Ai}(x_h)\circ x_h\}$, where $k_{Ai}(x_j)$ is equal to a number of lexical unit $x_j$ within the description of document $A_i$.

In the cases considered above, a multi-attribute object (symbol, document) $A_i$ is represented as a set of repeating elements (standard samples, lexical units) $x_j$ or as a multiset $\boldsymbol{A}_i$. Obviously, every graphic object $A_i$ may occur several times within the recognized text or image, and every document $A_i$ may exist in several versions within the file. So, there are groups, which combine different versions (copies) of one and the same objects. And each group of versions of multi-attribute object $A_i$, and the collection $\mathcal{A}=\{A_1,..., A_n\}$ wholly can also be represented as multisets over the correspondent set $X$.

A multiset $\boldsymbol{A}$ is said to be finite when all $k_A(x)$ are finite. A multiset $\boldsymbol{A}$ becomes a crisp set $A$ when $k_A(x)=\chi_A(x)$, where $\chi_A(x)=1$, if $x\in A$, and $\chi_A(x)=0$, if $x\notin A$. Multisets $\boldsymbol{A}$ and $\boldsymbol{B}$ are said to be equal ($\boldsymbol{A}=\boldsymbol{B}$), if $k_A(x)=k_B(x)$, and a multiset $\boldsymbol{B}$ is said to be contained or included in a multiset $\boldsymbol{A}$ ($\boldsymbol{B}\subseteq\boldsymbol{A}$), if $k_B(x)\leq k_A(x)$, $\forall x\in X$.

There are defined the following operations with multisets [17]: union $\boldsymbol{A}\cup\boldsymbol{B}$, $k_{A\cup B}(x)=\max[k_A(x), k_B(x)]$; intersection $\boldsymbol{A}\cap\boldsymbol{B}$, $k_{A\cap B}(x)=\min[k_A(x), k_B(x)]$; addition

$\boldsymbol{A}+\boldsymbol{B}$, $k_{A+B}(x)=k_A(x)+k_B(x)$; subtraction $\boldsymbol{A}-\boldsymbol{B}$, $k_{A-B}(x)=k_A(x)-k_{A\cap B}(x)$; symmetric difference $\boldsymbol{A}\Delta\boldsymbol{B}$, $k_{A\Delta B}(x)=|k_A(x)-k_B(x)|$; multiplication by a scalar (reproduction) $b\bullet\boldsymbol{A}$, $k_{b\cdot A}(x)=b\cdot k_A(x)$, $b\in\mathbf{N}$; multiplication $\boldsymbol{A}\bullet\boldsymbol{B}$, $k_{A\cdot B}(x)=k_A(x)\cdot k_B(x)$; arithmetic power $\boldsymbol{A}^n$; direct product $\boldsymbol{A}\times\boldsymbol{B}$, $k_{A\times B}(x_i,x_j)=k_A(x_i)\cdot k_B(x_j)$, $x_i\in\boldsymbol{A}$, $x_j\in\boldsymbol{B}$; direct power $(\times\boldsymbol{A})^n$.

Many features of operations under multisets are analogues to features of operations under ordinary sets. These are idempotency, involution, identity, commutativity, associativity, and distributivity. As well as for ordinary sets not all operations under multisets are mutually commutative, associative and distributive. In general, the operations of addition, multiplication by a scalar, multiplication, and raising to arithmetic powers are not defined in the theory of sets. When multisets are reduced to sets, the operations of multiplication and raising to an arithmetic power transform into a set intersection, but the operations of set addition and set multiplication by a scalar will be impracticable.

A collection $\mathcal{A}=\{A_1,...,A_n\}$ of multi-attribute objects can be considered as points in the multiset metric space $(\mathcal{A},d)$. Different types of metric spaces $(\mathcal{A},d)$ are defined by the following distances between multisets:

$$d_{1c}(\boldsymbol{A},\boldsymbol{B})=[m(\boldsymbol{A}\Delta\boldsymbol{B})]^{1/c};\quad d_{2c}(\boldsymbol{A},\boldsymbol{B})=[m(\boldsymbol{A}\Delta\boldsymbol{B})/m(\boldsymbol{Z})]^{1/c};$$

$$d_{3c}(\boldsymbol{A},\boldsymbol{B})=[m(\boldsymbol{A}\Delta\boldsymbol{B})/m(\boldsymbol{A}\cup\boldsymbol{B})]^{1/c}, \tag{2}$$

where $c\geq0$ is an integer, $m(\boldsymbol{A})$ is a measure of multiset $\boldsymbol{A}$, and the multiset $\boldsymbol{Z}$ is called as the maximal multiset with a multiplicity function $k_Z(x)=\max_{A\in\mathcal{A}}k_A(x)$. Multiset measure $m$ is a real-valued non-negative function defined on the algebra of multisets $L(\boldsymbol{Z})$. Multiset measure can be determined in the various ways, for instance, as a linear combination of multiplicity functions: $m(\boldsymbol{A})=\sum_j w_j k_A(x_j)$, $w_j>0$ is an importance of the element $x_j$. The distances $d_{2c}(\boldsymbol{A},\boldsymbol{B})$ and $d_{3c}(\boldsymbol{A},\boldsymbol{B})$ satisfy the normalization condition $0\leq d(\boldsymbol{A},\boldsymbol{B})\leq1$. The distance $d_{3c}(\boldsymbol{A},\boldsymbol{B})$ is undefined for $\boldsymbol{A}=\boldsymbol{B}=\varnothing$. So, $d_{3c}(\varnothing,\varnothing)=0$ by the definition. For any fixed $c$, the metrics $d_{1c}$ and $d_{2c}$ are the continuous and uniformly continuous functions, the metric $d_{3c}$ is the piecewise continuous function almost everywhere on the metric space [16-18].

The proposed spaces are the new types of metric spaces that differ from the well-known set-theoretical metric spaces [4]. The distance $d_{1c}(\boldsymbol{A},\boldsymbol{B})$ characterizes a difference between properties of two objects, and is analogues of the Hamming-type distance between objects that is traditional for many applications. The distance $d_{2c}(\boldsymbol{A},\boldsymbol{B})$ represents a difference between properties of two objects related to properties of the so called maximal object, and can be called as the completely averaged distance. And the distance $d_{3c}(\boldsymbol{A},\boldsymbol{B})$ reflects a difference between properties of two objects related to common properties of these objects, and can be called as the locally averaged distance. In the case of ordinary sets for $c=1$, $d_{11}(A,B)=m(A\Delta B)$ is called the Fréchet distance, $d_{31}(A,B)=m(A\Delta B)/m(A\cup B)$ is called the Steinhaus distance [4]. Various features of multisets and multiset metric spaces are considered and discussed in [17].

## 3 Representation and Aggregation of Multi-attribute Objects

The problem of group classification of multi-attribute objects is formulated generally as follows. There is a collection of objects $A=\{A_1,...,A_n\}$, which are described by $m$ attributes $Q_1,...,Q_m$. Every attribute $Q_s$ has its own scale $X_s=\{x_s^1,\ldots,x_s^{h_s}\}$, $s=1,\ldots,m$, gradations of which are numerical, symbolic or verbal values, discreet or continuos. Each object is represented in $k$ versions or copies, which are usually distinguished by the values of attributes. For example, object characteristics have been measured in different conditions or in different ways, either several experts independently evaluated objects upon many criteria. Need to divide objects into several groups (classes) $C_1,...,C_g$, describe and interpret the properties of these groups of objects. The number $g$ of object groups can be arbitrary or predefined, and the classes themselves can be ordered or unordered.

We first consider a few illustrative examples showing how one can represent multi-attribute objects.

Let the collection $A$ consists of 10 objects $A_1,...,A_{10}$, described with 8 attributes $Q_1,...,Q_8$. Each of the attributes may take one of the grades on five-point scale $X=\{1, 2, 3, 4, 5\}$. Assume that the objects $A_1,...,A_{10}$ are the school pupils, and the attributes of the objects are the annual scores to 8 disciplines (subjects): $Q_1$.Mathematics, $Q_2$.Physics, $Q_3$.Chemistry, $Q_4$.Biology, $Q_5$.Geography, $Q_6$.History, $Q_7$.Literature, $Q_8$.Foreign Language. The gradation of the estimate scale $X$ indicates the following: 1 – very bad, 2 – bad, 3 – satisfactory, 4 – good, 5 – excellent. Or, the objects $A_1,...,A_{10}$ are the answers to some questionnaire, with which the opinions of a group of people on some issues are studied. The attributes of objects will be the estimates given by 8 respondents $Q_1,...,Q_8$, which are coded as follows: 1 – strongly disagree, 2 – disagree, 3 – neutral, 4 – agree, 5 – strongly agree.

Associate each multi-attribute object $A_i$ with a vector or tuple $q_i=(q_{i1}^{e_1},...,q_{im}^{e_m})$ in the Cartesian $m$-space of attributes $Q=Q_1\times\ldots\times Q_m$. The collection $A$ of objects and their attributes can be represented by a table "Objects-Attributes" $Q=||q_{is}||_{n\times m}$. Rows of the matrix Q correspond to the objects, columns agree with the attributes, whereas entries $q_{is}$ are the components $q_{is}^{e_s}$ of the correspondent vectors/tuples. The data table Q for the above examples is shown in Table 1. This table is taken from [8] and is a part of questionnaire about data analysis course. For instance, the object $A_1$ is associated with the vector/tuple $q_1=(4, 5, 4, 5, 4, 5, 4, 5)$.

We point out another possible way to define multi-attribute objects using multisets. Let us consider the set of estimates $X=\{x_1,...,x_h\}$ as a generating set, and associate each multi-attribute object $A_i$ with a multiset $\boldsymbol{A}_i=\{k_{Ai}(x_1)\circ x_1,...,k_{Ai}(x_h)\circ x_h\}$ over the set $X$. Here the value of counting function $k_{Ai}(x_j)$ shows how many times the score $x_j$ occurs in the description of the object $A_i$. The collection $A$ of objects and their attributes can be represented by another table "Objects-Attributes" $K=||k_{ij}||_{n\times h}$, entries $k_{ij}$ of which are the multiplicities $k_{Ai}(x_j)$ of elements of the correspondent multisets. The data table K for the above objects $A_1,...,A_{10}$ is shown in Table 2. For instance, the object $A_1$ is associated with the multiset $\boldsymbol{A}_1=\{0\circ x_1, 0\circ x_2, 0\circ x_3, 4\circ x_4, 4\circ x_5\}$. This notation says that the object $A_1$ has 4 estimates $x_4$ meaning 'good' or 'agree', 4 estimates $x_5$ meaning 'excellent' or 'strongly agree', and other estimates are absent. In some cases, it is conveniently to place elements of the multiset in reverse order – from the best to the worst estimates – and write the multiset as $\boldsymbol{A}_1=\{4\circ x_5, 4\circ x_4, 0\circ x_3, 0\circ x_2, 0\circ x_1\}$. Notice, however, that, strictly speaking, the elements of multiset are considered as disordered.

**Table 1** Data table Q

| A\Q | $Q_1$ | $Q_2$ | $Q_3$ | $Q_4$ | $Q_5$ | $Q_6$ | $Q_7$ | $Q_8$ |
|---|---|---|---|---|---|---|---|---|
| $A_1$ | 4 | 5 | 4 | 5 | 4 | 5 | 4 | 5 |
| $A_2$ | 4 | 1 | 2 | 1 | 3 | 2 | 2 | 2 |
| $A_3$ | 1 | 1 | 3 | 1 | 4 | 1 | 1 | 4 |
| $A_4$ | 5 | 3 | 2 | 4 | 4 | 5 | 4 | 5 |
| $A_5$ | 4 | 4 | 4 | 4 | 4 | 5 | 4 | 4 |
| $A_6$ | 5 | 5 | 4 | 4 | 4 | 5 | 5 | 4 |
| $A_7$ | 4 | 1 | 2 | 3 | 3 | 3 | 1 | 2 |
| $A_8$ | 4 | 5 | 4 | 2 | 3 | 4 | 5 | 3 |
| $A_9$ | 3 | 2 | 3 | 1 | 3 | 3 | 2 | 2 |
| $A_{10}$ | 5 | 5 | 4 | 5 | 3 | 5 | 5 | 4 |

**Table 2** Data table K

| A\X | $x_1$ | $x_2$ | $x_3$ | $x_4$ | $x_5$ |
|---|---|---|---|---|---|
| $A_1$ | 0 | 0 | 0 | 4 | 4 |
| $A_2$ | 2 | 4 | 1 | 1 | 0 |
| $A_3$ | 5 | 0 | 1 | 2 | 0 |
| $A_4$ | 0 | 1 | 1 | 3 | 3 |
| $A_5$ | 0 | 0 | 0 | 7 | 1 |
| $A_6$ | 0 | 0 | 0 | 4 | 4 |
| $A_7$ | 2 | 2 | 3 | 1 | 0 |
| $A_8$ | 0 | 1 | 2 | 3 | 2 |
| $A_9$ | 1 | 3 | 4 | 0 | 0 |
| $A_{10}$ | 0 | 0 | 1 | 2 | 5 |

Now suppose that each object exists in several versions or copies. For example, two sets of semester (semi-year) scores are given for each of the pupils $A_1,...,A_{10}$ to the same 8 disciplines $Q_1,...,Q_8$. Or 8 respondents $Q_1,...,Q_8$ are interviewed twice, answering the same questions $A_1,...,A_{10}$. This means that each object is described by not one but two vectors/tuples of estimates, or that there are two versions of the same object. For instance, the object $A_1$ is associated with the group of vector/tuple consisting of $q_1^{(1)}=(4, 5, 4, 5, 4, 5, 4, 5)$ and $q_1^{(2)}=(5, 5, 5, 5, 4, 4, 4, 5)$.

So, in the general case, one and the same multi-attribute object $A_i$, $i=1,\ldots,n$ is represented as a group of $k$ vectors/tuples $\{q_i^{(1)},\ldots,q_i^{(k)}\}$ in the Cartesian $m$-space of attributes $Q=Q_1\times\ldots\times Q_m$. Here $q_i^{(f)}=(q_{i1}^{e_1(f)},\ldots,q_{im}^{e_m(f)})$, $f=1,\ldots,k$ is one of the object versions, and a component $q_{is}^{e_s(f)}$, $s=1,\ldots,m$ of the vector/tuple $q_i^{(f)}$ is one of the scale grade $x_j\in X$, $j=1,\ldots,h$.

Note that the group of vectors/tuples corresponding to the object $A_i$ is to be considered as a whole, despite the possible incomparability of individual vectors $q_i^{(f)}$ and/or inconsistency of their components $q_{is}^{e_s(f)}$. Components of vectors are numeric variables. Therefore, the group of vectors is often replaced by a single vector representing the whole group, whose components are determined by some additional considerations. For example, it may be a vector having as components the averaged values of the corresponding components of all group members. Then the object $A_1$ will be represented as the vector of 'averaged' components $q_1=(4.5, 5.0, 4.5, 5.0, 4.0, 4.5, 4.0, 5.0)$. However such a vector does not correspond to any concrete point of the $m$-dimensional attribute space $Q=Q_1\times\ldots\times Q_m$, formed by a discrete numerical scale $X=\{1, 2, 3, 4, 5\}$, in which there are no fractional numbers. To be able to operate with such vectors, one should either expand the scale $X$ of estimates by introducing the intermediate numerical gradations, for instance, as follows: $X=\{1.00, 1.25, 1.50, 1.75, 2.00,\ldots, 4.00, 4.25, 4.50, 4.75, 5.00\}$, or to consider $X$ as a continuous scale. And the first, and the second, strictly speaking, changes the initial original formulation of the problem.

When objects are represented as tuples with symbolic or verbal components, that is used the scale $X=\{1, 2, 3, 4, 5\}$, where, for example, 1 – very bad, 2 – bad, 3 – satisfactory, 4 – good, 5 – excellent, a group of similar tuples can no longer be replaced by a single tuple with 'averaged' components, because such an operation is mathematically incorrect.

These difficulties can be easily overcome with the help of multisets. Associate each version $A_i^{(f)}$, $i=1,\ldots,n$, $f=1,\ldots,k$ of the multi-attribute object $A_i$ with a multiset $\boldsymbol{A}_i^{(f)}=\{k_{Ai}^{(f)}(x_1)\circ x_1,\ldots,k_{Ai}^{(f)}(x_h)\circ x_h\}$ over the set $X=\{x_1,\ldots,x_h\}$, and each object $A_i$ with a multiset $\boldsymbol{A}_i=\{k_{Ai}(x_1)\circ x_1,\ldots,k_{Ai}(x_h)\circ x_h\}$ over the same set $X$. The value of multiplicity function of the multiset $\boldsymbol{A}_i$ is calculated, for instance, according to the rule: $k_{Ai}(x_j)=\sum_f k_{Ai}^{(f)}(x_j)$. Thus, two versions of the object $A_1$ are represented by two multisets $\boldsymbol{A}_1^{(1)}=\{0\circ x_1, 0\circ x_2, 0\circ x_3, 4\circ x_4, 4\circ x_5\}$ and $\boldsymbol{A}_1^{(2)}=\{0\circ x_1, 0\circ x_2, 0\circ x_3, 3\circ x_4, 5\circ x_5\}$, and the object $A_1$ itself – by the multiset $\boldsymbol{A}_1=\{0\circ x_1, 0\circ x_2, 0\circ x_3, 7\circ x_4, 9\circ x_5\}$.

If for some reason it is important not the total number of different values on all attributes $Q_1$-$Q_8$ but the number of different values for each attribute $Q_s$, for example, scores for each discipline or answers of each respondent, and these values should be distinguished, then a multiset can be defined in other way. Introduce the hyperscale of attributes $X=X_1\cup\ldots\cup X_m=\{x_1^{\,1},\ldots,x_1^{\,h_1};\ \ldots;\ x_m^{\,1},\ldots,x_m^{\,h_m}\}$, which combines together all gradations of the individual attribute scales. Then each object $A_i$ or its version $A_i^{(f)}$ can be associated with a multiset

$$\boldsymbol{A}_i=\{k_{Ai}(x_1^{\,1})\circ x_1^{\,1},\ldots,k_{Ai}(x_1^{\,h_1})\circ x_1^{\,h_1};\ \ldots;\ k_{Ai}(x_m^{\,1})\circ x_m^{\,1},\ldots,k_{Ai}(x_m^{\,h_m})\circ x_m^{\,h_m}\} \qquad (3)$$

over the set of values $X=\{x_1^{\,1},\ldots,x_1^{\,h_1};\ \ldots;\ x_m^{\,1},\ldots,x_m^{\,h_m}\}$. Here the value of multiplicity function $k_{Ai}(x_s^{\,e_s})$ shows the number of values $x_s^{\,e_s}\in X_s$, $e_s=1,\ldots,h_s$, $s=1,\ldots,m$ of each attribute $Q_s$ representing within the description of the object $A_i$. For example, the object $A_1$ will be now associated with the following multiset of individual values on all attributes:

$$\begin{aligned}\boldsymbol{A}_1=\{&0\circ x_1^{\,1},0\circ x_1^{\,2},0\circ x_1^{\,3},1\circ x_1^{\,4},1\circ x_1^{\,5};\ 0\circ x_2^{\,1},0\circ x_2^{\,2},0\circ x_2^{\,3},0\circ x_2^{\,4},2\circ x_2^{\,5};\\ &0\circ x_3^{\,1},0\circ x_3^{\,2},0\circ x_3^{\,3},1\circ x_3^{\,4},1\circ x_3^{\,5};\ 0\circ x_4^{\,1},0\circ x_4^{\,2},0\circ x_4^{\,3},0\circ x_4^{\,4},2\circ x_4^{\,5};\\ &0\circ x_5^{\,1},0\circ x_5^{\,2},0\circ x_5^{\,3},2\circ x_5^{\,4},0\circ x_5^{\,5};\ 0\circ x_6^{\,1},0\circ x_6^{\,2},0\circ x_6^{\,3},1\circ x_6^{\,4},1\circ x_6^{\,5};\\ &0\circ x_7^{\,1},0\circ x_7^{\,2},0\circ x_7^{\,3},2\circ x_7^{\,4},0\circ x_7^{\,5};\ 0\circ x_8^{\,1},0\circ x_8^{\,2},0\circ x_8^{\,3},0\circ x_8^{\,4},2\circ x_8^{\,5}\}.\end{aligned}$$

Despite the apparent awkwardness of notation, this representation is extremely convenient from the computational point of view when comparing multisets and operating with them, because one can perform operations simultaneously on all elements of multisets. Note that the expression (3) for any multiset can be easily written in the usual form (1) $\boldsymbol{A}_i=\{k_{Ai}(x_j)\circ x_j|x_j\in X=\{x_1,\ldots,x_h\}\}$, if in the set

**Table 3** Data table $Q^{\#}$

| A\Q | $Q_1$ | $Q_2$ | $Q_3$ | $Q_4$ | $Q_5$ | $Q_6$ | $Q_7$ | $Q_8$ |
|---|---|---|---|---|---|---|---|---|
| $A_1$ | 4 | 5 | 4 | 5 | 4 | 5 | 4 | 5 |
| | 5 | 5 | 5 | 5 | 4 | 4 | 4 | 5 |
| $A_2$ | 4 | 1 | 2 | 1 | 3 | 2 | 2 | 2 |
| | 3 | 2 | 1 | 1 | 4 | 3 | 3 | 2 |
| $A_3$ | 1 | 1 | 3 | 1 | 4 | 1 | 1 | 4 |
| | 1 | 2 | 3 | 1 | 5 | 2 | 1 | 3 |
| $A_4$ | 5 | 3 | 2 | 4 | 4 | 5 | 4 | 5 |
| | 4 | 4 | 3 | 5 | 4 | 5 | 3 | 4 |
| $A_5$ | 4 | 4 | 4 | 4 | 4 | 5 | 4 | 4 |
| | 5 | 5 | 3 | 4 | 4 | 4 | 5 | 4 |
| $A_6$ | 5 | 5 | 4 | 4 | 4 | 5 | 5 | 4 |
| | 4 | 5 | 4 | 4 | 4 | 4 | 5 | 5 |
| $A_7$ | 4 | 1 | 2 | 3 | 3 | 3 | 1 | 2 |
| | 3 | 2 | 1 | 4 | 2 | 4 | 2 | 3 |
| $A_8$ | 4 | 5 | 4 | 2 | 3 | 4 | 5 | 3 |
| | 5 | 4 | 5 | 3 | 4 | 5 | 4 | 4 |
| $A_9$ | 3 | 2 | 3 | 1 | 3 | 3 | 2 | 2 |
| | 4 | 3 | 2 | 2 | 2 | 3 | 3 | 2 |
| $A_{10}$ | 5 | 5 | 4 | 5 | 3 | 5 | 5 | 4 |
| | 5 | 5 | 5 | 4 | 2 | 4 | 4 | 4 |

**Table 4** Data table $K^{\#}$

| A\X | $x_1$ | $x_2$ | $x_3$ | $x_4$ | $x_5$ |
|---|---|---|---|---|---|
| $A_1$ | 0 | 0 | 0 | 4 | 4 |
| | 0 | 0 | 0 | 3 | 5 |
| $A_2$ | 2 | 4 | 1 | 1 | 0 |
| | 2 | 2 | 3 | 1 | 0 |
| $A_3$ | 5 | 0 | 1 | 2 | 0 |
| | 3 | 2 | 2 | 0 | 1 |
| $A_4$ | 0 | 1 | 1 | 3 | 3 |
| | 0 | 0 | 2 | 4 | 2 |
| $A_5$ | 0 | 0 | 0 | 7 | 1 |
| | 0 | 0 | 1 | 4 | 3 |
| $A_6$ | 0 | 0 | 0 | 4 | 4 |
| | 0 | 0 | 0 | 5 | 3 |
| $A_7$ | 2 | 2 | 3 | 1 | 0 |
| | 1 | 3 | 3 | 1 | 0 |
| $A_8$ | 0 | 1 | 2 | 3 | 2 |
| | 0 | 0 | 1 | 4 | 3 |
| $A_9$ | 1 | 3 | 4 | 0 | 0 |
| | 0 | 4 | 3 | 1 | 0 |
| $A_{10}$ | 0 | 0 | 1 | 2 | 5 |
| | 0 | 1 | 0 | 4 | 3 |

$X=\{x_1^{\,1},\ldots,x_1^{\,h_1};\ \ldots;\ x_m^{\,1},\ldots,x_m^{\,h_m}\}$ to make the following change of variables: $x_1^{\,1}=x_1,\ldots,\ x_1^{\,h_1}=x_{h1},\ x_2^{\,1}=x_{h1+1},\ldots,\ x_2^{\,h_2}=x_{h1+h2},\ldots,\ x_m^{\,h_m}=x_h,\ h=h_1+\ldots+h_m$.

The collection $A$ of objects, when each of objects is available in several versions, and their attributes can be again represented by a table "Objects-Attributes", but having a larger dimension. The table "Objects-Attributes" $Q^{\#}=||q_{is}||_{kn\times m}$ for objects $A_1,\ldots,A_{10}$ of the earlier examples, which are represented by vectors/tuples, is given in Table 3. Entries $q_{is}$ of the table $Q^{\#}$ are the components $q_{is}^{\,e_s(f)}$ of the correspondent vectors/tuples. The table "Objects-Attributes" $K^{\#}=||k_{ij}||_{kn\times h}$ for objects $A_1,\ldots,A_{10}$, which are represented by multisets of the type (1), is shown in Table 4. Entries $k_{ij}$ of the table $K^{\#}$ are the multiplicities $k_{Ai}{}^{(f)}(x_j)$ of elements of the multisets corresponding to versions of objects. The table "Objects-Attributes" $L=||k_{ij}||_{n\times h}$, $h=h_1+\ldots+h_m$ for objects, represented as multiset of the type (3), is given in Table 5. Entries $k_{ij}$ of the table L are the multiplicities $k_{Ai}(x_j)$ of elements of the multisets corresponding to objects.

Object grouping is one of the useful techniques for studying structure of object collection. By classifying objects, we have to assign each object to any class and use information about a membership of the objects to classes for an elaboration and correction of classification rules. In the most general sense, classification rules can be represented as logical statements of the following type:

**Table 5** Data table L

| A\X | $x_1^1\ x_1^2\ x_1^3\ x_1^4\ x_1^5$ | $x_2^1\ x_2^2\ x_2^3\ x_2^4\ x_2^5$ | $x_1^1\ x_3^2\ x_3^3\ x_3^4\ x_3^5$ | $x_4^1\ x_4^2\ x_4^3\ x_4^4\ x_4^5$ | $x_5^1\ x_5^2\ x_5^3\ x_5^4\ x_5^5$ | $x_6^1\ x_6^2\ x_6^3\ x_6^4\ x_6^5$ | $x_7^1\ x_7^2\ x_7^3\ x_7^4\ x_7^5$ | $x_8^1\ x_8^2\ x_8^3\ x_8^4\ x_8^5$ |
|---|---|---|---|---|---|---|---|---|
| $A_1$ | 0 0 0 1 1 | 0 0 0 0 2 | 0 0 0 1 1 | 0 0 0 0 2 | 0 0 0 2 0 | 0 0 0 1 1 | 0 0 0 2 0 | 0 0 0 0 2 |
| $A_2$ | 0 0 1 1 0 | 1 1 0 0 0 | 1 1 0 0 0 | 2 0 0 0 0 | 0 0 1 1 0 | 0 1 1 0 0 | 0 1 1 0 0 | 0 2 0 0 0 |
| $A_3$ | 2 0 0 0 0 | 1 1 0 0 0 | 0 0 2 0 0 | 2 0 0 0 0 | 0 0 0 1 1 | 1 1 0 0 0 | 2 0 0 0 0 | 0 0 1 1 0 |
| $A_4$ | 0 0 0 1 1 | 0 0 1 1 0 | 0 1 1 0 0 | 0 0 0 1 1 | 0 0 0 2 0 | 0 0 0 0 2 | 0 0 0 1 1 | 0 0 0 1 1 |
| $A_5$ | 0 0 0 1 1 | 0 0 0 1 1 | 0 0 0 1 1 | 0 0 0 2 0 | 0 0 0 2 0 | 0 0 0 1 1 | 0 0 0 1 1 | 0 0 0 2 0 |
| $A_6$ | 0 0 0 1 1 | 0 0 0 0 2 | 0 0 0 2 0 | 0 0 0 2 0 | 0 0 0 2 0 | 0 0 0 1 1 | 0 0 0 0 2 | 0 0 0 1 1 |
| $A_7$ | 0 0 1 1 0 | 1 1 0 0 0 | 1 1 0 0 0 | 0 0 1 1 0 | 0 1 1 0 0 | 0 0 1 1 0 | 1 1 0 0 0 | 0 1 1 0 0 |
| $A_8$ | 0 0 0 1 1 | 0 0 0 1 1 | 0 0 0 1 1 | 0 1 1 0 0 | 0 0 1 1 0 | 0 0 0 1 1 | 0 0 0 1 1 | 0 0 1 1 0 |
| $A_9$ | 0 0 1 1 0 | 0 1 1 0 0 | 0 1 1 0 0 | 1 1 0 0 0 | 0 1 1 0 0 | 0 0 2 0 0 | 0 1 1 0 0 | 0 0 2 0 0 |
| $A_{10}$ | 0 0 0 0 2 | 0 0 0 0 2 | 0 0 0 1 1 | 0 0 0 1 1 | 0 1 1 0 0 | 0 0 0 1 1 | 0 0 0 1 1 | 0 0 0 2 0 |

IF ⟨conditions⟩, THEN ⟨decision⟩. (4)

The antecedent term ⟨conditions⟩ specifies the requirements for selecting objects. For example, they may be names of objects, values or combinations of values of attributes describing objects, constraints on the values of attributes, relationships between objects, rules for comparing objects to each other or with some particular members of the classes, and the like. Objects are compared by similarity or difference of their properties, which are formalized usually with specially defined measures of closeness. To select a specific member of the class, for example, a typical representative of the class or a center of the class, impose certain requirements. The consequent term ⟨decision⟩ denotes the name of the generated class and/or the object membership to the predefined class if the required conditions are carried out.

Variety of operations with multisets allows us to use different ways for aggregating multi-attribute objects into classes. For instance, a class $C_t$ of objects $A_i$, $i$=1,...,$n$ can be constructed as a sum $\boldsymbol{Y}_t=\sum_i \boldsymbol{A}_i$, $k_{Yt}(x_j)=\sum_i k_{Ai}(x_j)$, union $\boldsymbol{Y}_t=\bigcup_i \boldsymbol{A}_i$, $k_{Yt}(x_j)=\max_i k_{Ai}(x_j)$ or intersection $\boldsymbol{Y}_t=\bigcap_i \boldsymbol{A}_i$, $k_{Yt}(x_j)=\min_i k_{Ai}(x_j)$ of multisets, which represent the objects considered. A class $C_t$ of objects can be also formed as a linear combination of corresponding multisets $\boldsymbol{Y}_t=\sum_i b_i \bullet \boldsymbol{A}_i$, $\boldsymbol{Y}_t=\bigcup_i b_i \bullet \boldsymbol{A}_i$ or $\boldsymbol{Y}_t=\bigcap_i b_i \bullet \boldsymbol{A}_i$. When a class $C_t$ is formed by adding multisets, all features of all members in the group $\boldsymbol{Y}_t$ of multisets (all values of all attributes) are combined. In the case of union or intersection of multisets, the best features (maximal values of all attributes) or the worth features (minimal values of all attributes) of individual members in the group $\boldsymbol{Y}_t$ of multisets are intensified. For example, the multiset $\boldsymbol{A}_i$ corresponding to the object $A_i$ has been formed with the addition of multisets $\boldsymbol{A}_i^{(f)}$, which are associated with versions of this object.

## 4 Group Clustering Multi-attribute Objects

Cluster analysis deals with a division of object collection $\boldsymbol{A}=\{A_1,...,A_n\}$ in several groups (clusters) $C_1,...,C_g$ basing on the notion of closeness between objects [1, 8, 14]. Two general approaches to constructing groups of objects are used usually in clustering techniques: (i) minimization of difference (maximization of similarity) between objects within a group; (ii) maximization of difference (minimization of similarity) between groups. A difference and similarity between the object features are determined by a value of distance between objects in the attribute space. It is assumed that special rules are given for computing the distance between any pair of objects and for combining two objects or groups of objects to build a new group.

Let us represent the object described by many qualitative (symbolic or verbal) attributes and groups of such objects as a multiset of the type (1) or (3). Taking into account the formula (2) for $c$=1 and $m(\boldsymbol{A})=\sum_j w_j k_{A}(x_j)$, where $w_j$>0 is an importance of the attribute $Q_j$, a difference between groups of objects can be defined as one of the following distances in the multiset space $(\boldsymbol{A},d)$:

$$d_{11}(\boldsymbol{Y}_p,\boldsymbol{Y}_q)=D_{pq};\quad d_{21}(\boldsymbol{Y}_p,\boldsymbol{Y}_q)=D_{pq}/W;\quad d_{31}(\boldsymbol{Y}_p,\boldsymbol{Y}_q)=D_{pq}/M_{pq}. \tag{5}$$

A similarity between groups of objects can be expressed by one of the following indexes:

$$s_1(\boldsymbol{Y}_p,\boldsymbol{Y}_q)=1-(D_{pq}/W);\quad s_2(\boldsymbol{Y}_p,\boldsymbol{Y}_q)=L_{pq}/W;\quad s_3(\boldsymbol{Y}_p,\boldsymbol{Y}_q)=L_{pq}/M_{pq}. \tag{6}$$

Here $D_{pq}=\sum_j w_j|k_{Yp}(x_j)-k_{Yq}(x_j)|$; $L_{pq}=\sum_j w_j\min[k_{Yp}(x_j),\ k_{Yq}(x_j)]$; $M_{pq}=\sum_j w_j\max[k_{Yp}(x_j),\ k_{Yq}(x_j)]$; $W=\sum_j w_j\sup_t k_{Yt}(x_j)$. The values of functions $k_{Yp}(x_j)$, $k_{Yq}(x_j)$ depend on the option used for combining multi-attribute objects into groups. In the case of multisets, the expressions $s_1$, $s_2$, $s_3$ (6) generalize the well-known nonmetric indexes of object similarity such as the simple matching coefficient, Russel and Rao measure of similarity, Jaccard coefficient or Tanimoto measure, respectively [1, 14].

Consider main ideas of cluster analysis for objects represented as multisets. Suppose, for simplicity, that the formulas (5) and (6) are determined a difference and similarity in a multi-attribute space between objects within any group, between any object $A_i$ and a group of objects $C_t$, and between groups of objects.

Hierarchical clustering such multi-attribute objects, when a number of the clusters generated is unknown beforehand, consists of the following principal stages.

Step 1. Set $g=n$, $g$ is the number of clusters, $n$ is the number of objects. Then each cluster $C_i$ consists of the single object $A_i$, and $\boldsymbol{Y}_i=\boldsymbol{A}_i$ for all $i=1,...,g$.

Step 2. Calculate distances $d(\boldsymbol{Y}_p,\boldsymbol{Y}_q)$ between pairs of multisets representing clusters $C_p$ and $C_q$ for all $1\leq p,q\leq g$, $p\neq q$ by using one of the metrics (5).

Step 3. Find a pair of the closest clusters $C_u$ and $C_v$, which satisfy the condition

$$d(\boldsymbol{Y}_u,\boldsymbol{Y}_v) = \min_{p,q} d(\boldsymbol{Y}_p,\boldsymbol{Y}_q), \tag{7}$$

and form a new cluster $C_r$ that is represented as a sum of multisets $\boldsymbol{Y}_r=\boldsymbol{Y}_u+\boldsymbol{Y}_v$, an union $\boldsymbol{Y}_r=\boldsymbol{Y}_u\cup\boldsymbol{Y}_v$, an intersection $\boldsymbol{Y}_r=\boldsymbol{Y}_u\cap\boldsymbol{Y}_v$, or as any linear combination of these operations with corresponding multisets.

Step 4. Reduce the number of clusters by unit: $g=n-1$. If $g=1$, then output the result and stop. If $g>1$, then go to the next step.

Step 5. Recalculate distances $d(\boldsymbol{Y}_p,\boldsymbol{Y}_r)$ between pairs of new multisets representing clusters $C_p$ and $C_r$ for all $1\leq p,r\leq g$, $p\neq r$. Go to step 3.

The hierarchical algorithm builds up a tree or dendrogram by adding objects step by step into groups. The new objects or clusters $C_p$ and $C_q$ are aggregated by branching down the tree from the root, at each step moving to one of the closest clusters $C_r$. Clustering objects is ended when all objects are merged in a single group or several groups depending on the problem considered. The process may be also terminated when an acceptable rule is carried out, for instance, the value of difference between objects overcame a certain threshold level.

Note that on the step 3 many pairs of the closest clusters $C_u$,$C_v$ may appear, which are equivalent with respect to the minimum of distance $d(\boldsymbol{Y}_p,\boldsymbol{Y}_q)$ in the multi-attribute space. So, various branch points of the algorithm (ways of successive aggregation of multisets) exist, and different final trees of objects can be built. Special tests showed that the smallest number of final clusters appears as a result of multisets' addition, and the biggest one – as a result of multisets' intersection. Using the distance $d_{31}$ leads to a smaller number of branch points of the algorithm in a comparison with the distances $d_{11}$ and $d_{21}$, the application of which give the similar results.

In order to diminish a number of possible variants of combining objects, one can use other additional criteria of object closeness, for instance, the criterion of cluster compactness, instead of the single criterion (7). In this case, a modified algorithm of hierarchical clustering objects looks as follows.

Step 3.1. Find all equivalent pairs of the closest clusters $C_u,C_v$ represented as multisets $\boldsymbol{Y}_u,\boldsymbol{Y}_v$ in accordance with formula (7) and form new clusters $C_{rl}$ ($l$=1,…,$t_r$) by using one of the operations under multisets mentioned above; $t_r$ is a number of equivalent clusters' pairs with the same distance $d(\boldsymbol{Y}_u,\boldsymbol{Y}_v)$.

Step 3.2. Find a cluster $C_r$* represented as a multiset $\boldsymbol{Y}_r$* that minimizes the cluster compactness

$$f(\boldsymbol{Y}_r^*) = \min_l \sum_{i,p\in C_{rl}} d(\boldsymbol{A}_i,\boldsymbol{A}_p)/N_{rl}, \tag{8}$$

where $N_{rl}$ is equal to a number of objects $A_i$ within the cluster $C_{rl}$. Go to step 4.

Using many criteria of the object closeness leads to better outcomes of hierarchical clustering objects for all options of cluster construction.

Consider as an illustrative example, the problem of nominal classification without a teacher. Need to divide the collection of objects $A=\{A_1,...,A_{10}\}$ with attributes $Q_1,...,Q_8$ in two classes. All attributes have one and the same five-point scale $X=\{x_1, x_2, x_3, x_4, x_5\}$ of qualitative (symbolic or verbal) grades. It is necessary also to give the interpretation of the obtained classes.

Represent each object as a multiset of the type (1) $\boldsymbol{A}_i=\{k_{Ai}(x_j)\circ x_j|x_j\in X\}$. Objects' descriptions are given in the table K "Objects-Attributes" (Table 2). To classify such multi-attribute objects we use the simplified algorithm of hierarchical clustering, which includes the following steps.

$1^0$. Consider every object as a separate cluster. The object/cluster corresponds to a multiset $\boldsymbol{Y}_i=\boldsymbol{A}_i$, $i$=1,...,10.

$2^0$. Choose as a measure of the difference between objects/clusters, one of the metrics (5) $d_{11}(\boldsymbol{Y}_p,\boldsymbol{Y}_q)=D_{pq}=\sum_j|k_{Yp}(x_j)-k_{Yq}(x_j)|$, $j$=1,...,5, assuming that all of the attributes $x_j$ are equally important ($w_j$=1). Compute the distances between all pairs of objects/clusters.

$3^0$. The nearest objects are $A_1$ and $A_6$. The distance between them in the multiset space is equal to $d_{11}(\boldsymbol{A}_1,\boldsymbol{A}_6)=0$, since the multiset $\boldsymbol{A}_1$ and $\boldsymbol{A}_6$ are the same. Combine the objects $A_1$ and $A_6$ in the cluster $C_{11}$. This cluster will correspond to a multiset $\boldsymbol{Y}_{11}$ that is formed through the operation of multisets' addition:

$$\boldsymbol{Y}_{11} = \boldsymbol{A}_1+\boldsymbol{A}_6 = \{0\circ x_1, 0\circ x_2, 0\circ x_3, 4\circ x_4, 4\circ x_5\}+\{0\circ x_1, 0\circ x_2, 0\circ x_3, 4\circ x_4, 4\circ x_5\} =$$
$$= \{0\circ x_1, 0\circ x_2, 0\circ x_3, 8\circ x_4, 8\circ x_5\}.$$

Objects $A_1$ and $A_6$ are removed from further consideration. The number of objects/clusters decreases by 1.

$4^0$. Compute the distances between all pairs of objects and the remaining pairs of object/new cluster. In this step, the closest objects would be $A_4$ and $A_8$ placed at the distance $d_{11}(\boldsymbol{A}_4,\boldsymbol{A}_8)=|0-0|+|1-1|+|1-2|+|3-3|+|3-2|=2$. Objects $A_4$ and $A_8$ form the cluster $C_{12}$, which corresponds to a multiset

$$\boldsymbol{Y}_{12} = \boldsymbol{A}_4+\boldsymbol{A}_8 = \{0\circ x_1,\ 1\circ x_2,\ 1\circ x_3,\ 3\circ x_4,\ 3\circ x_5\}+\{0\circ x_1,\ 1\circ x_2,\ 2\circ x_3,\ 3\circ x_4,\ 2\circ x_5\} = \\ = \{0\circ x_1,\ 2\circ x_2,\ 3\circ x_3,\ 6\circ x_4,\ 5\circ x_5\},$$

and are removed from further consideration.

$5^0$. Compute the distances between all pairs of the remaining objects and the obtained clusters. In this step, the closest objects would be two pairs $A_2$, $A_7$ and $A_7$, $A_9$, which are placed at the equal distances in the multiset space: $d_{11}(\boldsymbol{A}_2,\boldsymbol{A}_7)=4$ and $d_{11}(\boldsymbol{A}_7,\boldsymbol{A}_9)=4$. To form a new cluster $C_{13}$ choose the pair of objects $A_7$,$A_9$ and remove them from further consideration. Cluster $C_{13}$ corresponds to a multiset

$$\boldsymbol{Y}_{13} = \boldsymbol{A}_7+\boldsymbol{A}_9 = \{3\circ x_1,\ 5\circ x_2,\ 7\circ x_3,\ 1\circ x_4,\ 0\circ x_5\}.$$

$6^0$. Calculating step by step the distances between all pairs of objects/clusters and choosing at each step the closest pair, we obtain clusters $C_{14}$, $C_{15}$, $C_{16}$, $C_{17}$, $C_{18}$, represented by the following multisets:

$$\boldsymbol{Y}_{14} = \boldsymbol{A}_2+\boldsymbol{A}_3 = \{7\circ x_1,\ 4\circ x_2,\ 2\circ x_3,\ 3\circ x_4,\ 0\circ x_5\},\quad d_{11}(\boldsymbol{A}_2,\boldsymbol{A}_3)=8;$$
$$\boldsymbol{Y}_{15} = \boldsymbol{Y}_{11}+\boldsymbol{A}_5 = \{0\circ x_1,\ 0\circ x_2,\ 0\circ x_3,\ 15\circ x_4,\ 9\circ x_5\},\quad d_{11}(\boldsymbol{Y}_{11},\boldsymbol{A}_5)=8;$$
$$\boldsymbol{Y}_{16} = \boldsymbol{Y}_{12}+\boldsymbol{A}_{10} = \{0\circ x_1,\ 2\circ x_2,\ 4\circ x_3,\ 8\circ x_4,\ 10\circ x_5\},\quad d_{11}(\boldsymbol{Y}_{12},\boldsymbol{A}_{10})=8;$$
$$\boldsymbol{Y}_{17} = \boldsymbol{Y}_{13}+\boldsymbol{Y}_{14} = \{10\circ x_1,\ 9\circ x_2,\ 9\circ x_3,\ 4\circ x_4,\ 0\circ x_5\},\quad d_{11}(\boldsymbol{Y}_{13},\boldsymbol{Y}_{14})=12;$$
$$\boldsymbol{Y}_{18} = \boldsymbol{Y}_{15}+\boldsymbol{Y}_{16} = \{0\circ x_1,\ 2\circ x_2,\ 4\circ x_3,\ 23\circ x_4,\ 19\circ x_5\},\quad d_{11}(\boldsymbol{Y}_{15},\boldsymbol{Y}_{16})=14.$$

$7^0$. The procedure terminates when two clusters $C_{17}$ and $C_{18}$, which correspond to multiset $\boldsymbol{Y}_{17}$ and $\boldsymbol{Y}_{18}$, remain. The cluster $C_{17}$ aggregates the objects $A_2$, $A_3$, $A_7$, $A_9$, which have, basically, "low" estimates $x_1$, $x_2$, $x_3$, and the cluster $C_{18}$ combines the objects $A_1$, $A_6$, $A_5$, $A_4$, $A_8$, $A_{10}$, which have, in general, "high" estimates $x_4$, $x_5$. The tree shown in Figure 1 is built as a result of the classification procedures.

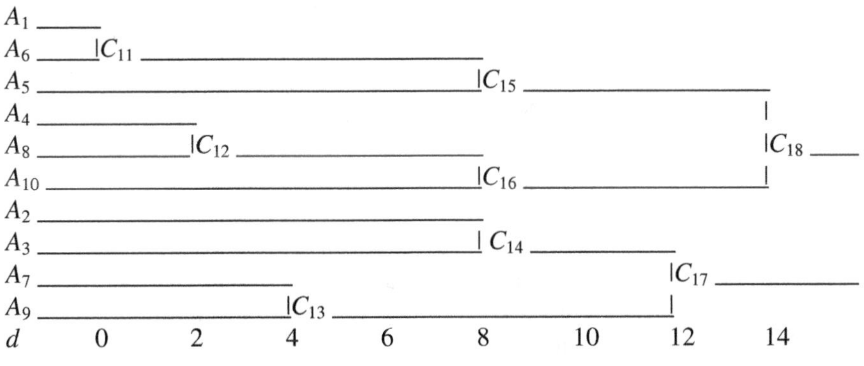


**Fig. 1** Output tree of the hierarchical clustering algorithm

If in step $5^0$ to form another cluster that combines a pair of the objects $A_2$ and $A_7$, then the final partition of the objects' collection will consist of the same clusters $C_{17}$ and $C_{18}$. The final result of classifying multi-attribute objects, obtained in our case, coincides with the statement in [8] where objects were represented as vectors of numeric values of attributes (Table 1), and aggregation into clusters proceeds with the adding algorithm. Note that the both algorithms have been successful in identifying similar groups of objects.

The above hierarchical clustering algorithm for solving the problem of nominal classification of objects, which are described by many qualitative attributes, can be expanded without difficulty to the cases, where objects can exist in several versions (copies) with different values of the attributes, and several classes are given.

In nonhierarchical clustering, a number $g$ of clusters is considered as fixed and determined beforehand. A general framework of nonhierarchical clustering objects described by many qualitative attributes and represented as multisets is as follows.

Step 1. Select any initial partition of the object collection $A=\{A_1,...,A_n\}$ into $g$ clusters $C_1,...,C_g$.

Step 2. Distribute all objects $A_1,...,A_n$ into clusters $C_1,...,C_g$ according to a certain rule. For instance, calculate distances $d(\boldsymbol{A}_i,\boldsymbol{Y}_t)$ between multisets $\boldsymbol{A}_i$ representing objects $A_i$ ($i=1,...,n$) and multisets $\boldsymbol{Y}_t$ representing clusters $C_t$ ($t=1,...,g$), and allocate the object $A_i$ into the nearest cluster $C_h$ with $d(\boldsymbol{A}_i,\boldsymbol{Y}_h)=\min_t d(\boldsymbol{A}_i,\boldsymbol{Y}_t)$. Or find a center $A_t{}^\circ$ for each cluster $C_t$ by solving the following minimization problem:

$$J(\boldsymbol{A}_t{}^\circ,\boldsymbol{Y}_t) = \min_p \Sigma_i d(\boldsymbol{A}_i,\boldsymbol{A}_p), \tag{9}$$

and allocate each object $A_i$ into the cluster $C_r$ with the closest center, that is $d(\boldsymbol{A}_i,\boldsymbol{A}_r{}^\circ)=\min_t d(\boldsymbol{A}_i,\boldsymbol{A}_t{}^\circ)$. The functional $J(\boldsymbol{A}_t{}^\circ,\boldsymbol{Y}_t)$ (9) is analogues by sense of the criterion of cluster compactness $f(\boldsymbol{Y}_r{}^*)$ (8). Remark that the cluster center $A_t{}^\circ$ may coincide with one of the real objects $A_i$ of the collection $A$ or be the so-called 'phantom' object, which is absent in the collection $A$ but constructed from attributes $x_j$ as a multiset $\boldsymbol{A}_t{}^\circ$ of the type (1) or (3).

Step 3. If all objects $A_i$ ($i=1,...,n$) do not change their membership given by the initial partition of objects into clusters, then output the result and stop. Otherwise go to step 2.

A result of the object classification is evaluated by a quality of partition. The best partition can be found, in particular, as a solution of the following optimization problem:

$$\Sigma_t J(\boldsymbol{A}_t{}^\circ,\boldsymbol{Y}_t) \rightarrow \min \tag{10}$$

where $J(\boldsymbol{A}_t{}^\circ,\boldsymbol{Y}_t)$ is defined, for example, by formula (9).

The condition of min $d(\boldsymbol{Y}_p,\boldsymbol{Y}_q)$ is to be replaced by the condition of max $s(\boldsymbol{Y}_p,\boldsymbol{Y}_q)$ when an index $s$ of object similarity (6) is used in a clustering algorithm.

In solving practical problems, the following approach to structuring a collection of objects may be useful. At first, objects are classified by any technique of hierarchical clustering, and several possible partitions of objects are formed. Then the collection of partitions is analyzed by a technique of nonhierarchical clustering, and the most preferable or optimal partition of objects is found.

## 5 Group Sorting Objects

We now consider the approach to solve the problem of group ordinal classification of multi-attribute objects $A_1,...,A_n$ with many teachers called also as the problem of group multicriteria sorting. Suppose, there are $k$ experts, and each of experts evaluates every object $A_i$, $i=1,...,n$ with respect to $m$ qualitative criteria $Q_1,...,Q_m$. Each criterion (attribute) $Q_s$, $s=1,...,m$ has its own symbolic or verbal scale $X_s=\{x_s^1,...,x_s^{h_s}\}$, which may be ordered, for example, from the worst to the best. In addition, each expert assigns every object $A_i$ to one of the classes $C_1,...,C_g$, which differ in their properties and may be ordered by preference. Thus, there are $k$ different versions of each object $A_i$ and $k$ individual expert rules for sorting objects, which are usually not agreed among themselves.

Need to find a sufficiently simple generalized rule for group multicriteria sorting of the type (4), which approximates a large family of discordant individual rules of expert classification of objects and can assign objects to a given class, not rejecting possible contradictions of objects' evaluations. The problem of group sorting objects described by many qualitative attributes, and especially in the cases, where objects can exist in several versions, is one of the most hard problems of classification. Difficulties are tied, generally, with a need to process simultaneously large amounts of symbolic and/or verbal data, the convolution of which is either impossible or mathematically incorrect.

The method MASKA (abbreviation of the Russian words Multi-Attribute Consistent Classification of Alternatives) has been developed for group sorting multi-attribute objects and based on the multiset theory [16-18]. Let us represent each multi-attribute object $A_i$, $i=1,...,n$ as the following multiset of the type (3):

$$A_i=\{k_{Ai}(x_1^1)\circ x_1^1,...,k_{Ai}(x_1^{h_1})\circ x_1^{h_1},..., k_{Ai}(x_m^1)\circ x_m^1,...,k_{Ai}(x_m^{h_m})\circ x_m^{h_m}, \\ k_{Ai}(r_1)\circ r_1,...,k_{Ai}(r_g)\circ r_g\}, \tag{11}$$

which is drawn from a set $X'=Q_1\cup...\cup Q_m\cup R=X\cup R$. The extended set of attributes $X'$ combines the subsets of multiple criteria estimates $X_s=\{x_s^1,...,x_s^{h_s}\}$ and the subset of sorting attributes $R=\{r_1,...,r_g\}$ where $r_t$ is an expert conclusion that any object belongs to the class $C_t$, $t=1,...,g$. Values of $k_{Ai}(x_s^{e_s})$ and $k_{Ai}(r_t)$ are equal correspondingly to numbers of experts who estimates the object $A_i$ with the attribute value $x_s^{e_s}$ and gives the conclusion $r_t$. Obviously, conclusions of many different experts may be similar, diverse, or contradictory. These inconsistencies express subjective preferences of individual experts and cannot be accidental errors.

The relations between the collection of multi-attribute objects $A=\{A_1,...,A_n\}$ and the set of attributes $X'$ are described by the extended data matrix $L'=||k_{ij}||_{n\times(h+g)}$, $h=h_1+...+h_m$. Each row of the matrix L' corresponds to any object $A_i$, each column agrees with a certain value of criterion $Q_s$ or sorting attribute $R$, whereas an entry $k_{ij}$ is a multiplicity $k_{Ai}(x_j')$ of the attribute value $x_j'\in X'$. The matrix L' is an analog of the so-called decision table or information table that is used often in data analysis, decision making, pattern recognition, and so on.

The representation (11) of the object $A_i$ can be considered also as a collective decision rule (4) of several experts for sorting this multi-attribute object. This rule is associated with multiset arguments in the formula (11) as follows. The antecedent term ⟨conditions⟩ includes the various combinations of criteria estimates $x_s^{e_s}$, which describe the object features, and expert conclusions $r_t$. The consequent term ⟨decision⟩ denotes that the object $A_i$ is assigned to the class $C_t$, if acceptable conditions are fulfilled. For instance, the object $A_i$ is said to belong to the class $C_t$ in accordance with one of the following rules of majority: if $k_{Ai}(r_t)>k_{Ai}(r_p)$ for all $p\neq t$, or if $k_{Ai}(r_t)>\sum_{p\neq t} k_{Ai}(r_p)$.

In order to simplify the classification problem, let us assume that the collection of objects $A=\{A_1,...,A_n\}$ is to be sorted out only two ordered classes $C_a$ (say, more preferable) and $C_b$ (less preferable). The division of objects into only two classes is not the principle restriction. Whenever objects are to be sorted out more than two classes, it is possible to divide the collection $A$ into two groups, then into subgroups, and so on. For instance, if it is necessary to select some groups of competitive projects, then, at first, these projects can be classified as projects approved and not approved; at second, the not approved projects can be sorted as projects rejected and can be considered later, and so on.

Let us form each class of multi-attribute objects as a sum of multisets of the type (11). So, the following multisets correspond to the classes $C_a$ and $C_b$:

$$\begin{aligned}\boldsymbol{Y}_t=\{&k_{Yt}(x_1^{\,1})\circ x_1^{\,1},...,k_{Yt}(x_1^{\,h_1})\circ x_1^{\,h_1},..., k_{Yt}(x_m^{\,1})\circ x_m^{\,1},...,k_{Yt}(x_m^{\,h_m})\circ x_m^{\,h_m},\\ &k_{Yt}(r_a)\circ r_a,\ k_{Yt}(r_b)\circ r_b\},\end{aligned} \tag{12}$$

where $t=a,b$, $k_{Yt}(x_s^{e_s})=\sum_{i\in I_t} k_{Ai}(x_s^{e_s})$, $k_{Yt}(r_t)=\sum_{i\in I_t} k_{Ai}(r_t)$, $I_a\bigcup I_b=\{1,...,n\}$, $I_a\bigcap I_b=\varnothing$. The relations between the classes $C_a$, $C_b$ and the set of attributes $X'$ are described now by the reduced decision table – data matrix $\mathrm{M}=||k_{ij}'||_{2\times(h+2)}$, which consists of 2 rows and $h$+2 columns. Each row of the matrix M corresponds to any class $C_t$, each column agrees with a certain value of criterion $Q_s$ or sorting attribute $R$, whereas an entry $k_{ij}'$ is a multiplicity $k_{Yt}(x_j')$ of the attribute value $x_j'\in X'$. The expression (12) represents the collective decision rule (4) of all experts for sorting the collection of multi-attribute object out the class $C_t$.

Consider the inverted decision table – data matrix $\mathrm{M}^{-1}=||k_{ji}'||_{(h+2)\times 2}$, which consists of $h$+2 rows and 2 columns. Each row of the matrix $\mathrm{M}^{-1}$ corresponds to one of values of criterion $Q_s$ or sorting attribute $R$, each column agrees with any class $C_t$, whereas an entry $k_{ji}'$ is a multiplicity $k_{Yt}(x_j')$ of the attribute value $x_j'\in X'$. Let us introduce a set of new attributes $Y'=\{y_a,y_b\}$, which elements related to the classes $C_a$ and $C_b$. Then rows of the matrix $\mathrm{M}^{-1}$ form a collection $B$ of new objects represented as the following new multisets:

$$\boldsymbol{R}_a=\{k_{\boldsymbol{R}a}(y_a)\circ y_a,\ k_{\boldsymbol{R}a}(y_b)\circ y_b\},\ \boldsymbol{R}_b=\{k_{\boldsymbol{R}b}(y_a)\circ y_a,\ k_{\boldsymbol{R}b}(y_b)\circ y_b\},$$

$$\boldsymbol{Q}_j=\{k_{\boldsymbol{Q}j}(y_a)\circ y_a,\ k_{\boldsymbol{Q}j}(y_b)\circ y_b\} \tag{13}$$

drawn from the set $Y'$. Here $k_{\boldsymbol{R}a}(y_t)=k_{Yt}(r_a)$, $k_{\boldsymbol{R}b}(y_t)=k_{Yt}(r_b)$, $k_{\boldsymbol{Q}j}(y_t)=k_{Yt}(x_j)$, $j=1,\dots,h$. We shall call the multisets $\boldsymbol{R}_a$, $\boldsymbol{R}_b$ as 'categorical' and the multisets $\boldsymbol{Q}_j$ as 'substantial' multisets.

Note that the categorical multisets $\boldsymbol{R}_a$ and $\boldsymbol{R}_b$ correspond to the best binary decomposition of the objects collection $\boldsymbol{A}$ into the given classes $C_a$ and $C_b$ according to the following primary sorting rules of experts:

$$\text{IF } \langle (k_{Ai}(r_a) > k_{Ai}(r_b)) \rangle, \text{ THEN } \langle \text{Object } A_i \in C_a \rangle,$$
$$\text{IF } \langle (k_{Ai}(r_a) < k_{Ai}(r_b)) \rangle, \text{ THEN } \langle \text{Object } A_i \in C_b \rangle. \tag{14}$$

And the distance $d^*=d(\boldsymbol{R}_a,\boldsymbol{R}_b)$ between the categorical multisets $\boldsymbol{R}_a$ and $\boldsymbol{R}_b$ is the maximal distance in a new multiset metric space $(\boldsymbol{B},d)$. In the case of ideal initial sorting object without inconsistencies of individual expert rules, the maximal distance (5) is equal to $d_{11}^*=kn$, $d_{12}^*=1/(h+g)$, $d_{13}^*=1$.

The main idea of how to find a simple generalized decision rule, which the mostly coincides with initial inconsistent sorting rules of many individual experts, is formulated as follows. It is necessary to construct a pair of new substantial multisets $\boldsymbol{Q}_{sa}$ and $\boldsymbol{Q}_{sb}$ for every group $Q_s$, $s=1,\ldots,m$ of attributes such that the multisets $\boldsymbol{Q}_{sa}$, $\boldsymbol{Q}_{sb}$ within each pair, which are considered as points of multiset metric space $(\boldsymbol{B},d)$, are to be placed at the maximal distance. The multisets $\boldsymbol{Q}_{sa}$ and $\boldsymbol{Q}_{sb}$ are determined as the following sums of multisets: $\boldsymbol{Q}_{sa}=\sum_{j\in Jsa}\boldsymbol{Q}_j$, $\boldsymbol{Q}_{sb}=\sum_{j\in Jsb}\boldsymbol{Q}_j$, where index subsets $J_{sa}\cup J_{sb}=\{1,\ldots,h_s\}$, $J_{sa}\cap J_{sb}=\varnothing$.

The substantial multisets $\boldsymbol{Q}_{sa}{}^*$ and $\boldsymbol{Q}_{sb}{}^*$, which correspond to the best binary decomposition of the objects collection $\boldsymbol{A}$ for the $s$-th attribute $Q_s$, are a solution of the following optimization problem:

$$d(\boldsymbol{Q}_{sa},\boldsymbol{Q}_{sb}) \to \max d(\boldsymbol{Q}_{sa},\boldsymbol{Q}_{sb}) = d(\boldsymbol{Q}_{sa}{}^*,\boldsymbol{Q}_{sb}{}^*). \tag{15}$$

Note that each substantial multiset $\boldsymbol{Q}_{st}$ aggregates a group of multisets $\boldsymbol{Q}_j$, which corresponds to a subgroup of attribute values $Q_{st}=\cup_{j\in Jst}Q_j$, $t=a,b$. We shall call an attribute value $x_j\in Q_{st}{}^*$, $j\in J_{st}$ that characterizes the class $C_t$ as a classifying attribute for the correspondent class.

The set of attributes $Q_1,\ldots,Q_m$ can be ranged by the value of distance $d(\boldsymbol{Q}_{sa}{}^*,\boldsymbol{Q}_{sb}{}^*)$ or by the level of approximation rate $V_s=d(\boldsymbol{Q}_{sa}{}^*,\boldsymbol{Q}_{sb}{}^*)/d(\boldsymbol{R}_a,\boldsymbol{R}_b)$. The classifying attribute $x_j\in Q_{st}{}^*$ that provides the acceptable level of approximation rate $V_s\geq V_0$ is to be included in the generalized decision rule for group multicriteria sorting objects. The level of approximation rate $V_s$ shows a relative significance of the $s$-th property $Q_s$ within the generalized decision rule.

Various combinations of the classifying attributes produce the generalized decision rules for group classifying objects into the classes $C_a$ and $C_b$ as follows:

$$\text{IF } \langle (x_j\in Q_{ua}{}^*) \text{ AND } (x_j\in Q_{va}{}^*) \text{ AND } (x_j\in Q_{wa}{}^*) \text{ AND} \ldots \rangle, \quad \text{THEN } \langle \text{Object } A_i\in C_a \rangle, \tag{16}$$

$$\text{IF } \langle (x_j\in Q_{ub}{}^*) \text{ AND } (x_j\in Q_{vb}{}^*) \text{ AND } (x_j\in Q_{wb}{}^*) \text{ AND} \ldots \rangle, \quad \text{THEN } \langle \text{Object } A_i\in C_b \rangle. \tag{17}$$

Remark that these rules are quite different, generally speaking.

Among the objects, which have been assigned to the given class $C_a$ or $C_b$ in accordance with the generalized decision rule (16) or (17), there are correctly and not correctly classified objects. So, we need to find such attribute values that maximize numbers $N_a$ and $N_b$ of correctly classified objects, and minimize numbers $N_{ac}$ and $N_{bc}$ of not correctly classified objects.

We first find a single criterion $Q_{ua}$*, then a couple of criteria $Q_{ua}$* and $Q_{va}$*, then three criteria $Q_{ua}$*, $Q_{va}$*, $Q_{wa}$*, four criteria and so on, step by step, which are included in the generalized decision rules (16) or (17), and provide the minimal difference $N_a - N_{ac}$ or $N_b - N_{bc}$. Finally, we obtain the following decision rules:

$$\text{IF } \langle(\textstyle\sum_{x\in Qua^*} k_{Ai}(x_j) > \sum_{x\in Qub^*} k_{Ai}(x_j)) \text{ AND } (\sum_{x\in Qva^*} k_{Ai}(x_j) > \sum_{x\in Qvb^*} k_{Ai}(x_j)) \text{ AND}\ldots\text{AND } (k_{Ai}(r_a) > k_{Ai}(r_b))\rangle, \text{ THEN } \langle \text{Object } A_i \in C_a \backslash C_{ac} \rangle, \tag{18}$$

$$\text{IF } \langle(\textstyle\sum_{x\in Qua^*} k_{Ai}(x_j) < \sum_{x\in Qub^*} k_{Ai}(x_j)) \text{ AND } (\sum_{x\in Qva^*} k_{Ai}(x_j) < \sum_{x\in Qvb^*} k_{Ai}(x_j)) \text{ AND}\ldots\text{AND } (k_{Ai}(r_a) < k_{Ai}(r_b))\rangle, \text{ THEN } \langle \text{Object } A_i \in C_b \backslash C_{bc} \rangle. \tag{19}$$

These consistent generalized decision rules define the specified classes $C_a \backslash C_{ac}$ (say, completely preferable) and $C_b \backslash C_{bc}$ (completely not preferable) of correctly classified objects and the mostly approximate the family of initial sorting rules of many individual experts.

Simultaneously the specified class $C_c = C_{ac} \cup C_{bc}$ of contradictory classified objects is built. Such objects satisfy the following decision rules for inconsistent sorting:

$$\text{IF } \langle[(\textstyle\sum_{x\in Qua^*} k_{Ai}(x_j) > \sum_{x\in Qub^*} k_{Ai}(x_j)) \text{ AND } (\sum_{x\in Qva^*} k_{Ai}(x_j) > \sum_{x\in Qvb^*} k_{Ai}(x_j)) \text{ AND}\ldots\text{AND } (k_{Ai}(r_a) < k_{Ai}(r_b))]$$
$$\text{OR } [(\textstyle\sum_{x\in Qua^*} k_{Ai}(x_j) < \sum_{x\in Qub^*} k_{Ai}(x_j)) \text{ AND } (\sum_{x\in Qva^*} k_{Ai}(x_j) < \sum_{x\in Qvb^*} k_{Ai}(x_j)) \text{ AND}\ldots\text{AND } (k_{Ai}(r_a) > k_{Ai}(r_b))]\rangle, \text{ THEN } \langle \text{Object } A_i \in C_c \rangle \tag{20}$$

and have to be analyzed additionally. This rule helps us to discover possible inconsistencies of individual expert rules.

The algorithm for constructing the generalized decision rules, which provide consistent group sorting multi-attribute objects, includes the following stages.

Step 1. Compute the decision table $L' = \|k_{ij}\|_{n\times(h+g)}$ that represents the collection of multi-attribute objects $A = \{A_1, ..., A_n\}$ and the extended set of attributes $X' = \{x_1^1, \ldots, x_1^{h_1}; \ldots; x_m^1, \ldots, x_m^{h_m}, r_1, \ldots, r_g\}$. Aggregate columns $r_1, \ldots, r_g$ in two columns $r_a$ and $r_b$ according to any requirement.

Step 2. Combine the objects $A_1, ..., A_n$ with respect to the given classes $C_a$, $C_b$ by adding the correspondent multisets in accordance with any collective decision rule (4) that integrates individual sorting rules of several experts. Compute the reduced decision table $M = \|k_{ij}'\|_{2\times(h+2)}$ and the inverted decision table $M^{-1} = \|k_{ji}'\|_{(h+2)\times 2}$, which represent classes $C_a$, $C_b$ of objects and values of criteria and sorting attributes.

Step 3. Solve the optimization problem (15) for every attribute $Q_s$, $s=1, \ldots, m$ and find the classifying attributes $x_j \in Q_{sa}$* and $x_j \in Q_{sb}$*. Range the classifying attributes $x_j \in Q_{st}$*, $t=a,b$ by the value of distance $d(\boldsymbol{Q}_{sa}^*, \boldsymbol{Q}_{sb}^*)$ or by the level of approximation rate $V_s = d(\boldsymbol{Q}_{sa}^*, \boldsymbol{Q}_{sb}^*)/d(\boldsymbol{R}_a, \boldsymbol{R}_b)$.

Step 4. Select the classifying attributes $x_j \in Q_{st}$*, which provide the acceptable level of approximation rate $V_s \geq V_0$. Form the generalized decision rules (16) and (17) for group sorting objects.

Step 5. Form the consistent decision rules (18) for group sorting objects out of the completely preferable class $C_a \backslash C_{ac}$ and determine the correctly classified objects.

Step 6. Form the consistent decision rules (19) for group sorting objects out of the completely not preferable class $C_b \backslash C_{bc}$ and determine the correctly classified objects.

Step 7. Form the decision rules (20) for determination of the specified class $C_c = C_{ac} \bigcup C_{bc}$ of contradictory classified objects.

The generalized decision rules for group multicriteria sorting objects can easily be written in natural language, using formulations of the verbal values of the classifying attributes.

Consider how this algorithm works on a simple illustrative example of the problem of ordinal classification. It is required to sort objects $A_1,...,A_{10}$ out of two classes $C_a$ (more preferable) and $C_b$ (less preferable). Objects are characterized by attributes of $Q_1,...,Q_8$, each of them takes qualitative estimates (scores) on a five-point scale $X=\{x_1, x_2, x_3, x_4, x_5\}$. Recall that these estimates are symbolic or verbal variables. For example, $x_1$ – very bad, strongly disagree, $x_2$ – bad, disagree, $x_3$ – satisfactory, neutral, $x_4$ – good, agree, $x_5$ – excellent, strongly agree. Objects are presented in single versions, descriptions of which are given in the Table 2 $K=||k_{ij}||_{10\times5}$ "Objects-Attributes".

Build a decision table $K'=||k_{ij}||_{10\times(5+2)}$ that is the expanded data table K with two additional columns corresponding to the sorting attributes $r_a$ and $r_b$. These attributes characterize an expert opinions ("votes" of disciplines or respondents) about a membership of the object to one of the classes $C_a$ or $C_b$, for instance, as follows: the attribute $r_a$ is associated with the best estimates ($x_5$ и $x_4$), and the attribute $r_b$ – with the worst estimates ($x_3$, $x_2$, $x_1$). The a decision table K', rows of which are multiset of the type (11).

$\boldsymbol{A}_i=\{k_{Ai}(x_1)\circ x_1,\ k_{Ai}(x_2)\circ x_2,\ k_{Ai}(x_3)\circ x_3,\ k_{Ai}(x_4)\circ x_4,\ k_{Ai}(x_5)\circ x_5,\ k_{Ai}(r_a)\circ r_a,\ k_{Ai}(r_b)\circ r_b\}$, is presented in Table 6.

**Table 6** Decision table K'

| A\X' | $x_1$ | $x_2$ | $x_3$ | $x_4$ | $x_5$ | $r_a$ | $r_b$ |
|---|---|---|---|---|---|---|---|
| $A_1$ | 0 | 0 | 0 | 4 | 4 | 8 | 0 |
| $A_2$ | 2 | 4 | 1 | 1 | 0 | 1 | 7 |
| $A_3$ | 5 | 0 | 1 | 2 | 0 | 2 | 6 |
| $A_4$ | 0 | 1 | 1 | 3 | 3 | 6 | 2 |
| $A_5$ | 0 | 0 | 0 | 7 | 1 | 8 | 0 |
| $A_6$ | 0 | 0 | 0 | 4 | 4 | 8 | 0 |
| $A_7$ | 2 | 2 | 3 | 1 | 0 | 1 | 7 |
| $A_8$ | 0 | 1 | 2 | 3 | 2 | 5 | 3 |
| $A_9$ | 1 | 3 | 4 | 0 | 0 | 0 | 8 |
| $A_{10}$ | 0 | 0 | 1 | 2 | 5 | 7 | 1 |

**Table 7** Reduced decision table M

| C\X' | $x_1$ | $x_2$ | $x_3$ | $x_4$ | $x_5$ | $r_a$ | $r_b$ |
|---|---|---|---|---|---|---|---|
| $\boldsymbol{Y}_a$ | 0 | 2 | 4 | 23 | 19 | 42 | 6 |
| $\boldsymbol{Y}_b$ | 10 | 9 | 9 | 4 | 0 | 4 | 28 |

Combine objects in two classes $C_a$ and $C_b$, for example, in accordance with such a collective rule of 'majority of voices': if the number of the best estimates exceeds the number of the worst estimates, then the object is to be assigned to the class $C_a$. In the opposite case, the object is to be assigned to the class $C_b$. Thus, the class $C_a$ will include the objects $A_1$, $A_4$, $A_5$, $A_6$, $A_8$, $A_{10}$, and the class $C_b$ – the objects $A_2$, $A_3$, $A_7$, $A_9$. The classes $C_a$ and $C_b$ are described by the following multisets of the type (12):

$$\boldsymbol{Y}_a=\{0\circ x_1, 2\circ x_2, 4\circ x_3, 23\circ x_4, 19\circ x_5, 42\circ r_a, 6\circ r_b\},$$
$$\boldsymbol{Y}_b=\{10\circ x_1, 9\circ x_2, 9\circ x_3, 4\circ x_4, 0\circ x_5, 4\circ r_a, 28\circ r_b\}.$$

which are rows of the reduced decision table $M=\|k_{ij}'\|_{2\times(5+2)}$ (Table 7).

Consider the inverted decision table $M^{-1}=\|k_{ji}'\|_{(5+2)\times 2}$ and construct the following a categorical and substantial multisets:

$$\boldsymbol{Q}_1=\{0\circ y_a, 10\circ y_b\}, \boldsymbol{Q}_2=\{2\circ y_a, 9\circ y_b\}, \boldsymbol{Q}_3=\{4\circ y_a, 9\circ y_b\},$$

$$\boldsymbol{Q}_4=\{23\circ y_a, 4\circ y_b\}, \boldsymbol{Q}_5=\{19\circ y_a, 0\circ y_b\},$$

$$\boldsymbol{R}_a=\{42\circ y_a, 4\circ y_b\}, \boldsymbol{R}_b=\{6\circ y_a, 28\circ y_b\}.$$

Let us take one of the metrics (5) $d_{11}(\boldsymbol{Q}_i,\boldsymbol{Q}_j)=D_{ij}=\sum_t |k_{Qi}(y_t)-k_{Qj}(y_t)|$, $t=a,b$ as the measure of the multisets' closeness, considering all of attributes $y_t$ as equally important ($w_t$=1). The distance between categorical multisets is equal to

$$d(\boldsymbol{R}_a,\boldsymbol{R}_b) = |42-6|+|4-28| = 60.$$

The solution of the single optimization problem (15) is the following substantial multisets: $\boldsymbol{Q}_a^*=\boldsymbol{Q}_4+\boldsymbol{Q}_5$ and $\boldsymbol{Q}_b^*=\boldsymbol{Q}_1+\boldsymbol{Q}_2+\boldsymbol{Q}_3$, which are placed at the maximal distance

$$d(\boldsymbol{Q}_a^*,\boldsymbol{Q}_b^*) = |19+23-4-2|+|4-9-9-10| = 36+24 = 60$$

among all the possible combinations of pairs of multisets $\boldsymbol{Q}_a^*$ and $\boldsymbol{Q}_b^*$. The multiset $\boldsymbol{Q}_a^*$ defines a group of the classifying attributes $Q_a^*=Q_4\cup Q_5$, which characterizes the class $C_a$ and includes the attributes $x_4$, $x_5$. The multiset $\boldsymbol{Q}_b^*$ defines a group of the classifying attributes $Q_b^*=Q_1\cup Q_2\cup Q_3$, which characterizes the class $C_b$ and includes the attributes $x_1$, $x_2$, $x_3$.

The generalized decision rules for group classification of multi-attribute objects, expressed in terms of natural language, look as follows:

"The more preferable class $C_a$ consists of objects that have estimates 'good' and 'excellent' ($x_4$, $x_5$). The less preferable class $C_b$ consists of objects that have estimates 'very bad', 'bad', 'satisfactory' ($x_1$, $x_2$, $x_3$)".

Note that the values of distances $d(\boldsymbol{R}_a,\boldsymbol{R}_b)$ and $d(\boldsymbol{Q}_a*,\boldsymbol{Q}_b*)$ are equal, and the above generalized decision rules for group classification of objects does not differ practically from the collective rules of 'majority of voices', according which the objects are assigned to a certain class. These circumstances are explained by absence of contradictions between the individual decision rules for initial sorting objects. In general, it will not. It is interesting to remark that the constructed classification of multi-attribute objects coincides with the result obtained previously using the algorithm of hierarchical clustering.

It is also not so difficult to solve the problem of group ordinal classification of objects, which are described by many qualitative (symbolic or verbal) attributes, in the cases, when these objects are represented in several versions (copies) with different values of the attributes, a few classes of decisions are given, and individual decision rules for initial sorting of objects are contradictory.

## 6 Applications

Let us describe briefly some applications of the suggested techniques.

Competitive selection of proposals is a very popular means in various business areas. Often such selection is based on multiple criteria estimates and conclusions given by several experts. Rules for group sorting applications have been constructed by using the MASKA method. These decision rules are based on many contradictory individual decision rules, described by non-numerical data, and could not been found with other known techniques of multicriteria decision aiding.

For example, in the process of elaborating the State Research Program on High-Temperature Superconductivity five subgroups of experts considered more than 250 applications and approved about 170 projects. Three experts estimated every application with respect to 6 criteria with verbal scales. The questionnaire for project estimation included the following criteria: $Q_1$.The project contribution to the program goals; $Q_2$.Long-range value of the project; $Q_3$.Novelty of the approach to solve the task; $Q_4$.Team qualification; $Q_5$.Resources available for the project realization; $Q_6$.Profiles of the project results.

Each criterion has a nominative or ordered scale of verbal estimates. The scale of the criterion $Q_4$.'Team qualification' is as follows:

$x_4^1$ – the team has the best qualification and experience;

$x_4^2$ – the team has a qualification and experience sufficient for the project realization;

$x_4^3$ – the team has a qualification and experience insufficient for the project realization;

$x_4^4$ – the team has an unknown qualification and experience.

Experts evaluated each application by all criteria $Q_1$-$Q_6$ and made one of the following conclusions for sorting projects: $r_1$ – to approve the project; $r_2$ – to reject the project; $r_3$ – to consider the project later after improving. Obviously, different experts can evaluate differently one and the same project. The individual expert recommendations for the project approval may also coincide or not as well.

The Table 8 includes (i) a part of the decision table L', which presents a collection of projects; (ii) the reduced decision table M, which presents classes $C_a$ and $C_b$ of approved and not approved projects; (iii) the distances between the categorical multisets $\boldsymbol{R}_a$, $\boldsymbol{R}_b$, the substantial multisets $\boldsymbol{Q}_{sa}$, $\boldsymbol{Q}_{sb}$, and the levels of approximation rate $V_s$. Here the column $r_a$ agrees with the expert conclusion $r_1$, and the column $r_b$ integrates the expert conclusions $r_2$ and $r_3$.

As Table 8 is shown, the criterion $Q_4$ that characterizes a qualification and experience of team is the most important for the project selection. The next ones are the criteria $Q_1$ expressed the project contribution to the program goals, and $Q_5$ evaluated resources available for the project realization.

**Table 8** Decision tables L' and M of competitive applications

(i)

| A/X' | $x_1^1$ | $x_1^2$ | $x_1^3$ | $x_2^1$ | $x_2^2$ | $x_2^3$ | $x_3^1$ | $x_3^2$ | $x_3^3$ | $x_4^1$ | $x_4^2$ | $x_4^3$ | $x_4^4$ | $x_5^1$ | $x_5^2$ | $x_5^3$ | $x_5^4$ | $x_6^1$ | $x_6^2$ | $x_6^3$ | $r_a$ | $r_b$ |
|---|---|---|---|---|---|---|---|---|---|---|---|---|---|---|---|---|---|---|---|---|---|---|
| $\boldsymbol{A}_1$ | 1 | 2 | 0 | 2 | 1 | 0 | 3 | 0 | 0 | 2 | 1 | 0 | 0 | 0 | 2 | 1 | 0 | 2 | 1 | 0 | 3 | 0 |
| … | | | | | | | | | | | | | | | | | | | | | | |
| $\boldsymbol{A}_i$ | 1 | 1 | 1 | 0 | 2 | 1 | 1 | 2 | 0 | 0 | 2 | 1 | 0 | 0 | 1 | 2 | 0 | 0 | 0 | 3 | 2 | 1 |
| $\boldsymbol{A}_{i+1}$ | 1 | 1 | 1 | 0 | 2 | 1 | 1 | 2 | 0 | 0 | 2 | 1 | 0 | 0 | 1 | 2 | 0 | 0 | 0 | 3 | 1 | 2 |
| … | | | | | | | | | | | | | | | | | | | | | | |
| $\boldsymbol{A}_n$ | 0 | 2 | 1 | 0 | 1 | 2 | 0 | 3 | 0 | 0 | 1 | 1 | 1 | 0 | 0 | 2 | 1 | 0 | 3 | 0 | 0 | 3 |

(ii)

| | | | | | | | |
|---|---|---|---|---|---|---|---|
| $\boldsymbol{Y}_a$ | 144 360 21 | 81 324 120 | 99 336 90 | 219 297 9 0 | 72 435 18 0 | 126 300 99 | 510 15 |
| $\boldsymbol{Y}_b$ | 45 156 51 | 27 93 132 | 36 111 105 | 51 132 63 6 | 60 147 30 15 | 45 135 72 | 78 174 |

(iii)

| | | | | | | | |
|---|---|---|---|---|---|---|---|
| $d_{11}$ | 333 | 297 | 303 | 393 | 327 | 273 | 591 |
| $V_s$ | 0,563 | 0,503 | 0,517 | 0,665 | 0,553 | 0,462 | |

The use of multisets to describe and sort objects allowed us to build several generalized classification rules, which approximate a large family of individual sorting rules, and to find correctly and contradictory classified projects. The generalized decision rules for group selecting the approved projects were as follows:

"The project is to be included in the Program if the team has the best or sufficient qualification and experience for the project realization" (the criterion estimates $x_4^1$ or $x_4^2$; the approximation rate $V_s \geq 0{,}65$);

"The project is to be included in the Program if the team has the best or sufficient qualification and experience for the project realization, the project is very important or important for achievement of the major program goals, and resources are fully sufficient or sufficient for the project realization" (the criteria estimates $x_4^1$ or $x_4^2$, and $x_1^1$ or $x_1^2$, and $x_5^1$ or $x_5^2$; the approximation rate $V_s \geq 0{,}55$).

The last generalized decision rule coincides with the empirical sorting rule that has been formed in the real-life situation [16].

Another case study is aimed to evaluating credibility of cardholders [18]. Every year banks and credit card companies lose millions of dollars due to excess expenditures of credit cardholders. In order to diminish such losses banks try to predict a financial behavior of potential cardholders. Decision rules for multiple criteria categorization of cardholder credibility can be produced, for instance, by extracting knowledge from data bases that are collected today in banks. Such data bases describe real-life financial histories of cardholders and include sex, age, residence, area of activity, income, balance, payments, purchase, cash, credit rating, and other cardholder characteristics.

A large size of files and variety of attributes, which are numerical and verbal, ordinal and nominal, cause difficulties in constructing decision rules for classifying cardholder credibility based on their personal and financial data. Moreover, there exist persons among thousands of real cardholders who have the same or very closed personal attributes but the opposite financial behavior. It means, these persons belong to the different categories, thus they have contradictory descriptions. A presence of inconsistencies of such kinds in data samples may deteriorate a quality of learning procedures, and leads to poor robustness of results.

The cardholders' representations in the decision table of multisets and sorting multisets allow us to construct a set of general decision rules of the type (4) for credit card portfolio management. Decision rules for categorization of cardholder credibility are expressed in terms of personal and financial data.

The proposed methods for group classifying multi-attribute objects operate with arbitrary-size files of qualitative and/or quantitative attributes without mixing data, do not need of previous adjustment on a sampling collection, and take into account various inconsistent and contradictory expert judgments without forcing one to find a compromise among them. Note that such group decision rules could not been found with any other known classification techniques [1, 5, 7].

## 7 Conclusion

An analysis of the problem considered is an important stage of decision aiding. The investigation of object groups and possible relations between the objects can help decision maker to formulate choice strategies and decision rules, which will be more adequate with the reality, to make his solutions more substantial and reasonable. The multi-aspect analysis and structuring alternatives allow us to gain an insight into the problem nature and to find better decisions. However, there are situations when the known methods can not be applied to analysis and solution of the problem. The most important features of these problems are plurality and redundancy of data that characterize objects, alternatives, options, and their features.

In this paper, we suggested techniques for processing and structuring a collection of objects described by many quantitative and/or qualitative attributes when several versions of objects may exist. These techniques are based on the theory of multiset metric spaces and more suitable than the well-known approaches such as utility functions, outranking, fuzzy or rough set methods [5, 7, 11, 19, 20, 21, 22, 23]. The multiset approach gives us new tools to solve traditional classification problems in more simple and constructive manner, and solve new types of problems, which are never sold earlier due to their peculiarities.

Multiset methods provide effective group multiple criteria decision aiding in a wide variety of problems, for instance, for selection of R&D projects [16], evaluation of cardholder credibility [18]. These tools are useful also for many-sided analysis of multi-attribute objects in pattern recognition [2], text processing [15], data mining, and other areas. Another promising application of multiset techniques may be content based analysis, retrieval, arrangement and classification of multimedia information. Underline once more that available information, which is contained in object descriptions, is used only in the native forms without any transformation of qualitative (symbolic, verbal) data into numerical ones.

**Acknowledgements.** This work is partially supported by the Russian Academy of Sciences, Research Programs 'Intelligent Information Technologies, Mathematical Modeling, Systems Analysis, and Automation', 'Information Technologies and Methods for Complex Systems Analysis', the Russian Foundation for Basic Research (projects 08-01-00247, 08-07-13532, 09-07-00009, 09-07-12111).

## References

1. Anderberg, M.R.: Cluster Analysis for Applications. Academic Press, New York (1973)
2. Arlazarov, V.L., Loginov, A.S., Slavin, O.L.: Characteristics of programs for optic recognition of texts. Programming 3, 45–63 (2002) (in Russian)
3. Blizard, W.: Multiset theory. Notre Dame Journal of Formal Logic 30, 36–66 (1989)
4. Deza, M.M., Laurent, M.: Geometry of Cuts and Metrics. Springer, Berlin (1997)
5. Doumpos, M., Zopounidis, C.: Multicriteria Decision Aid Classification Methods. Kluwer Academic Publishers, Dordrecht (2002)
6. Furems, E.: Knowledge-based multi-attribute classification problems structuring. In: Ruan, D., Montero, J., Lu, J., Martinez, L., D'hondt, P., Kerre, E. (eds.) Computational Intelligence in Decision and Control, pp. 465–470. World Scientific Publisher, Singapore (2008)
7. Greco, S., Matarazzo, B., Slowinski, R.: Rough sets methodology for sorting problems in presence of multiple attributes and criteria. European Journal of Operational Research 138, 247–259 (2002)
8. Hartigan, J.A.: Clustering Algoritms. Wiley, New York (1975)
9. Hwang, C.L., Lin, M.J.: Group Decision Making under Multiple Criteria. Springer, Berlin (1987)
10. Knuth, D.E.: The Art of Computer Programming. Seminumerical Algorithms, vol. 2. Addison-Wesley, Reading (1998)
11. Köksalan, M., Ulu, C.: An interactive approach for placing alternatives in preference classes. European Journal of Operational Research 144, 429–439 (2003)
12. Larichev, O.I.: Verbal Decision Analysis. Nauka, Moscow (2006) (in Russian)

13. Larichev, O.I., Olson, D.L.: Multiple Criteria Analysis in Strategic Siting Problems. Kluwer Academic Publishers, Boston (2001)
14. Miyamoto, S.: Cluster analysis as a tool of interpretation of complex systems. Working Paper WP-87-41. IIASA, Laxenburg, Austria (1987)
15. Petrovsky, A.B.: Structuring techniques in multiset spaces. In: Fandel, G., Gal, T. (eds.) Multiple Criteria Decision Making, pp. 174–184. Springer, Berlin (1997)
16. Petrovsky, A.B.: Multi-attribute sorting of qualitative objects in multiset spaces. In: Koksalan, M., Zionts, S. (eds.) Multiple Criteria Decision Making in the New Millenium, pp. 124–131. Springer, Berlin (2001)
17. Petrovsky, A.B.: Spaces of Sets and Multisets. Editorial URSS, Moscow (2003) (in Russian)
18. Petrovsky, A.B.: Multi-attribute classification of credit cardholders: multiset approach. Int. J. Manag. and Dec. Making 7(2/3), 166–179 (2006)
19. Roubens, M.: Ordinal multiattribute sorting and ordering in the presence of interacting points of view. In: Bouyssou, D., Jacquet-Lagrèze, E., Perny, P., Slowinski, R., Vanderpooten, D., Vincke, P. (eds.) Aiding Decisions with Multiple Criteria: Essays in Honor of Bernard Roy, pp. 229–246. Kluwer Academic Publishers, Dordrecht (2001)
20. Roy, B.: Multicriteria Methodology for Decision Aiding. Kluwer Academic Publishers, Dordrecht (1996)
21. Roy, B., Bouyssou, D.: Aide Multicritère à la Décision: Méthodes et Cas. Economica, Paris (1993)
22. Saaty, T.: Multicriteria Decision Making: The Analytic Hierarchy Process. – RWS Publications, Pittsburgh (1990)
23. Vincke, P.: Multicriteria Decision Aid. Wiley, Chichester (1992)
24. Yager, R.R.: On the theory of bags. Int. J. General Systems 13, 23–37 (1986)
25. Zadeh, L.A.: From computing with numbers to computing with words – from manipulation of measurements to manipulation of perceptions. IEEE Transactions on Circuits and Systems 45(1), 105–119 (1999)

# Empirical Evaluation of Selected Algorithms for Complexity-Based Classification of Software Modules and a New Model

Jian Han Wang, Nizar Bouguila, and Taoufik Bdiri

**Abstract.** Software plays a major role in many organizations. Organizational success depends partially on the quality of softwares used. In recent years, many researchers have recognized that statistical classification techniques are well-suited to develop software quality prediction models. Different statistical software quality models, using complexity metrics as early indicators of software quality, have been proposed in the past. At a high-level the problem of software categorization is to classify software modules into fault prone and non-fault prone. Indeed, a learner is given a set of training modules and the corresponding class labels (i.e fault prone or non-fault-prone), and outputs a classifier. Then, the classifier takes an unlabeled module (i.e hitherto-unseen module) and assigns it to a class. The focus of this paper is to study some selected classification techniques widely used for software categorization. Indeed, practitioners are faced with a body of approaches and literature that give several conflicting advices about the usefulness of these classification approaches. The techniques evaluated in this paper include: principal component analysis, linear discriminant analysis, multiple linear regression, logistic regression, support vector machine and finite mixture models. Moreover, we propose a Bayesian approach based on finite Dirichlet mixture models. We evaluate experimentally these approaches using a real data set. Our experimental results show that different algorithms lead to different statistically significant results.

## 1 Introduction

With the increasing need of complex computer systems, the advance in hardware performance, the size and complexity of softwares used is inevitably growing rapidly. Thus, more and more energy and investigation are devoted to the software quality field to seek techniques that can accurately reflect software performance and

Jian Han Wang · Nizar Bouguila · Taoufik Bdiri
Concordia Institute for Information Systems Engineering (CIISE)
Concordia University, Montreal, QC H3G 1T7, Canada
e-mail: {bouguila,jian,bdiri}@ciise.concordia.ca

V. Sgurev et al. (Eds.): Intelligent Systems: From Theory to Practice, SCI 299, pp. 99–131.
springerlink.com 

reliability [61]. Indeed, developing and maintaining a given software system is a challenging problem that has a lot of difficulties such as complexity, conformity, changeability and invisibility (See [14, 12], for instance, for more discussions and details about these major essential difficulties in software engineering). Software is composed of a great number of relatively independent units called modules (i.e a set of source-code files) which perform certain functions [31]. One way to test software quality is to determine the number of faults in each module. These faults may be related, for instance, to changes [1] to source code happening while the software is executing [105, 109] and are in general in a small portion [2] of the modules [67, 90]. Most of the time, people are not concerned about the exact number of changes, rather than setting a threshold. If the number of faults (i.e defects in a program that can cause incorrect execution [31]) found in certain module exceeds this previously set criterion, it is regarded as fault-prone, otherwise non fault-prone [80, 95]. For example, if a threshold of two faults is set, each module having two or more changes will be assigned to the fault-prone group and considered unstable, with high-risk and might cause failure.

A software prediction model is viewed as an empirical tool using a certain algorithm to forecast modules types (i.e fault-prone or non fault-prone) [81] and should be easy to interpret [36]. A key common characteristic of these prediction models is that they establish a relationship between the measures of modules attributes and the types [75, 30]. The fundamental construction of the predictive models is based upon the faults and corresponding measures collected from past similar program development and maintenance scenarios. When the model is built, we can determine the quality and reliability of new modules, if the measures of their attributes are in hand. The understanding of the modules through prediction models helps to target high-risk modules which need priority attention, extensive testing, redesign and improvement in early life cycle [34, 36], which is very valuable [3], cost-effective, and improve the efficiency of inspection efforts [21]. It is not acceptable to postpone the assurance of software quality until the product's release [4] [111, 71]. For instance, in telecommunication or military systems [80, 95, 81, 32, 29, 59], if faults are not early identified, but found later in operational phase, any slightly changed signal or message used to communicate will likely cause expensive consequences [5]. Software quality models are very valuable for software engineering of embedded systems, too [92]. In addition, delaying correction in testing and operational phase may result in higher cost [5]. Conversely, knowing the troublesome modules in time will guide the designers to optimize the development process and allocate the effects to the right modules in dire need of being enhanced [74, 94]. For example, predicting the high-risk modules during the design phase allows designers to refine or restructure the

[1] See [65] for a discussion about the types and classes of changes that may occur.

[2] According to the 80/20 rule (i.e Pareto rule), about 20 percent of a software system is responsible for 80 percent of its errors, costs and rework [1, 7].

[3] Note that some studies have been devoted to assess the return on investment of software classification models (See [82], for instance).

[4] An evaluation of software quality models, based on classification trees, over several releases is presented in [101].

system to reduce its complexity. And if those are identified in the implementation phase, the majority of the test resources will be assigned to which are most likely to cause quality problems. Thus, a software predictive model, which can categorize program modules into fault-prone or non fault-prone, not only locate the troublesome modules earlier, but also benefits the designers to effectively use the resources to the accurate ones, which have internal faults, with the utmost probability. In addition, these models may even be used to guide maintenance activities during the operations phase [28, 100].

Since software categorization plays a critical role in the software quality field, nowadays more and more modeling pattern recognition [35], statistical analysis, and machine learning [68, 12] techniques are employed in building predictive models and have proven to be of great practical value such as neural networks [99], optimal set reduction [34], discriminant power [54, 55], discriminant coordinates [58], fuzzy classification [7], classification trees [103, 93, 66, 102], regression trees [79], random forests [37], boolean discriminant functions [52], combined classification models [83], support vector machine, discriminant analysis, logistic regression, and finite mixture models. Although the algorithms used differ, they all employ complex metrics as an input predictor, and a prediction of fault-prone or non fault-prone as an output response variable and aim at reducing the cost of the misclassification [89]. Software complexity metrics act as partial measures of software attributes and represent quantitative description of internal program characteristics (i.e each module is considered as a multidimensional vector). Indeed, it has been shown that the quantity of software faults found during execution has direct relationship with the complexity metrics obtained in early software life cycle and it also has a great impact on software quality [31, 73, 27].

Despite the great interest in prediction models and compared with other software engineering areas such as inspections and use-cases, only few studies (See, [15, 16, 8], for instance) have been devoted to compare and evaluate the different techniques used [41]. In this paper, we perform a survey and comparison of some selected software categorization techniques by using real data set. Moreover, we propose a novel Bayesian approach based on finite Dirichlet mixture models previously presented in [50]. The rest of this paper is organized as follows. Section 2 provides some information on software complexity metrics. Section 3 describes the key ideas behind the different selected techniques that we have compared in our study and covering principal component analysis, discriminant analysis, multiple linear regression, logistic regression, support vector machine and finite mixture models. In Section 4, we present our finite Bayesian Dirichlet mixture model and its complete learning algorithm. The experimental evaluation is detailed in Section 5. Finally, Section 6 provides a discussion and some concluding remarks.

## 2 Modules Representation Using Complexity Metrics

The different classification approaches that we will describe in this paper represent each software module using complexity metrics which have been developed to measure software quality and capture modules features [31, 109, 96, 18, 87]. Indeed, each module is considered to be a multidimensional vector in the complexity metrics space. These metrics are not only a part of measurable [5] software attributes which can be gathered in the early life cycle of software design, but also are proven indicators which describe the software complexity and analyze its improvement [77, 76]. In many previous studies, it was observed that the software complexity is directly related to software quality and fault-correction activity [31, 4, 38, 24], which means, for instance, that fault counts and change counts are highly correlated. Thus, measures of software complexity are good indicators to understand and model the quality of software. Software metrics are constructed by a variety of measures of program codes. Many product metrics and techniques to evaluate them have been proposed [72, 110, 13]. In particular, Lines of Code (LOC) is generally closely related to the number of faults found later when executing software. In addition to the Lines of Code, there are other well-known and widely used measures of product metrics. For instance, Halstead's software science [42] is an approach dedicated to build software complexity measures by identifying a set of basic elements describing the modules, such as operands and operators [108]. Operands refer to variables and constants, and operators indicate symbols or combination of symbols that affect the values of operands. The basic measures of this approach is based upon four scalar numbers derived directly from the module's source code: (1) the number of unique operators (2) the number of unique operands (3) the total number of operators (4) the total number of operands. Furthermore, the basic Halstead complexity measures are combined in a number of ways to produce additional measures, which are widely adopted as indicators in vast majority cases. Halstead's complex metrics are popularly employed in evaluating mainstream programming, such as Pascal.

During the past decade, object-oriented approaches have been extensively used in software development environments. The conceptual and structural nature of these approaches, have created new challenges in the software quality field such as exploring new and special metrics [106, 39], and assessing software quality in object-oriented environments [63]. A well-known example is Chidamber and Kemerer's metrics suite proposed in [78] and widely studied and evaluated in the literature [106, 43]. The main reason that software complexity metrics are widely used, is that they can be collected in the very early software life cycle. Some of them are obtained directly from measuring the source codes and high-level design, and some are even taken from the software specifications. However, since part of the components of complexity metrics is the combination of some of the others, there are potential linear relationship within them [2]. Besides, some of the metrics used may be redundant with marginal contribution [55]. Thus, it is necessary to explore the structure of observations to understand the mutual collinearity existing within the components. In this paper, we adopt principal components analysis technique to investigate the underlying relationship between every two predictors, and process observations beforehand.

[5] See [33, 25, 57], for instance, for interesting discussions about measurement theory in software engineering.

## 3 Statistical Methods

Different techniques have been proposed to develop a predictive relationship between software complexity metrics and the categorization of the modules into fault-prone and non fault-prone[6]. These predictive models are built generally from examples [68] using training data sets composed of labeled observations (i.e modules taken, for instance, from historical projects). Then, according to these built models new unlabeled modules can be identified as fault-prone or non fault-prone which allow software engineers to detect troublesome modules in the early life-cycle of a software product. Before building a quality prediction model, an important step generally implemented is validating [52] and analyzing the software metrics used, to examine the interrelationship among them, to reduce the dimensionality of the observations describing the modules and then simplifying the quantity of calculations. Note that the validated metrics can then be applied on multiple projects [53]. Principal Components Analysis (PCA) is the most used technique for this task and allows the extraction of the most relevant information brought by the used metrics. In this section, we will introduce these most successful techniques in details, mainly centering around their usefulness in software quality prediction.

### 3.1 Principal Components Analysis

Principal Components Analysis (PCA) is a widely used exploratory multivariate technique [6]. Suppose we have a set of $N$ modules $\mathscr{X} = (x_1, x_2, \ldots, x_N)$, where each module is represented by a $d$-dimensional vector, of complexity metrics $x_{il}, l = 1, \ldots, d$, $x_i = (x_{i1}, x_{i2}, x_{i3}, \ldots, x_{id}) \in \mathbb{R}^d$, $i = 1, \ldots, N$, where two or more metrics have high degree of linear correlation. This is called multicollinearity, and it is a major problem in many models such as regression analysis built on the basic assumption that selected variables are independent [96]. When multicollinearity exists among some metrics, the established statistical model become unstable, and coefficients parameters estimated by training data sets are very sensitive [32]. Besides, the model will not be robust enough to forecast response variables of new observations. A solution to this problem is the application of PCA to transform correlated metric data into orthogonal variables. As in practice, software complexity metrics are often found highly correlated to each other and are a linear combination of a small number of orthogonal metric domains [26], PCA has been applied in many works [90, 81, 32, 88, 98].

PCA finds a linear transformation $W^T$ which maps the $d$-dimensional metrics vectors space into a new space with lower dimension $d^{new} < d$. The $d^{new}$-dimensional vectors $x_i^{new}$ are given by:

$$x_i^{new} = W^T x_i \tag{1}$$

[6] Note that some studies have re-examined the analysis under the assumption that only two classes can be distinguished by considering a number of differentiable groups instead of two (See [58, 40], for instance).

With PCA we try to find the optimal projection $E$ which maximizes the determinant of the scatter matrix $W^T \Sigma W$ of the new projected samples $\mathscr{X}^{new} = (x_1^{new}, \ldots, x_N^{new})$

$$E = \arg\max_W |W^T \Sigma W| \tag{2}$$

where $\Sigma$ is the scatter matrix of the original data

$$\Sigma = \sum_{i=1}^{N} (x_i - \bar{x})(x_i - \bar{x})^T \tag{3}$$

$\bar{x}$ is the mean vector of $\mathscr{X}$

$$\bar{x} = \frac{1}{N} \sum_{i=1}^{N} x_i \tag{4}$$

and $E = [E_1, \ldots, E_{d^{new}}]$ is composed of the $d$-dimensional eigenvectors of $S$ corresponding to the $d^{new}$ largest eigenvalues [64].

## 3.2 *Discriminant Analysis*

Discriminant Analysis technique [6] is applied when we attempt to build a predictive model of groups membership based upon observed characteristics of each observation (i.e module). In software categorization case, this technique generates a discriminant function which can classify software modules as either high or low risk according to the software complexity metrics [80, 95, 32, 90]. This discriminant function, generated from a set of observations of labeled modules, can then be applied to new observations with software measurements but unknown groups membership. There are several discriminant analysis models (i.e linear, non linear and logistic discriminant model) that can be chosen depending on the data type of predictive variables such as all quantitative, all qualitative or mixed [112].

### 3.2.1 Linear Discriminant Analysis

Linear Discriminant Analysis (LDA) was used extensively by software engineering researchers to both assess software quality [81, 32] and evaluate software metrics [107, 51]. LDA assumes that the classes are linearly separable and follow homoscedastic gaussian distributions. Under this assumption, one can show that the optimal subspace where we can perform the classification is given by the vectors $W$ which are the solution of the following generalized eigenvalue problem

$$\Sigma_b W = \lambda \Sigma_w W \tag{5}$$

where $\Sigma_w$ is the within-class scatter matrix and given by

$$\Sigma_w = \sum_{j=1}^{M} \sum_{i=1}^{n_j} (x_i - \bar{x}_j)(x_i - \bar{x}_j)^T \tag{6}$$

where $n_j$ is the number of vectors in class $j$ and $\bar{x}_j$ is the mean of class $j$. $\Sigma_b$ is the between-class scatter matrix and given by

$$\Sigma_b = \sum_{j=1}^{M} (\bar{x}_j - \bar{x})(\bar{x}_j - \bar{x})^T \tag{7}$$

where $M$ is the total number of classes. The linear discriminant model generally used, to differentiate fault-prone from non fault-prone modules, is based on the following generalized squared distance:

$$D_j^2(x) = (x - \bar{x}_j)^T \Sigma_p^{-1} (x - \bar{x}_j) \tag{8}$$

where, $\bar{x}_j$ represents the mean vector of class $j \in \{1,2\}$ and $\Sigma_p$ is the so-called pooled covariance matrix given by:

$$\Sigma_p = \frac{\sum_{j=1}^{2} n_j \Sigma_j}{\sum_{j=1}^{2} n_j} \tag{9}$$

where $\Sigma_j$ is the covariance matrix of class $j$ and $n_j$ represents the number of modules in class $j$. Thus, the posterior probability of membership of $x$ in class $j$ is :

$$p_j(x) = \frac{e^{-1/2 D_j^2(x)}}{\sum_{j=1}^{2} e^{-1/2 D_j^2(x)}} \tag{10}$$

According to the discriminant function given by the previous equation, a vector $x$ is assigned to the class $j$ yielding to the greater posterior probability. Despite its effectiveness, a major inconvenient of LDA is the Gaussian assumption which is not the best choice [88]. A solution to this problem is nonparametric discriminant analysis. Another major drawback of LDA is the linearity of the classification surface. To overcome this problem, SVMs can be used to offer both linear and non-linear flexible classification surfaces. Moreover, discriminant analysis is less appropriate, when many of the metrics are discrete and an alternative approach in this case is logistic regression [97].

### 3.2.2 Nonparametric Discriminant Analysis

Nonparametric Discriminant Analysis (NDA) does not make assumptions about the distribution of the data and was widely used for classification in the case of software quality modeling [95, 89, 90, 80, 88, 91]. Let $f_j$ be the multivariate probability density function representing class $j$. Nonparametric discriminant analysis is based on the empiric estimation of the densities $f_j$ which gives an approximation $\hat{f}_j$ to it as the following:

$$\hat{f}_j(x_i|\lambda) = \frac{1}{n_j} \sum_{i=1}^{n_k} K_j(x_i|x_{ji}, \lambda) \tag{11}$$

where $K_j(x_i|x_{ji},\lambda)$ is a multivariate normal kernel on vector $x_i$, with modes at $x_{ji}$ which is a vector in class $j$, and given by

$$K_j(x_i|x_{ji},\lambda) = (2\pi\lambda^2)^{-n_j/2}|\Sigma_j|^{-1/2}\exp((\frac{-1}{2\lambda^2})(x_i - x_{ji})^T\Sigma_j^{-1}(x_i - x_{ji})) \quad (12)$$

where $\lambda$ is a smoothing parameter chosen by optimizing the misclassification rates of cross validation on the training data set [89]. Then, the classification is based, in the case of our problem, on the following rule

$$class(x_i) = \begin{cases} 1 & \text{if} \quad \frac{\hat{f}_1(x_i)}{\hat{f}_2(x_i)} > \frac{n_2}{n_1} \\ 2 & \text{otherwise} \end{cases} \quad (13)$$

## 3.3 Multiple Linear Regression

Multiple linear regression [10] performs a summary of the relationship between the module types, fault prone or non-fault prone, and the software complexity metrics, which is represented as a multivariate linear regression model. Here the determined module type, so-called response variable or dependent variable, is denoted as $Y$, and the software complexity metrics, which are composed of independent indicators, are represented as $x_i$. Written mathematically, the standard multiple regression is,

$$Y_i = \beta_0 + \beta_1 x_{i1} + \beta_2 x_{i2} + ... + \beta_i x_{id} + \varepsilon_i \quad (14)$$

where $\beta_i$ are the coefficient parameters, and $\varepsilon_i$ are normally distributed random variables, called error terms on the assumption that mean equals to 0 and variance is unknown and constant. Some approaches are widely employed to estimate the parameters, such as least square estimations, least absolute value estimation, relative least squares and minimum relative error procedures [85]. Least square estimation is the most used method among them, and the estimated regression model parameters are yielded by minimizing $\sum_{i=1}^{N}(\hat{Y}_i - Y_i)^2$, where $\hat{Y}_i = \hat{\beta}_0 + \hat{\beta}_1 x_{i1} + \hat{\beta}_2 x_{i2} + ... + \hat{\beta}_i x_{id}$, $\hat{Y}_i$ and $\hat{\beta}_i$ represent estimated values.

Multiple linear regression models are built from a set of potential large number of predictive terms, and a subset of significant independent terms should be determined to enter into the multiple regression models [84]. Some techniques are employed for adding or removing explanatory variables from the model: forward selection, backward elimination and stepwise regression, which are all iterative procedures. Forward selection starts with an empty subset in the model and add one explanatory variable (which most contribute to the model) at a time, continuing the iterations until reaching a certain stop criterion. On the contrary, backward elimination beginning with all the predictors, removes one of them (considered the most redundant) in every iterative procedure. Stepwise regression [84] can be referred as a forward selection with replacement. In each subsequent iterative step, the model is evaluated, using computed statistical significance, with or without a potential predictor to see if it contributes to the explanatory power of the model. After determining the most significant complexity metrics and estimating the model parameters, the linear combination of the predictors can be used to predict if the future modules are high-risk or not.

## 3.4 Logistic Regression

Logistic regression [11] was extensively used in software engineering for both metrics validation [106] and modules classification [34, 36, 100, 97, 56]. It is a widely applied statistical modeling technique when the dependant (i.e response) variable has only two possible values which is the case in our studied problem (fault-prone vs. non fault-prone). The independent variables (software metrics in our case), however, may be categorical, discrete or continuous. The logistic regression model is given by the following form [11]

$$\ln\left(\frac{\pi(x_i)}{1-\pi(x_i)}\right) = \beta_0 + \beta_1 x_{i1} + \ldots + \beta_d x_{id} \tag{15}$$

where $\pi(x_i)$ is the probability of the event: *the module $x_i$ is fault-prone*, and has the following multivariate exponential form

$$\pi(x_i) = \frac{\exp(\beta_0 + \beta_1 x_{i1} + \ldots + \beta_d x_{id})}{1 + \exp(\beta_0 + \beta_1 x_{i1} + \ldots + \beta_d x_{id})} \tag{16}$$

The ratio $\frac{\pi(x_i)}{1-\pi(x_i)}$ is usually interpreted as odds of occurrence, which compares the probability of the event *fault-prone* to the probability of the *non fault-prone* one. This odds ranges form zero to infinity, whereas its logarithm ranges from 0 to 1, and called the log odds or the logit. From Eq. 15, we can see that the $\frac{\pi(x_i)}{1-\pi(x_i)}$ has a linear relationship with $x_i$, and the parameters $\beta_1, \beta_2, \ldots, \beta_d$, so-called regression coefficients, embody the changes in the log odds. The estimation of these coefficients is in general based on the maximum likelihood approach and can be carried out with a wide variety of statistical software packages [11].

In practical applications, after the logistic regression model is set up, a threshold is experimentally designated to determine if the new modules are troublesome or not. For example, the threshold can be determined through a classification rule that minimize the expected cost of misclassification [100, 89, 97]:

$$class(x_i) = \begin{cases} 1 : fault-prone & \text{if } \frac{\hat{\pi}(x_i)}{1-\hat{\pi}(x_i)} > \frac{C_I}{C_{II}} \frac{\pi_{nfp}}{\pi_{fp}} \\ 2 : non\ fault-prone & \text{otherwise} \end{cases} \tag{17}$$

where $C_I$ and $C_{II}$ are respectively the cost of type I (a non-fault prone is classified as fault-prone) and type II (a fault-prone is classified as non fault-prone) misclassification; and $\pi_{nfp}$ and $\pi_{fp}$ represent the prior probabilities of non fault-prone and fault-prone, respectively.

## 3.5 Support Vector Machine

Support Vector Machine (SVM) [104] is a two-class classification method that has been used successfully in many applications dealing with data classification in general and software modules in particular [17]. In the following, we briefly summarize the theory of SVM. For two-class pattern recognition, we try to estimate a function $f : \mathbb{R}^d \rightarrow \{\pm 1\}$ using $l$ training $d$-dimensional vectors $x_i$ and class labels $y_i$,

$$(x_1, y_1), \ldots, (x_l, y_l) \in \mathbb{R}^d \times \{\pm 1\} \tag{18}$$

after the training the function $f$ should be able to correctly classify new test vectors $x$ into one of the two classes. Suppose that we have a hyperplane separating the first class (positive class) for the the second class (negative class). The idea behind SVM is to find the optimal hyperplane permitting a maximal margin of separation between the two classes and defined by

$$w.x + b = 0 \quad w \in \mathbb{R}^d, b \in \mathbb{R} \tag{19}$$

corresponding to decision function

$$f(x) = sign(w.x + b) \tag{20}$$

where $b$ is the distance to the hyperplane from the origin and $w$ is the normal of the hyperplane which can be estimated through the use of training data by solving a quadratic optimization problem [104]. $w$ can be estimated by

$$w = \sum_{i=1}^{l} v_i x_i \tag{21}$$

where $v_i$ are coefficient weights.

In general, classes are not linearly separable. In order to overcome this problem, SVM can be extended by introducing a kernel $K$ to map the data into another dot product space $F$ using a nonlinear map

$$\Phi : \mathbb{R}^d \rightarrow F \tag{22}$$

In this new space $F$, the classes will be linearly separable. The kernel $K$ is given by

$$K(x, x_i) = (\Phi(x).\Phi(x_i)) \tag{23}$$

and measures the similarity between data vectors $x$ and $x_i$. Then, the decision rule is

$$f(x) = sign\left(\sum_{i=1}^{l} v_i K(x_i, x) + b\right) \tag{24}$$

An important issue here is the choice of the kernel function and some well-known classic choices are: Polynomial with degree $d$, $K(x_i, x) = (x_i^T x + 1)^d$, Radial basis function (RBF) with parameter $\sigma$, $K(x_i,x) = \frac{\exp(-\|x_i-x\|^2)}{2\sigma^2}$, and Sigmoid with parameters $\kappa$ and $\theta$, $K(x_i,x) = \tanh(\kappa x_i^T x + \theta)$.

## 3.6 Finite Mixture Models

Finite mixture models are among the most applied and accepted statistical approaches [22]. Finite mixture models have several clear attractions: they have a solid grounding in the theory of probability and statistics, they are flexible enough to approximate any other statistical model and they are a natural choice when the data to model is heterogenous [22]. Moreover, finite mixtures permit a formal approach to unsupervised learning. The use of finite mixture models as a statistical tool for early prediction of fault-prone program modules has been investigated, for instance, in [62]. Finite mixtures can be viewed as a superimposition of a finite number of component densities and thus adequately model situations in which each data element is assumed to have been generated by one (unknown) component. More formally, a finite mixture model with $M$ components is defined as

$$p(x_i|\Theta) = \sum_{j=1}^{M} p(x_i|\theta_j)p_j \tag{25}$$

The parameters of a mixture for $M$ clusters are denoted by $\Theta = (\theta_1,\ldots,\theta_M,P)$, where $P = (p_1,\cdots,p_M)$ is the mixing parameter vector. Of course, being probabilities, the $p_j$ must satisfy $0 < p_j \leq 1, \quad j = 1,\ldots,M$ and $\sum_{j=1}^{M} p_j = 1$. The choice of the component model $p(x_i|\theta_j)$ is very critical in mixture decomposition. The number of components required to model the mixture and the modeling capabilities are directly related to the component model used [22]. In the past two decades, much effort has been devoted to Gaussian mixture models estimation and selection (i.e determination of the number of components). The multivariate Gaussian probability density function is the common assumption when using finite mixture models and is given by

$$p(x_i|\theta_j) = \frac{\exp[\frac{1}{2}(x_i-\mu_j)^T\Sigma_j^{-1}(x_i-\mu_j)]}{(2\pi)^{d/2}|\Sigma_j|^{1/2}} \tag{26}$$

Thus, in the case of a finite Gaussian mixture model, we have $\theta_j = (\mu_j,\Sigma_j)$.

An important problem in the case of finite mixture models is the estimation of the parameters. During the last two decades, the method of maximum likelihood (ML) has become the most common approach to this problem [22]. It is well known that the maximum likelihood (ML) estimate:

$$\hat{\Theta}_{ML} = \arg\max_{\Theta}\{L(\Theta,\mathscr{X})\} \tag{27}$$

where $L(\Theta,\mathscr{X})$ is the log-likelihood corresponding to a $M$-component is:

$$L(\Theta,\mathscr{X}) = \log\prod_{i=1}^{N} p(x_i|\Theta) = \sum_{i=1}^{N}\log\sum_{j=1}^{M} p(x_i|\theta_j)p_j \tag{28}$$

The maximization defining the ML estimates is subject to the constraints over the mixing parameters and cannot be found analytically [22]. However, the ML estimates of the mixture parameters can be obtained using expectation maximization (EM) and related techniques [22]. The EM algorithm is a general approach to maximum likelihood in the presence of incomplete data. In EM, the "complete" data are considered to be $y_i = \{x_i, z_i\}$, where $z_i = (z_{i1}, \ldots, z_{iM})$, with:

$$z_{ij} = \begin{cases} 1 \text{ if } x_i \text{ belongs to class } j \\ 0 \quad\quad \text{otherwise} \end{cases} \tag{29}$$

constituting the "missing" data. The relevant assumption is that the density of an observation $x_i$, given $z_i$, is given by $\prod_{j=1}^{M} p(x_i|\theta_j)^{z_{ij}}$. The resulting *complete-data log-likelihood* is:

$$L(\Theta, \mathscr{Z}, \mathscr{X}) = \sum_{i=1}^{N} \sum_{j=1}^{M} z_{ij} \log(p(x_i|\theta_j)p_j) \tag{30}$$

Where $\mathscr{Z} = (z_1, \ldots, z_N)$. The EM algorithm produces a sequence of estimates $\{\Theta^t, t = 0, 1, 2 \ldots\}$ by applying two steps in alternation until some convergence criterion is satisfied:

1. **E-step:** Compute $\hat{z}_{ij}$ given the parameter estimates from the initialization:
$$\hat{z}_{ij} = \frac{p(x_i|\theta_j)p_j}{\sum_{l=1}^{M} p(x_i|\theta_l)p_l}$$
2. **M-step:** Update the parameter estimates according to: $\hat{\Theta} = \arg\max_{\Theta} L(\Theta, \mathscr{Z}, \mathscr{X})$

The quantity $\hat{z}_{ij}$ is the conditional expectation of $z_{ij}$ given the observation $x_i$ and parameter vector $\Theta$. The value $z^*_{ij}$ of $\hat{z}_{ij}$ at a maximum of equation 30 is the conditional probability that observation $i$ belongs to class $j$ (the *a posteriori* probability); the classification of an observation $x_i$ is taken to be $\{k/z^*_{ik} = max_j z^*_{ij}\}$, which is the Bayes rule. When we maximize the function given by equation 30, we obtain:

$$p_j^{(t+1)} = \frac{1}{N} \sum_{i=1}^{N} \hat{z}_{ij} \tag{31}$$

$$\mu_j^{(t+1)} = \frac{\sum_{i=1}^{N} \hat{z}_{ij} x_i}{\sum_{i=1}^{N} \hat{z}_{ij}} \tag{32}$$

$$\Sigma_j^{(t+1)} = \frac{\sum_{i=1}^{N} \hat{z}_{ij} [(x_i - \mu_j^{(t)})(x_i - \mu_j^{(t)})^T]}{\sum_{i=1}^{N} \hat{z}_{ij}} \tag{33}$$

Another important problem now is the selection of the number of components $M$ which best describes the data. For this purpose, many approaches have been suggested. From a computational point of view, these approaches can be classified into three classes: deterministic, stochastic, and resampling methods [22]. The most used approaches, however, are the deterministic methods which can themselves be classified in two main classes: in the first, we have approximate Bayesian criteria like the Schwarz's Bayesian information criterion (BIC) [20] and the Laplace-empirical criterion (LEC) [22]. The second class contains approaches based on information/coding theory concepts such as the minimum message length (MML) [9], Akaike's information criterion (AIC) [23], and minimum description length (MDL) criterion [70]. A more detailed survey of selection criteria approaches can be found in [22]. For instance, the authors in [66, 62], have used the AIC criterion given by

$$AIC(M) = -2L(\Theta, \mathscr{X}) + 2N_p \tag{34}$$

where $N_p$ is the number of parameters in the model and is equal to $Md + (M-1) + Md(d+1)/2$ in the case of finite Gaussian mixture models. The selection of the optimal number of clusters $M^*$ is done by $M^* = \arg\min_M AIC(M)$.

# 4 The Bayesian Finite Dirichlet Mixture

## 4.1 Finite Dirichlet Mixture

The Dirichlet mixture is the multivariate generalization of the Beta mixture. Finite Beta mixtures were studied by Bouguila et al. in [47] which highlights some difficulties when performing the likelihood approach and proposes Bayesian inference to estimate the parameters. Despite the fact that this distribution plays an important role in statistical inference, and in contrast to the vast amount of theoretical work that exists on the Dirichlet distribution, however, very little work has been done on its practical applications. The majority of the studies either consider a single Dirichlet distribution [3, 19] or use it as a prior to the multinomial [46]. Indeed, the majority of the researchers consider finite Gaussian mixtures for data modeling. The Dirichlet mixture, however, could offer better modeling capabilities as shown in [48, 49, 45] where it is applied, as a parent distribution and not as a prior, for different image processing tasks.

If the random vector $\mathbf{x} = (x_1, \ldots, x_d)$ follows a Dirichlet distribution with parameters $\alpha = (\alpha_1, \ldots, \alpha_d)$ the joint density function is given by

$$p(\mathbf{x}|\alpha) = \frac{\Gamma(|\alpha|)}{\prod_{i=1}^{d} \Gamma(\alpha_i)} \prod_{i=1}^{d} x_i^{\alpha_i - 1} \tag{35}$$

where $\sum_{i=1}^{d} x_i = 1$[7], and $|\alpha| = \sum_{i=1}^{d} \alpha_i, \quad \alpha_i > 0 \quad \forall i = 1 \ldots d$. This distribution is the multivariate extension of the 2-parameter Beta distribution [47]. Unlike the normal distribution, the Dirichlet does not have separate parameters describing the mean and variation. The mean and the variance, however, can be calculated through $\alpha$ and are given by

[7] The Dirichlet distribution can be extended easily to be defined in any $d$-dimensional rectangular domain $[a_1, b_1] \times \ldots \times [a_d, b_d]$ where $(a_1, \ldots, a_d) \in \mathbb{R}^d$ and $(b_1, \ldots, b_d) \in \mathbb{R}^d$.

$$\mu_i = E(x_i) = \frac{\alpha_i}{|\alpha|} \tag{36}$$

$$Var(x_i) = \frac{\alpha_i(|\alpha| - \alpha_i)}{|\alpha|^2(|\alpha| + 1)} \tag{37}$$

By substituting Eq. 36 in Eq. 35, the Dirichlet distribution can be written as the following

$$p(\mathbf{x}||\alpha|, \mu) = \frac{\Gamma(|\alpha|)}{\prod_{i=1}^{d} \Gamma(\mu_i|\alpha|)} \prod_{i=1}^{d} x_i^{\mu_i|\alpha|-1} \tag{38}$$

where $\mu = (\mu_1, \ldots, \mu_d)$. Note that this alternative parameterization was also adopted in the case of the Beta distribution by Bouguila et al. [47] and provides interpretable parameters. Indeed, $\mu$ and $|\alpha|$ represent the mean and measures the sharpness of the distribution, respectively. A large value of $|\alpha|$ produces a sharply peaked distribution around the mean $\mu$. And when $|\alpha|$ decreases, the distributions becomes broader. An additional advantage of this parameterization is the fact that $\mu$ lies within a bounded space, increase computational efficiency. Then, this parameterization will be adopted.

Let $p(\mathbf{X}|\xi)$ be an $M$-component finite Dirichlet mixture model. The symbol $\xi$ refers to the entire set of parameters to be estimated:

$$\xi = (\mu_1, \ldots, \mu_M, |\alpha_1|, \ldots, |\alpha_M|, p(1), \ldots, p(M))$$

This set of parameters can be divided into three subsets $\xi_1 = (|\alpha_1|, \ldots, |\alpha_M|)$, $\xi_2 = (\mu_1, \ldots, \mu_M)$, and $\xi_3 = (p_1, \ldots, p_M)$. Then, the different parameters $\xi_1$, $\xi_2$ and $\xi_3$ can be calculated independently.

Most statistical estimation work is done using deterministic approaches. With deterministic approaches, a random sample of observations is drawn from a distribution or a mixture of distributions with unknown parameters assumed to be fixed. The estimation is performed through EM and related techniques. In contrast with deterministic approaches, Bayesian approaches consider parameters as random variables and allow probability distributions to be associated with them. Bayesian estimation is feasible now thanks to the development of simulation-based numerical integration techniques such as Markov chain Monte Carlo (MCMC) [60]. MCMC methods simulate requires estimates by running appropriate Markov Chains using specific algorithms such as Gibbs sampler.

## 4.2 Bayesian Estimation

### 4.2.1 Parameters Estimation

Bayesian estimation is based on learning from data by using Bayes's theorem in order to combine both prior information with the information brought by the data to produce the posterior distribution. The prior represents our prior belief about the parameter before looking at the data. The posterior distribution summarizes our belief about the parameters after we have analyzed the data. The posterior distribution can be expressed as

$$p(\xi|\mathscr{X}) \propto p(\mathscr{X}|\xi)p(\xi) \tag{39}$$

From the previous equation, we can see that Bayesian estimation requires a prior distribution $p(\xi)$ specification for the mixture parameters.

We start by the distribution $p(\xi_3|\mathscr{Z})$ and we have

$$p(\xi_3|\mathscr{Z}) \propto p(\xi_3)p(\mathscr{Z}|\xi_3) \tag{40}$$

We determine now $p(\xi_3)$ and $p(\mathscr{Z}|\xi_3)$. We know that the vector $\xi_3$ is defined on the simplex $\{(p_1,\dots,p_M) : \sum_{j=1}^{M-1} p_j < 1\}$, then a natural choice, as a prior, for this vector is the Dirichlet distribution [60]

$$p(\xi_3) = \frac{\Gamma(\sum_{j=1}^{M}\eta_j)}{\prod_{j=1}^{M}\Gamma(\eta_j)} \prod_{j=1}^{M} p_j^{\eta_j-1} \tag{41}$$

where $\eta = (\eta_1,\dots,\eta_M)$ is the parameter vector of the Dirichlet distribution. Moreover, we have

$$p(\mathscr{Z}|\xi_3) = \prod_{i=1}^{N} p(z_i|\xi_3) = \prod_{i=1}^{N} p_1^{z_{i1}} \dots p_M^{z_{iM}} = \prod_{i=1}^{N}\prod_{j=1}^{M} p_j^{z_{ij}} = \prod_{j=1}^{M} p_j^{n_j} \tag{42}$$

where $n_j = \sum_{i=1}^{N} \mathbb{I}_{z_{ij}=j}$. Then

$$\begin{aligned} p(\xi_3|\mathscr{Z}) &= \frac{\Gamma(\sum_{j=1}^{M}\eta_j)}{\prod_{j=1}^{M}\Gamma(\eta_j)} \prod_{j=1}^{M} p_j^{\eta_j-1} \prod_{j=1}^{M} p_j^{n_j} = \frac{\Gamma(\sum_{j=1}^{M}\eta_j)}{\prod_{j=1}^{M}\Gamma(\eta_j)} \prod_{j=1}^{M} p_j^{\eta_j+n_j-1} \\ &\propto \mathscr{D}(\eta_1+n_1,\dots,\eta_M+n_M) \end{aligned} \tag{43}$$

where $\mathscr{D}$ is a Dirichlet distribution with parameters $(\eta_1+n_1,\dots,\eta_M+n_M)$. We note that the prior and the posterior distributions, $p(\xi_3)$ and $p(\xi_3|\mathscr{Z})$, are both Dirichlet. In this case we say that the Dirichlet distribution is a conjugate prior for the mixture proportions. We held the hyperparameters $\eta_j$ fixed at 1 which is a classic and reasonable choice. For a mixture of Dirichlet distributions, it is therefore possible to associate with each $|\alpha_j|$ a prior $p_j(|\alpha_j|)$ and with each $\mu_j$ a prior $p_j(\mu_j)$. For the same reasons as the mixing proportion, we can select a Dirichlet prior for $\mu_j$

$$p(\mu_j) = \frac{\Gamma(\sum_{l=1}^{d}\vartheta_l)}{\prod_{l=1}^{M}\Gamma(\vartheta_l)} \prod_{l=1}^{d} \mu_{jl}^{\vartheta_l-1} \tag{44}$$

For $|\alpha_j|$, we adopt a vague prior of inverse Gamma shape $p(|\alpha_j|^{-1}) \sim \mathscr{G}(1,1)$:

$$p(|\alpha_j|) \propto |\alpha_j|^{-3/2} \exp\big(-1/(2|\alpha_j|)\big) \tag{45}$$

Having these priors, the posterior distribution is

$$p(|\alpha_j|,\mu_j|\mathscr{Z},\mathscr{X}) \propto p(|\alpha_j|)p(\mu_j) \prod_{z_{ij}=1} p(\mathbf{X}_i||\alpha_j|,\mu_j)$$

$$\propto |\alpha_j|^{-3/2} \exp\left(-1/(2|\alpha_j|)\right) \frac{\Gamma(\sum_{l=1}^{d} \vartheta_l)}{\prod_{l=1}^{d} \Gamma(\vartheta_l)} \prod_{l=1}^{d} \mu_{jl}^{\vartheta_l - 1}$$
$$\times \left( \frac{\Gamma(|\alpha_j|)}{\prod_{l=1}^{d} \Gamma(\mu_{jl}|\alpha_j|)} \right)^{n_j} \prod_{l=1}^{d} \left( \prod_{Z_{ij}=1} X_{il} \right)^{\mu_{jl}|\alpha_j| - 1}$$

The hyperparameters $\vartheta_l$ are chosen to be equal to 1. Having all the posterior probabilities in hand, the steps of the Gibbs sampler are

1. Initialization
2. Step t: For t=1,...
   a. Generate $Z_i^{(t)} \sim \mathcal{M}(1; \hat{Z}_{i1}^{(t-1)}, \dots, \hat{Z}_{iM}^{(t-1)})$
   b. Compute $n_j^{(t)} = \sum_{i=1}^{N} \mathbb{I}_{Z_{ij}^{(t)}=j}$
   c. Generate $P^{(t)}$ from Eq. 43
   d. Generate $(|\alpha_j|, \mu_j)^{(t)}$ $(j = 1, \dots, M)$ from Eq. 46 using the Metropolis-Hastings (M-H) algorithm [60].

Having our algorithm in hand, an important problem is the determination of the number of iterations needed to reach convergence which is discussed in [50].

### 4.2.2 Selection of the Number of Clusters

The choice of the number of components $M$ affect the flexibility of the model. For the selection of the number of clusters we use integrated (or marginal) likelihood defined by

$$p(\mathcal{X}|M) = \int \pi(\xi|\mathcal{X}, M) d\xi = \int p(\mathcal{X}|\xi, M)\pi(\xi|M) d\xi \tag{46}$$

Where $\xi$ is the vector of parameters of a finite mixture model, $\pi(\xi|M)$ is its prior density, and $p(\mathcal{X}|\xi, M)$ is the likelihood function. The main problem now is how to compute the integrated likelihood. In order to resolve this problem, let $\hat{\xi}$ denotes the posterior mode, satisfying $\frac{\partial \log(\pi(\hat{\xi}|\mathcal{X},M))}{\partial \xi} = 0$, where $\frac{\partial \log(\pi(\hat{\xi}|\mathcal{X},M))}{\partial \xi}$ denotes the gradient of $\log(\pi(\hat{\xi}|\mathcal{X},M))$ evaluated at $\xi = \hat{\xi}$. The Hessian matrix of minus the $\log(\pi(\hat{\xi}|\mathcal{X},M))$ evaluated at $\xi = \hat{\xi}$ is denoted by $H(\hat{\xi})$. To approximate the integral given by (46), the integrand is expanded in a second-order Taylor series about the point $\xi = \hat{\xi}$, and the Laplace approximation gives

$$p(\mathcal{X}|M) = \int \exp\left( \log(\pi(\hat{\xi}|\mathcal{X}, M) - \frac{1}{2}(\Theta - \hat{\xi})^T H(\hat{\xi}) \right.$$
$$\left. \times (\xi - \hat{\xi}) \right) d\xi = p(\mathcal{X}|\hat{\xi}, M)\pi(\hat{\xi}|M)(2\pi)^{\frac{N_p}{2}} \sqrt{|H(\hat{\xi})|} \tag{47}$$

where $N_p$ is the number of parameters to be estimated and is equal to $(d+2)M$ in our case, and $|H(\hat{\xi})|$ is the determinant of the Hessian matrix. Note that the Laplace approximation is very accurate as shown in [69]. Indeed, the relative error of this approximation, given by

$$\frac{p(\mathscr{X}|M)_{Laplace} - p(\mathscr{X}|M)_{correct}}{p(\mathscr{X}|M)_{correct}} \tag{48}$$

is $O_p(1/N)$. For numerical reasons, it is better to work with the Laplace approximation on the logarithm scale. Taking logarithms, we can rewrite ( 47) as

$$\begin{aligned}\log(p(\mathscr{X}|M)) &= \log(p(\mathscr{X}|\hat{\xi},M)) + \log(\pi(\hat{\xi}|M)) \\ &+ \frac{N_p}{2}\log(2\pi) + \frac{1}{2}\log(|H(\hat{\xi})|)\end{aligned} \tag{49}$$

In order to compute the Laplace approximation, we have to determine $\hat{\xi}$ and $H(\hat{\xi})$. However, in many practical situations an analytic solution is not available. Besides, the computation of $|H(\hat{\xi})|$ is difficult especially for high-dimensional data. Then, we use another efficient approximation [69] which can be deduced from (49) by retaining only the terms which increase linearly with $N$ and we obtain

$$\log(p(\mathscr{X}|M)) = \log(p(\mathscr{X}|\hat{\xi},M)) - \frac{N_p}{2}\log(N) \tag{50}$$

which is the Bayesian information criterion (BIC) [20]. Having (50) in hand, the number of components in the mixture model is taken to be $\{M/\log(p(\mathscr{X}|M)) = \max_M \log(p(\mathscr{X}|M)), M = M_{min},\ldots,M_{max}\}$.

## 5 Experimental Results

In this section, we experimentally evaluate the performance of the different approaches presented in the previous section on a real data set called Medical Imaging System (MIS) [31]. In the following, we first describe the data set, the metrics used and the experimental methodology, then we give and analyze the experimental results.

### 5.1 *The MIS Data Set, Metrics and the Experimental Methodology*

MIS is a widely used commercial software system consisting of about 4500 routines written in approximate 400,000 lines of Pascal, FORTRAN, and PL/M assembly code. The practical number of changes (faults) as well as 11 software complexity metrics of each module in this program were determined during three-years system testing and maintenance. Basically, the MIS data set used in this paper, is composed of 390 modules and each module is described by 11 complexity metrics acting as variables:

- LOC is the number of lines of code, including comments.
- CL is the number of lines of code, excluding comments.
- TChar is the number of characters
- TComm is the number of comments.
- MChar is the number of comment characters.
- DChar is the number of code characters
- $N = N_1 + N_2$ is the program length, where $N_1$ is the total number of operators and $N_2$ is the total number of operands.
- $\hat{N} = \eta_1 \log_2 \eta_1 + \eta_2 \log_2 \eta_2$ is an estimated program length, where $\eta_1$ is the number of unique operators and $\eta_2$ is the number of unique operands.
- $N_F = (\log_2 \eta_1)! + (\log_2 \eta_2)!$ is Jensen's [24] estimator of program length.
- V(G), McCabe's cyclomatic number, is one more than the number of decision nodes in the control flow graph.
- *BW* is Belady's bandwidth metric, where

$$BW = \frac{1}{n} \sum_i iL_i \tag{51}$$

  and $L_i$ represents the number of nodes at level $i$ in a nested control flow graph of $n$ nodes [24]. This metric indicates the average level of nesting or width of the control flow graph representation of the program.

Figure 1 shows the number of faults found in the software as a function of the different complexity metrics. According to this figure, it is clear that the number of changes (or faults) increases as the metrics values increase.

In documented MIS data set, modules 1 to 114 are regarded as non fault-prone (number of faults less than 2), and those with 10 to 98 faults are considered to be fault-prone. Thus, there are 114 non fault-prone and 89 fault-prone modules.

Resampling is an often used technique to test classification algorithms by generating training and test sets. The training set is used to build the software quality prediction model, and the test set is used to validate the predictive accuracy of the model. In our experiments, we have used $k$-fold cross validation where original data sets are divided into $k$ subsamples of approximately equal size. Each time one of the $k$ subsamples is selected as test data set to validate the model, and the remaining $k-1$ subsamples acts as training data sets. Then, the process is repeated $k$ times, with each of the $k$ subsamples used exactly once as test data set. The $k$ results are averaged to produce a misclassification error. Our specific resampling choice was 10-fold cross validation. In the case of our problem, there are two types of misclassification, type I and type II. Type I misclassification occurs when a non fault-prone module is wrongly classified as fault-prone and type II misclassification occurs when a fault-prone modules is mistakenly classified as non fault-prone. In our experiments, type I and type II misclassification rates are used as the measure of effectiveness and efficiency to compare the different selected classification algorithms.

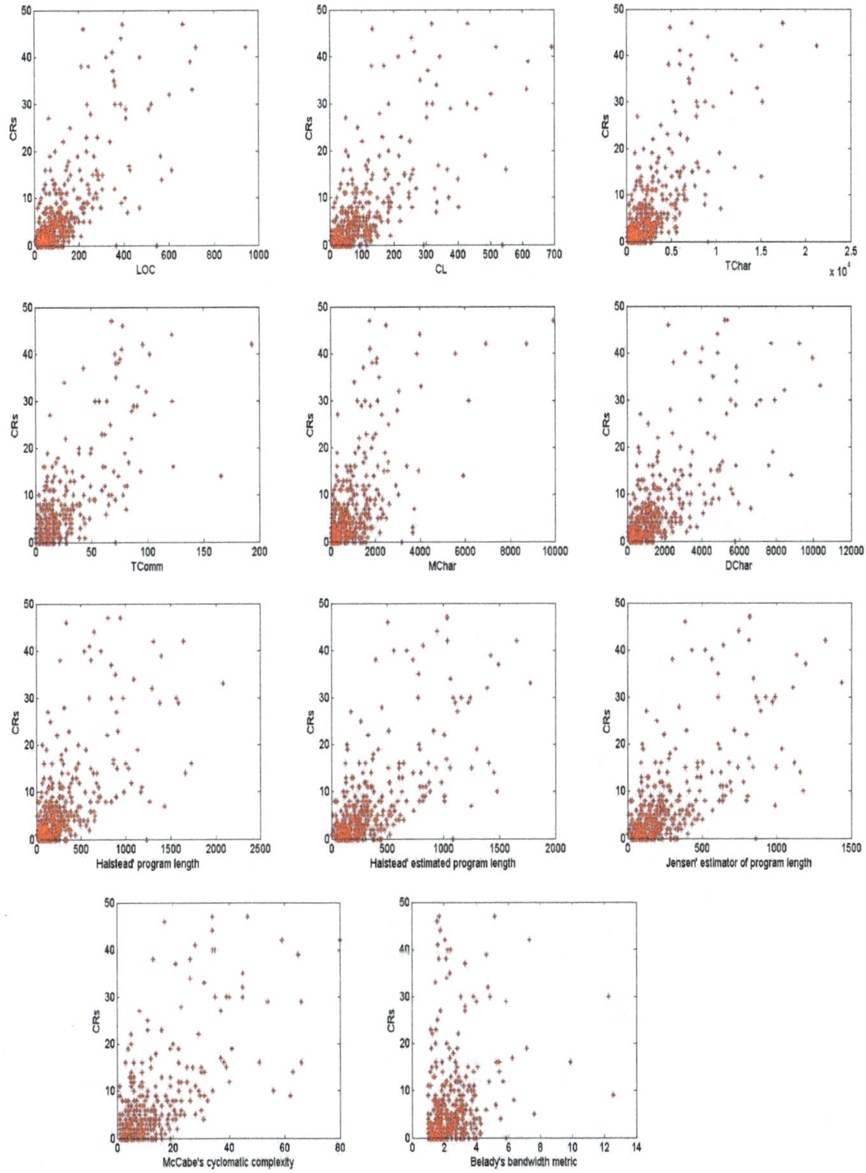


**Fig. 1** The relationship between the metrics and number of CRs

In order to assess the statistical significance of the different results achieved by supervised algorithms, we have used Student's $t$ test; and for unsupervised one (i.e finite mixture model), a test for the difference of two proportions has been employed [86]. To conduct Student's $t$ test, let $p_A^{(i)}$ be the misclassification rate of test data set $i$ ($i$ from 1 to 10) by algorithm A, and $p_B^{(i)}$ represents the same meaning. If we suppose 10 differences $p^{(i)} = p_A^{(i)} - p_B^{(i)}$ are achieved independently, then we can use Student's $t$ test to compute the statistic $t = \bar{p}\sqrt{n}/\sqrt{\frac{\sum_{i=1}^{n}(p^{(i)}-\bar{p})^2}{n-1}}$, where $n$=10 and $\bar{p} = \frac{1}{n}\sum_{i=1}^{n} p^{(i)}$. Using the null hypothesis, this Student's distribution has 9 ($n$-1) degrees of freedom. In this case, the null hypothesis can be rejected if $|t| > t_{9,0.975} = 2.262$. To compare the results achieved by unsupervised algorithms, we adopt another statistical test to measure the difference. Let $p_A$ represents the proportion of misclassified modules by algorithm A, so does $p_B$. Suppose $p_A$ and $p_B$ are normally distributed, so that their quantity of difference $(p^A - p^B)$ is normally distributed as well. The null hypothesis is rejected if $|z| = |(p^A - p^B)/\sqrt{2p(1-p)/n}| > Z_{0.975} = 1.96$, where $p = (p^A + p^B)/2$.

## 5.2 *Experimental Results and Analysis*

### 5.2.1 PCA Results

As a first step in our experiments, we have applied PCA to the MIS data set. Table 1 shows highest five eigenvalues as well as their corresponding eigenvectors and they express 98.57% of the features of the datasets in all. The columns from domain 1 to domain 5 are the principal component scores. According to this table, we can see that the first two largest eigenvalues express up to 90.8% information of the original dataset and then could be considered as comprehensive to some extent to describe the MIS dataset. Fig. 2 shows the PCA results by considering the first two components. Each of the eleven predictors is represented in this figure by a vector, the direction and length of which denote how much contribution to the two principal components the predictor provides. The first principal component, represented by the horizontal axis, has positive coefficients for all components. The second principal component, represented by the vertical axis, has positive coefficients for the components $BW', V(G)$, almost no coefficients for the components $DChar, N, \hat{N}, NF$, and negative coefficients for the remaining five. Note that in Fig. 2, the components $BW', V(G)$ and $MChar$ are standing out, which indicate that they have less correlation with other indicators. On the contrary, the indicators $DChar, N, \hat{N}$, and $NF$ are highly correlated.

### 5.2.2 Classification Result

In this subsection, we present the results obtained using the different classification approaches that we have presented in the previous section. Table 2 shows these results with and without PCA pretreatment. In this table, type I and type II errors, and the accuracy rates which represent the ratio of the corrective classification modules

**Table 1** Principal Components Analysis for MIS

| Complexity Matrix | Domain 1 | Domain 2 | Domain 3 | Domain 4 | Domain 5 |
|---|---|---|---|---|---|
| LOC | 0.3205 | 0.0903 | -0.0526 | -0.2928 | -0.4016 |
| CL | 0.3159 | 0.0270 | 0.1029 | 0.3481 | -0.5452 |
| TChar | 0.3226 | 0.1287 | -0.1794 | 0.1792 | 0.1572 |
| TComm | 0.2992 | 0.1577 | -0.2484 | -0.6689 | -0.0144 |
| MChar | 0.2729 | 0.2911 | -0.6850 | 0.2304 | 0.2915 |
| DChar | 0.3230 | 0.0191 | 0.2246 | 0.0319 | 0.0341 |
| N | 0.3176 | 0.0056 | 0.2785 | -0.0172 | 0.0772 |
| $\hat{N}$ | 0.3167 | 0.0092 | 0.3312 | -0.0078 | 0.3617 |
| NF | 0.3166 | 0.0120 | 0.3358 | 0.0007 | 0.3589 |
| V(G) | 0.3052 | -0.2011 | -0.0763 | -0.4917 | -0.3771 |
| BW | 0.1751 | -0.9077 | -0.2593 | 0.1319 | 0.1502 |
| Eigenvalue | 9.1653 | 0.8224 | 0.4662 | 0.2330 | 0.1546 |
| %Variance | 83.32 | 7.48 | 4.24 | 2.12 | 1.41 |
| %Cumulative | 83.32 | 90.8 | 95.04 | 97.16 | 98.57 |

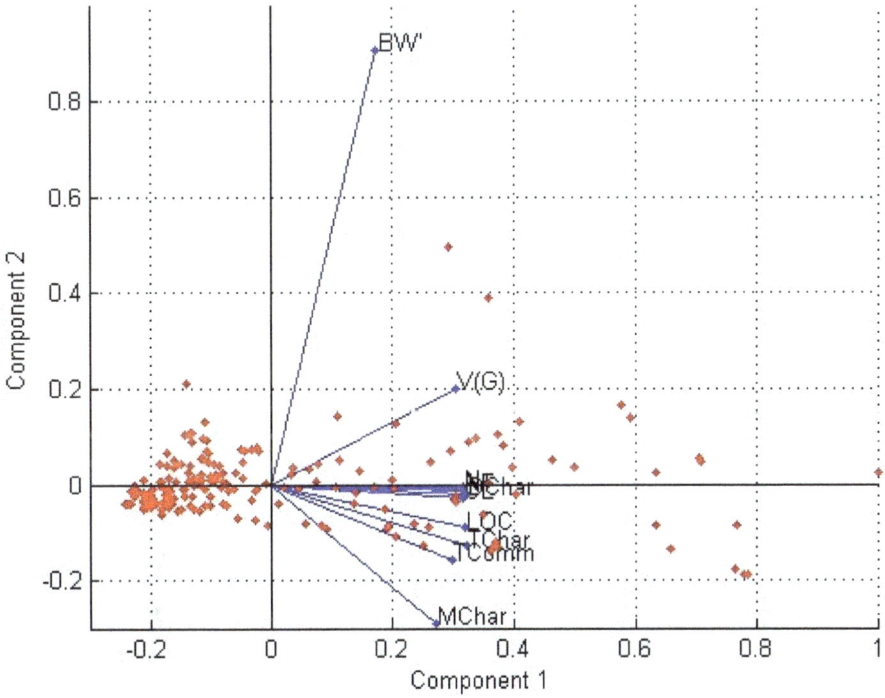


**Fig. 2** Two-dimensional plot of variable coefficients and scores

to the total, are employed as the indicators to compare the overall classification capabilities. Comparing the different approaches using the accuracy rates, it is clear that the PCA pre-process improves generally the results and that LDA with PCA pretreatment performs best in our case, achieving highest accuracy rate 88.76%.

Table 3 lists the results achieved by using only the first two principal components as input to the selected algorithms except multiple linear regression (it is

**Table 2** Type I and Type II errors, and the accuracy rates using different approaches with and without PCA

| | Type I error | Type II error | Accuracy Rate |
|---|---|---|---|
| LDA | 9.22% | 16.47% | 87.24% |
| LDA + PCA | 6.92% | 17.40% | 88.76% |
| NDA | 1.60% | 30.79% | 85.71% |
| NDA + PCA | 2.43% | 24.12% | 88.17% |
| Logistic Regression | 9.08% | 18.51% | 87.21% |
| Logistic Regression + PCA | 6.43% | 20.43% | 88.14% |
| Multiple Linear Regression | 22.37% | 10.64% | 82.69% |
| Multiple Linear Regression + PCA | 12.04 % | 13.08% | 85.69% |
| SVM (Polynomial) | 10.67% | 28.11% | 81.59% |
| SVM (Polynomial) + PCA | 12.94% | 24.61% | 81.76% |
| SVM (RBF) | 14.33 | 27.22 | 79.95% |
| SVM (RBF) + PCA | 13.14% | 20.55% | 82.88% |
| SVM (Sigmoid) | 30.25% | 26.63% | 70.88% |
| SVM (Sigmoid) + PCA | 33.10% | 25.31% | 72.28% |
| Gaussian Mixture Model | 1.75% | 41.57% | 80.78% |
| Gaussian Mixture Model + PCA | 1.75% | 41.57% | 80.78% |

**Table 3** Type I and Type II errors by processing first two principal components

| | Type I error | Type II error | Accuracy Rate |
|---|---|---|---|
| LDA | 1.77% | 37.53% | 82.81% |
| LDA + PCA | 1.55% | 36.37% | 83.23% |
| NDA | 2.25% | 38.83% | 82.85% |
| NDA + PCA | 2.25% | 38.00% | 83.36% |
| Logistic Regression | 5.63% | 22.27% | 86.67% |
| Logistic Regression + PCA | 3.76% | 24.48% | 87.09% |
| SVM (Polynomial) | 9.09% | 88.17% | 60.54% |
| SVM (Polynomial) + PCA | 13.49% | 23.63% | 81.74% |
| SVM (RBF) | 24.17% | 18.23% | 78.38% |
| SVM (RBF) + PCA | 16.46% | 19.09% | 82.26% |
| SVM (Sigmoid) | 71.73% | 35.32% | 45.28% |
| SVM (Sigmoid) + PCA | 51.27% | 31.69% | 58.24% |
| Gaussian Mixture Model | 20.17% | 15.73% | 81.77% |
| Gaussian Mixture Model + PCA | 20.17% | 15.73% | 81.77% |

inappropriate to evaluate multiple linear regression by only two predictors). By comparing these results with the results shown in table 2, it is clear that in most of the cases, the results are better when we consider all principal components. The only exception is the results achieved by Gaussian finite mixture model. When tracking the intermediate variables, it occurs that, for each module, the two procedures with or without PCA pretreatment, respectively, arrive at the same probability of being fault-prone, as well as being non fault-prone. Table 3 also shows that, Logistic Regression with PCA performs best, orderly followed by NDA with PCA and LDA with PCA. SVM technique still functions here, but when classifying with Sigmoid kernel function, the accuracy rate decreases a lot.

Tables from 4 to 8 show the absolute value results of Student's $t$ test when using different approaches with and without PCA. The statistical significance tests are conducted in order to make extensive comparisons under various circumstances. Tables 4 and 5 show comparisons between the different classification methods

**Table 4** Absolute-value results of resampled paired $t$ test with 11 software complexity metrics without PCA pretreatment

| | $t$ test (type I error) | $t$ test (type II error) |
|---|---|---|
| | vs. LDA | vs. LDA |
| NDA | 2.0775 | 2.0360 |
| Logistic Regression | 0.0358 | 0.2456 |
| Multiple Linear Regression | 2.0962 | 1.1881 |
| SVM (Polynomial) | 0.3233 | 1.3729 |
| SVM (RBF) | 1.6156 | 1.1884 |
| SVM (Sigmoid) | 3.8010 | 1.2808 |
| | vs. NDA | vs. NDA |
| Logistic Regression | 1.9266 | 5.7418 |
| Multiple Linear Regression | 3.9220 | 3.7158 |
| SVM (Polynomial) | 3.2742 | 0.4011 |
| SVM (RBF) | 3.8099 | 0.8718 |
| SVM (Sigmoid) | 4.9427 | 0.6865 |
| | vs. Logistic Regression | vs. Logistic Regression |
| Multiple Linear Regression | 1.7345 | 1.1734 |
| SVM (Polynomial) | 0.4001 | 1.1290 |
| SVM (RBF) | 1.2063 | 2.1146 |
| SVM (Sigmoid) | 4.3904 | 1.1760 |
| | vs. Multiple Linear Regression | vs. Multiple Linear Regression |
| SVM (Polynomial) | 1.7914 | 2.7224 |
| SVM (RBF) | 1.3103 | 2.5823 |
| SVM (Sigmoid) | 0.8530 | 2.4620 |
| | vs. SVM (Polynomial) | vs. SVM (Polynomial) |
| SVM (RBF) | 0.7176 | 0.1114 |
| SVM (Sigmoid) | 3.6469 | 0.2349 |
| | vs. SVM (RBF) | vs. SVM (RBF) |
| SVM (Sigmoid) | 2.4779 | 0.0879 |

**Table 5** Absolute-value results of resampled paired *t* test with PCA pretreatment

| | *t* test (type I error) | *t* test (type II error) |
|---|---|---|
| | vs. LDA | vs. LDA |
| NDA | 1.6656 | 1.3882 |
| Logistic Regression | 0.1304 | 0.5349 |
| Multiple Linear Regression | 1.0977 | 1.0609 |
| SVM (Polynomial) | 1.1543 | 1.9027 |
| SVM (RBF) | 1.1567 | 0.5309 |
| SVM (Sigmoid) | 2.6381 | 1.4429 |
| | vs. NDA | vs. NDA |
| Logistic Regression | 1.3731 | 0.5315 |
| Multiple Linear Regression | 2.2245 | 3.4619 |
| SVM (Polynomial) | 2.7321 | 0.0803 |
| SVM (RBF) | 3.0483 | 0.6148 |
| SVM (Sigmoid) | 2.9724 | 0.2211 |
| | vs. Logistic Regression | vs. Logistic Regression |
| Multiple Linear Regression | 1.3211 | 1.3369 |
| SVM (Polynomial) | 1.7492 | 0.7191 |
| SVM (RBF) | 0.0405 | 0.1845 |
| SVM (Sigmoid) | 0.0405 | 0.1845 |
| | vs. Multiple Linear Regression | vs. Multiple Linear Regression |
| SVM (Polynomial) | 0.6748 | 2.0658 |
| SVM (RBF) | 1.6699 | 0.0130 |
| SVM (Sigmoid) | 2.8554 | 0.5899 |
| | vs. SVM (Polynomial) | vs. SVM (Polynomial) |
| SVM (RBF) | 0.0452 | 0.5443 |
| SVM (Sigmoid) | 2.3825 | 0.0881 |
| | vs. SVM (RBF) | vs. SVM (RBF) |
| SVM (Sigmoid) | 1.9211 | 0.7198 |

using the total eleven software complexity metrics with and without PCA, respectively. Table 6 shows also cross-comparisons, but with the first two significant principal components. In table 7 and 8, we investigate the statistical significance of the difference between the results achieved by each approach when we apply it with and without PCA by considering all the principal components and the first two most important components, respectively. The results in these four tables are computed using the outputs of every two algorithms, and any absolute value larger than $t_{9,0.975} = 2.262$ represents a statistical difference. The inspection of these two tables reveals, that on the one hand the disparity do exist between some of the algorithms, being wise to select simpler algorithm if the classification accuracy is not significantly different between the two; on the other hand we must point out that using merely the evaluation results with MIS data sets to measure the candidate algorithms is inappropriate to reach an absolute conclusion about the performance of the different approaches. Recent studies show that some factors seriously affect the performance of the classification algorithms [44]. Data set characteristics and training

**Table 6** Absolute-value results of resampled paired t test with PCA pretreatment and using the first two significant principal components

| | *t* test (type I error) | *t* test (type II error) |
|---|---|---|
| | vs. LDA | vs. LDA |
| NDA | 1.6656 | 1.3882 |
| Logistic Regression | 1.6560 | 2.2456 |
| SVM (Polynomial) | 3.3642 | 1.0491 |
| SVM (RBF) | 3.0496 | 1.0105 |
| SVM (Sigmoid) | 4.2935 | 0.4204 |
| | vs. NDA | vs. NDA |
| Logistic Regression | 1.1709 | 3.1772 |
| SVM (Polynomial) | 3.2828 | 1.2730 |
| SVM (RBF) | 3.0664 | 1.3567 |
| SVM (Sigmoid) | 4.0206 | 0.4859 |
| | vs. Logistic Regression | vs. Logistic Regression |
| SVM (Polynomial) | 3.3619 | 0.1842 |
| SVM (RBF) | 2.6313 | 0.2801 |
| SVM (Sigmoid) | 3.6404 | 0.7881 |
| | vs. SVM (Polynomial) | vs. SVM (Polynomial) |
| SVM (RBF) | 0.3338 | 0.0288 |
| SVM (Sigmoid) | 0.0525 | 0.6106 |
| | vs. SVM (RBF) | vs. SVM (RBF) |
| SVM (Sigmoid) | 0.3903 | 0.7060 |

**Table 7** Absolute-value results of resampled paired t test with 11 software complexity metrics with and without PCA-pretreatment

| | t test(type I error) | t test(type II error) |
|---|---|---|
| LDA | 0.5167 | 0.1782 |
| NDA | 1.0000 | 2.6938 |
| Logistic Regression | 0.5295 | 0.2353 |
| Multiple Linear Regression | 3.1840 | 0.6866 |
| SVM (Polynomial) | 0.5374 | 0.3915 |
| SVM (RBF) | 0.2134 | 1.0780 |
| SVM (Sigmoid) | 0.2854 | 0.2130 |

**Table 8** Absolute-value results of resampled paired t test with 11 and 2 software complexity metrics

| | t test(type I error) | t test(type II error) |
|---|---|---|
| LDA | 2.2048 | 2.4728 |
| NDA | 0.1329 | 2.0252 |
| Logistic Regression | 0.2088 | 0.3632 |
| SVM (Polynomial) | 2.7016 | 0.0303 |
| SVM (RBF) | 2.7200 | 0.4772 |
| SVM (Sigmoid) | 1.5541 | 0.7272 |

data set size are dominating factors. According to some empirical research,"best" prediction technique, depends on the context or data set characteristics. For example, generally LDA outperforms for data sets coming from Gaussian Distribution or with some outliers. Moreover, increasing the size of training data sets is always welcomed and improve the prediction results.

To sum up what we mentioned above, even if all the candidate classification algorithms have certain ability to partition data sets, choosing proper classifier strongly depends on the data sets characteristics and the comparative advantages of each classifier. LDA is more suitable for data sets following Gaussian distribution and with unequal within-class proportion. The essence of this algorithm is trying to find the linear combination of the predictors which most separate the two populations, by maximizing the between-class variance, and at same time minimizing the within-class variance. However, to analyze non-Gaussian data which are not linearly related or without common covariance within all groups, the logistic regression is preferred. But, logistic regression has its own underlying assumptions and inherent restrictions. In empirical applications, logistic regression is better for discrete outcomes. Besides, under the circumstances of continuous responses, multiple regression is more powerful. As logistic regression NDA has no requirement concerning the distribution of data. Multiple Linear Regression is very effective when dealing with small quantity of independent variables, but easily being affected by outliers, so identifying and removing these outliers before building the model is necessary. The linear regression model generates unique coefficient parameters to every predictor. So, if the undesired outliers exist in training data sets, the built model with these fixed parameters can not provide reliable prediction for the software modules to test. In this context, other methods robust to the contamination of data (e.g. robust statistics) should be used. Regarding SVM, it transfers data sets into another high-dimensional feature space by means of kernel function, and finds support vectors as the determinant boundary to separate data set. Because the classification is achieved by maximizing the margin between the two classes, that process maximizes the generalization ability of this learning machine, which will not be deteriorated even if the data are somewhat changed within their original range. However, a drawback of SVM is that the kernel function computation is time-consuming. Finite mixture model is an unsupervised statistical approach which permits the partition of data without the training procedure. This technique should be favored when historical data is very costly or hard to collect.

We have also tested our finite Dirichlet mixture model estimated by both the Bayesian approach and a deterministic one based on EM algorithm. To monitor the convergence of the Bayesian algorithm, we run 5 parallel chains of 9000 iterations each. The values of the multiple potential scale reduction factor are shown in Fig. 3. According to this figure convergence occurs around the 7500 iteration. Table 9 shows the values of Type I and Type II errors when using both the deterministic and Bayesian algorithms. According to this table the two approaches give the same type I error which corresponds to 4 misclassified modules. The Bayesian approach outperforms the deterministic one in the case of type II error. Table 10 shows the classification probabilities, of the 4 misclassified modules causing Type I

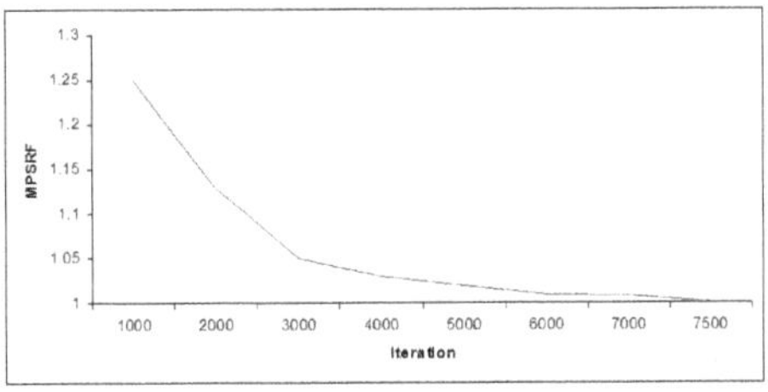


**Fig. 3** Plot of multiple potential scale reduction factor values

**Table 9** Type I and Type II errors using both the deterministic and Bayesian approaches

| | Type I error | Type II error |
|---|---|---|
| Maximum Likelihood | 3.51% | 28.08% |
| Bayesian | 3.51% | 26.96% |

errors. From this table, we can see clearly that the Bayesian approach has increased estimated probabilities, associated with the misclassified data samples of belonging to the correct class.

**Table 10** Classification probabilities (probabilities to be in the non fault-prone class) of the misclassified modules causing type I errors

| Module Number | Bayesian | Maximum Likelihood |
|---|---|---|
| 6 | 0.31 | 0.27 |
| 41 | 0.34 | 0.29 |
| 69 | 0.41 | 0.42 |
| 80 | 0.37 | 0.32 |

## 6 Conclusion

In this paper, we have shown that learning to classify software modules is a fundamental problem in software engineering and has been attacked using different approaches. Software quality prediction models can point to 'hot spots' modules that are likely to have a high error rate or that need high development effort and further attention. In this paper we studied several approaches for software modules classification. Our study consisted of a detailed experimental evaluation. In general, it is difficult to say with conviction, which algorithm is better than the others and judgement is generally subjective and data set-dependent. In fact, in any classification problem there are many issues that should be taken into account and on which classification algorithms can be compared. Besides, all the approaches that we have presented have success stories in a variety of software engineering tasks. Despite the success of these approaches, same problems still exists such as the choice of the number of metrics to describe a given module. Indeed, the description of the modules may include attributes based on subjective judgements which may give rise to errors in the values of the metrics, so we should select the most relevant attributes to build our classification models. Indeed, most results obtained for the software modules categorization problem start by assuming that the data are observed with no noise which is not justified by the reality of software engineering. Besides, the collection of historical modules used for training may include some modules for which an incorrect classification was made. Another important problem is the lack of sufficient historical data for learning in some cases.

**Acknowledgements.** The completion of this research was made possible thanks to the Natural Sciences and Engineering Research Council of Canada (NSERC), a NATEQ Nouveaux Chercheurs Grant, and a start-up grant from Concordia University.

## References

1. Porter, A.A., Selby, R.W.: Empirically guided software development using metric-based classification trees. IEEE Software 7(2), 46–54 (1990)
2. Mayer, A., Sykes, A.M.: Statistical Methods for the Analysis of Software Metrics Data. Software Quality Journal 1(4), 209–223 (1992)
3. Narayanan, A.: A Note on Parameter Estimation in the Multivariate Beta Distribution. Computer Mathematics and Applications 24(10), 11–17 (1992)
4. Curtis, B., Sheprad, S.B., Milliman, H., Borst, M.A., Love, T.: Measuring the Psychlogical Complexity of Software Maintenance Tasks with the Halstead and McCabe Metrics. IEEE Transactions on Software Engineering SE-5(2), 96–104 (1979)
5. Boehm, B.W., Papaccio, P.N.: Understanding and Controlling Software Costs. IEEE Transactions on Software Engineering 14(10), 1462–1477 (1988)
6. Ripley, B.D.: Pattern Recognition and Neural Networks. Cambridge University Press, Cambridge (1996)
7. Ebert, C.: Classification Techniques for Metric-Based Development. Software Quality Journal 5(4), 255–272 (1996)
8. Ebert, C., Baisch, E.: Industrial Application of Criticality Predictions in Software Development. In: Proc. of the 8th IEEE International Symposium on Software Reliability Engineering, pp. 80–89 (1998)
9. Wallace, C.S.: Statistical and Inductive Inference by Minimum Message Length. Springer, Heidelberg (2005)
10. Montgomery, D.C., Peck, E.A., Vining, G.G.: Introduction to Linear Regression Analysis, 3rd edn. Wiley-Interscience, Hoboken (2001)
11. Hosmer, D.W., Lemeshow, S.: Applied Logistic Regression. Wiley-Interscience Publication, Hoboken (2000)
12. Zhang, D., Tsai, J.J.P.: Machine Learning and Software Engineering. Software Quality Journal 11(2), 87–119 (2003)
13. Weyuker, E.J.: Evaluating software complexity measures. IEEE Transactions on Software Engineering 14(9), 1357–1365 (1988)
14. Brooks, F.: No Silver Bullet-Essense and Accidents of Software Engineering. IEEE Computer 20(4), 10–19 (1987)

15. Lanubile, F.: Why Software Reliability Predictions Fail. IEEE Software 13(4), 131–132, 137 (1996)
16. Lanubile, F., Visaggio, G.: Evaluating Predictive Quality Models Derived from Software Measures: Lessons Learned. Journal of Systems and Software 38(3), 225–234 (1997)
17. Xing, F., Guo, P., Lyu, M.R.: A Novel Method for Early Software Quality Prediction Based on Support Vector Machine. In: Proc. of the 16th IEEE International Symposium on Software Reliability Engineering, pp. 213–222 (2005)
18. Le Gall, G., Adam, M.-F., Derriennic, H., Moreau, B., Valette, N.: Studies on Measuring Software. IEEE Journal on Selected Areas in Communications 8(2), 234–246 (1990)
19. Ronning, G.: Maximum Likelihood Estimation of Dirichlet Distributions. Journal of Statistical Computation and Simulation 32, 215–221 (1989)
20. Schwarz, G.: Estimating the Dimension of a Model. The Annals of Statistics 6(2), 461–464 (1978)
21. Russel, G.W.: Experience With Inspection in Ultralarge-Scale Developments. IEEE Software 8(1), 25–31 (1991)
22. McLachlan, G.J., Peel, D.: Finite Mixture Models. Wiley, New York (2000)
23. Akaike, H.: A New Look at the Statistical Model Identification. IEEE Transactions on Automatic Control AC-19(6), 716–723 (1974)
24. Jensen, H., Vairavan, K.: An Experimental Study of Software Metrics for Real-time Software. IEEE Transaction on Software Engineering SE-11(4), 231–234 (1994)
25. Zuse, H.: Comments to the Paper: Briand, Eman, Morasca: On the Application of Measurement Theory in Software Engineering. Empirical Software Engineering 2(3), 313–316 (1997)
26. Munson, J.C., Khoshgoftaar, T.M.: The Dimensionality of Program Complexity. In: Proc. of Eleventh International Conference on Software Engineering, pp. 245–253 (1989)
27. Gaffney, J.: Estimating the Number of Faults in Code. IEEE Transactions on Software Engineering 10(4), 459–464 (1984)
28. Henry, J., Henry, S., Kafura, D., Matheson, L.: Improving Software Maintenance at Martin Marietta. IEEE Software 11(4), 67–75 (1994)
29. Mayrand, J., Coallier, F.: System Acquisition Based on Software Product Assessment. In: Proc. of 18th International Conference on Software Engineering, pp. 210–219 (1996)
30. Troster, J., Tian, J.: Measurement and Defect Modeling for a Legacy Software System. Annals of Software Engineering 1(1), 95–118 (1995)
31. Munson, J.C.: Handbook of Software Reliability Engineering. IEEE Computer Society Press/McGraw-Hill Book Company (1999)
32. Munson, J.C., Khoshgoftaar, T.M.: The Detection of Fault-Prone Programs. IEEE Transactions on Software Engineering 18(5), 423–433 (1992)
33. Briand, L., EL Emam, K., Morasca, S.: On the Application of Measurement Theory in Software Engineering. Empirical Software Engineering 1(1), 61–88 (1996)
34. Briand, L.C., Basili, V.R., Hetmanski, C.J.: Developing Interpretable Models with Optimized Set Reduction for Identifying High-Risk Software Components. IEEE Transactions on Software Engineering 19(11), 1028–1044 (1993)
35. Briand, L.C., Basili, V.R., Thomas, W.M.: A Pattern Recognition Approach for Software Engineering Data Analysis. IEEE Transactions on Software Engineering 18(11), 931–942 (1992)
36. Briand, L.C., Thomas, W.M., Hetmanski, C.J.: Modeling and Managing Risk Early in Software Development. In: Proc. of 15th International Conference on Software Engineering, pp. 55–65 (1993)

37. Guo, L., Ma, Y., Cukic, B., Singh, H.: Robust Prediction of Fault-Proneness by Random Forests. In: Proc. of the 15th IEEE International Symposium on Software Reliability Engineering, pp. 417–428 (2004)
38. Ottenstein, L.M.: Quantitative Estimates of Debugging Requirements. IEEE Transactions on Software Engineering SE-5(5), 504–514 (1979)
39. Mark, L., Jeff, K.: Object-Oriented Software Metrics. Prentice-Hall, Englewood Cliffs (1994)
40. Ohlsson, M.C., Wohlin, C.: Identification of Green, Yellow and Red Legacy Components. In: Proc. of the International Conference on Software Maintenance, pp. 6–15 (1998)
41. Ohlsson, M.C., Runeson, P.: Experience from Replicating Empirical Studies on Prediction Models. In: Proc. of the Eighth IEEE Symposium on Software Metrics, pp. 217–226 (2002)
42. Halstead, M.H., Leroy, A.M.: Elements of Software Science. Elseviser, New York (1977)
43. Hitz, M., Montazeri, B.: Chidamber and Kemerer's Metrics Suite: A Measurement Theory Perspective. IEEE Transactions on Software Engineering 22(4), 267–271 (1996)
44. Shepperd, M., Kadoda, G.: Comparing Software Prediction Techniques Using Simulation. IEEE Transactions on Software Engineering 27(11), 1014–1022 (2001)
45. Bouguila, N., Ziou, D.: Unsupervised Selection of a Finite Dirichlet Mixture Model: An MML-Based Approach. IEEE Transactions on Knowledge and Data Engineering 18(8), 993–1009 (2006)
46. Bouguila, N., Ziou, D.: Unsupervised Learning of a Finite Discrete Mixture: Applications to Texture Modeling and Image Databases Summarization. Journal of Visual Communication and Image Representation 18(4), 295–309 (2007)
47. Bouguila, N., Ziou, D., Monga, E.: Practical Bayesian Estimation of a Finite Beta Mixture Through Gibbs Sampling and its Applications. Statistics and Computing 16(2), 215–225 (2006)
48. Bouguila, N., Ziou, D., Vaillancourt, J.: Novel Mixtures Based on the Dirichlet Distribution: Application to Data and Image Classification. In: Perner, P., Rosenfeld, A. (eds.) MLDM 2003. LNCS (LNAI), vol. 2734, pp. 172–181. Springer, Heidelberg (2003)
49. Bouguila, N., Ziou, D., Vaillancourt, J.: Unsupervised Learning of a Finite Mixture Model Based on the Dirichlet Distribution and its Application. IEEE Transactions on Image Processing 13(11), 1533–1543 (2004)
50. Bouguila, N., Wang, J.H., Ben Hamza, A.: A Bayesian Approach for Software Quality Prediction. In: Proc. of the IEEE International Conference on Intelligent Systems, pp. 49–54 (2008)
51. Schneidewind, N.F.: Validating Software Metrics: Producing Quality Discriminators. In: Proc. of Second International Symposium on Software Reliability Engineering, pp. 225–232 (1991)
52. Schneidewind, N.F.: Methodology For Validating Software Metrics. IEEE Transactions on Software Engineering 18(5), 410–422 (1992)
53. Schneidewind, N.F.: Minimizing risk in applying metrics on multiple projects. In: Proc. of Third International Symposium on Software Reliability Engineering, pp. 173–182 (1992)
54. Schneidewind, N.F.: Software metrics validation: Space Shuttle flight software example. Annals of Software Engineering 1(1), 287–309 (1995)
55. Schneidewind, N.F.: Software metrics model for integrating quality control and prediction. In: Proc. of the Eighth International Symposium on Software Reliability Engineering, pp. 402–415 (1997)

56. Schneidewind, N.F.: Investigation of Logistic Regression as a Discriminant of Software Quality. In: Proc. of the Seventh IEEE Symposium on Software Metrics, pp. 328–337 (2001)
57. Fenton, N.: Software Measurement: A Necessary Scientific Basis. IEEE Transactions on Software Engineering 20(3), 199–206 (1994)
58. Ohlisson, N., Zhao, M., Helander, M.: Application of Multivariate Analysis for Software Fault Prediction. Software Quality Journal 7(1), 51–66 (1998)
59. Ohlsson, N., Alberg, H.: Predicting Fault-Prone Software Modules in Telephone Switches. IEEE Transactions on Software Engineering 22(12), 886–894 (1996)
60. Congdon, P.: Applied Bayesian Modelling. John Wiley and Sons, Chichester (2003)
61. Frankl, P., Hamlet, D., Littlewood, B., Strigini, L.: Evaluating Testing Methods by Delivered Reliability. IEEE Transactions on Software Engineering 24(8), 586–601 (1998)
62. Guo, P., Lyu, M.R.: Software Quality Prediction Using Mixture Models with EM Algorithm. In: Proc. First Asia-Pacific Conference on Quality Software, pp. 69–78 (2000)
63. Szabo, R.M., Khoshgoftaar, T.M.: An assessment of software quality in a C++ environment. In: Proc. of the Sixth International Symposium on Software Reliability Engineering, pp. 240–249 (1995)
64. Duda, R.O., Hart, P.E., Stork, D.G.: Pattern Classification. Wiley, New York (2001)
65. Pressman, R.S.: Software Engineering: A Practioner's Approach, 5th edn. McGraw-Hill, New York (2001)
66. Takahashi, R., Muraoka, Y., Nakamura, Y.: Building Software Quality Classification Trees: Approach, Experimentation, Evaluation. In: Proc. of the 8th IEEE International Symposium on Software Reliability Engineering, pp. 222–233 (1997)
67. Selby, R.W.: Empirically based analysis of failures in software systems. IEEE Transactions on Reliability 39(4), 444–454 (1990)
68. Selby, R.W., Porter, A.A.: Learning From Examples: Generation and Evaluation of Decision Trees for Software Ressource Analysis. IEEE Transactions on Software Engineering 14(12), 1743–1757 (1988)
69. Kass, R.E., Raftery, A.E.: Bayes Factors. Journal of the American Statistical Association 90, 773–795 (1995)
70. Rissanen, J.: Modeling by Shortest Data Description. Automatica 14, 465–471 (1978)
71. Biyani, S., Santhanam, P.: Exploring Defect Data from Development and Customer Usage on Software Modules over Multiple Releases. In: Proc. of the 8th IEEE International Symposium on Software Reliability Engineering, pp. 316–320 (1998)
72. Conte, S.D.: Metrics and Models in Software Quality Engineering. Addison-Wesley Professional, Reading (1996)
73. Crawford, S.G., McIntosh, A.A., Pregibon, D.: An Analysis of Static Metrics and Faults in C Software. Journal of Systems and Software 15(1), 37–48 (1985)
74. Stockman, S.G., Todd, A.R., Robinson, G.A.: A Framework for Software Quality Measurement. IEEE Journal on Selected Areas in Communications 8(2), 224–233 (1990)
75. Henry, S., Wake, S.: Predicting maintainability with software quality metrics. Journal of Software Maintenance: Research and Practice 3(3), 129–143 (1991)
76. Pfleeger, S.L.: Lessons Learned in Building a Corporate Metrics Program. IEEE Software 10(3), 67–74 (1993)
77. Pfleeger, S.L., Fitzgerald, J.C., Rippy, D.A.: Using multiple metrics for analysis of improvement. Software Quality Journal 1(1), 27–36 (1992)
78. Chidamber, S.R., Kemerer, C.F.: A Metrics Suite for Object-Oriented Design. IEEE Transactions on Software Engineering 20(6), 476–493 (1994)
79. Gokhale, S.S., Lyu, M.R.: Regression Tree Modeling for the Prediction of Software Quality. In: Proc. of the third ISSAT International Conference on Reliability and Quality in Design, pp. 31–36 (1997)

80. Khoshgoftaar, T.M., Allen, E.B.: Early Quality Prediction: A Case Study in Telecommunications. IEEE Software 13(4), 65–71 (1996)
81. Khoshgoftaar, T.M., Lanning, D.L., Pandya, A.S.: A Comparative Study of Pattern Recognition Techniques for Quality Evaluation of Telecommunications Software. IEEE Journal on Selected Areas in Communications 12(2), 279–291 (1994)
82. Khoshgoftaar, T.M., Allen, E.B., Jones, W.D., Hudepohl, J.P.: Return on Investment of Software Quality Predictions. In: Proc. of the IEEE Workshop on Application-Specific Software Engineering Technology, pp. 145–150 (1998)
83. Khoshgoftaar, T.M., Geleyn, E., Nguyen, L.: Empirical Case Studies of Combining Software Quality Classification Models. In: Proc. of the Third International Conference on Quality Software, pp. 40–49 (2003)
84. Khoshgoftaar, T.M., Munson, J.C., Lanning, D.L.: A comparative Study of Predictive Models for Program Changes During System Testing and Maintenance. In: Proc. of the IEEE Conference on Software Maintenance, pp. 72–79 (1993)
85. Khoshgoftaar, T.M., Munson, J.C., Bhattacharya, B.B., Richardson, G.D.: Predictive Modeling Techniques of Software Quality from Software Measures. IEEE Transactions on Software Engineering 18(11), 979–987 (1992)
86. Dietterich, T.G.: Approximate Statistical Test For Comparing Supervised Classification Learning Algorithms. Neural Computation 10(7), 1895–1923 (1998)
87. McCabe, T.J.: A Complexity Measure. IEEE Transactions on Software Engineering SE-2(4), 308–320 (1976)
88. Khoshgoftaar, T.M., Allen, E.B.: Multivariate Assessment of Complex Software Systems: A comparative Study. In: Proc. of First International Conference on Engineering of Complex Computer Systems, pp. 389–396 (1995)
89. Khoshgoftaar, T.M., Allen, E.B.: The Impact of Costs of Misclassification on Software Quality Modeling. In: Proc. of Fourth International Software Metrics Symposium, pp. 54–62 (1997)
90. Khoshgoftaar, T.M., Allen, E.B.: Classification of Fault-Prone Software Modules: Prior Probabilities, Costs, and Model Evaluation. Empirical Software Engineering 3(3), 275–298 (1998)
91. Khoshgoftaar, T.M., Allen, E.B.: A Comparative Study of Ordering and Classification of Fault-Prone Software Modules. Empirical Software Engineering 4(2), 159–186 (1999)
92. Khoshgoftaar, T.M., Allen, E.B.: Predicting Fault-Prone Software Modules in Embedded Systems with Classification Trees. In: Proc. of High-Assurance Systems Engineering Workshop, pp. 105–112 (1999)
93. Khoshgoftaar, T.M., Allen, E.B.: Controlling Overfitting in Classification-Tree Models of Software Quality. Empirical Software Engineering 6(1), 59–79 (2001)
94. Khoshgoftaar, T.M., Allen, E.B.: Ordering Fault-Prone Software Modules. Software Quality Journal 11(1), 19–37 (2003)
95. Khoshgoftaar, T.M., Allen, E.B.: A Practical Classification-Rule for Software-Quality Models. IEEE Transactions on Reliability 49(2), 209–216 (2000)
96. Khoshgoftaar, T.M., Munson, J.C.: Predicting Software Development Errors Using Software Complexity Metrics. IEEE Journal on Selected Areas in Communications 8(2), 253–261 (1990)
97. Khoshgoftaar, T.M., Halstead, R.: Process Measures for Predicting Software Quality. In: Proc. of High-Assurance Systems Engineering Workshop, pp. 155–160 (1997)
98. Khoshgoftaar, T.M., Allen, E.B., Goel, N.: The Impact of Software Evolution and Reuse on Software Quality. Empirical Software Engineering 1(1), 31–44 (1996)

99. Khoshgoftaar, T.M., Allen, E.B., Hudepohl, J.P., Aud, S.J.: Applications of Neural Networks to Software Quality Modeling of a Very Large Telecommunications System. IEEE Transactions on Neural Networks 8(4), 902–909 (1997)
100. Khoshgoftaar, T.M., Allen, E.B., Jones, W.D., Hudepohl, J.P.: Which Software Modules have Faults which will be Discovered by Customers? Journal of Software Maintenance: Research and Practice 11, 1–18 (1999)
101. Khoshgoftaar, T.M., Allen, E.B., Jones, W.D., Hudepohl, J.P.: Classification-Tree Models of Software-Quality Over Multiple Release. IEEE Transactions on Reliability 49(1), 4–11 (2000)
102. Khoshgoftaar, T.M., Yuan, X., Allen, E.B.: Balancing Misclassification Rates in Classification-Tree Models of Software Quality. Empirical Software Engineering 5(4), 313–330 (2000)
103. Khoshgoftaar, T.M., Yuan, X., Allen, E.B., Jones, W.D., Hudepohl, J.P.: Uncertain Classification of Fault-Prone Software Modules. Empirical Software Engineering 7(1), 297–318 (2002)
104. Vapnik, V.N.: The Nature of Statistical Learning Theory, 2nd edn. Springer, Heidelberg (1999)
105. Basili, V.R., Hutchens, D.H.: An Empirical Study of a Syntactic Complexity Family. IEEE Transactions on Software Engineering SE-9(6), 664–672 (1983)
106. Basili, V.R., Briand, L.C., Melo, W.L.: A Validation of Object-Oriented Design Metrics as Quality Indicators. IEEE Transactions on Software Engineering 22(10), 751–761 (1996)
107. Rodriguez, V., Tsai, W.T.: Evaluation of Software Metrics Using Discriminant Analysis. Information and Software Technology 29(3), 245–251 (1987)
108. Shen, V.Y., Conte, S.D., Dunsmore, H.E.: Software Science Revisited: A Critical Analysis of the Theory and its Empirical Support. IEEE Transactions on Software Engineering SE-9(2), 155–165 (1983)
109. Shen, V.Y., Yu, T.-J., Thebaut, S.M., Paulsen, L.R.: Identifying Error-Prone Software-An Empirical Study. IEEE Transactions on Software Engineering 11(4), 317–324 (1985)
110. Li, W., Henry, S.: Object-Oriented Metrics that Predict Maintainability. Journal of Systems and Software 23(2), 111–122 (1993)
111. Evanco, W.M., Agresti, W.M.: A Composite Complexity Approach for Software Defect Modeling. Software Quality Journal 3(1), 27–44 (1994)
112. Dillon, W.R., Goldstein, M.: Multivariate Analysis. Wiley, New York (1984)

# A System Approach to Agent Negotiation and Learning

František Čapkovič and Vladimir Jotsov

**Abstract.** The Petri nets (PN)-based analytical approach to describing the agent behaviour in MAS (multi agent systems) is presented. PN yield the possibility to express the MAS behaviour by means of the vector state equation in the form of linear discrete system. Thus, the modular approach to the creation of the MAS model can be successfully used too. Three different interconnections of modules (agents, interfaces, environment) expressed by PN subnets are introduced. Special attention is paid to conflict resolution and machine learning methods based on modelling tools, detection and resolution of conflicts. The approach makes possible to use methods of linear algebra. In addition, it can be successfully used at the system analysis (e.g. the reachability of states), at testing the system properties, and even at the system control synthesis.

**Keywords:** Agent, Petri Net, Negotiation, Conflict Resolution, Machine Learning.

## 1 Introduction

Agents are usually understood (Fonseca et al. 2001) to be persistent (software, but not only software) entities that can perceive, reason, and act in their environment and communicate with other agents. Multi agent systems (MAS) can be apprehended as a composition of collaborative agents working in shared environment. The agents together perform a more complex functionality. Communication enables the agents in MAS to exchange information. Thus, the agents can coordinate their actions and cooperate with each other. The agent behaviour has both internal

František Čapkovič
Institute of Informatics, Slovak Academy of Sciences, Dúbravská cesta 9, 845 07
Bratislava, Slovak Republic, Tel.: +421 2 59411244; Fax: +421 2 54773271
e-mail: Frantisek.Capkovic@savba.sk

Vladimir Jotsov
Institute of Information Technologies, Bulgarian Academy of Sciences,
P.O. Box 161, Sofia, Bulgaria, Tel.: +359 898828678
e-mail: jotsov@ieee.org

V. Sgurev et al. (Eds.): Intelligent Systems: From Theory to Practice, SCI 299, pp. 133–160.
springerlink.com 

and external attributes. From the external point of view the agent is (Demazeau 2003) a real or virtual entity that (i) evolves in an environment; (ii) is able to perceive this environment; (iii) is able to act in this environment; (iv) is able to communicate with other agents; (v) exhibits an autonomous behaviour. From the internal point of view the agent is a real or virtual entity that encompasses some local control in some of its perception, communication, knowledge acquisition, reasoning, decision, execution, and action processes. While the internal attributes characterize rather the agent inherent abilities, different external attributes of agents are manifested themselves in different measures in a rather wide spectrum of MAS applications - like e.g. computer-aided design, decision support, manufacturing systems, robotics and control, traffic management, network monitoring, telecommunications, e-commerce, enterprise modelling, society simulation, office and home automation, etc. It is necessary to distinguish two groups of agents and/or agent societies, namely, human and artificial. The principle difference between them consists especially in the different internal abilities. These abilities are studied by many branches of sciences including those finding themselves out of the technical branches - e.g. economy, sociology, psychology, etc. MAS are also used in intelligent control, especially for a cooperative problem solving (Yen et al. 2001). In order to describe the behaviour of discrete event systems (DES) Petri nets (PN) (Peterson 1981, Murata 1989) are widely used - in flexible manufacturing systems, communication systems of different kinds, transport systems, etc. Agents and MAS can be (observing their behaviour) understood to be a kind of DES. PN yield both the graphical model and the mathematical one, they have a formal semantics. There are many techniques for proving their basic properties (Peterson 1981, Murata 1989) like reachability, liveness, boundedness, conservativeness, reversibility, coverability, persistence, fairness, etc. Consequently, PN represent the enough general means to be able to model a wide class of systems. These arguments are the same like those used in DES modelling in general in order to prefer PN to other approaches. In addition, there were developed many methods in PN-theory that are very useful at model checking - e.g. like the methods of the deadlocks avoidance, methods for computing P (place)-invariants and T (transition)-invariants, etc. Moreover, PN-based models dispose of the possibility to express not only the event causality, but also of the possibility to express analytically the current states expressing the system dynamics development. Even, linear algebra and matrix calculus can be utilized on this way. This is very important especially at the DES control synthesis. The fact that most PN properties can be tested by means of methods based on the reachability tree (RT) and invariants is indispensable too. Thus, the RT and invariants are very important in PN-based modelling DES. In sum, the modelling power of PN consists especially in the facts that (i) PN have formal semantics. Thus, the execution and simulation of PN models are unambiguous; (ii) notation of modelling a system is event-based. PN can model both states and events; (iii) there are many analysis techniques associated with PN. Especially, the approach based on place/transition PN (P/T PN) enables us to use linear algebra and matrix calculus - exact and in practice verified approaches. This makes possible the MAS analysis in analytical terms, especially, by computing RT, invariants, testing properties, model checking, even the efficient model-based control synthesis. Moreover, PN can be used not only for handling software agents but also for 'material' agents - like robots and other technical devices. PN are suitable also at modelling, analysing and control of any modular DES and they are able to deal with any problem on that way. Mutual interactions of agents are considered within the framework of the global model. Such an approach is sufficiently general in order to allow us to create the model that yields the possibility to analyse any situation. Even, the environment behaviour can be modelled as an agent of the agent system too. Thus, the model can acquire arbitrary structure and generate different situations.

## 2 The Petri Net-Based Model

Use the analogy between the DES atomic activities $a_i \in \{a_1, \cdots, a_n\}$ and the PN places $p_i \in \{p_1, \cdots, p_n\}$ as well as between the discrete events $e_j \in \{e_1, \cdots, e_m\}$ occurring in DES and the PN transitions $t_j \in \{t_1, \cdots, t_m\}$. Then, DES behaviour can be modelled by means of P/T PN. The analytical model has the form of the linear discrete system as follows

$$\mathbf{x}_{k+1} = \mathbf{x}_k + \mathbf{B}.\mathbf{u}_k, k = 0, \cdots, K$$
$$\mathbf{B} = \mathbf{G}^T - \mathbf{F}$$
$$\mathbf{F}.\mathbf{u}_k \le \mathbf{x}_k$$

where k is the discrete step of the dynamics development; $\mathbf{x}_k = (\sigma^k_{p_1}, \cdots, \sigma^k_{p_n})^T$ is the n-dimensional state vector in the step k; $\sigma^k_{p_i} \in \{0,1,\cdots,c_{p_i}\}$, i=1,...,n, express the states of the DES atomic activities, namely the passivity is expressed by $\sigma^k_{p_i} = 0$ and the activity is expressed by $0 \le \sigma^k_{p_i} \le c_{p_i}$ ; $c_{p_i}$ is the capacity as to the activities - e.g. the passivity of a buffer means the empty buffer, the activity means a number of parts stored in the buffer and the capacity is understood to be the maximal number of parts which can be put into the buffer; $\mathbf{u}_k = (\gamma^k_{t_1}, \cdots, \gamma^k_{t_m})^T$ is the m-dimensional control vector of the system in the step k; its components $\gamma^k_{t_j} \in \{0,1\}$, j=1,...,m, represent occurring of the DES discrete events (e.g. starting or ending the atomic activities, occurrence of failures, etc.) - when the j-th discrete event is enabled $\gamma^k_{t_j} = 1$, when the event is disabled $\gamma^k_{t_j} = 0$; **B**, **F**, **G** are structural matrices of constant elements; $\mathbf{F} = \{f_{ij}\}$, $f_{ij} \in \{0, M_{f_{ij}}\}$, i=1,...,n, j=1,...,m, express the causal relations between the states of the DES (in the role of causes) and the discrete events occurring during the DES operation (in the role of consequences) – nonexistence of the corresponding relation is expressed by $M_{f_{ij}} = 0$, existence and multiplicity of the relation are expressed by $M_{f_{ij}} > 0$; $G = \{g_{ij}\}$, $g_{ij} \in \{0, M_{g_{ij}}\}$, i=1,...,m, j=1,...,n, express very analogically the causal relations between the discrete events (as the causes) and the DES states (as the consequences); the structural matrix $\mathbf{B}$ is given by means of the arcs incidence matrices $\mathbf{F}$ and $\mathbf{G}$ according to the above introduced relation; $(.)^T$ symbolizes the matrix or vector transposition. The PN marking, which in PN theory is usually denoted by $\boldsymbol{\mu}$, was denoted here by the letter $\mathbf{x}$ usually denoting the state in system theory. The expressive power of the PN-based approach consists in the ability to describe in details (by states and events) how agents behave and/or how agents collaborate. The deeper is the model abstraction level the greater is the model dimensionality (n, m). It is a limitation, so a compromise between the model grain and its dimensionality has to be done. However, such a 'curse' occurs in any kind of systems. There are many papers interested in PN-based modelling of agents and MAS from different reasons – (Hung and Mao 2002, Nowostawski et al. 2001) and a copious amount of other papers. However, no systematic modular approach in analytical terms occurs there. An attempt at forming of such an approach is presented in this chapter. It arises from the author's previous results (Čapkovič 2005, 2007).

## 3 Modular Approaches to Modelling

The modular approach makes possible to model and analyse each module separately as well as the global composition of modules. In general, three different kinds of the model creation can be distinguished according to the form of the interface connecting the modules (PN subnets), namely (i) the interface consisting exclusively of PN transitions; (ii) the interface consisting exclusively of PN places; (iii) the interface in the form of a PN subnet with an arbitrary structure containing both positions and transitions. Let us introduced the structure of the PN model of MAS with agents $A_i$, i = 1, 2, ..., $N_A$, for these three different forms of the interface among agents.

### *3.1 The Transition-Based Interface*

When the interface contains only $m_c$ additional PN transitions, the structure of the actual contact interface among the agents $A_i$, i = 1, 2, ..., $N_A$, is given by the $(n \times m_c)$-dimensional matrix $\mathbf{F}_c$ and $(m_c \times n)$-dimensional matrix $\mathbf{G}_c$ as follows

$$\mathbf{F} = \begin{pmatrix} \mathbf{F}_1 & \mathbf{0} & \cdots & \mathbf{0} & \mathbf{0} & \mathbf{F}_{c_1} \\ \mathbf{0} & \mathbf{F}_2 & \cdots & \mathbf{0} & \mathbf{0} & \mathbf{F}_{c_2} \\ \vdots & \vdots & \ddots & \vdots & \vdots & \vdots \\ \mathbf{0} & \mathbf{0} & \cdots & \mathbf{F}_{N_A-1} & \mathbf{0} & \mathbf{F}_{c_{N_A-1}} \\ \mathbf{0} & \mathbf{0} & \cdots & \mathbf{0} & \mathbf{F}_{N_A} & \mathbf{F}_{c_{N_A}} \end{pmatrix} =$$
$$= \left( blockdiag(\mathbf{F}_i)_{i=1,\cdots,N_A} \mid \mathbf{F}_c \right)$$

$$\mathbf{G} = \begin{pmatrix} \mathbf{G}_1 & \mathbf{0} & \cdots & \mathbf{0} & \mathbf{0} \\ \mathbf{0} & \mathbf{G}_2 & \cdots & \mathbf{0} & \mathbf{0} \\ \vdots & \vdots & \ddots & \vdots & \vdots \\ \mathbf{0} & \mathbf{0} & \cdots & \mathbf{G}_{N_A-1} & \mathbf{0} \\ \mathbf{0} & \mathbf{0} & \cdots & \mathbf{0} & \mathbf{G}_{N_A} \\ \mathbf{G}_{c_1} & \mathbf{G}_{c_2} & \cdots & \mathbf{G}_{c_{N_A-1}} & \mathbf{G}_{c_{N_A}} \end{pmatrix} = \begin{pmatrix} blockdiag(\mathbf{G}_i)_{i=1,\cdots N_A} \\ \mathbf{G}_c \end{pmatrix}$$

$$\mathbf{B} = \begin{pmatrix} \mathbf{B}_1 & \mathbf{0} & \cdots & \mathbf{0} & \mathbf{0} & \mathbf{B}_{c_1} \\ \mathbf{0} & \mathbf{B}_2 & \cdots & \mathbf{0} & \mathbf{0} & \mathbf{B}_{c_2} \\ \vdots & \vdots & \ddots & \vdots & \vdots & \vdots \\ \mathbf{0} & \mathbf{0} & \cdots & \mathbf{B}_{N_A-1} & \mathbf{0} & \mathbf{B}_{c_{N_A-1}} \\ \mathbf{0} & \mathbf{0} & \cdots & \mathbf{0} & \mathbf{B}_{N_A} & \mathbf{B}_{c_{N_A}} \end{pmatrix} =$$
$$= \left( blockdiag(\mathbf{B}_i)_{i=1,\cdots,N_A} \mid \mathbf{B}_c \right)$$

where $\mathbf{B}_{\mathrm{i}} = \mathbf{G}_{\mathrm{i}}^{\mathrm{T}} - \mathbf{F}_{\mathrm{i}}$; $\mathbf{B}_{\mathrm{c_i}} = \mathbf{G}_{\mathrm{c_i}}^{\mathrm{T}} - \mathbf{F}_{\mathrm{c_i}}$; i=1,...., $\mathrm{N_A}$; $\mathbf{F}_{\mathrm{c}} = \left(\mathbf{F}_{\mathrm{c_1}}^{\mathrm{T}}, \cdots, \mathbf{F}_{\mathrm{c_{N_A}}}^{\mathrm{T}}\right)^{\mathrm{T}}$; $\mathbf{G}_{\mathrm{c}} = \left(\mathbf{G}_{\mathrm{c_1}}, \cdots, \mathbf{G}_{\mathrm{c_{N_A}}}\right)$; $\mathbf{B}_{\mathrm{c}} = \left(\mathbf{B}_{\mathrm{c_1}}^{\mathrm{T}}, \cdots, \mathbf{B}_{\mathrm{c_{N_A}}}^{\mathrm{T}}\right)^{\mathrm{T}}$ with $\mathbf{F}_{\mathrm{i}}, \mathbf{G}_{\mathrm{i}}, \mathbf{B}_{\mathrm{i}}$, i=1,... $\mathrm{N_A}$, representing the parameters of the PN-based model of the agent $\mathrm{A_i}$ and with $\mathbf{F}_{\mathrm{c}}, \mathbf{G}_{\mathrm{c}}, \mathbf{B}_{\mathrm{c}}$ representing the structure of the interface between the agents cooperating in MAS.

## 3.2 The Place-Based Interface

When the interface contains only $n_d$ additional PN places the structure of the actual contact interface among the agents $A_i$, i=1,..., $N_A$, is given by the $(n_d \times m)$-dimensional matrix $\mathbf{F}_d$ and $(m \times n_d)$-dimensional matrix $\mathbf{G}_d$ as follows

$$\mathbf{F} = \begin{pmatrix} \mathbf{F}_1 & \mathbf{0} & \cdots & \mathbf{0} & \mathbf{0} \\ \mathbf{0} & \mathbf{F}_2 & \cdots & \mathbf{0} & \mathbf{0} \\ \vdots & \vdots & \ddots & \vdots & \vdots \\ \mathbf{0} & \mathbf{0} & \cdots & \mathbf{F}_{N_A-1} & \mathbf{0} \\ \mathbf{0} & \mathbf{0} & \cdots & \mathbf{0} & \mathbf{F}_{N_A} \\ \mathbf{F}_{d_1} & \mathbf{F}_{d_2} & \cdots & \mathbf{F}_{d_{N_A-1}} & \mathbf{F}_{d_{N_A}} \end{pmatrix} = \begin{pmatrix} blockdiag(\mathbf{F}_i)_{i=1,\cdots N_A} \\ \mathbf{F}_d \end{pmatrix}$$

$$\mathbf{G} = \begin{pmatrix} \mathbf{G}_1 & \mathbf{0} & \cdots & \mathbf{0} & \mathbf{0} & \mathbf{G}_{d_1} \\ \mathbf{0} & \mathbf{G}_2 & \cdots & \mathbf{0} & \mathbf{0} & \mathbf{G}_{d_2} \\ \vdots & \vdots & \ddots & \vdots & \vdots & \vdots \\ \mathbf{0} & \mathbf{0} & \cdots & \mathbf{G}_{N_A-1} & \mathbf{0} & \mathbf{G}_{d_{N_A-1}} \\ \mathbf{0} & \mathbf{0} & \cdots & \mathbf{0} & \mathbf{G}_{N_A} & \mathbf{G}_{d_{N_A}} \end{pmatrix} =$$

$$= \left( blockdiag(\mathbf{G}_i)_{i=1,\cdots,N_A} \mid \mathbf{G}_d \right)$$

$$\mathbf{B} = \begin{pmatrix} \mathbf{B}_1 & \mathbf{0} & \cdots & \mathbf{0} & \mathbf{0} \\ \mathbf{0} & \mathbf{B}_2 & \cdots & \mathbf{0} & \mathbf{0} \\ \vdots & \vdots & \ddots & \vdots & \vdots \\ \mathbf{0} & \mathbf{0} & \cdots & \mathbf{B}_{N_A-1} & \mathbf{0} \\ \mathbf{0} & \mathbf{0} & \cdots & \mathbf{0} & \mathbf{B}_{N_A} \\ \mathbf{B}_{d_1} & \mathbf{B}_{d_2} & \cdots & \mathbf{B}_{d_{N_A-1}} & \mathbf{B}_{d_{N_A}} \end{pmatrix} = \begin{pmatrix} blockdiag(\mathbf{B}_i)_{i=1,\cdots N_A} \\ \mathbf{B}_d \end{pmatrix}$$

where $\mathbf{B}_i = \mathbf{G}_i^T - \mathbf{F}_i$; $\mathbf{B}_{d_i} = \mathbf{G}_{d_i}^T - \mathbf{F}_{d_i}$ ; i=1,..., $N_A$; $\mathbf{F}_d = (\mathbf{F}_{d_1}, \cdots, \mathbf{F}_{d_{N_A}})$ ; $\mathbf{G}_d = (\mathbf{G}_{d_1}^T, \cdots, \mathbf{G}_{d_{N_A}}^T)^T$; $\mathbf{B}_d = (\mathbf{B}_{d_1}, \cdots, \mathbf{B}_{d_{N_A}})$ with $\mathbf{F}_i, \mathbf{G}_i, \mathbf{B}_i$, i=1,... $N_A$, representing the parameters of the PN-based model of the agent $A_i$ and with

$\mathbf{F}_d, \mathbf{G}_d, \mathbf{B}_d$ representing the structure of the interface between the agents cooperating in MAS.

### 3.3 The Interface in the Form of a PN Subnet

When the interface among the agents $A_i$, $i=1,...,N_A$, has a form of the PN subnet containing $n_d$ additional places and $m_c$ additional transitions its structure is given by the $(n_d \times m_c)$-dimensional matrix $\mathbf{F}_{d\leftrightarrow c}$ and $(m_c \times n_d)$-dimensional matrix $\mathbf{G}_{c\leftrightarrow d}$. However, it is only the structure of the PN subnet. Moreover, the row and the column consisting of corresponding blocks have to be added in order to model the contacts of the interface with the elementary agents. Hence, we have the following structural (incidence) matrices

$$\mathbf{F} = \begin{pmatrix} \mathbf{F}_1 & \mathbf{0} & \cdots & \mathbf{0} & \mathbf{0} & | & \mathbf{F}_{c_1} \\ \mathbf{0} & \mathbf{F}_2 & \cdots & \mathbf{0} & \mathbf{0} & | & \mathbf{F}_{c_2} \\ \vdots & \vdots & \ddots & \vdots & \vdots & | & \vdots \\ \mathbf{0} & \mathbf{0} & \cdots & \mathbf{F}_{N_A-1} & \mathbf{0} & | & \mathbf{F}_{c_{N_A-1}} \\ \mathbf{0} & \mathbf{0} & \cdots & 0 & \mathbf{F}_{N_A} & | & \mathbf{F}_{c_{N_A}} \\ \mathbf{F}_{d_1} & \mathbf{F}_{d_2} & \cdots & \mathbf{F}_{d_{N_A-1}} & \mathbf{F}_{d_{N_A}} & | & \mathbf{F}_{d\leftrightarrow c} \end{pmatrix} =$$

$$= \begin{pmatrix} blockdiag(\mathbf{F}_i)_{i=1,\dots,N_A} & | & \mathbf{F}_c \\ \mathbf{F}_d & | & \mathbf{F}_{d\leftrightarrow c} \end{pmatrix}$$

$$\mathbf{G} = \begin{pmatrix} \mathbf{G}_1 & \mathbf{0} & \cdots & \mathbf{0} & \mathbf{0} & | & \mathbf{G}_{d_1} \\ \mathbf{0} & \mathbf{G}_2 & \cdots & \mathbf{0} & \mathbf{0} & | & \mathbf{G}_{d_2} \\ \vdots & \vdots & \ddots & \vdots & \vdots & | & \vdots \\ \mathbf{0} & \mathbf{0} & \cdots & \mathbf{G}_{N_A-1} & \mathbf{0} & | & \mathbf{G}_{d_{N_A-1}} \\ \mathbf{0} & \mathbf{0} & \cdots & \mathbf{0} & \mathbf{G}_{N_A} & | & \mathbf{G}_{d_{N_A}} \\ \mathbf{G}_{c_1} & \mathbf{G}_{c_2} & \cdots & \mathbf{G}_{c_{N_A-1}} & \mathbf{G}_{c_{N_A}} & | & \mathbf{G}_{c\leftrightarrow d} \end{pmatrix} =$$

$$= \begin{pmatrix} blockdiag(\mathbf{G}_i)_{i=1,\dots,N_A} & | & \mathbf{G}_d \\ \mathbf{G}_c & | & \mathbf{G}_{c\leftrightarrow d} \end{pmatrix}$$

$$
\mathbf{B} = \begin{pmatrix}
\mathbf{B}_1 & \mathbf{0} & \cdots & \mathbf{0} & \mathbf{0} & | & \mathbf{B}_{c_1} \\
\mathbf{0} & \mathbf{B}_2 & \cdots & \mathbf{0} & \mathbf{0} & | & \mathbf{B}_{c_2} \\
\vdots & \vdots & \ddots & \vdots & \vdots & | & \vdots \\
\mathbf{0} & \mathbf{0} & \cdots & \mathbf{B}_{N_A-1} & \mathbf{0} & | & \mathbf{B}_{c_{N_A-1}} \\
\mathbf{0} & \mathbf{0} & \cdots & \mathbf{0} & \mathbf{B}_{N_A} & | & \mathbf{B}_{c_{N_A}} \\
\mathbf{B}_{d_1} & \mathbf{B}_{d_2} & \cdots & \mathbf{B}_{d_{N_A-1}} & \mathbf{B}_{d_{N_A}} & | & \mathbf{B}_{d\leftrightarrow c}
\end{pmatrix} =
$$

$$
= \begin{pmatrix}
blockdiag(\mathbf{B}_i)_{i=1,\ldots,N_A} & | & \mathbf{B}_c \\
\mathbf{B}_d & | & \mathbf{B}_{d\leftrightarrow c}
\end{pmatrix}
$$

where $\mathbf{B}_{\mathrm{i}} = \mathbf{G}_{\mathrm{i}}^{\mathrm{T}} - \mathbf{F}_{\mathrm{i}}$; $\mathbf{B}_{\mathrm{d_i}} = \mathbf{G}_{\mathrm{d_i}}^{\mathrm{T}} - \mathbf{F}_{\mathrm{d_i}}$; $\mathbf{B}_{\mathrm{c_i}} = \mathbf{G}_{\mathrm{c_i}}^{\mathrm{T}} - \mathbf{F}_{\mathrm{c_i}}$; $\mathrm{i}=1,\ldots,\mathrm{N_A}$; $\mathbf{B}_{\mathrm{d\leftrightarrow c}} = \mathbf{G}_{\mathrm{c\leftrightarrow d}}^{\mathrm{T}} - \mathbf{F}_{\mathrm{d\leftrightarrow c}}$. It can be seen that the matrices **F, G, B** acquire a special structure. Each of them has the big diagonal block (with the smaller blocks in its diagonal describing the structure of the elementary agents $A_i$, $i=1,\ldots,N_A$) and the part in the form of a special structure of the matrix (like the capitol letter L turned over to the left for $180^o$) containing the small blocks representing interconnections among the agents. In these matrices the smaller blocks $\mathbf{F}_{\mathrm{d\leftrightarrow c}}$, $\mathbf{G}_{\mathrm{c\leftrightarrow d}}$, $\mathbf{B}_{\mathrm{d\leftrightarrow c}}$ are situated in their diagonals just in the breakage of the turned L, but outwards.

### 3.4 The Reason of the Modular Models

The modular models described above are suitable for modelling the very wide spectrum of the agent cooperation in MAS. The elementary agents can have either the same structure or the mutually different one. The modules can represent not only agents but also different additional entities including the environment behaviour. Three examples will be presented below in the section 5 to illustrate the usage of the proposed models.

## 4 Analysing the Agent Behaviour

The agent behaviour can be analysed by means of testing the model properties. This can be done by using the graphical tool and/or analytically. The graphical tool was developed in order to draw the PN model of the system to be analysed, to simulate its dynamics development for a chosen initial state (PN marking), to compute its P-invariants and T-invariants, to compute and draw its RT, etc. The RT is the most important instrument for analysing the dynamic behaviour of agents and MAS. Because the same leaves can occur in the RT repeatedly, it is suitable to connect them in such a case into one node. Consequently, the reachability graph (RG) is obtained from RT. Both RT and RG have the same adjacency matrix. RG is very useful not only at the system analysis (model checking), but also at the control synthesis. However, the RG-based control synthesis is not handled in this paper. It was presented in (Čapkovič 2005, 2007).

## 5 Illustrative Examples

The advantage of the modular approach to modelling MAS is that different kinds of the agent communication can be analysed in different evoking dynamic situations occurring at the agent cooperation, negotiation, etc. The models can be created flexibly by means of arbitrary aggregating or clustering elementary agents into MAS. To illustrate using the modular models of MAS, let us introduce three examples as follows. In spite of their simplicity they give us the basic conception of the utility of such models.

### *5.1 Example 1*

Consider the simple MAS with two agents $A_1$, $A_2$ of the same structure. Its PN-based model is in Figure 1. The system parameters are the following

$$\mathbf{F} = \begin{pmatrix} \mathbf{F}_1 & \mathbf{0} & \mathbf{F}_{c_1} \\ \mathbf{0} & \mathbf{F}_2 & \mathbf{F}_{c_2} \end{pmatrix} \quad ; \quad \mathbf{G} = \begin{pmatrix} \mathbf{G}_1 & \mathbf{0} \\ \mathbf{0} & \mathbf{G}_2 \\ \mathbf{G}_{c_1} & \mathbf{G}_{c_2} \end{pmatrix}$$

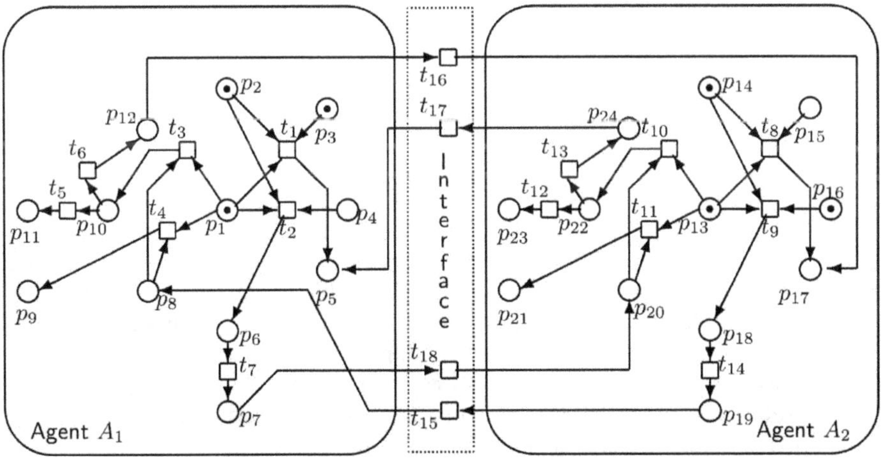


**Fig. 1** The PN-based model of the two agents cooperation - the transitions-based interface

$$\mathbf{F}_1 = \mathbf{F}_2 = \begin{pmatrix} 1 & 1 & 1 & 1 & 0 & 0 & 0 \\ 1 & 1 & 0 & 0 & 0 & 0 & 0 \\ 1 & 0 & 0 & 0 & 0 & 0 & 0 \\ 0 & 1 & 0 & 0 & 0 & 0 & 0 \\ 0 & 0 & 0 & 0 & 0 & 0 & 0 \\ 0 & 0 & 0 & 0 & 0 & 0 & 1 \\ 0 & 0 & 0 & 0 & 0 & 0 & 0 \\ 0 & 0 & 1 & 1 & 0 & 0 & 0 \\ 0 & 0 & 0 & 0 & 0 & 0 & 0 \\ 0 & 0 & 0 & 0 & 1 & 1 & 0 \\ 0 & 0 & 0 & 0 & 0 & 0 & 0 \\ 0 & 0 & 0 & 0 & 0 & 0 & 0 \end{pmatrix}$$

$$\mathbf{G}_1^T = \mathbf{G}_2^T = \begin{pmatrix} 0 & 0 & 0 & 0 & 0 & 0 & 0 \\ 0 & 0 & 0 & 0 & 0 & 0 & 0 \\ 0 & 0 & 0 & 0 & 0 & 0 & 0 \\ 0 & 0 & 0 & 0 & 0 & 0 & 0 \\ 1 & 0 & 0 & 0 & 0 & 0 & 0 \\ 0 & 1 & 0 & 0 & 0 & 0 & 0 \\ 0 & 0 & 0 & 0 & 0 & 0 & 1 \\ 0 & 0 & 0 & 0 & 0 & 0 & 0 \\ 0 & 0 & 0 & 1 & 0 & 0 & 0 \\ 0 & 0 & 1 & 0 & 0 & 0 & 0 \\ 0 & 0 & 0 & 0 & 1 & 0 & 0 \\ 0 & 0 & 0 & 0 & 0 & 1 & 0 \end{pmatrix}$$

The interpretation of the PN places is the following: $p_1$ = the agent ($A_1$) is free; $p_2$ = a problem has to be solved by $A_1$; $p_3 = A_1$ is able to solve the problem ($P_{A1}$ ); $p_4 = A_1$ is not able to solve $P_{A1}$ ; $p_5 = P_{A1}$ is solved; $p_6 = P_{A1}$ cannot be solved by $A_1$ and another agent(s) should be contacted; $p_7 = A_1$ asks another agent(s) to help him to solve $P_{A1}$ ; $p_8 = A_1$ is asked by another agent(s) to solve a problem $P_B$; $p_9 = A_1$ refuses the help; $p_{10} = A_1$ accepts the request of another agent(s) for help; $p_{11} = A_1$ is not able to solve $P_B$; $p_{12} = A_1$ is able to solve $P_B$. In case of the agent $A_2$ the interpretation is the same only the indices of the places are shifted for 12 (in the MAS in order to distinguish $A_2$ from $A_1$). The same is valid also for the agent transitions, only the shifting is for 8 (because the number of transitions in the model of $A_1$ is 7). The transitions represent the starting or ending the atomic activities. The interface between the agents is realized by the additional transitions $t_{15}$-$t_{18}$. The number of the transitions follows from the situation to be modelled and analysed. Here, the situation when $A_2$ is not able to solve its own problem $P_{A2}$ and $A_2$ asks $A_1$ for help to solve it is modelled. Because $A_1$ accept the request and it is able to solve the problem $P_{A2}$, finally the $P_{A2}$ is resolved by the agent $A_1$. Consequently, the interface can be described by the structural matrices

$$\mathbf{F}_{c_1} = \begin{pmatrix} 0 & 0 & 0 & 0 \\ 0 & 0 & 0 & 0 \\ 0 & 0 & 0 & 0 \\ 0 & 0 & 0 & 0 \\ 0 & 0 & 0 & 0 \\ 0 & 0 & 0 & 0 \\ 0 & 0 & 0 & 1 \\ 0 & 0 & 0 & 0 \\ 0 & 0 & 0 & 0 \\ 0 & 0 & 0 & 0 \\ 0 & 0 & 0 & 0 \\ 0 & 1 & 0 & 0 \end{pmatrix} \quad \mathbf{F}_{c_2} = \begin{pmatrix} 0 & 0 & 0 & 0 \\ 0 & 0 & 0 & 0 \\ 0 & 0 & 0 & 0 \\ 0 & 0 & 0 & 0 \\ 0 & 0 & 0 & 0 \\ 0 & 0 & 0 & 0 \\ 1 & 0 & 0 & 0 \\ 0 & 0 & 0 & 0 \\ 0 & 0 & 0 & 0 \\ 0 & 0 & 0 & 0 \\ 0 & 0 & 0 & 0 \\ 0 & 0 & 1 & 0 \end{pmatrix}$$

$$\mathbf{G}_{c_1}^{\mathrm{T}} = \begin{pmatrix} 0 & 0 & 0 & 0 \\ 0 & 0 & 0 & 0 \\ 0 & 0 & 0 & 0 \\ 0 & 0 & 0 & 0 \\ 0 & 0 & 1 & 0 \\ 0 & 0 & 0 & 0 \\ 0 & 0 & 0 & 0 \\ 1 & 0 & 0 & 0 \\ 0 & 0 & 0 & 0 \\ 0 & 0 & 0 & 0 \\ 0 & 0 & 0 & 0 \\ 0 & 0 & 0 & 0 \end{pmatrix} \quad \mathbf{G}_{c_2}^{\mathrm{T}} = \begin{pmatrix} 0 & 0 & 0 & 0 \\ 0 & 0 & 0 & 0 \\ 0 & 0 & 0 & 0 \\ 0 & 0 & 0 & 0 \\ 0 & 1 & 0 & 0 \\ 0 & 0 & 0 & 0 \\ 0 & 0 & 0 & 0 \\ 0 & 0 & 0 & 1 \\ 0 & 0 & 0 & 0 \\ 0 & 0 & 0 & 0 \\ 0 & 0 & 0 & 0 \\ 0 & 0 & 0 & 0 \end{pmatrix}$$

We can analyse arbitrary situations. For the initial state e.g. in the form $\mathbf{x}_0 = \left({}^{A_1}\mathbf{x}_0^T, {}^{A_2}\mathbf{x}_0^T\right)^T$ where ${}^{A_1}\mathbf{x}_0^T = (1,1,1,0,0,0,0,0,0,0,0,0,0)^T$, ${}^{A_2}\mathbf{x}_0^T = (1,1,0,1,0,0,0,0,0,0,0,0,0)^T$, we can compute the parameters of the RG, i.e. the adjacency matrix and the feasible state vectors being the nodes of the RT/RG. The feasible state vectors are given in the form of the columns of the matrix $\mathbf{X}_{\text{reach}}$. Hence, the RG for the modelled situation is displayed in Figure 2. There, the RG nodes are the feasible state vectors $\mathbb{X}_1, \ldots, \mathbb{X}_{13}$. They are expressed by the columns of the matrix as follows $\mathbf{X}_{\text{reach}} = \left({}^1\mathbf{X}_{\text{reach}}^T, {}^2\mathbf{X}_{\text{reach}}^T\right)^T$ where

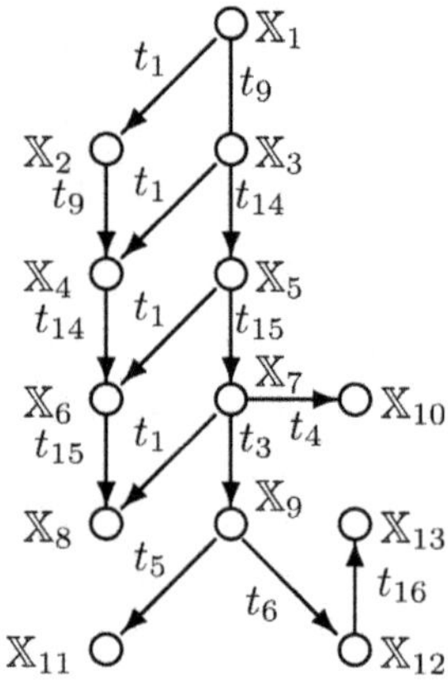


**Fig. 2** The RG of the two agents negotiation corresponding to the given initial state $\mathbf{x}_0$

$$
{}^1\mathbf{X}_{\text{reach}} = \begin{pmatrix}
1 & 0 & 1 & 0 & 1 & 0 & 1 & 0 & 0 & 0 & 0 & 0 & 0 \\
1 & 0 & 1 & 0 & 1 & 0 & 1 & 0 & 1 & 1 & 1 & 1 & 1 \\
1 & 0 & 1 & 0 & 1 & 0 & 1 & 0 & 1 & 1 & 1 & 1 & 1 \\
0 & 0 & 0 & 0 & 0 & 0 & 0 & 0 & 0 & 0 & 0 & 0 & 0 \\
0 & 1 & 0 & 1 & 0 & 1 & 0 & 1 & 0 & 0 & 0 & 0 & 0 \\
0 & 0 & 0 & 0 & 0 & 0 & 0 & 0 & 0 & 0 & 0 & 0 & 0 \\
0 & 0 & 0 & 0 & 0 & 0 & 0 & 0 & 0 & 0 & 0 & 0 & 0 \\
0 & 0 & 0 & 0 & 0 & 0 & 1 & 1 & 0 & 0 & 0 & 0 & 0 \\
0 & 0 & 0 & 0 & 0 & 0 & 0 & 0 & 0 & 1 & 0 & 0 & 0 \\
0 & 0 & 0 & 0 & 0 & 0 & 0 & 0 & 1 & 0 & 0 & 0 & 0 \\
0 & 0 & 0 & 0 & 0 & 0 & 0 & 0 & 0 & 0 & 1 & 0 & 0 \\
0 & 0 & 0 & 0 & 0 & 0 & 0 & 0 & 0 & 0 & 0 & 1 & 0
\end{pmatrix}
$$

$$
{}^{2}\mathbf{X}_{\text{reach}} = \begin{pmatrix}
1 & 1 & 0 & 0 & 0 & 0 & 0 & 0 & 0 & 0 & 0 & 0 & 0 \\
1 & 1 & 0 & 0 & 0 & 0 & 0 & 0 & 0 & 0 & 0 & 0 & 0 \\
0 & 0 & 0 & 0 & 0 & 0 & 0 & 0 & 0 & 0 & 0 & 0 & 0 \\
1 & 1 & 0 & 0 & 0 & 0 & 0 & 0 & 0 & 0 & 0 & 0 & 0 \\
0 & 0 & 0 & 0 & 0 & 0 & 0 & 0 & 0 & 0 & 0 & 0 & 1 \\
0 & 0 & 1 & 1 & 0 & 0 & 0 & 0 & 0 & 0 & 0 & 0 & 0 \\
0 & 0 & 0 & 0 & 1 & 1 & 0 & 0 & 0 & 0 & 0 & 0 & 0 \\
0 & 0 & 0 & 0 & 0 & 0 & 0 & 0 & 0 & 0 & 0 & 0 & 0 \\
0 & 0 & 0 & 0 & 0 & 0 & 0 & 0 & 0 & 0 & 0 & 0 & 0 \\
0 & 0 & 0 & 0 & 0 & 0 & 0 & 0 & 0 & 0 & 0 & 0 & 0 \\
0 & 0 & 0 & 0 & 0 & 0 & 0 & 0 & 0 & 0 & 0 & 0 & 0 \\
0 & 0 & 0 & 0 & 0 & 0 & 0 & 0 & 0 & 0 & 0 & 0 & 0
\end{pmatrix}
$$

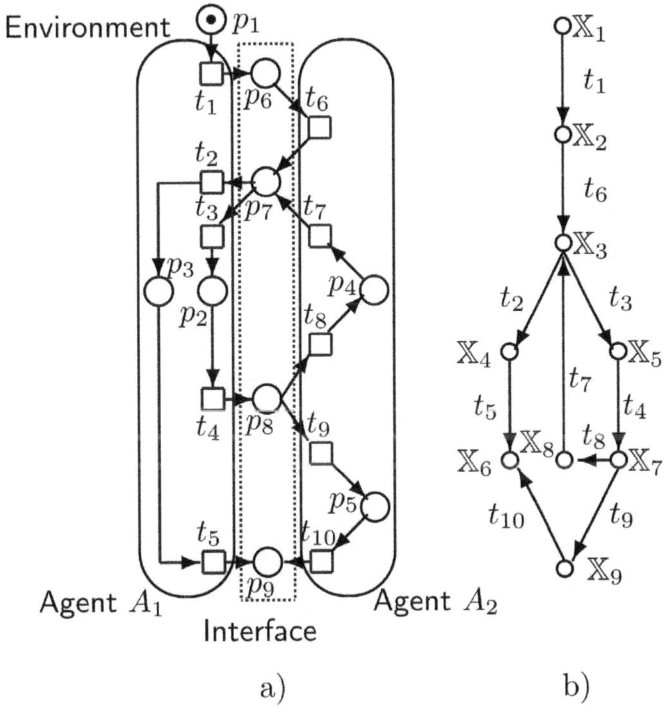


**Fig. 3** Two agents negotiation in an environment - the places-based interface. a) The PN-based model. b) The RG corresponding to the given initial state $\mathbf{x}_0$ .

## 5.2 Example 2

Consider two virtual companies A, B (Lenz et al. 2001). In the negotiation process the company A creates an information document containing the issues of the project (e.g. which the mutual software agents already agreed on) and those that are still unclear. The PN-based model and the corresponding RG are displayed in Figure 3. The interpretations of the PN places and the PN transitions are: $p_1$ – the start depending on the state of environment; $p_2$, $p_4$ - the updated proposal; $p_3$, $p_5$ - the unchanged proposal; $p_6$ - the information document; $p_7$ - the proposal to A; $p_8$ - the proposal to B; $p_9$ - the contract; $t_1$ - creating the information document; $t_2$, $t_9$ – checking the proposal and the agreement with it; $t_3$, $t_8$ - checking the proposal and asking changes; $t_4$ - sending the updated proposal; $t_5$, $t_{10}$ - accepting the unchanged proposal; $t_6$ - preparing the proposal; $t_7$ - sending the updated proposal. For the initial state $\mathbf{x}_0 \equiv X_1 = (1,0,0,0,0,0,0,0,0)^T$ we have the RG nodes $X_1$, ..., $X_9$ stored as the columns of the matrix

$$\mathbf{X}_{reach} = \begin{pmatrix} 1 & 0 & 0 & 0 & 0 & 0 & 0 & 0 & 0 \\ 0 & 0 & 0 & 0 & 1 & 0 & 0 & 0 & 0 \\ 0 & 0 & 0 & 1 & 0 & 0 & 0 & 0 & 0 \\ 0 & 0 & 0 & 0 & 0 & 0 & 0 & 1 & 0 \\ 0 & 0 & 0 & 0 & 0 & 0 & 0 & 0 & 1 \\ 0 & 1 & 0 & 0 & 0 & 0 & 0 & 0 & 0 \\ 0 & 0 & 1 & 0 & 0 & 0 & 0 & 0 & 0 \\ 0 & 0 & 0 & 0 & 0 & 0 & 1 & 0 & 0 \\ 0 & 0 & 0 & 0 & 0 & 1 & 0 & 0 & 0 \end{pmatrix}$$

Consider the situation when the terminal state $\mathbf{x}_t$ is achieving the contract represented by the feasible state $X_9 = (0,0,0,0,1,0,0,0,0)^T$ , i.e. the 9-th column of the matrix $\mathbf{X}_{reach}$. To reach the terminal state $X_9 \equiv \mathbf{x}_t$ from the initial state $X_1 \equiv \mathbf{x}_0$ (the first column of the matrix $\mathbf{X}_{reach}$) we can utilize the in-house graphical tool GraSim for control synthesis built on the base of the DES control synthesis-method proposed in (Čapkovič, 2005). Then, the state trajectory of the system is $X_1 \rightarrow X_2 \rightarrow X_3 \rightarrow X_5 \rightarrow X_7 \rightarrow X_9$.

## 5.3 Example 3

Consider two agents with the simple structure defined in (Saint-Voirin et al. 2003). The agents are connected by the interface in the form of the PN

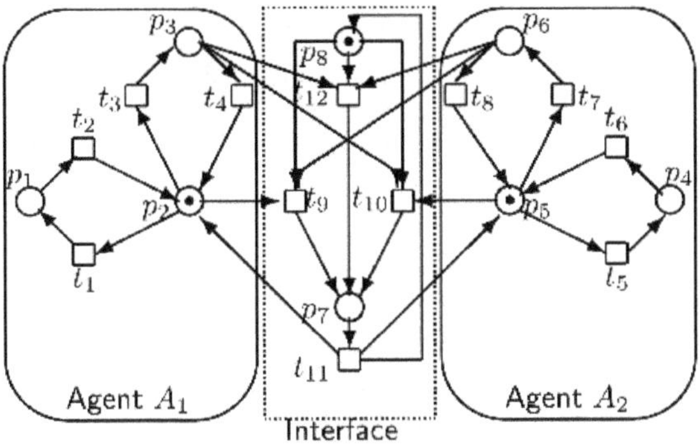


**Fig. 4** The PN-based model of the two agents communication by way of the interface in the form of the PN-subnet

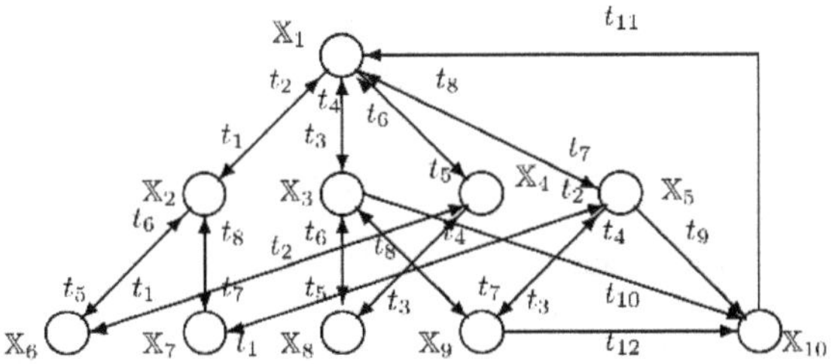


**Fig. 5** The PN-based model of the two agents communication by way of its RG

subnet – see Figure 4 (upper part). The interpretation of the places is the following: $p_1$ - $A_1$ does not want to communicate; $p_2$ - $A_1$ is available; $p_3$ - $A_1$ wants to communicate; $p_4$ - $A_2$ does not want to communicate; $p_5$ - $A_2$ is available; $p_6$ - $A_2$ wants to communicate; $p_7$ - communication; $p_8$ - availability of the communication channel(s) Ch (representing the interface). The PN transition $t_9$ fires the communication when $A_1$ is available and $A_2$ wants to communicate with $A_1$, $t_{10}$ fires the communication when $A_2$ is available and $A_1$ wants to communicate with $A_2$, and $t_{12}$ fires the communication when both $A_1$ and $A_2$ wants to communicate each other. For the initial state $\mathbb{X}_1 \equiv \mathbf{x}_0^T = (0,1,0,0,1,0,0,1)^T$ we have the RG given in the lower part of Figure 4. The RG nodes are the state vectors stored as the columns of the matrix

$$\mathbf{X}_{reach} = \begin{pmatrix} 0 & 1 & 0 & 0 & 0 & 1 & 1 & 0 & 0 & 0 \\ 1 & 0 & 0 & 1 & 1 & 0 & 0 & 0 & 0 & 0 \\ 0 & 0 & 1 & 0 & 0 & 0 & 0 & 1 & 1 & 0 \\ 0 & 0 & 0 & 1 & 0 & 1 & 0 & 1 & 0 & 0 \\ 1 & 1 & 1 & 0 & 0 & 0 & 0 & 0 & 0 & 0 \\ 0 & 0 & 0 & 0 & 1 & 0 & 1 & 0 & 1 & 0 \\ 0 & 0 & 0 & 0 & 0 & 0 & 0 & 0 & 0 & 1 \\ 1 & 1 & 1 & 1 & 1 & 1 & 1 & 1 & 1 & 0 \end{pmatrix}$$

In order to illustrate the control synthesis let us synthesize control able to transfer the system from the initial state $\mathbf{x}_0 \equiv X_1$ to the terminal one, consider the terminal state $\mathbf{x}_t \equiv X_{10} = (0,0,0,0,0,0,1,0)^T$ representing the communication of the agents. There exist two feasible trajectories $X_1 \rightarrow X_3 \rightarrow X_{10}$ and $X_1 \rightarrow X_5 \rightarrow X_{10}$.

## 6 Detection of Conflicts

Any lack of collaboration in a group of agents or intrusion could be found as an information conflict with existing models. Many methods exist where a model is given and everything non-matching it knowledge is assumed a contradiction. Say, in anomaly intrusion detection system, if the traffic has been increased, it is a contradiction to the existing statistical data and intrusion alert has been issued. The considered approach is to discover and trace different logical connections to reveal and resolve conflict information (Jotsov 2008). Constant inconsistency resolution process gradually improves the system DB and KB and leads to better intrusion detection and prevention.

Models for conflicts are introduced and used, they represent different forms of ontologies.

Let the strong (classical) negation is denoted by '¬' and the weak (conditional, paraconsistent (Arruda 1982)) negation is '~'. In the case of an evident conflict (inconsistency) between the knowledge and its ultimate form–the contradiction– the conflict situation is determined by the direct comparison of the two statements (the *conflicting sides*) that differ one form another just by a definite number of symbols '¬' or '~'. For example A and ¬A; B and not B (using ¬ equivalent to 'not'), etc.

The case of implicit (or hidden) negation between two statements A and B can be recognized only by an analysis of a preset models of the type of (1).

$$\{U\}[\eta: A, B]. \tag{1}$$

where $\eta$ is a type of negation, U is a statement with a validity including the validities of the concepts A and B and it is possible that more than two conflicting sides may be present. Below it is accepted that the contents in the figure brackets U is called *an unifying feature*. In this way it is possible to formalize not only the features that separate the conflicting sides but also the unifying concepts joining the sides. For example, the intelligent detection may be either automated or of a human-machine type but the conflict cannot be recognized without the investigation of the following model.

$$\{\text{detection procedures}\}[\neg\text{: automatic, interactive}].$$

The formula (1) formalizes a model of the conflict the sides of which unconditionally negate one another. In the majority of situations the sides participate in the conflict only under definite conditions: $\chi_1, \chi_2, \ldots \chi_z$.

$$\{U\}[\eta: A_1,A_2,\ldots A_p] <\tilde{\chi}_1 * \tilde{\chi}_2 * \ldots * \tilde{\chi}_z>. \tag{2}$$

where $\tilde{\chi}$ is a literal of $\chi$, i.e. $\tilde{\chi}\equiv\chi$ or $\tilde{\chi}\equiv\eta\chi$, * is the logical operation of conjunction, disjunction or implication.

The present research allows switch from models of contradictions to ontologies (Jotsov 2007) in order to develop new methods for revealing and resolving contradictions and also to expand the basis for cooperation with the Semantic Web community and with other research groups. This is the way to consider the suggested models from (1) or (2) as one of the forms of static ontologies.

The following factors have been investigated:

T – time factor: non-simultaneous events do not bear a contradiction.

M – place factor: events that have taken place not on one and the same place do not bear a contradiction. In this case the concept of place may be expanded up to a coincidence or to differences in possible worlds.

N – a disproportion of concepts emits a contradiction. For example if one of the parts of the contradiction is a small object and the investigated object is huge then and only then it is the case with a contradiction.

O – one and the same object. If the parts of the contradiction are referred to different objects then there is no contradiction.

P – the feature should be the same. If the parts of the contradiction are referred to different features then there is no contradiction.

S – simplification factor. If the logic of user actions is executed in a sophisticated manner then there is a contradiction.

W – mode factor. For example if the algorithms are applied in different modes then there is no contradiction.

MO – contradiction to the model. The contradiction exists if and only if ( *iff* ) when at least one of the measured parameters does not correspond to the meaning from the model. For example the traffic is bigger than the maximal value from the model.

Example. We must isolate errors that are done due to lack of attention from tendentious faults. In this case we introduce the following model (3):

$$\{\ \text{user} : \text{faults}\ \}[\sim: \text{accidental, tendentious}\ ] \langle T, \neg M, O; \neg S \rangle \qquad (3)$$

It is possible that one and the same person does sometimes accidental errors and in other cases tendentious faults; these failures must not be simultaneous on different places and must not be done by one and the same person. On the other side if there are multiple errors (e.g. more than three) for short time intervals (e.g. 10 minutes), for example during authentications or in various subprograms of the security software then we have a case of a violence, nor a series of accidental errors. In this way it is possible to realize comparisons, juxtapositions and other logical operations to form security policies thereof.

## 7 Conflict Resolution Methods

Recently we shifted conflict or contradiction models with ontologies (Jotsov 2008) that give us possibility to apply new resolution methods. For pity, the common game theoretic form of conflict detection and resolution is usually heuristic-driven and too complex. We concentrate on ultimate conflict resolution forms using contradictions. For the sake of brevity the resolution groups of methods are described schematically.

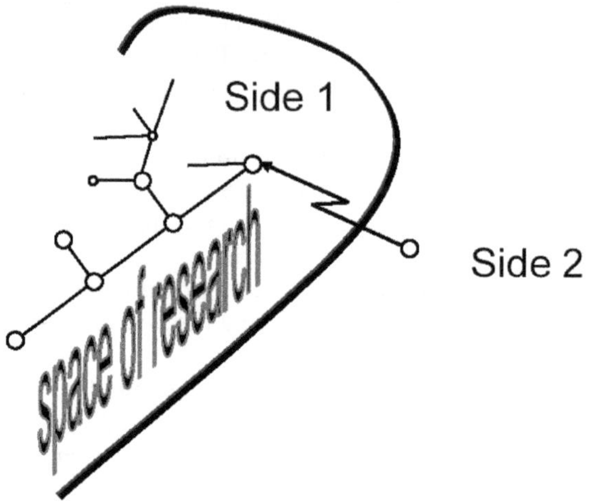


**Fig. 6** Avoidable (postponed) conflicts when Side 2 is outside the research space

The conflict recognition is followed by its resolution (Jotsov 2007). The schemas of different groups of resolution methods have been presented in Fig. 5 to Fig. 10.

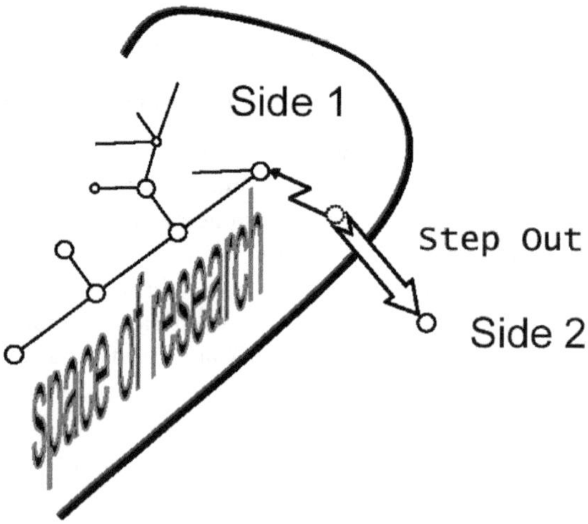


**Fig. 7** Conflict resolution by stepping out of the research space (postponed or resolved conflicts)

In situations from Fig. 6 one of the conflicting sides doesn't belong to the considered research space. Hence the conflict may be not resolved just in the moment, only a conflict warning is to be issued in future. Say, if we are looking for an intrusion attack, and side 2 matches printing problems, then the system could avoid the resolution to this problem. This conflict is not necessary to be resolved automatically, experts may resolve it later using the saved information. In Fig. 7 a situation is depicted where the conflict is resolvable by stepping out from the conflict area. This type of resolution is frequently used in multi-agent systems where conflicting sides step back to the pre-conflict positions and one or both try to avoid the conflict area. In this case a warning on the conflict situation has been issued.

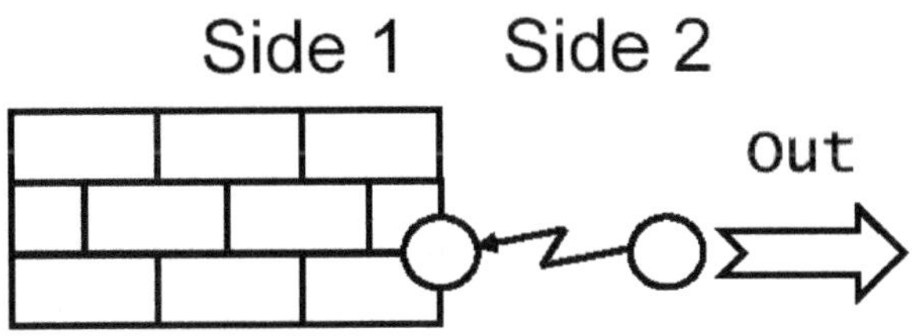


**Fig. 8** Automatically resolvable conflicts

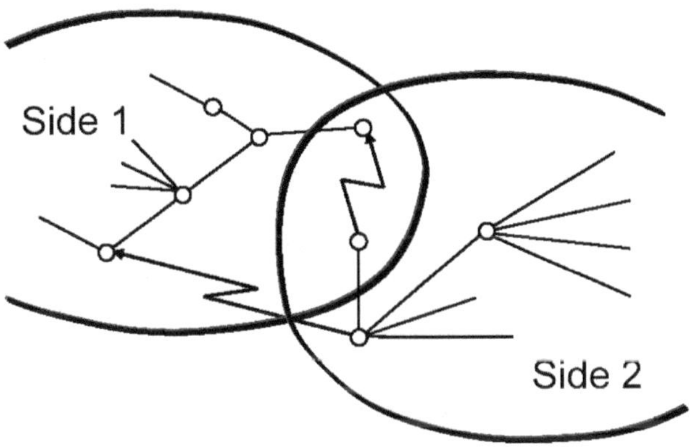


**Fig. 9** Conflicts resolvable using human-machine interaction

The situation from Fig. 8 is automatically resolvable without issuing a warning message. Both sides have different priorities, say side 1 is introduced by a security expert, and side 2 is introduced by a non-specialist. In this case side 2 has been removed immediately.

A situation is depicted in Fig. 9 where both sides have been derived by an inference machine, say by using deduction. In this case the origin for the conflict could be traced, and the process is using different human-machine interaction methods.

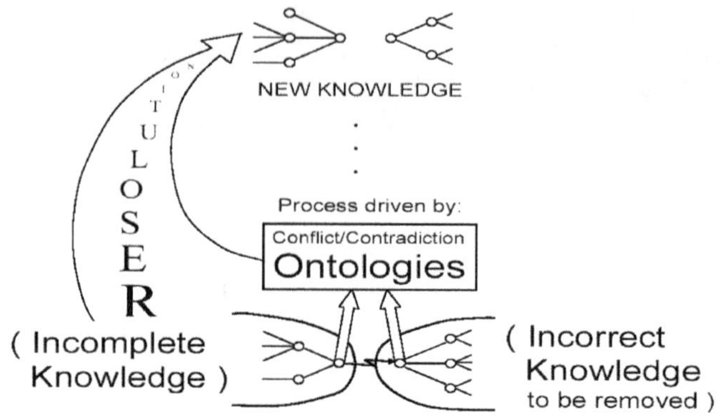


**Fig. 10** Learning via resolving conflicts or with contradictions

At last, the conflict may be caused by incorrect or incomplete conflict models (1) or (2), in this case they have been improved after same steps shown in Fig. 6 to Fig. 9.

The resolution leads to knowledge improvement as shown in Fig. 10, and this improvement gradually builds a machine self-learning. Of course in many situations from Fig. 6 to Fig. 9 the machine prompts the experts just in the moment or later, but this form of interaction isn't so boring, because it appears in time and it is a consequence of a logical reasoning based on knowledge (1) or (2).

Our research shows that automatic resolution processes may constantly stay resident using free computer resources OR in other situations they may be activated by user's command. In the first case the knowledge and data bases will be constantly improved by continuous elimination of incorrect information or by improving the existing knowledge as a result of revealing and resolving contradictions. As a result our contradiction resolution methods have been upgraded to a machine self-learning method i.e. learning without teacher which is very effective in the case of ISS. This type of machine learning is novel and original in both theoretical and applied aspects.

## 8 Essential Advantages to Machine Learning, Collective Evolution and Evolutionary Computation

Knowledge bases (KBs) are improved after isolating and resolving contradictions in the following way. One set is replaced by another while other knowledge is supplemented or specified. The indicated processes are not directed by the elaborator or by the user. The system functions autonomously and it requires only a preliminary input of models and the periodical updates of strategies for resolving contradictions. Competitions to the stated method may be methods for machine supervised – or unsupervised – learning. During supervised learning, for example by using artificial neural networks, training is a long and a complicated expensive process and the results from the applications outside the investigated matter are non-reliable. The 'blind' reproduction of teacher's actions is not effective and it has no good prospects except if cases when it is combined with other unsupervised methods. In cases of unsupervised training via artificial neural networks the system is overloaded by heuristic information and algorithms for processing heuristics and it cannot be treated as autonomous. The presented method contains autonomous unsupervised learning based on the doubt-about-everything principle or on the doubt-about-a-subset-of-knowledge principle. The contradiction-detecting procedure can be resident; it is convenient to use computer resources except for peak hours of operation.

The unsupervised procedure consists of three basic steps. During the first step the contradiction is detected using models from (1) to (3). During the second step the contradiction is resolved using one of the presented above resolution schemes depending on the type of conflict situation. As a result from the undertaken actions, after the second stage the set $K$ is transformed into $K'$ where it is possible to eliminate from $K$ the subset of incorrect knowledge $W \subseteq K$, to correct the subset of knowledge with an incomplete description of the object domain $I \subseteq K$, to add a subset of new knowledge for specification $U \subseteq K$. The latter of cited subsets includes postponed problems that are discussed in the version described in Fig. 5, knowledge with a possible discrepancy of the expert estimates (problematic knowledge) and other knowledge for future research which is detected based on the heuristic information.

In cases of ontologies, metaknowledge or other sophisticated forms of management strategies, the elimination of knowledge and the completion of KBs becomes a non-trivial problem. For this reason the concepts of orchestration and choreography of ontologies are introduced in the Semantic Web and especially for WSMO. The elimination of at least one of the relations inside the knowledge can lead to discrepancies in one or in several subsets of knowledge in $K$. That is why after the presented second stage and on the third stage a check-up of relations is performed including eliminated of modified knowledge and the new knowledge from subsets $W$, $N$, $I$, $U$ are tested for non-discrepancies via an above described procedure. After the successful finish of the process a new set of knowledge $K'$ is formed that is more qualitative than that in $K$; according to this criterion it is a result from a machine unsupervised learning managed by a priori defined models of contradictions and by the managing strategies with or without using metaknowledge.

The relations concerning parts of the contradiction or other knowledge from set $K$ cannot always be presented via implications, standard links of is_a type and lots of other known links. Two sets of objects tied to both sides of the contradiction are shown in Fig. 11. A part of the objects are related to the respective side via implications, is_a or via some other relation. The bigger part of the objects does not possess this type of links but it enters the area round the part of the contradiction marked by a dotted line. For example let the following relation between two concepts or two objects is confirmed statistically. Attacks are often accompanied by scanning ports but from the fact that a port is scanned it does not follow that there will be an attack. The interpretation in the case is 'the object is related somehow to the investigated part of the contradiction'; in other words the nature of the relation is not fully clarified. Therefore it is necessary to determine the goal 'clarification of reasons and the nature of the relation'. In a great number of cases the investigation of the goal is reduced to one of known relations fixed above or the link is of a defeasible type; if the exception from the rules is true then a part of the relations in Fig. 11 are eliminated or they change their direction. If the goal 'clarification of reasons and the nature of the relation' is not resolvable to the moment then the knowledge linked by the relation form a set of heuristic confirmations or denials of knowledge in the figure; they can be used not only to resolve a contradiction but also in the classical case of knowledge conflict.

Fig. 9 contains an inference tree where by definition all logical links are in the form of a classical implication. In Fig. 11 it is possible one of the relations in the zone round the part in the left section of the figure to be of is_a type and the other one to be an implication. In such cases the detection of reasons led to a contradiction is a problem without solutions in the sense of using classically formally logical approaches, so it is necessary to use non-classical applications (here original methods and applications to zero the information, etc. are introduced).

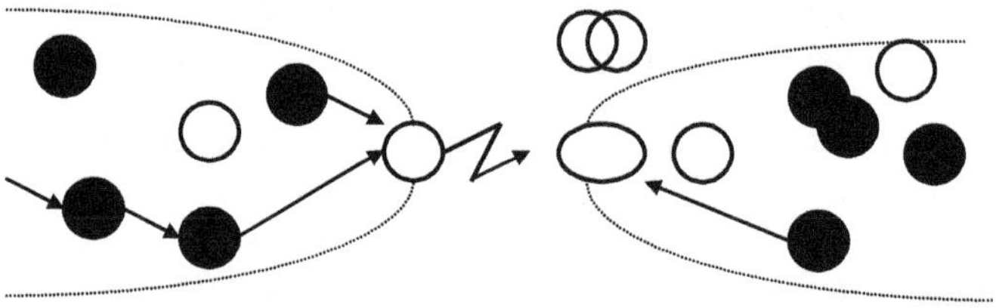

**Fig. 11** Set of concepts linked to parts of the contradiction

Any KB with more than a thousand knowledge items contains inevitably at least one contradiction. Contradictions are detected and resolved in accordance with the presented above methods. Each of the methods is described by exact but the result from their application is not algorithmically forecastable. In this way the presented solutions possess the characteristic properties of data mining: the results depend on the data or the knowledge, i.e. they are *data driven* and the applications are *knowledge driven*. Processing contradictions improves the quality of the gathered information according to criteria that are introduced above. At that the insignificant elements most often are eliminated; this leads to an additional increase of the quality. So an effect is realized from a principle that has been formulated in XII$^{th}$ century, that has been investigated since the dawn of informatics and that has not been successfully realized. Occam's razor means the following: choose simple, elegant, reject complex solutions. The method of inference from contradictory information allows remove inadequacies from situations that are similar to the one depicted in Fig. 12 and also to correct the insignificant elements and/or to present the situation in a more clear way as it is done with the transition from Fig. 12 to Fig. 13.

Fig. 12 presents a generalized graph showing the tendencies in the development of information security systems (ISS). The values are mean for contemporary ISS concerning the parameters of price ($), time periods (years) and quality (%). The last parameter operates via the fuzzy concept of 'quality' so the construction of the figure resides not on

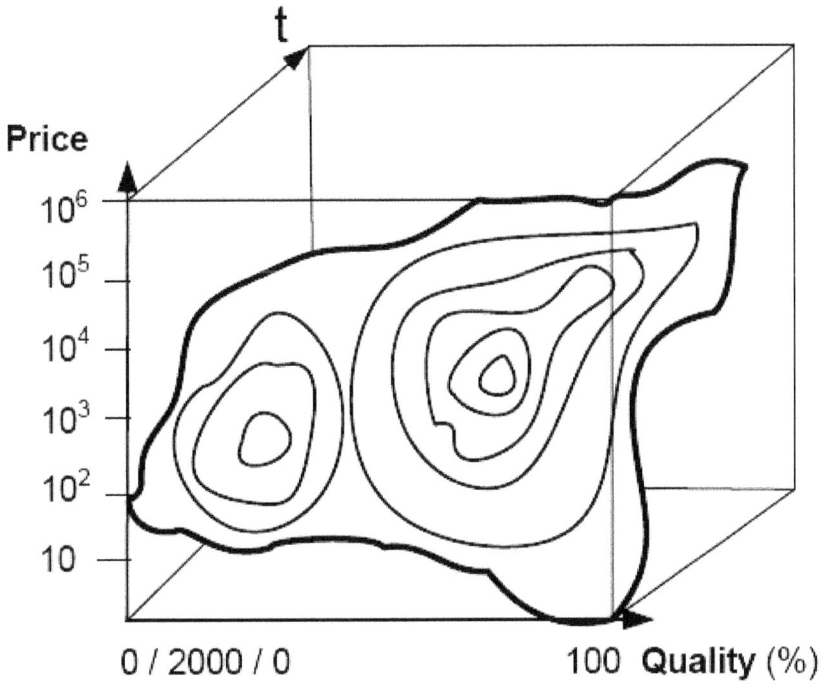


**Fig. 12** Tendencies in the development of information security systems (ISS)

Fig. 13 presents the same information with elimination of various insignificant data and knowledge aiming at clarifying the strategic picture of the domain.

The statistical data from Fig. 13 show that in course of time from the general direction gradually fall out more non-qualitative or hard to adapt systems and that the crisis accelerates the process. The price of qualitative systems permanently grows but even the most expensive from them offer smaller securities compared to the ones before several years. Therefore significant changes are necessary in the security software to manage the situation.

Contemporary realizations of statistical methods are effective and convenient to use but the comfort must be paid: information is encapsulated. In other words it is not possible to build tools to acquire new knowledge or to solve other problems of logical nature. If we partition the methods for intelligent processing in two groups (quantitative and qualitative) then statistical methods belong to the first group and

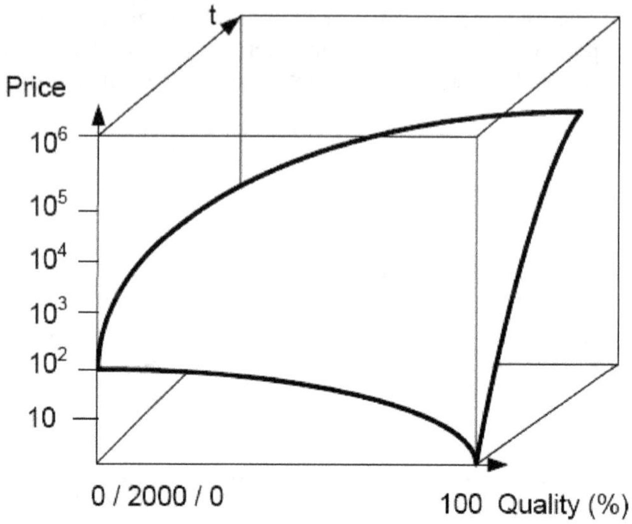


**Fig. 13** Tendencies after removing insignificant elements

the logical methods belong to the second group. For this reason their mechanical fusion is non-perspective. At the same time in the present chapter new methods to extract information from KBs or via human-machine dialogue are presented. Under the common management of the logical metamethod for intelligent analysis of information it is possible successfully to apply also statistical methods for SMM.

The very conflict resolution does not lead to clarifying all problems related to it. Answers to questions like 'what did the contradiction lead to' or 'why did the contradiction appear', or 'how to correct knowledge that led to a contradiction' are constructed depending on the way of detecting and resolving the contradiction. Stating questions like 'are there other conflicts and contradictions', 'is the contradiction caused by an expert error or by the expert style of work', 'since when the quality of knowledge is worsened', 'till when agents will contradict to a given situation', 'why do agents conflict' and many other similar questions as goals, agents possess more possibilities to improves KBs and to improve themselves by unsupervised learning though that in certain cases it is possible to ask also expert's assistance. In such cases our ontological elaborations will help us concerning the inclusion of new relations of types ORIGIN, HOW, WHY, etc. After terminating the research of the stated above subgoals it is possible automatically to add the presented relations to the existing in the system ontologies. So the relation agent-ontology is closed and new possibilities for in-depth processing of knowledge by the agents are revealed.

Conflicts or contradictions between various parts of knowledge assist the construction of constraints during the formation of ontologies; simultaneously it is possible to specify also separate attributes of the ontologies.

Parallel processing in cases of distributed artificial intelligence, e.g. via Petri nets, introduces substantial advantages in realizations of the already described methods. Fig.14 shows results from self estimates of agent's actions based on its used knowledge. Elements of quadruple logic are applied. It is clear from the figure that intermediate calculations do not lead to the searched final result, the answers 'YES' or 'NO' to the stated goal or at least to one of the four possible answers. Let at the same time the other agent 2 participating in the conflict has an estimate of knowledge from Fig.15. Let also the results from Fig. fig. 14 and 15 are compared one to another after a predefined period of time to solve the conflict, e.g. after 5 minutes.

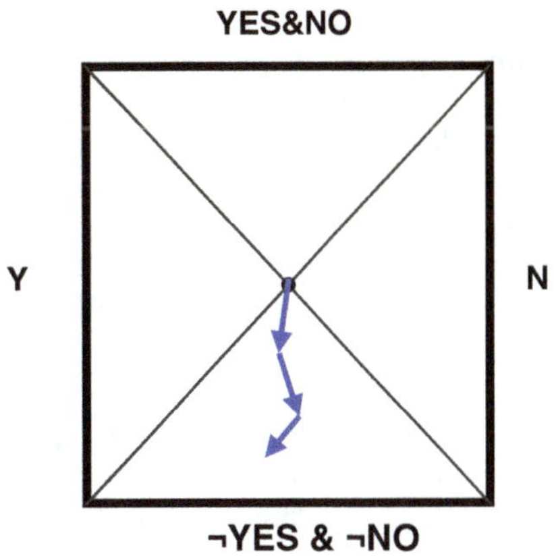


**Fig. 14** Parallel processing via knowledge juxtaposition

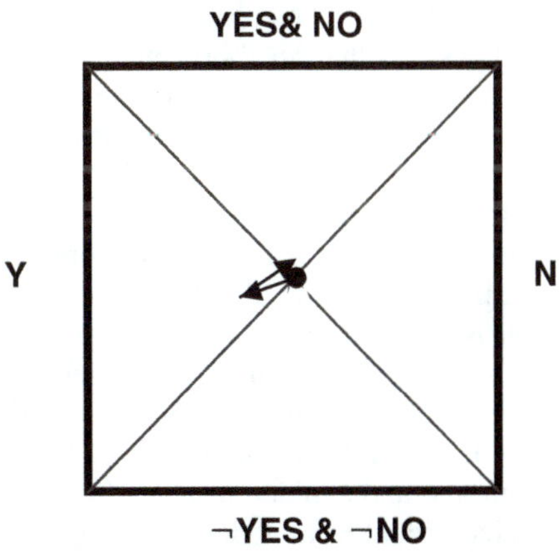


**Fig. 15** Corresponding calculations to Agent 2

The comparison shows that agent 1 has a substantially bigger progress in the conflict resolution compared to agent 2. The additional information related to the comparative analysis of the parallel activities of the agents offers much greater possibilities to resolve the conflict in comparison with the analysis by the agents one by one.

The obtained results most often are in the form of fuzzy estimates; the chances to construct logical proofs based on similar comparisons are much rarer. They allow substantially reduce the algorithmic complexity of the problem to resolve conflicts or of contradictions in multiagent systems.

Processes for resolving conflicts between two software agents most often are evolutionary. In the general case the goal (the conflict resolution) is denoted as a fitness function and the necessary solution is found applying genetic algorithms. At the same time the investigated above methods for conflict resolution introduce substantial possibilities to improve existing methods for evolutionary computations.

The evolution of agents is analogous to the evolution of living organisms where significant changes do not happen by chance (vaguely) but only of necessity. A contradiction with normal conditions for existence (work) is a bright example for necessity of changes, i.e. a motor of evolutionary changes. In cases when the fitness function is a problem for conflict resolution then the introduced above methods offer the following advantages compared to other methods of evolutionary computations:

**1.** The conflict-resolving fitness function cannot be defined a priori because nobody is able to foresee all imperfections of knowledge or data that can lead to a formation of the defined function. The set of fitness functions appears and it is changed automatically in this way during the process of operation of the proposed application.

**2.** Mutations, crossovers and other factors influencing the evolutionary process are changed in the following way. The probabilities for the events, e.g. 2% for mutations and 8% for crossovers, can be substantially changed and the most important is the fact that they may not be defined randomly because this leads to a rather adjusted character of the evolution. The constraints for ontologies of other types of constraints will be used instead that are born by the appearance and/or conflict resolutions. Constraints form zones with high probabilities for mutations or crossovers respectively. In cases of crossovers individuals avoid mutations or mutageneous zones and these principles are changed only in exceptional cases, e.g. in cases of irresolvable contradictions.

**3.** The fitness function (or the set of fitness functions) can be variable or it can change the direction during the process of solving the problem (a variable fitness function). The variation is realized using additional information. For example, if an agent has in its goal list 'to make a contact with agent *X* aiming at a mutual defense in the computer network' and after the last attack it is evident that agent *X* has contributed to the attack then this agent *X* must not be searched for collaborations but it must be monitored and avoided instead.

**4.** In cases when an individual or a model is applied successfully for a definite fitness function then the individual, the model or the fitness function may be replaced to provoke a new conflict. Collaborative behavior can be replaced by a competitive behavior in the same way, etc. Artificial changes for provoking and resolving new conflicts via the stated above tools serve to control the solutions and for adaptations of successful solutions in other domains. One of the effects of evolutionary transition through conflicts, crises and contradictions is the elimination from insignificant details and unnecessary complications; in other words it is possible to realize a new application or to achieve Occam's Razor effects.

## 9 Conclusion

The PN-based modular approach was utilized in order to model the agent cooperation in MAS in the form of the vector linear discrete dynamic system. Such a system approach is based on the analogy with DES. It is applicable for both the wide class of agents and the wide class of forms of agent cooperation in MAS. Three possible forms of the module representing the interface among agents were proposed, described and illustrated - namely, the interface: (i) based on additional PN transitions; (ii) based on additional PN places; (iii) in the form of the additional PN-subnet. The dynamic behaviour of the systems was tested for arbitrarily chosen initial states by means of corresponding RG. Using the PN-based approach enable us to find feasible states in analytical terms and to insight into their causality. This allows to observe the system dynamics and to find fit control strategies.

Conflict resolution methods are considered. It is shown that the resolution process lead to novel machine learning methods.

**Acknowledgments.** The research was partially supported by the Slovak Grant Agency for Science VEGA under grant # 2/0075/09 and the contract between Intstitute of Informatiocs - Slovak Academy of Sciences and Institute of Information Technologies - Bulgarian Academy of Sciences. The authors thank VEGA for this support.

## References

Čapkovič, F.: An Application of the DEDS Control Synthesis Method. Journal of Universal Computer Science 11(2), 303–326 (2005)

Čapkovič, F.: DES Modelling and Control vs. Problem Solving Methods. International Journal of Intelligent Information and Database Systems 1(1), 53–78 (2007)

Čapkovič, F.: DES Modelling and Control vs. Problem Solving Methods. International Journal of Intelligent Information and Database Systems 1(1), 53–78 (2007)

Demazeau, Y.: MAS methodology. In: Tutorial at the 2nd French-Mexican School of the Rep-artee Cooperative Systems - ESRC 2003, September 29-October 4. IRISA, Rennes (2003)

Demazeau, Y.: MAS methodology. In: Tutorial at the 2nd French-Mexican School of the Rep-artee Cooperative Systems - ESRC 2003, September 29-October 4. IRISA, Rennes (2003)

Fonseca, S., Griss, M., Letsinger, R.: Agent Behavior Architectures - A MAS Framework Comparison. HP Labs Technical Report HPL-2001-332. HP, Palo Alto (2001)

Hung, P.C.K., Mao, J.Y.: Modeling e-negotiation activities with Petri nets. In: Spraguer Jr., R.H. (ed.) Proceedings of 35th Hawaii International Conference on System Sciences, HICSS 2002, Big Island, Hawaii, vol. 1, p. 26. IEEE Computer Society Press, Piscataway (2002) (CD ROM)

Lenz, K., Oberweis, A., Schneider, S.: Trust based contracting in virtual organizations: A concept based on contract workflow management systems. In: Schmid, B., Stanoevska-Slabeva, K., Tschammer, V. (eds.) Towards the E-Society – E-Commerce, E-Business, and E-Government, pp. 3–16. Kluwer Academic Publishers, Boston (2001)

Murata, T.: Petri Nets: Properties, Analysis and Applications. Proceedings of the IEEE 77(4), 541–588 (1989)

Nowostawski, M., Purvis, M., Cranefield, S.: A layered approach for modelling agent conver-sations. In: Wagner, T., Rana, O.F. (eds.) Proceedings of 2nd International Workshop on Infra-structure for Agents, MAS, and Scalable MAS, 5th Int. Conference on Autonomous Agents - AA, Montreal, Canada, May 28-June 1, pp. 163–170. AAAI Press, Menlo Park (2001)

Peterson, J.L.: Petri Net Theory and Modeling the Systems. Prentice Hall Inc., Englewood Cliffs (1981)

Saint-Voirin, D., Lang, C., Zerhouni, N.: Distributed cooperation modelling for maintenance using Petri nets and multi-agents systems. In: Proceedings of 5th IEEE Int. Symposium on Computational Intelligence in Robotics and Automation, CIRA 2003, Kobe, Japan, July 16-20, vol. 1, pp. 366–371. IEEE Press, Piscataway (2003)

Yen, J., Yin, J., Ioerger, T.R., et al.: CAST: Collaborative agents for simulating teamwork. In: Nebel, B. (ed.) Proceedings of 17th Int. Joint Conference on Artificial Intelligence - IJCAI 2001, Seatle, USA, vol. 2, pp. 1135–1142. Morgan Kaufmann Publishers, San Francisco (2001)

Jotsov, V.: Dynamic Ontologies in Information Security Systems. J. Information Theory and Applications 15(4), 319–329 (2008)

Arruda, A.: A survey on paraconsistent logic. In: Arruda, A., Chiaqui, C., Da Costa, N. (eds.) Math. Logic in Latin America, pp. 1–41. North-Holland, Berlin (1982)

Jotsov, V.: Semantic Conflict Resolution Using Ontologies. In: Proc. 2nd Intl. Conference on System Analysis and Information Technologies, SAIT 2007, RAS, Obninsk, September 11-14, vol. 1, pp. 83–88 (2007)

# An Application of Mean Shift and Adaptive Control to Active Face Tracking

Ognian Boumbarov, Plamen Petrov, Krasimir Muratovski, and Strahil Sokolov

**Abstract.** This paper considers the problem of face detection and tracking with an active camera. An algorithm using Haar-like face detector and Mean-shift method for face tracking has been presented. We propose an adaptive algorithm that provides automated control of a Pan-Tilt camera (PTC) to follow a person's face and keep their image centered in the camera view. An error vector defined in the image plane and representing the face offset with respect to the center of the image is used for camera control. Adaptive feedback control design and stability analysis are performed via Lyapunov techniques. Simulation results are presented to illustrate the effectiveness of the proposed algorithms.

## 1 Introduction

In recent years, face detection and tracking has been an important research topic in computer vision. Applications of people tracking are, for example, surveillance systems where motion is the most critical feature to track. Also, the identification of a person, facial expression recognition, and behavior recognition may require head tracking as a pre-processing procedure. The use of autonomous pan-tilt camera as opposed to fixed cameras extends the range of sensing and effectiveness of surveillance systems. However, the task of person tracking with moving camera is much more complex than the person tracking from a fixed camera and effective tracking in a pan-tilt scenario remains a challenge.

Ognian Boumbarov · Krasimir Muratovski · Strahil Sokolov
Faculty of Telecommunications, Technical University of Sofia,
8, Kl.Ohridski Str., 1000 Sofia, Bulgaria
e-mail: olb@tu-sofia.bg, kraskoval@abv.bg, s_sokolov@tu-sofia.bg

Plamen Petrov
Faculty of Mechanical Engineering, Technical University of Sofia,
8, Kl.Ohridski Str., 1000 Sofia, Bulgaria
e-mail: ppetrov@tu-sofia.bg

V. Sgurev et al. (Eds.): Intelligent Systems: From Theory to Practice, SCI 299, pp. 161–179.
springerlink.com 

While there has been significant amount of work on person tracking from the single static camera, there has been much less work on people tracking using active cameras, [1, 2]. The adequate control of the pan-tilt camera is an essential phase of the tracking process. In [3], a method for the detection and tracking of human face and facial features was presented. The located head is searched and its centroid is fed back to a camera motion control algorithm which tries to keep the face cantered in the image using a pan-tilt camera unit. In [4], a PI-type controller was proposed for a pan-tilt camera. An image-based pan-tilt camera control for automated surveillance systems with multiple cameras was proposed in [5].

A robust face detector should be able to find the faces regardless of their number, color, positions, occlusions, orientations, facial expressions, etc. Additionally, color and motion, when available, may be characteristics in face detection. Even if the disadvantages of color based methods like sensitivity on varying lighting conditions make them not as robust methods, they can still be easily used as a pre-processing step in face detection.

In [6] is described an efficient method for face detection. Viola and Jones use a set of computationally efficient "rectangle" features which act on pairs of input images. The features compare regions within the input images at different locations, scales, and orientations. They use the AdaBoost algorithm to train the face similarity function by selecting features. Given a large face database, the set of face pairs is too large for effective training.

A number of authors describe methods for object detection and tracking based on color. Such methods generate more false-positives than Viola-Jones and are not as robust to illumination changes.

In [7, 8] the Mean-Shift algorithm is presented. It represents object tracking by region matching. Between frames the tracking region can change location and size. The Mean-Shift algorithm is used as a way to converge from an initial supposition for object location and scale to the best match based on color histogram similarity. The contributions of each pixel are Gaussian weighted based on their distance from the window's center. Similarity is measured as Bhattacharyya distance.

In this paper, we present an algorithm that provides automated control of a pan-tilt camera to follow a person's face and keep his image cantered in the camera view. Our approach to the face tracking is to design a control scheme using visual information only. The target motion is unknown and may be characterized by sudden changes. An offset (error) vector defined in the image plane and representing the coordinates of the target with respect to the image frame is used for camera feedback control. The proposed adaptive control achieves asymptotic stability of the closed-loop system. The control design and stability analysis are performed via Lyapunov techniques.

In last years, several methods for face tracking have been proposed. The objective is to develop robust algorithms working in real conditions, invariant to noise in the image, the face pose and orientation, occlusion, changes in lighting conditions and background. These algorithms can be implemented on the base of specific features as: the face skin colour [9, 10, 11], particular features of the face /nose, eyes, mouth/ and their relative position [12, 13], skin texture [14], as well as a combination between them. In this paper, we use Haar-like face detector and Mean-shift method combined with active camera control for face detection and tracking.

The rest of the paper is organized as follows. In Section 2, the face detection and tracking procedures are described. In Section 3, we present the adaptive feedback tracking controller for the pan-tilt camera and its visual implementation. We provide real and simulation results in Section 4. Conclusions are presented in Section 5.

## 2 Face Detection and Tracking

The goal of the face detection and tracking algorithm is to determine and transmit the position of the face center to the camera control system. These coordinates serve as information for camera control in a way to keep face center near to center of current frame.

The correct face detection and tracking depends on the variations of pose, illumination, facial expressions at different locations, scales, and orientations. In this section we address the problems one by one using computationally efficient algorithms.

For solving of the more general problems in tracking, we use Mean-shift combined with active camera control. The more subtle problem of different locations, scales and head orientations is solved by using efficient face detection based on Haar-like features. Viola and Jones [6] proposed a detection scheme based on combined Haar-like features and cascaded classifiers. This method is very fast to search through different scales because of the simple rectangular features and the integral image and is robust against varying background and foreground. The training of Viola-Jones detector takes time, but once the detector is trained and stored in memory, it works very fast and efficiently. This detector works on the Intensity (Y) component only (from color space YCbCr).

After a face is detected we proceed to the stage of tracking by using Mean-shift. For construction of the tracking model we use the color components Cb and Cr only that are extracted from the detected face's window. This gives us the possibility to use a 2D color histogram with size 255x255 bins in order to achieve more precise representation of the color distribution of the detected face's model.

### 2.1 Face Detection Using Haar-Like Features

The face detection component is a cascaded structure. It contains several strong classifiers. Each strong classifier consists of several weak classifiers. A weak classifier has its own specific weight in the structure. It processes a specific feature and uses thresholds for rejection or acceptance of that feature. We can generate many sub-images from an image by various positions and scales. Each strong classifier rejects numbers of sub-images; the rejected sub-images are no longer being processed.

Features used by our system can be computed very fast through integral images. In this chapter, we will introduce integral images, feature types and the way to use integral images to calculate a value from a feature. The approach for face detection using Haar-like features requires the computation of an integral image. Integral image is also called Summed Area Table (SAT). It represents a sum of grayscale pixel values for a particular area in an image. The main steps of the initial face detection procedure are presented on Fig.2.

*Computation of the integral frame*

Rectangle Integral image RII is introduced in [7]. The value at position (i, j) in a RII represents the sum of pixels above and left to (i, j) in the original image (Fig.1):

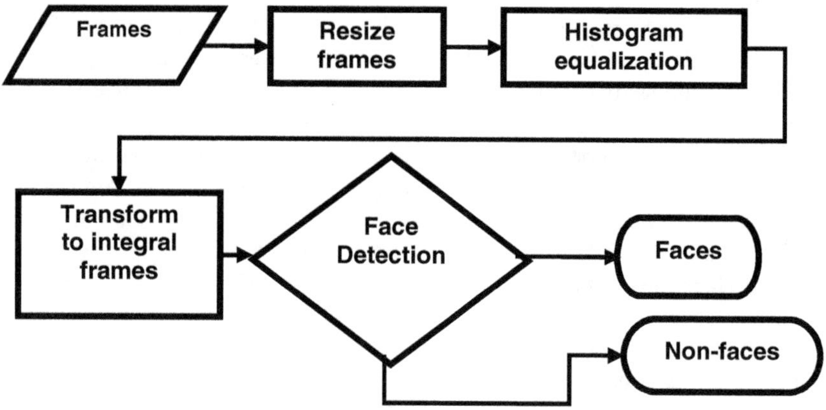


**Fig. 1** Flow chart of the initial face detection procedure used in our system

$$RII(i,j) = \sum_{i\le i', j\le j'} I(i',j') \tag{1}$$

where RII(i, j) is the value of RII at position (i, j) and I(i', j') is the grayscale pixel value of the original image at position (i', j'). For each frame, we compute its RII through only one pass of scanning all its pixels. In practice, first we compute the cumulative row sum and then add the sum to RII at previous row and the same column to get the sum of pixels above and left:

$$\begin{aligned} \text{ROW\_SUM}(i,j) &= \text{ROW\_SUM}(i-1,j) + I(i,j) \\ \text{RII}(i,j) &= \text{RII}(i,j-1) + \text{ROW\_SUM}(i,j) \end{aligned} \tag{2}$$

*Selection of a characteristic set of features for face* ***detection***
There are some characteristics of faces which are useful for detecting faces. We can observe that some areas are darker then other areas. For example, the areas of two eyes are usually darker than the area of the bridge of a nose; the area across eyes is usually darker than the area of cheeks. Thus, we can use the differences of sums of dark areas and sums of light areas to detect faces.

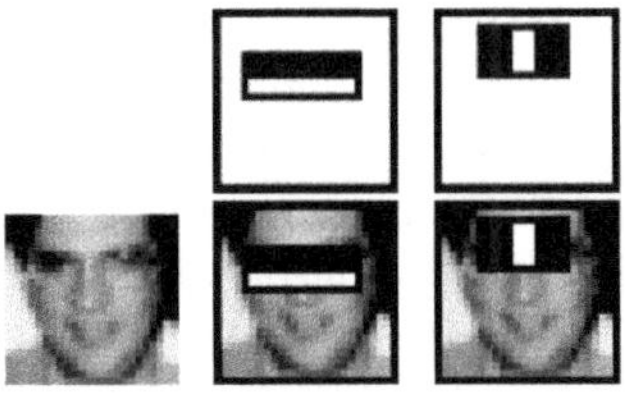

**Fig. 2** Face characteristics

Lienhart et al. introduces four classes of feature types [15] as shown in Fig.2. Features calculate values by subtracting the sum of pixels of the white area from the sum of pixels of the black area in an image. We extend those feature types to be more generalized (Fig.3):

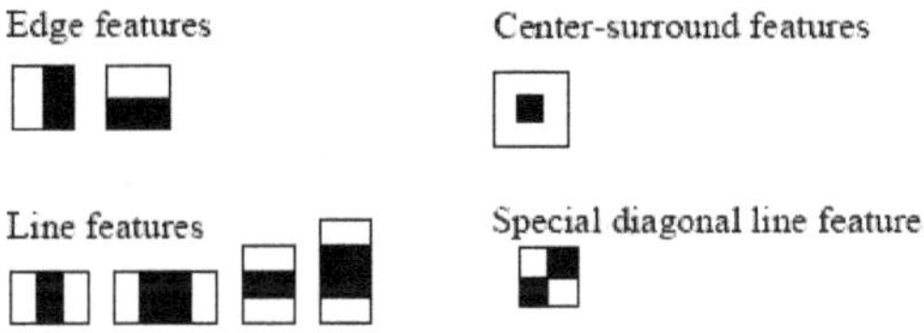


**Fig. 3** Extended feature types

*Application of a machine-learning technique for creation of a cascade of strong classifiers*
In our work we use boosting as the basic learning algorithm to learn a strong classifier. Freund and Schapire introduce boosting algorithm and its application "AdaBoost" [16] which stands for Adaptive Boosting. AdaBoost is used to both select features and to train a strong classifier. A strong classifier contains several weak classifiers. A weak classifier is used for determining faces and non-faces and only required to be slightly better than random sampling. A weak classifier is composed of a feature, a weight and two thresholds. For each feature, the weak learner determines a high threshold and a low threshold such that the minimum numbers of samples are misclassified.

A weak classifier $h_m(n)$ determines an image as a face if the value calculated by the feature is between those two thresholds:

$$h_m(n) = \begin{cases} 1, & low\_threshold \le f_m(n) \le high\_threshold \\ 0, & otherwise \end{cases} \quad (3)$$

where, $n$ is the n-th image and $h_m(n)$ is the output generated by it from the m-th weak classifier. In most works, the weak classifiers contain only one threshold. Some authors recently tend to use two thresholds for raising detection rate and lowering false positives. The structure contains several strong classifiers (denoted with circles "1, 2, 3 …") as shown on Fig.4. Viola-Jones face detector is known to operate in online mode for small image sizes. In the cascaded structure, each strong classifier rejects a number of non-face images. This is efficient for detecting large numbers of sub-images. In the early stages, most of the non-face parts of the frame are rejected; only few numbers of non-faces (hard samples) are needed to be processed in the late stages.

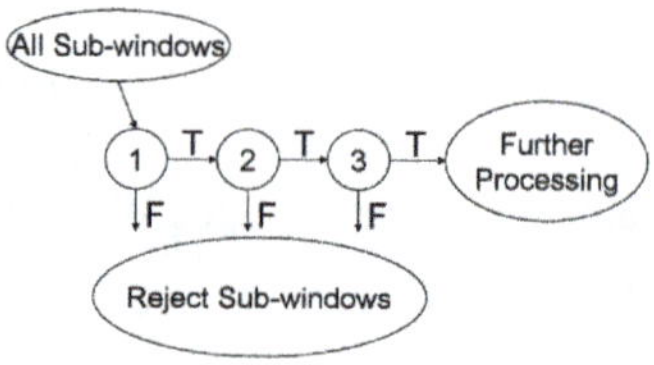


**Fig. 4** Cascaded structure

In order to reduce false detections from V-J FD it is necessary to add a stage for validation. We decided to implement the Convolutional Neural Network (CNN) approach [17, 22]. The stage is depicted on Fig.5.

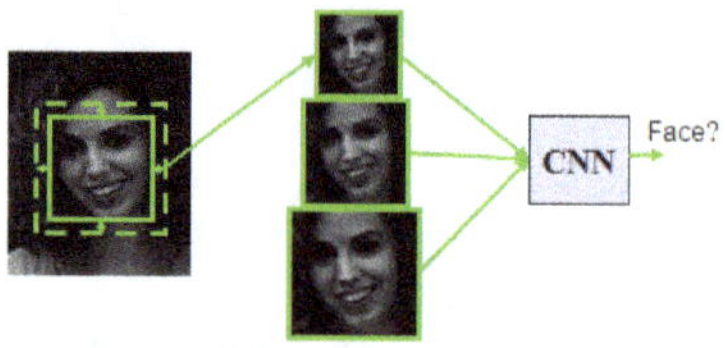


**Fig. 5** Face validation using CNN

We take the output of Viola-Jones face detector and apply the pretrained CNN to validate the face. According to the proposed verification method the face-like object's outer containing window is processed - a pyramid of images across different scales is built for this extended area. The pyramid consists of the extended face-like object's images scaled by factor 0.8. The size of the last pyramid's image has to be not less than 32x36 (default CNN's input size). Each image from the pyramid is being processed by the CNN at once instead of being scanned by a fixed-size window. After the processing of all pyramids' images by a CNN, the number of multiple detections is evaluated and an object is accepted as a face when the number of multiple detections is equal or greater than the threshold value which is chosen experimentally.

## 2.2 *Kernel Density Estimation*

In this paper, we use the "Mean Shift" algorithm [18, 19] for face tracking. The Mean Shift method is a nonparametric iterative technique with gradient descending for finding mode of probability distribution. In our algorithm kernel density estimation (KDE) is used in order to make the color feature more robust and discriminative for the mean-shift based tracker. KDE is a kind of nonparametric technique, which is also widely used in the area of data analysis. Combining advantages of both colour histograms and spatial kernel, Comaniciu [20] proposes a nonparametric tracking approach based on mean shift analysis.

### 2.2.1 Description of Mean Shift

To realize the face detection and tracking, a preliminary training stage is necessary, i.e., we utilize preliminary information in the form of tracking object model for the target, which has to be tracked. In our case, as tracking object model we use the human skin region obtained with initial target localization: initial face gravity center and enclosing kernel size-based on Viola-Jones face detector.

For constructing of tracking model, in this paper, we use the *Cb and Cr* color components of the *YCbCr* color space. By using a 2-dimensional color histogram with size 256x256 bins it is possible to achieve better accurate representation of the face color distribution model.

Let $\{m_{i,j}\}_{i=1...Nm,j=1..Mm}$ be the pixel positions of the first image of the model with size *Nm x Mm* centered at 0. The histogram of the model is given by:

$$\hat{mh}_{lCrCb} = C_m \sum_{i=1}^{Nm} \sum_{j=1}^{Mm} k\left(\left\|\frac{m_{i,j}}{w_m}\right\|^2\right)\delta_m(Cb,Cr), \tag{4}$$

$$Cr,Cb=0,1...255,$$

where $\delta_m(Cb,Cr) = \delta\left[b(m_{i,j})-Cr\right]\delta\left[b(m_{i,j})-Cb\right]$ and $\delta$ is the Kroneker delta function. To obtain a general face model we have to integrate the histograms of used face images contained in the training set. This is done by simple averaging, i.e:

$$\hat{mh}_{CrCb} = \frac{1}{p}\sum_{l=1}^{p} \hat{mh}_{lCrCb},\quad Cr,Cb=0,1...255 \tag{5}$$

Let $\{tr_{i,j}\}_{i=1...Nt, j=1..Mt}$ are the pixel locations for an image of the target centered at $tr_0$. In the tracking process, computation of the target histogram is performed into the search window centered at $tr_0$:

$$\hat{th}CrCb(tr_0) = C_t \sum_{i=1}^{Nt} \sum_{j=1}^{Mt} k\left(\left\|\frac{tr_0 - tr_{i,j}}{w_t}\right\|^2\right)\delta_t(Cb,Cr) \tag{6}$$

$$Cr, Cb = 0,1...255$$

where $\delta_t(Cb,Cr) = \delta\left[b(tr_{i,j}) - Cr\right]\delta\left[b(tr_{i,j}) - Cb\right]$.

In Equations (4) and (7), $k$ is the kernel profile. The model and target size of the kernel profile is denoted by $w_m$ and $w_t$, respectively, and $b$ is histogram index for pixel position $tr_{i,j}$. $Cm$ and $Ct$ are normalization constants given by the following expressions:

$$C_m = \frac{1}{\sum_{i=1}^{Nm} \sum_{j=1}^{Mm} k\left(\left\|\frac{m_{i,j}}{w_m}\right\|^2\right)} ; C_t = \frac{1}{\sum_{i=1}^{Nt} \sum_{j=1}^{Mt} k\left(\left\|\frac{tr_0 - tr_{i,j}}{w_t}\right\|^2\right)} \tag{7}$$

The goal of using the kernel in process of histogram computing is to give bigger weights for pixels near to object of interest. To estimate the similarity of the target and model histograms, we use the Bhattacharyya coefficient:

$$\begin{aligned} \hat{\rho}(tr_0) &= \rho[\hat{th}CrCb(tr_0), \hat{mh}CrCb] = \\ &= \sum_{C_b=0}^{255} \sum_{C_r=0}^{255} \sqrt{\hat{th}CrCb(tr_0)\hat{mh}CrCb} \end{aligned} \tag{8}$$

Using the Taylor expansion, the Bhattacharyya coefficient is approximated as follows:

$$\begin{aligned} &\rho[\hat{th}CbCr(tr_0), \hat{mh}CbCr] \approx \\ &\approx \frac{1}{2} \sum_{C_b=0}^{255} \sum_{C_r=0}^{255} \sqrt{\hat{th}CbCr(tr_0)\hat{mh}\mathrm{CbCr}} + \frac{C_t}{2} \sum_{j=1}^{Mt} \sum_{i=1}^{Nt} w_{i,j}\left(\left\|\frac{tr_0 - tr_{i,j}}{w_t}\right\|^2\right) \end{aligned} \tag{9}$$

where $w_{i,j} = \sum_{C_b=0}^{255} \sum_{C_r=0}^{255} \delta(Cb,Cr)\sqrt{\frac{\hat{m}hCbCr}{\hat{th}CbCr(tr_o)}}$ with size $NtxMt$. (9a)

The process of tracking is realized with iterative execution of Mean-Shift algorithm which is iterative maximization of (9) for previous pixel location (xc0, yc0) given by vector $X = [x_c, y_c]^T$ with coordinates:

$$x_c = \frac{\sum_{i=1}^{Nt}\sum_{j=1}^{Mt} i w_{i,j} g\left(\left\|\frac{tr_0 - tr_{i,j}}{w_t}\right\|^2\right)}{\sum_{i=1}^{Nt}\sum_{j=1}^{Mt} w_{i,j} g\left(\left\|\frac{tr_0 - tr_{i,j}}{w_t}\right\|^2\right)}, \quad y_c = \frac{\sum_{i=1}^{Nt}\sum_{j=1}^{Mt} j w_{i,j} g\left(\left\|\frac{tr_0 - tr_{i,j}}{w_t}\right\|^2\right)}{\sum_{i=1}^{Nt}\sum_{j=1}^{Mt} w_{i,j} g\left(\left\|\frac{tr_0 - tr_{i,j}}{w_t}\right\|^2\right)} \qquad (10)$$

where $g\left(\frac{tr_0 - tr_{i,j}}{w_t}\right) = -k'\left(\frac{tr_0 - tr_{i,j}}{w_t}\right)$, assuming that the derivative of k(x) exists for all $k \in [0, \infty)$, except for a finite set of points.

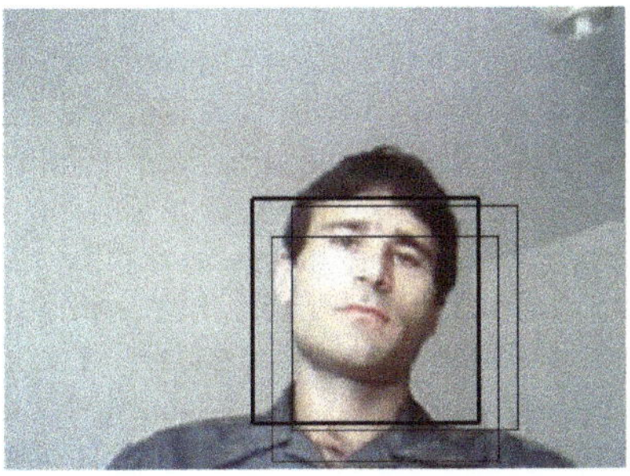

**Fig. 6** Illustration of the process of convergence using Mean Shift in 3 iterations

In order to increase the speed of the proposed algorithm, it is necessary to reduce in accurate way the size of search window, where histogram calculation is proceeded. An example of window adaptation is proposed in [9], where the zero moment is calculated from the probability distribution. In our algorithm, except calculation of zero moment for $w_{i,j}$ we use additional criterion, who takes into account similarity between the histograms of face model and tracking object. That is because in variable light conditions variations of the correlation coefficients are unavoidable and in case of constant distance between camera and tracking object this can result in change of zero moment and thence incorrect change in window size.

At start of the algorithm searching is perform in the whole image (frame). The coordinates of the face center for first frame (xc0, yc0) are used as start values for searching in next frame, which results in minimizing of computing complexity. After complete search procedure final coordinates (xc,yc) of face center for given set of frame are pass to block for Adaptive Camera Control.

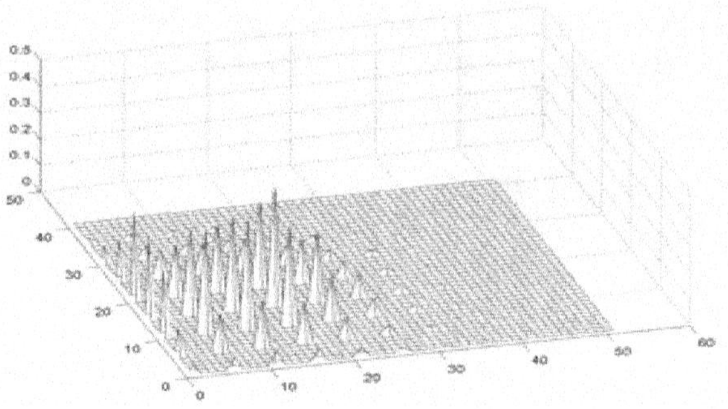

**Fig. 7** Representation of Bhattacharyya coefficients for search window with size 40 x 50 pixels for the face center from Fig. 6

### *2.3 Description of Proposed Algorithm for Face Detection and Tracking*

The proposed algorithm contains two stages – initialization and working stage.

The initialization uses the Viola-Jones face detector. It starts the Mean-shift tracker at the location and size of the tracking window *in frame k=1*. The algorithm begins with search through the whole frame for a face. After face localization with Viola-Jones face detector, the face gravity center and the size of the enclosing face kernel are defined. The human skin region, belonging to the enclosing face kernel, is used as tracking object model. The main parameters of the model are the histogram bins of the Cb and Cr components of the face detected. When the histogram model of the object for tracking is obtained, the nearest mode in the probability density is located by using Mean Shift. The Mean-Shift estimated face position is sent to the active camera control unit.

The working stage is realized for each subsequent video frame. It locates the search window position equal to the face position of previous frame. The search window has area bigger in twice than the initial/previous kernel's area. After that, the Mean-shift tracker is started to locate nearest mode in the probability density in the search window and to send the Mean-Shift estimated face position to the active camera control unit. The current size and location of the tracked object are reported and used to set the size and location of the search window in the next video image. The process is then repeated for continuous tracking.

## 3 Adaptive Camera Control

### *3.1 Control Algorithm Design*

In this Section, we consider the problem of controlling the motion of a pan-tilt camera. We are dealing with a dynamic environment, and the control objective is to maintain the person's face being tracked in the center of the camera view.

During this process, the coordinates of the face center with respect to the image plane (face offset) is retrieved from processed acquired images. A kinematic model of the camera is developed which is used for the design of an adaptive camera feedback tracking controller. The proposed feedback control makes use of visual information only without prediction of the target motion motivated by the potential application in which the person motion is unknown and may be characterized by sudden changes.

Since we want to track the center of the face, let $e_c = [x_c, y_c]^T$ be the face offset with respect to the center of the image, i.e., the coordinates of the centroid of the segmented image inside the window where the face is located. The following simplifying assumptions are made: a) The intersection of the pan and camera axes coincides with the focus of the camera; b) The coordinates $x_c$ and $y_c$ of the blob in the image plane depend only on the pan angle $\theta$ and the tilt angle $\varphi$, respectively.

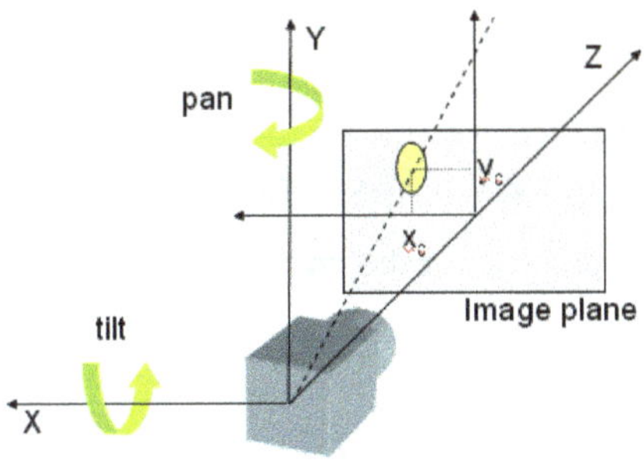


**Fig. 8** Pan-tilt camera unit and the face offset in the image plane

In that follows, we consider in detail the control design for the loop concerning the pan motion of the camera.

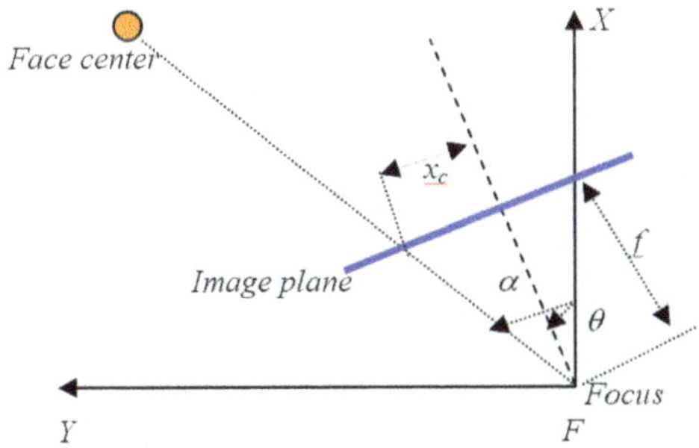


**Fig. 9** Geometrical representation of the face $x_c$-coordinate and the pan angle $\alpha$

Let $f$ be the camera focal length. The following equation holds for the dependence of image coordinate $x_c$ on the pan angle $\theta$, (Fig. 9):

$$x_c = f \tan(\alpha - \theta) \tag{11}$$

Differentiating (1), a kinematic model for the $x_c$ offset is obtain in form:

$$\dot{x}_c = \frac{f}{\cos^2(\alpha - \theta)}(\dot{\alpha} - \dot{\theta}) \tag{12}$$

where the pan velocity $\dot{\theta}$ is considered as a control input. The angle $\alpha$ can be determined from (11) in terms of the pan angle $\theta$. The term $\dot{\alpha}$ depends on the instantaneous motion of the face with respect to the frame FXY. In this paper, we assume that $\omega_\alpha := \dot{\alpha}$ is piece-wise unknown constant parameter and is not available for feedback control.

The control objective is to asymptotically stabilize to zero the system (12) in the presence of the unknown constant parameter $\dot{\alpha}$ . The control problem consists in finding an adaptive feedback control law for the system (12) with control input $\omega_\theta := \dot{\theta}$ such that $\lim_{t \to \infty} x_c(t) = 0$ . The estimate $\hat{\omega}_\alpha$ of $\omega_\alpha$ used in the control law is obtained from the dynamic part of the proposed controller, which is designed as a parameter update law.

We propose the following feedback control

$$\omega_\theta = \hat{\omega}_\alpha + kx_c \tag{13}$$

where $k$ is a positive gain.

Consider the following Lyapunov function candidate

$$V = \frac{1}{2}x_c^2 + \frac{1}{2\gamma}\tilde{\omega}_\alpha^2 \tag{14}$$

Where

$$\tilde{\omega}_\alpha = \omega_\alpha - \hat{\omega}_\alpha \tag{15}$$

Using (12) and (13), for the derivative of $V$ one obtains

$$\dot{V} = -\frac{kfx_c^2}{\cos^2(\alpha - \theta)} + \tilde{\omega}_\alpha \left[ \frac{fx_c}{\cos^2(\alpha - \theta)} - \frac{1}{\gamma}\dot{\hat{\omega}} \right] \tag{16}$$

where all the terms containing $\tilde{\omega}_\alpha$ have been groped together. To eliminate them, the update law is chosen as

$$\dot{\hat{\omega}}_\alpha = \gamma \frac{f x_c}{\cos^2(\alpha-\theta)} \tag{17}$$

and we obtain for the derivative of $V$

$$\dot{V} = -\frac{k f x_c^2}{\cos^2(\alpha-\theta)} \leq 0 \tag{18}$$

The resulting closed-loop adaptive system becomes

$$\begin{aligned} \dot{x}_c &= \frac{f}{\cos^2(\alpha-\theta)}(\tilde{\omega}_\alpha - k x_c) \\ \dot{\tilde{\omega}}_\alpha &= -\gamma \frac{f x_c}{\cos^2(\alpha-\theta)} \end{aligned} \tag{19}$$

*Remark*: It should be noted that the system (19) is autonomous since from (15), the difference ($\alpha$-$\theta$) can be expressed in terms of $x_c$. □

**Proposition 1.** Assume that $\omega_\alpha := \dot{\alpha}$ in (12) is unknown constant parameter. If the control law given by (13) is applied to (12), where the estimate $\hat{\omega}_\alpha$ of $\omega_\alpha$ is obtained from the parameter update law (17), the origin $x=[x_c,\tilde{\omega}_\alpha]^T = 0$ of the closed-loop system (19) is asymptotically stable.

*Proof.* Based on the LaSalle's invariance principle [6], the stability properties of (19) follow from (14) and (18). Let $D = \{ x=[x_c,\tilde{\omega}_\alpha]^T \in \Re^2 \mid |\alpha - \theta| < \pi/2 \}$. The system has an equilibrium point at the origin. The function $V(t)$: D $\rightarrow \Re$ is continuously differentiable and positive definite. From (18), it follows that (14) is no increasing, ($V(t) \leq V(0)$), and this in turn implies that $x_c(t)$ and $\tilde{\omega}_\alpha(t)$ are bounded and converge to the largest invariant set $M$ of (19) contained in the set $E = \{x \in D \mid \dot{V} = -\frac{k f x_c^2}{\cos^2(\alpha-\theta)} = 0\}$. Suppose $x(t)$ is a trajectory that belongs identically to $E$. But, $x_c(t) \equiv 0$, $\Rightarrow$ $\dot{x}_c \equiv 0$ which, in turn implies (from the first and the second equations of (23)) that $\tilde{\omega}_\alpha(t) \equiv 0$, $\dot{\tilde{\omega}}_\alpha \equiv 0$. Therefore, from [21], (Corollary 3.1, p.116), the only solution that can stay identically in $E$ is the trivial solution $x(t) = 0$. Thus, the origin is asymptotically stable, i.e.

$$\lim_{t\to\infty} x = [x_c(t), \tilde{\omega}_\alpha(t)]^T = 0 \tag{20}$$

A similar construction in the vertical plane holds for the dependence of the other face coordinate $y_c$ on the tilt angle $\varphi$, (Fig. 10). The adaptive control algorithm for the tilt angle $\varphi$ in order to place the face centroid in the center of the image plane is derived in analogous fashion.

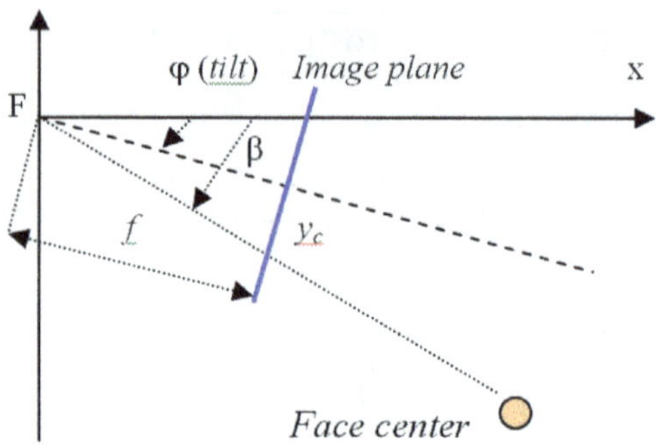


**Fig. 10** Geometrical representation of the face $y_c$-coordinate and the tilt angle $\varphi$

## 3.2 Visual Implementation of the Adaptive Control Law

In this Section, we show how to implement the proposed adaptive control law in order to predict the position of the working window in the mean-shift algorithm (Section 2) at the next discrete time interval. We use data for pan/tilt angle the face offset expressed in pixels in the camera plane, which is obtained at discrete time instants from the camera.

The quantities needed for control computation can be approximated trough finite differences as follows. At every discrete time instant $t_k$ ($k$ =0,1,2,….) the pan angle $\theta_k$ and $x_{ck}$ are measured, and $\alpha_k$ is computed using (11). In the following, the subscript $k$ will refer to the time instant in which the quantity is defined. The sampling interval is denoted by $\Delta t = t_{k+1} - t_k$ and is equal to 40ms. The incremental displacements can be estimated via the Euler method. At $t=0$ ($k=0$), the first estimate for $\omega_\alpha, \hat{\omega}_{\alpha 0}$, is set arbitrary. The control law at $t = t_k$ is computed as follows

$$\omega_{\theta k} = \hat{\omega}_{\alpha k} + kx_{ck} \tag{21}$$

This allows us to compute $\theta_{k+1}$ as

$$\theta_{k+1} = \theta_k + \omega_{\theta k} \Delta t_k \tag{22}$$

We can also compute an estimate of $\hat{\alpha}_{k+1}$ at $t_{k+1}$ as

$$\hat{\alpha}_{k+1} = \hat{\alpha}_k + \hat{\omega}_{\alpha k} \Delta t_k \tag{23}$$

Finally, using Eq. (11), an easy calculation yields the estimate $\hat{x}_{ck+1}$ of the offset $x_c$ at $t_{k+1}$

$$\hat{x}_{ck+1} = f.\tan(\hat{\alpha}_{k+1} - \theta_{k+1}) \tag{24}$$

The estimate of $\omega_{\alpha k+1}$, $\hat{\omega}_{\alpha k+1}$, at $t_{k+1}$ (for k=1,2,3...) is computed from (12) as follows

$$\hat{\omega}_{\alpha k+1} = \hat{\omega}_{\alpha k} + \gamma \frac{f x_{ck}}{\cos^2(\alpha_k - \theta_k)} \tag{25}$$

## 4 Results

In this section, we present our experimental results in order to evaluate the effectiveness of the proposed control method. MATLAB was used to assess its performance. The available information is the $x_c$ offset observed by the camera in the image plane coordinate system. We assume a camera with a focal length of *3mm.*We also assume that the point where the pan axis intersects the camera axis coincides with the focus of the camera. For the simulation purposes, the offset $x_c$ is evaluated directly in millimeters (*mm*) instead of pixel representation in order to avoid the transformation procedure related to the scaling factors (for a concrete camera) in the intrinsic camera calibration matrix.

Our System was tested on live video, size 640x480 on a 2GHz Intel® Pentium® Dual Core laptop. Our algorithm showed better results, in comparison to using Viola-Jones face detector framewise. The initial frame is scanned and validated by the first initialization procedure in 0.6 seconds. On Fig.11. we have shown the initial face detection result using Viola-Jones face detector (red point) and Mean-Shift (white point) in the initial frame, and the two kernels: the enclosing kernel with size - based on Viola-Jones face detector with CNN validation and the external kernel (search window).

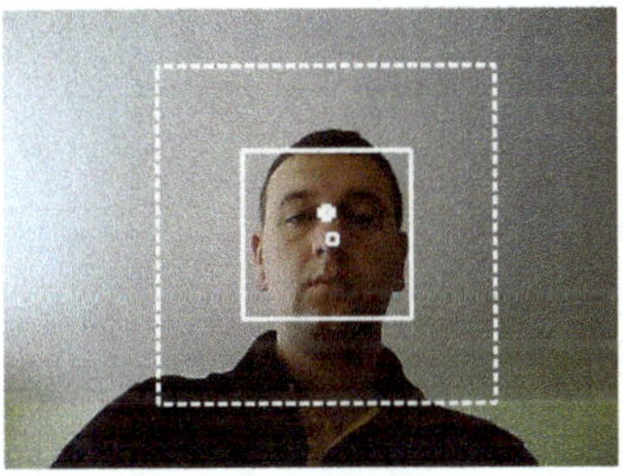

**Fig. 11** Initial face detection

The tracking procedure is realized in 30 fps, for one face in a frame, moving linearly with constant velocity.

The human face is situated a distance of 5m from the camera. Initially the angle α = 0.5rad = cte (Fig. 10) which corresponds to the initial value of xc(0) = 0.0164m (16.4mm). Initially, the pan angle θ(0) = 0. The control law was in the form (13) and (17).

Several scenarios were considered. The face path consists of circular motion of three different constant velocity movements of fixed duration (10 seconds each): turning both to the left and to the right followed by a motion less phase. The corresponding face velocities are as follows (Table 1), where in the case of circular motion of the face with radius R = 5m, the velocities of 1m/s = cte and -0.5m/s = cte corresponds to the rate of change of angle α, $\dot{\alpha} = 0.2 rad/s$ and $\dot{\alpha} = -0.1 rad/s$, respectively.

**Table 1** Succeeded constant velocity movements of fixed duration (5 seconds each) of the human face

| Face motion | Velocity |
|---|---|
| 1) Circular motion , *R=5m* | *1m/s* |
| 2) Circular motion, *R=5m* | *-0.5m/s* |
| 3) Motionless, *R=5m* | *0m/s* |

The initial estimate $\hat{\omega}_{\alpha 0}$ of $\omega_\alpha$ used in the control law is *0rad/s*. Simulation results are shown in Figures 12, 13 and 14. In the first simulation, from Figure 14, we can see the evolution of the angles $\alpha$ and $\theta$ in time. The camera successfully hunts the face and places it at the center of the image plane zeroing the difference $(\alpha - \theta)$. Figure 13 shows the evolution of the face offset $x_c$ along the x-axis in the image plane. As shown in Figure 14, the estimate $\hat{\omega}_\alpha$ of $\omega_\alpha$ tends asymptotically to its actual value. The results of the simulation verify the validity of the proposed controller.

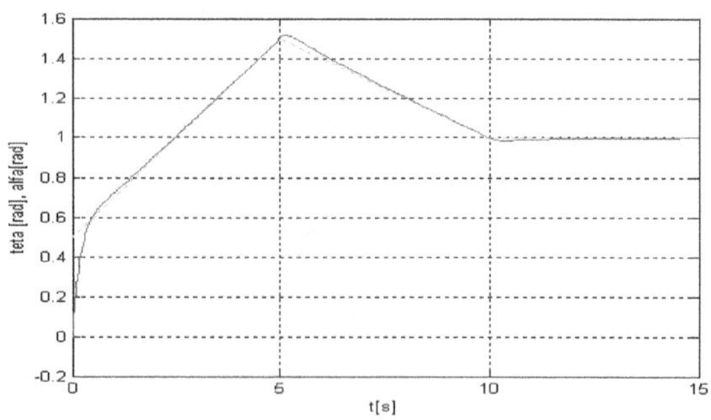


**Fig. 12** Evolution of the pan angle $\theta$ (blue solid line) and angle $\alpha$ (green dashed line); $\alpha(0)$=*0.5rad*; $\theta(0)$=*0rad* according to the succeeded constant velocity movements of fixed duration (5 seconds each) of the human face, (Table 1)

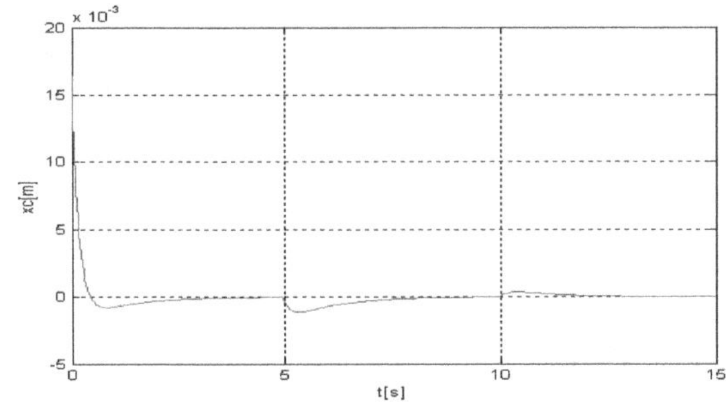


**Fig. 13** Evolution of the face offset $x_c$ along the x-axis in the image plane; $x_c(0) = 0.0164m$ (*16.4mm*) according to the succeeded constant velocity movements of fixed duration (5 seconds each) of the human face, (Table 1)

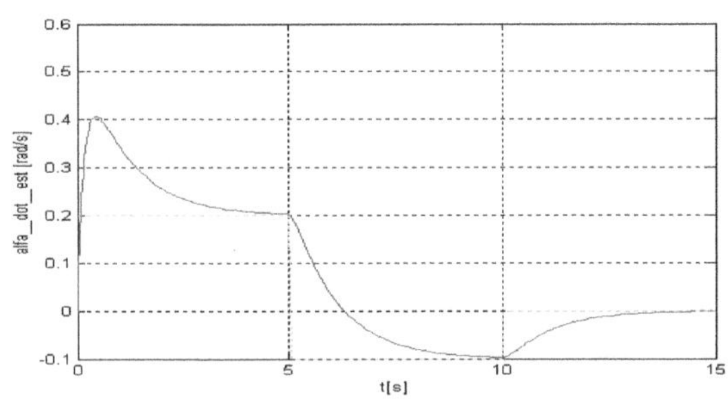


**Fig. 14** Evolution of $\hat{\omega}_\alpha$ (blue continuous line) and its actual value $\omega_\alpha$ (green dashed line); $\hat{\omega}_\alpha(0) = 0$, according to the succeeded constant velocity movements of fixed duration (5 seconds each) of the human face, (Table 1)

## 4 Conclusion

In this paper, we address the problem of face detection and tracking with a pan-tilt camera. An adaptive algorithm that provides automated control of a pan-tilt camera to follow a person's face and keep his image centered in the camera vie has been proposed. We assume that the face velocity during circular motion (forward or backward) is unknown constant parameters. The control scheme for face tracking uses visual information only (an offset vector defined in the image plane). The target motion is unknown and may be characterized by sudden changes. The pan-tilt camera control velocities are computed using velocity estimates. For piecewise constant face velocity, at steady state, the camera successfully hunts the face and places it at the center of the image plane. The results of the simulation verify the validity of the proposed controller. Future research will address the problem of controlling the camera in the case of unknown *time-varying* velocities of the human face.

**Acknowledgments.** This work was supported by National Ministry of Education and Science of Bulgaria under contract DO02-41/2008 "Human Biometric Identification in Video Surveillance Systems", Ukrainian-Bulgarian R&D joint project.

## References

1. Murray, D., Basu, A.: Motion tracking with an active camera. IEEE Transactions on Pattern Analysis and Machine Intelligence, vol. 16(5), pp.449-459 (1994)
2. Jeong, D., Yang, Y.K., Kang, D.G., Ra, J.B.: Real-Time Head Tracking Based on Color and Shape Information. Image and Video Communications and Processing. In: Proc. of SPIE-IS&T Electronic Imaging, SPIE, vol. 5685, pp. 912–923 (2005)
3. Jordao, L., Perrone, M., Costeira, J.P., Santos-Victor, J.: Active face and feature tracking. In: Proc. Int. Conf. Image Analysis and Processing, pp. 572–577 (1999)
4. Oh, P., Allen, P.: Performance of partitioned visual feedback controllers. In: Proc. IEEE Int. Conf. Rob. and Automation, pp. 275–280 (1999)
5. Lim, S., Elgammal, A., Davis, L.S.: Image-based pan - tilt camera control in a multi - camera surveillance environment. In: Proc. Int. Conf. Mach. Intelligence, pp. 645–648 (2003)
6. Viola, P., Jones, M.J.: Rapid Object Detection using a Boosted Cascade of Simple Features. In: Proceedings of IEEE Computer Society Conference on CVPR, pp. 511–518 (2001)
7. Comaniciu, D., Ramesh, V., Meer, P.: Real-Time Tracking of Non-Rigid Objects Using Mean Shift. In: Proc. IEEE Conf. on CVPR, pp. 142–149 (2000)
8. Xu, D., Wang, Y., An, J.: Applying a New Spatial Color Histogram in Mean-Shift Based Tracking Algorithm. College of Automation, Northwestern Polytechnical University, Xi'an 710072, China (2001)
9. Bradski, G.: Computer Vision Face Tracking For Use in a Perceptual User Interface. Microcomputer Research Lab, Santa Clara, CA, Intel Corporation (1998)
10. Yang, G., Huang, T.S.: Human Face Detection in Complex Background. Patt. Recognition 27(1), pp. 53–63 (1994)
11. Kotropoulos, C., Pitas, I.: Rule-Based Face Detection in Frontal Views. In: Proc. Int'l Conf. Acoustics, Speech and Signal Processing, vol. 4, pp. 2537–2540 (1997)
12. Dai, Y., Nakano, Y.: Face-Texture Model Based on SGLD and Its Application in Face Detection in a Color Scene. Pattern Recognition vol. 29(6), pp. 1007–1017 (1996)
13. Yang, M.H., Ahuja, N.: Detecting Human Faces in Color Images. In: Proc. IEEE Int. Conf. Image Processing, vol. 1, pp. 127–130 (1998)
14. McKenna, S., Raja, Y., Gong, S.: Tracking Colour Objects Using Adaptive Mixture Models. Image and Vision Computing 17(3/4), pp. 225–231 (1999)
15. Lienhart, R., Maydt, J.: An Extended Set of Haar-like Features for Rapid Object Detection. In: Proceedings of The IEEE International Conference on Image Processing, vol. 1, pp. 900–903 (2002)
16. Freund, Y., Schepire, R.: A Short Introduction to Boosting. Journal of Japanese Society for Artificial Intelligence 14(5), pp. 771–780 (1999)
17. Garcia, C., Delakis, M.: Convolution Face Finder: A Neural Architecture for Fast and Robust Face Detection. IEEE Transactions on Pattern Analysis and Machine Intelligence 26(11), pp. 1408–1423 (2004)

18. Comaniciu, D., Meer, P.: Mean shift: A robust approach toward feature space analysis. IEEE Trans. Patt. Anal. Mach. Intelligence 24(5), pp. 603–619 (2002)
19. Comaniciu, D., Meer, P.: Mean Shift Analysis and Applications. In: IEEE Int. Conf. Computer Vision (ICCV 1999), Kerkyra, Greece, pp. 1197–1203 (1999)
20. Comaniciu, D., Ramesh, V., Meer, P.: Kernel-based object tracking. IEEE Transactions on Pattern Analysis and Machine Intelligence vol. 25(5), pp. 56–577 (2003)
21. Khalil, H.: Nonlinear Systems. Prentice-Hall, Englewood Cliffs (1996)
22. Paliy, I., Kurylyak, Y., Boumbarov, O., Sokolov, S.: Combined Approach to Face Detection for Biometric Identification Systems. In: Proceedings of the IEEE 5th International Workshop on Intelligent Data Acquisition and Advanced Computing Systems: Technology and Applications (IDAACS 2009), Rende (Cosenza), Italy, pp. 425–429 (2009)

# Time Accounting Artificial Neural Networks for Biochemical Process Models

Petia Georgieva, Luis Alberto Paz Suárez, and Sebastião Feyo de Azevedo

**Abstract.** This paper is focused on developing more efficient computational schemes for modeling in biochemical processes. A theoretical framework for estimation of process kinetic rates based on different temporal (time accounting) Artificial Neural Network (ANN) architectures is introduced. Three ANNs that explicitly consider temporal aspects of modeling are exemplified: i) Recurrent Neural Network (RNN) with global feedback (from the network output to the network input); ii) Time Lagged Feedforward Neural Network (TLFN) and iii) Reservoir Computing Network (RCN). Crystallization growth rate estimation is the benchmark for testing the methodology. The proposed hybrid (dynamical ANN & analytical submodel) schemes are promising modeling framework when the process is strongly nonlinear and particularly when input-output data is the only information available.

## 1 Introduction

The dynamics of chemical and biochemical processes are usually described by mass and energy balance differential equations. These equations combine two elements, the phenomena of conversion of one reaction component into another (i.e. the reaction kinetics) and the transport dynamics of the components through the reactor. The identification of such mathematical models from experimental input/output data is still a challenging issue due to the inherent nonlinearity and complexity of this class of processes (for example polymerization or fermentation

Petia Georgieva
Signal Processing Lab, IEETA, DETI
University of Aveiro, 3810-193 Aveiro, Portugal
e-mail: petia@ua.pt

Luis Alberto Paz Suárez · Sebastião Feyo de Azevedo
Department of Chemical Engineering, Faculty of Engineering,
University of Porto, 4200-465 Porto, Portugal
e-mail: sfeyo@fe.up.pt

V. Sgurev et al. (Eds.): Intelligent Systems: From Theory to Practice, SCI 299, pp. 181–199.
springerlink.com 

reactors, distillation columns, biological waste water treatment, ect.) The most difficult problem is how to model the reaction kinetics and more particularly, the reaction rates. The traditional way is to estimate the reaction rates in the form of analytical expressions, Bastin and Dochain 1990. First, the parameterized structure of the reaction rate is determined based on data obtained by specially designed experiments. Then the respective parameters of this structure are estimated. Reliable parameter estimation is only possible if the proposed model structure is correct and theoretically identifiable, Wolter and Pronzato 1997. Therefore the reaction rate analytical structure is usually determined after a huge number of expensive laboratory experiments. It is further assumed that the initial values of the identified parameters are close to the real process parameters, Noykove et al. 2002, which is typically satisfied only for well known processes. The above considerations motivated a search for alternative estimation solutions based on computationally more attractive paradigms as are the Artificial Neural Networks (ANNs). The interest in ANNs as dynamical system models is nowadays increasing due to their good non-linear time-varying input-output mapping properties. The balanced network structure (parallel nodes in sequential layers) and the non-linear transfer function associated with each hidden and output nodes allows ANNs to approximate highly non-linear relationships without a priori assumption. Moreover, while other regression techniques assume a functional form, ANNs allow the data to define the functional form. Therefore, ANNs are generally believed to be more powerful than many other nonlinear modeling techniques.

The objective of this work is to define a computationally efficient framework to overcome difficulties related with poorly known kinetics mechanistic descriptors of biochemical processes. Our main contribution is the analytical formulation of a modeling procedure based on time accounting artificial neural networks (ANNs), for kinetic rates estimation. A hybrid (ANN & phenomenological) model and a procedure for ANN supervised training when target outputs are not available are proposed. The concept is illustrated on a sugar crystallization case study where the hybrid model outperforms the traditional empirical expression for the crystal growth rate.

The paper is organized as follows. In the next section a hybrid model of a general chemical or biochemical process is introduced, where a time accounting ANN is assumed to model the process kinetic rates in the framework of a nonlinear state space analytical process model. In section 3 three temporal ANN structures are discussed.

In section 4 a systematic ANN training procedure is formulated assuming that all kinetics coefficients are available but not all process states are measured. The proposed methodology is illustrated in section 5 for crystallization growth rate estimation.

## 2 Knowledge Based Hybrid Models

The generic class of reaction systems can be described by the following equations, Bastin and Dochain 1990

$$\frac{dX}{dt} = K\varphi(X,T) - DX + U_x \tag{1}$$

$$\frac{dT}{dt} = b\varphi(X,T) - d_0 T + U_T \tag{2}$$

where, for $n, m \in \mathrm{N}$ , the constants and variables denote

| | |
|---|---|
| $X = (x_1(t),.......x_n(t)) \in R^n$ | Concentrations of total amounts of $n$ process components |
| $K = [k_1,.......k_m] \in R^{n \times m}$ | kinetics coefficients (yield, stoichiometric, or other) |
| $\varphi = (\varphi_1,.......\varphi_m)^T \in R^m$ | Process kinetic rates |
| $T$ | Temperature |
| $b \in R^m$ | Energy related parameters |
| $q_{in} / V$ | Feeding flow/Volume |
| D | Dilution rate |
| $d_o$ | Heat transfer rate related parameter |

$U_x$ and $U_T$ are the inputs by which the process is controlled to follow a desired dynamical behavior. The nonlinear state-space model (1) proved to be the most suitable form of representing several industrial processes as crystallization and precipitation, polymerization reactors, distillation columns, biochemical fermentation and biological systems. Vector ($\varphi$) defines the rate of mass consumption or production of components. It is usually time varying and dependent of the stage of the process. In the specific case of reaction process systems $\varphi$ represents the reaction rate vector typical for chemical or biochemical reactions that take place in several processes, such as polymerization, fermentation, biological waste water treatment, etc. In nonreaction processes as for example crystallization and precipitation, $\varphi$ represents the growth or decay rates of chemical species. In both cases (reaction or nonreaction systems) $\varphi$ models the process kinetics and is the key factor for reliable description of the components concentrations. In this work, instead of an exhaustive search for the most appropriate parameterized reaction rate structure, three temporal (time accounting) ANN architectures are applied to estimate the vector of kinetic rates. The ANN sub-model is incorporated in the general dynamical model (1) and the mixed structure is termed knowledge-based hybrid model (KBHM), see Fig.1. A systematic procedure for ANN-based estimation of reaction rates is discussed in the next section.

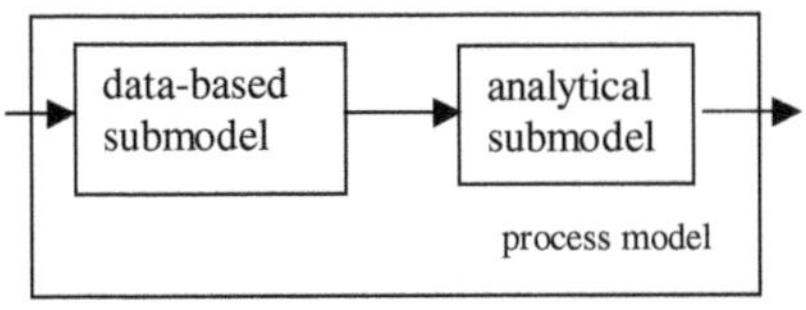


**Fig. 1** Knowledge-based hybrid model (KBHM)

## 3 Time Accounting Artificial Neural Networks

The Artificial Neural Network (ANN) is a computational structure inspired by the neurobiology. An ANN is characterized by its architecture (the network topology and pattern of connections between the nodes), the method of determining the connection weights, and the activation functions that employs. The multi-layer perceptron (MLP), which constitute the most widely used network architecture, is composed of a hierarchy of processing units organized in a parallel-series sets of neurons and layers. The information flow in the network is restricted to only one direction from the input to the output, therefore a MLP is also called Feedforward Neural Network (FNN). FNNs have been extensively used to solve static problems as classification, feature extraction, pattern recognition. In contrast to the FNN, the Recurrent Neural Network (RNN) processes the information in both (feedforward and feedback) directions due to the recurrent relation between network outputs and inputs, Mandic and Chambers 2001. Thus the RNN can encode and learn time dependent (past and current) information which is interpreted as memory. This paper specifically focuses on comparison of three different types of RNNs, namely, i) RNN with global feedback (from the network output to the network input); ii) Time lagged feedforward neural network (TLFN), and iii) Reservoir Computing Network (RCN).

*Recurrent Neural Network (RNN) with global feedback*
An example of RNN architecture where past network outputs are fed back as inputs is depicted in Fig. 2. It is similar to Nonlinear Autoregressive Moving Average with eXogenios input (NARMAX) filters, Haykin 1999.The complete RNN input consists of two vectors formed by present and past network exogenous inputs (r) and past fed back network outputs (p) respectively.

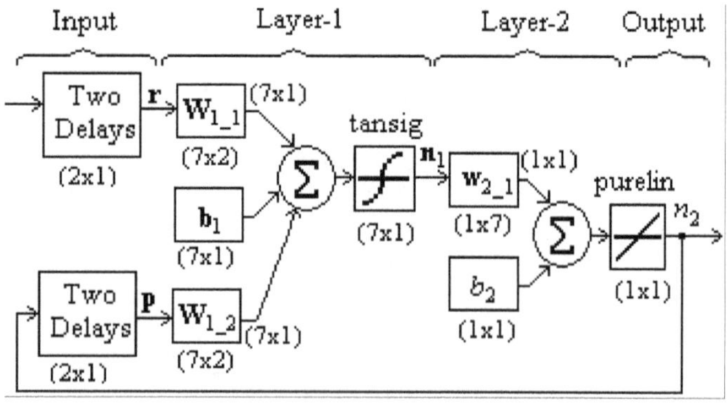


**Fig. 2** RNN architecture

The RNN model implemented in this work is the following

$$\mathbf{u}_{NN} = [\mathbf{r}, \mathbf{p}] \text{ (complete network input)} \quad (3)$$

$$\mathbf{r} = [r_1(k), \ldots. r_1(k-l), \ldots.. r_c(k), \ldots. r_c(k-l)] \text{ (network exogenous inputs)} \quad (4)$$

$$\mathbf{p} = [n_2(k-1),....n_2(k-h)] \quad \text{(recurrent network inputs)} \tag{5}$$

$$\mathbf{x} = \mathbf{W}_{11} \cdot \mathbf{r} + \mathbf{W}_{12} \cdot \mathbf{p} + \mathbf{b}_1 \quad \text{(network states)} \tag{6}$$

$$\mathbf{n}_1 = \left(e^{\mathbf{x}} - e^{-\mathbf{x}}\right) / \left(e^{\mathbf{x}} + e^{-\mathbf{x}}\right) \quad \text{(hidden layer output)} \tag{7}$$

$$n_2 = \mathbf{w}_{21} \cdot \mathbf{n}_1 + b_2 \quad \text{(network output)}, \tag{8}$$

where $\mathbf{W}_{11} \in R^{m\times 2}$, $\mathbf{W}_{12} \in R^{m\times 2}$, $\mathbf{w}_{21} \in R^{1\times m}$, $\mathbf{b}_1 \in R^{m\times 1}$, $b_2 \in R$ are the network weights (in matrix form) to be adjusted during the ANN training, $m$ is the number of nodes in the hidden layer. $l$ is the number of past exogenous input samples and $h$ is the number of past network output samples fed back to the input. The RNNs are a powerful technique for nonlinear dynamical system modeling, however their main disadvantage is that they are difficult to train and stabilize. Due to the simultaneous spatial (network layers) and temporal (past values) aspects of the optimization, the static Backpropagation (BP) learning method has to be substituted by the Backpropagation through time (BPTT) learning. BPTT is a complex and costly training method, which does not guarantee convergence and often is very time consuming, Mandic and Chambers 2001.

*Time lagged feedforward neural network (TLFN)*

TLFN is a dynamical system with a feedforward topology. The dynamic part is a linear memory, Principe et al. 2000. TLFN can be obtained by replacing the neurons in the input layer of an MLP with a memory structure, which is sometimes called a tap delay-line (see Fig. 3). The size of the memory layer (the tap delay) depends on the number of past samples that are needed to describe the input characteristics in time and it has to be determined on a case-by-case basis. When the memory is at the input the TLFN is also called Focused Time delay Neural Network (TDNN). There are other TLFN topologies where the memory is not focused only at the input but can be distributed over the next network layers. The main advantage of the TDNN is that it can be trained with the static BP method.

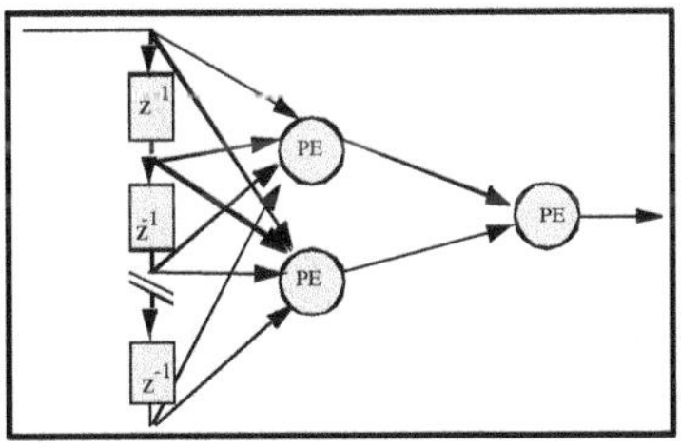


**Fig. 3** RNN architecture

*Reservoir Computing Network (RCN).*

RCN is a concept in the field of machine learning that was introduced independently in three similar descriptions, namely, Echo State Networks (Jaeger 2001), Liquid State Machines (Maass et al. 2002) and Backpropagation-Decorelation learning rule (Steil 2004). All three techniques are characterized by having a fixed hidden layer usually with randomly chosen weights that is used as a reservoir of rich dynamics and a linear output layer (termed also readout layer), which maps the reservoir states to the desired outputs (see Fig. 4). Only the output layer is trained on the response to input signals, while the reservoir is left unchanged (except when making a reservoir re-initialization). The concepts behind RCN are similar to ideas from both kernel methods and RNN theory. Much like a kernel, the reservoir projects the input signals into a higher dimensional space (in this case the state space of the reservoir), where a linear regression can be performed. On the other hand, due to the recurrent delayed connections inside the hidden layer, the reservoir has a form of a short term memory, called the fading memory which allows temporal information to be stored in the reservoir. The general state update equation for the nodes in the reservoir and the readout output equation are as follows:

$$x(k+1) = f\left(W_{res}^{res} x(k) + W_{inp}^{res} u(k) + W_{out}^{res} y(k) + W_{bias}^{res}\right) \tag{9}$$

$$y(k+1) = W_{res}^{out} x(k+1) + W_{inp}^{out} u(k) + W_{out}^{outr} y(k) + W_{bias}^{out} \tag{10}$$

Where: $u(k)$ denotes the input at time $k$; $x(k)$ represents the reservoir state; $y(k)$ is the output; and $f()$ is the activation function (with the hyperbolic tangent $tanh()$ as the most common type of activation function). The initial state is usually set to $x(0)=0$. All weight matrices to the reservoir (denoted as $W^{res}$) are initialized randomly , while all connections to the output (denoted as $W^{out}$) are trained. In the general state update equation (3), it is assumed a feedback not only between the reservoir neurons expressed by the term $W_{res}^{res} x(k)$, but also a feedback from the output to the reservoir accounted by $W_{out}^{res} y(k)$. The first feedback is considered as the short-term memory while the second one as a very long term memory. In order to simplify the computations Following the idea of Antonelo et al. 2007, for the present study the second feedback is discarded and a scaling factor $\alpha$ is introduced in the state update equation

$$x(k+1) = f\left((1-\alpha)x(k) + \alpha W_{res}^{res} x(k) + W_{inp}^{res} u(k) + W_{bias}^{res}\right) \tag{11}$$

Parameter $\alpha$ serves as a way to tune the dynamics of the reservoir and improve its performance. The value of $\alpha$ can be chosen empirically or by an optimization. The output calculations are also simplified (Antonelo et al. 2007) assuming no direct connections from input to output or connections from output to output

$$y(k+1) = W_{res}^{out} x(k+1) + W_{bias}^{out} \tag{12}$$

Each element of the connection matrix $W_{res}^{res}$ is drawn from a normal distribution with mean 0 and variance 1. The randomly created matrix is rescaled so that the spectral radius $|\lambda_{\max}|$ (the largest absolute eigenvalue) is smaller than 1. Standard settings of $|\lambda_{\max}|$ lie in a range between 0.7 and 0.98. Once the reservoir topology is set and the weights are assigned, the reservoir is simulated and optimized on the training data set. It is usually done by linear regression (least squares method) or ridge regression, Bishop 2006. Since the output layer is linear, regularization can be easily applied by adding a small amount of Gaussian noise to the RCN response.

The main advantage of RCN is that it overcomes many of the problems of traditional RNN training such as slow convergence, high computational requirements and complexity. The computational efforts for training are related to computing the transpose of a matrix or matrix inversion. Once trained, the resulting RCN-based system can be used for real time operation on moderate hardware since the computations are very fast (only matrix multiplications of small matrices).

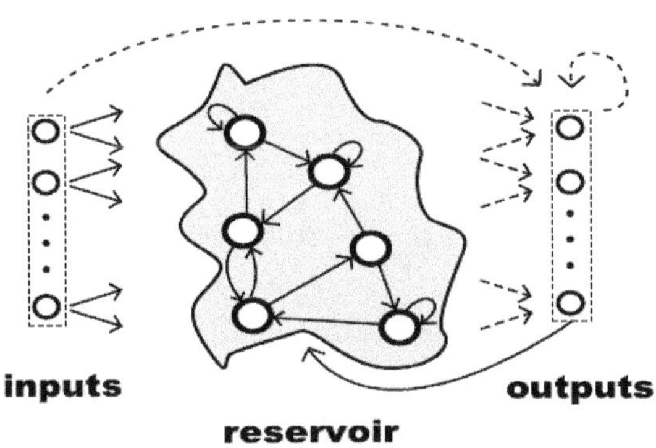


**Fig. 4** Reservoir Computing (RC) network with fixed connections (solid lines) and adaptable connections (dashed lines)

## 4 Kinetic Rates Estimation by Time Accounting ANN

The ANNs are a data-based modeling technique where during an optimization procedure (termed also learning) the network parameters (the weights) are updated based on error correction principle. At each iteration, the error between the network output and the corresponding reference has to be computed and the weights are changed as a function of this error. This principle is also known as supervised learning. However, the process kinetic rates are usually not measured variables, therefore targets (references) are not available and the application of any data-based modeling technique is questionable. A procedure is proposed in the present work to solve this problem. The idea is to propagate the ANN output through a fixed partial analytical model (*Anal. model*) until it comes to a measured process variable (see Fig.5). The proper choice of this *Anal. model* and the formulation of the error signal for network updating are discussed below. The procedure is based on the following assumptions:

**(A1)** Not all process states of model (1) are measured.

**(A2)** All kinetics coefficients are known, that is $b$ and all entries of matrix $K$ are available.

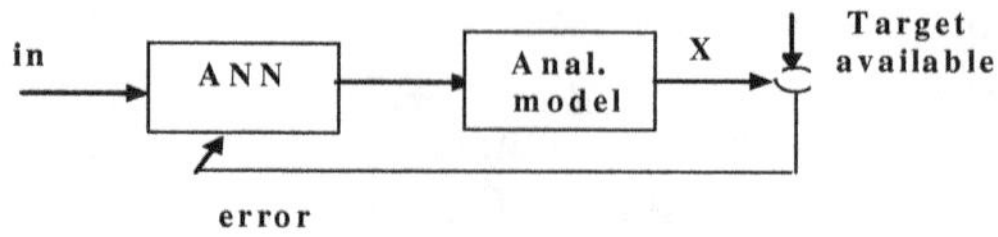


**Fig. 5** Hybrid ANN training structure

For more convenience, the model (1) is reformulated based on the following augmented vectors

$$X_{aug} = \begin{bmatrix} X \\ T \end{bmatrix}, \quad X_{aug} \in R^{n+1}, \quad K_{aug} = \begin{bmatrix} K \\ b \end{bmatrix}, \quad K_{aug} \in R^{(n+1)\times m}. \tag{13}$$

Then (1) is rewritten as

$$\frac{dX_{aug}}{dt} = K_{aug}\varphi(X_{aug}) - \overline{D}X_{aug} + U \quad \text{with} \quad \overline{D} = \begin{bmatrix} D & 0 \\ 0 & d_0 \end{bmatrix}, \quad U = \begin{bmatrix} U_x \\ U_T \end{bmatrix} \tag{14}$$

*Step 1: State vector partition A*

The general dynamical model (8) represents a particular class of nonlinear state-space models. The nonlinearity lies in the kinetics rates $\varphi(X_{aug})$ that are nonlinear functions of the state variables $X_{aug}$. These functions enter the model in the form $K_{aug}\varphi(X_{aug})$, where $K_{aug}$ is a constant matrix, which is a set of linear combinations of the same nonlinear functions $\varphi_1(X_{aug}),.......\varphi_m(X_{aug})$. This particular structure can be exploited to separate the nonlinear part from the linear part of the model by a suitable linear state transformation. More precisely, the following nonsingular partition is chosen, Chen and Bastin 1996.

$$LK_{aug} = \begin{bmatrix} K_a \\ K_b \end{bmatrix}, \quad rank(K_{aug}) = l, \tag{15}$$

where $L \in R^{n\times n}$ is a quadratic permutation matrix, $K_a$ is a $l x m$ full row rank submatrix of $K_{aug}$ and $K_b \in R^{(n-l)\times m}$. The induced partitions of vectors $X_{aug}$ and $U$ are

$$LX_{aug} = \begin{bmatrix} X_a \\ X_b \end{bmatrix}, \quad LU = \begin{bmatrix} U_a \\ U_b \end{bmatrix}, \tag{16}$$

with $X_a \in R^l$, $U_a \in R^l$, $X_b \in R^{n-l}$, $U_b \in R^{n-l}$.

According to (9), model (8) is also partitioned into two submodels

$$\frac{dX_a}{dt} = K_a\varphi(X_a, X_b) - \overline{D}X_a + U_a \tag{17}$$

$$\frac{dX_b}{dt} = K_b\varphi(X_a, X_b) - \overline{D}X_b + U_b \tag{18}$$

Based on (9), a new vector $Z \in R^{n+1-l}$ is defined as a linear combination of the state variables

$$Z = A_0 X_a + X_b, \tag{19}$$

where matrix $A_0 \in R^{(n+1-l)\times l}$ is the unique solution of

$$A_0 K_a + K_b = 0, \tag{20}$$

that is

$$A_0 = -K_b K_a^{-1}, \tag{21}$$

Note that, a solution for $A_0$ exist if and only if $K_a$ is not singular. Hence, a necessary and sufficient condition for the existence of a desired partition (9), is that $K_a$ is a *pxm* full rank matrix, which was the initial assumption. Then, the first derivative of vector Z is

$$\begin{aligned}\frac{dZ}{dt} &= A_0\frac{dX_a}{dt} + \frac{dX_b}{dt}\\ &= A_0\left[K_a\varphi(X_a, X_b) - \overline{D}X_a + U_a\right] + K_b\varphi(X_a, X_b) - \overline{D}X_b + U_b\\ &= (A_0K_a + K_b)\varphi(X_a, X_b) - \overline{D}(A_0X_a + X_b) + A_0U_a + U_b\end{aligned} \tag{22}$$

Since matrix $A_0$ is chosen such that eq. (13) holds, the term in (15) related with $\varphi$ is cancelled and we get

$$\frac{dZ}{dt} = -\overline{D}Z + A_0U_a + U_b \tag{23}$$

The state partition A results in a vector Z whose dynamics, given by eq. (15), is independent of the kinetic rate vector $\varphi$. In general, (9) is not an unique partition and for any particular case a number of choices are possible.

*Step 2: State vector partition B (measured & unmeasured states)*

Now a new state partition is defined as sub-vectors of measured and unmeasured states $X_1$, $X_2$, respectively. The model (8) is also partitioned into two submodels

$$\frac{dX_1}{dt} = K_1\varphi(X_1, X_2) - \overline{D}X_1 + U_1 \tag{24}$$

$$\frac{dX_2}{dt} = K_2\varphi(X_1, X_2) - \overline{D}X_2 + U_2 \tag{25}$$

From state partitions A and B, vector Z can be represented in the following way

$$Z = A_0 X_a + X_b = A_1 X_1 + A_2 X_2 . \tag{26}$$

The first representation is defined in (12), then applying linear algebra transformations $A_1$ and $A_2$ are computed to fit the equality (18). **The purpose of state partitions A and B is to estimate the unmeasured states (vector $X_2$) independently of the kinetics rates (vector $\varphi$).** The recovery of $X_2$ is defined as state observer .

<u>*Step 3: State observer*</u>

Based on (16) and starting with known initial conditions, Z can be estimated as follow (in this work the estimations are denoted by hat )

$$\frac{d\hat{Z}}{dt} = -D\hat{Z} + A_0(F_{in\_a} - F_{out\_a}) + (F_{in\_b} - F_{out\_b}) \tag{27}$$

Then according to (18) the unmeasured states $X_2$ are recovered as

$$\hat{X}_2 = A_2^{-1}(\hat{Z} - A_1 X_1) \tag{28}$$

Note that, estimates $\hat{X}_2$ exist if and only if $A_2$ is not singular, Bastin and Dochain 1990. Hence, **a necessary and sufficient condition for observability of the unmeasured states is that $A_2$ is a full rank matrix.**

<u>*Step 4: Error signal for NN training*</u>

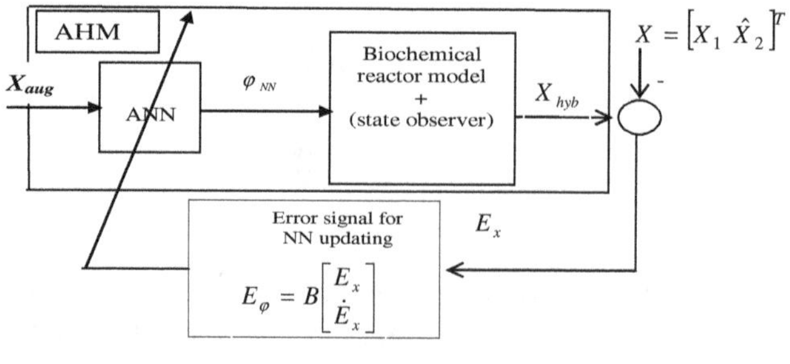


**Fig. 3** Hybrid NN-based reaction rates identification structure

The hybrid structure for NN training is shown in Fig. 3, where the adaptive hybrid model (AHM) is formulated as

$$\frac{dX_{hyb}}{dt} = K_{aug}\varphi_{NN} - \overline{D}X_{hyb} + U + \Omega(X_{aug} - X_{hyb}) \tag{29}$$

The true (but unknown) process behavior is assumed to be represented by (8). Then the error dynamics is modeled as the difference between (8) and (21)

$$\frac{d(X_{aug} - X_{hyb})}{dt} = K_{aug}(\varphi - \varphi_{NN}) - \overline{D}(X_{aug} - X_{hyb}) + \Omega(X_{aug} - X_{hyb}) \tag{30}$$

The following definitions take place:

$E_x = (X_{aug} - X_{hyb})$ is termed as the observation error,

$E_\varphi = \varphi - \varphi_{NN}$ is the error signal for updating the ANN parameters.

$X_{aug}$ consists of the measured ($X_1$) and the estimated ($\hat{X}_2$) states. Thus, (22) can be rearranged as follows

$$\frac{dE_x}{dt} = K_{aug}E_\varphi - (\overline{D} - \Omega)E_x \tag{31}$$

and from (23) the error signal for NN training is

$$E_\varphi = K_{aug}^{-1}\begin{bmatrix}\overline{D} - \Omega & 1\end{bmatrix}\begin{bmatrix}E_x \\ \dot{E}_x\end{bmatrix} = B\begin{bmatrix}E_x \\ \dot{E}_x\end{bmatrix}, \quad B = K_{aug}^{-1}\begin{bmatrix}\overline{D} - \Omega & 1\end{bmatrix} \tag{32}$$

$\Omega$ is a design parameter which defines the speed of the observation error convergence. The necessary identifiability condition for the kinetic rate vector is the non singularity of matrix $K_{aug}$. Note that, the error signal for updating the network parameters is a function of the observation error ($E_x$) and the speed of the observation error ($\dot{E}_x$). The intuition behind is that the network parameters are changed proportionally to their effect on the prediction of the process states and the prediction of their dynamics.

Step 5: Optimization porcesure - Levenberg-Marquardt Quasi-Newton algorithm

The cost function to be minimized at each iteration of network training is the sum of squared errors, where *N* is the time instants over which the optimization is performed (batch mode of training)

$$J_k = \frac{1}{N}\sum_{i=1}^{N}\left[E_\varphi(i)\right]^2 \tag{33}$$

A number of algorithms have been proposed to update the network parameters (*w*). For this study the Levenberg-Marquardt (LM) Quasi Newton method is the chosen algorithm due to its faster convergence than the steepest descent or conjugate gradient methods, Hagan et al. 1996. One ($k$) iteration of the classical Newton's method can be written as

$$w_{k+1} = w_k - H_k^{-1} g_k, \; g_k = \frac{\partial J_k}{\partial w_k}, \; H_k = \frac{\partial J_k^2}{\partial w_k \partial w_k} \tag{34}$$

where $g_k$ is the current gradient of the performance index (25) $H_k$ is the Hessian matrix (second derivatives) of the performance index at the current values ($k$) of the weights and biases. Unfortunately, it is complex and expensive to compute the Hessian matrix for a dynamical ANN. The LM method is a modification of the classical Newton method that does not require calculation of the second derivatives. It is designed to approach second-order training speed without having to compute directly the Hessian matrix. When the performance function has the form of a sum of error squares (25), at each iteration the Hessian matrix is approximated as

$$H_k = J_k^T J_k \tag{35}$$

where $J_k$ is the Jacobian matrix that contains first derivatives of the network errors ($e_k$) with respect to the weights and biases

$$J_k = \frac{\partial E_{\varphi k}}{\partial w_k}, \tag{36}$$

The computation of the Jacobian matrix is less complex than computing the Hessian matrix. The gradient is then computed as

$$g_k = J_k E_{\varphi k} \tag{37}$$

The LM algorithm updates the network weights in the following way

$$w_{k+1} = w_k - \left[ J_k^T J_k + \mu I \right]^{-1} J_k^T E_{\varphi k} \tag{38}$$

When the scalar $\mu$ is zero, this is just Newton's method, using the approximate Hessian matrix. When $\mu$ is large, this becomes gradient descent with a small step size. Newton's method is faster and more accurate near an error minimum, so the aim is to shift towards Newton's method as quickly as possible. Thus, $\mu$ is decreased after each successful step (reduction in performance function) and is increased only when a tentative step would increase the performance function. In this way, the performance function will always be reduced at each iteration of the algorithm.

## 5 Case Study - Estimation of Sugar Crystallization Growth Rate

Sugar crystallization occurs through mechanisms of nucleation, growth and agglomeration that are known to be affected by several not well-understood operating conditions. The search for efficient methods for process description is linked both to the scientific interest of understanding fundamental mechanisms of the crystallization process and to the relevant practical interest of production requirements. The sugar production batch cycle is divided in several phases. During the first phase the pan is partially filled with a juice containing dissolved sucrose. The liquor is concentrated by evaporation, under vacuum, until the supersaturation reaches a predefined value. At this point seed crystals are introduced into the pan to induce the production of crystals (crystallization phase). As evaporation takes place further liquor or water is added to the pan. This maintains the level of supersaturation and increases the volume contents. The third phase consists of tightening which is controlled by the evaporation capacity, see Georgieva et al. 2003 for more details. Since the objective of this paper is to illustrate the technique introduced in section 4, the following assumptions are adopted:

*i)* Only the states that explicitly depend on the crystal growth rate are extracted from the comprehensive mass balance process model;
*ii)* The population balance is expressed only in terms of number of crystals;
*iii)* The agglomeration phenomenon is neglected.

The simplified process model is then

$$\frac{dM_s}{dt} = -k_1 G + F_f \rho_f B_f Pur_f \tag{39}$$

$$\frac{dM_c}{dt} = k_1 G \tag{40}$$

$$\frac{dT_m}{dt} = k_2 G + bF_f + cJ_{vap} + d \tag{41}$$

$$\frac{dm_0}{dt} = k_3 G \tag{42}$$

where $M_s$ is the mass of dissolved sucrose, $M_c$ is the mass of crystals, $T_m$ is the temperature of the massecuite, $m_0$ is the number of crystals. $Pur_f$ and $\rho_f$ are the purity (mass fraction of sucrose in the dissolved solids) and the density of the incoming feed. $F_f$ is the feed flowrate, $J_{vap}$ is the evaporation rate and *b, c, d* are parameters incorporating the enthalpy terms and specific heat capacities. They are derived as functions of physical and thermodynamic properties. The full state vector is $X_{aug} = [M_s \quad M_c \quad T_m \quad m_0]^T$, with $K_{aug} = [-k_1 \quad k_1 \quad k_2 \quad k_3]^T$. Now we are in a position to apply the formalism developed in sections 2.2 and 2.3 for this particular reaction process.

We chose the following state partition A : $X_a = M_c$, $X_b = [M_s \quad T_m \quad m_0]^T$ and the solution of equation (13) is

$$A_0 = \begin{bmatrix} 1 & -\frac{k_2}{k_1} & -\frac{k_3}{k_1} \end{bmatrix}^T \tag{43}$$

$M_c$ and $T_m$ are the measured states, then the unique state partition B is

$$X_1 = [M_c \quad T_m]^T, \quad X_2 = [M_s \quad m_0]^T,$$

Taking into account (32), the matrices of the second representation of vector Z in (18) are computed as

$$A_1 = \begin{bmatrix} 1 & -\frac{k_2}{k_1} & -\frac{k_3}{k_1} \\ 0 & 1 & 0 \end{bmatrix}^T, \quad A_2 = \begin{bmatrix} 1 & 0 & 0 \\ 0 & 0 & 1 \end{bmatrix}^T$$

For this case D=0, then the estimation of the individual elements of Z are

$$\hat{Z}_1 = M_c + \hat{M}_s, \quad \hat{Z}_2 = -\frac{k_2}{k_1} M_c + T_m, \quad \hat{Z}_3 = -\frac{k_3}{k_1} M_c + \hat{m}_0 \tag{44}$$

The analytical expression for the estimation of the unmeasured states is then

$$\begin{bmatrix} \hat{M}_s \\ \hat{m}_0 \end{bmatrix} = \begin{bmatrix} 1 & 0 & 0 \\ 0 & 0 & 1 \end{bmatrix} \left( \begin{bmatrix} \hat{Z}_1 \\ \hat{Z}_2 \\ \hat{Z}_3 \end{bmatrix} - \begin{bmatrix} 1 & 0 \\ -\frac{k_2}{k_1} & 1 \\ -\frac{k_3}{k_1} & 0 \end{bmatrix} \begin{bmatrix} M_c \\ T_m \end{bmatrix} \right) \tag{45}$$

The observation error is defined as

$$E_x = \begin{bmatrix} \hat{M}_s - M_{shyb} \\ M_c - M_{chyb} \\ \hat{m}_0 - m_{0hyb} \\ T_m - T_{mhyb} \end{bmatrix} \tag{46}$$

In the numerical implementation the first derivative of the observation error is computed as the difference between the current $E_x(k)$ and the previous value $E_x(k-1)$ of the observation error divided by the integration step ($\Delta t$)

$$\dot{E}_x = \frac{E_x(k) - E_x(k-1)}{\Delta t} \tag{47}$$

The three types of time accounting ANNs were trained with the same training data coming from six industrial batches (training batches). The physical inputs to all networks are ($M_c, T_m, m_0, M_s$), the network output is $G_{NN}$. Two of the inputs ($M_c, T_m$) are measurable, the others ($m_0, M_s$) are estimated. In order to improve the comparability between the different networks a linear activation function is located at the single output node (see Fig. 2, Layer 2- purelin) and hyperbolic tangent functions are chosen for the hidden nodes (Fig. 2, Layer 1 - tansig). Though other S-shaped activation functions can be also considered for the hidden nodes, our choice was determined by the symmetry of the hyperbolic tangent function into the interval (-1, 1).

The hybrid models are compared with an analytical model of the sugar crystallization, reported in Oliveira et al. 2009, where $G$ is computed by the following empirical correlation

$$G = K_g \exp\left[-\frac{57000}{R(T_m + 273)}\right](S-1)\exp[-13.863(1-P_{sol})]\left(1+2\frac{V_c}{V_m}\right), \tag{48}$$

where S is the supersaturation, $P_{sol}$ is the purity of the solution and $V_c / V_m$ is the volume fraction of crystals. $K_g$ is a constant, optimized following a non-linear least-squares regression.

The performance of the different models is examined with respect to prediction quality of the crystal size distribution (CSD) at the end of the process which is quantified by two parameters - the final average (in mass) particle size (AM) and the final coefficient of particle variation (CV). The predictions given by the models are compared with the experimental data for the CSD (Table 1), coming from 8 batches not used for network training (validation batches). The results with respect to different configurations of the networks are summarized in Tables 2, 3, and 4. All hybrid models (eqs. 31 +RNN/TLNN/RCN) outperform the empirical model (37) particularly with respect to predictions of CV. The predictions based on TLFN and TCN are very close especially for higher reservoir dimension. Increasing the RCN hidden nodes (from 100 to 200) reduces the AM and CV prediction errors, however augmenting the reservoir dimension from 200 to 300 does not bring substantial improvements. The hybrid models with RNN exhibit the best performance though the successful results reported in Table 2 were preceded by a great number of unsuccessful (not converging) trainings. As with respect to learning efforts the RCN training takes in average few seconds on an Intel Core2 Duo processor based computer and by far is the easiest and fastest dynamical regressor.

**Table 1** Final CSD – experimental data versus analytical model predictions

| batch No. | experimental data | | analytical model (eqs. 31+ eq. 37) | |
|---|---|---|---|---|
| | AM[mm] | CV [%] | AM[mm] | CV [%] |
| 1 | 0.479 | 32.6 | 0.583 | 21.26 |
| 2 | 0.559 | 33.7 | 0.542 | 18.43 |
| 3 | 0.680 | 43.6 | 0.547 | 18.69 |
| 4 | 0.494 | 33.7 | 0.481 | 14.16 |
| 5 | 0.537 | 32.5 | 0.623 | 24.36 |
| 6 | 0.556 | 35.5 | 0.471 | 13.642 |
| 7 | 0.560 | 31.6 | 0.755 | 34.9 |
| 8 | 0.530 | 31.2 | 0.681 | 27.39 |
| **av. err** | | | **13.7%** | **36.1%** |

**Table 2** Final CSD – hybrid model predictions (eqs. 31+RNN)

| RNN | batch No. | AM[mm] | CV[%] |
|---|---|---|---|
| exogenous input | 1 | 0.51 | 29.6 |
| delay: 2 | 2 | 0.48 | 30.7 |
| | 3 | 0.58 | 33.6 |
| Recurrent input | 4 | 0.67 | 31.7 |
| delay: 2 | 5 | 0.55 | 29.5 |
| | 6 | 0.57 | 34.5 |
| Total N° of inputs: 14 | 7 | 0.59 | 29.6 |
| hidden neurons:5 | 8 | 0.53 | 32.2 |
| **Average error (%)** | | **4.1** | **7.5** |
| exogenous input | 1 | 0.59 | 30.7 |
| delay: 1 | 2 | 0.55 | 41.5 |
| | 3 | 0.59 | 39.3 |
| Recurrent input | 4 | 0.51 | 35.9 |
| delay: 3 | 5 | 0.49 | 32.1 |
| | 6 | 0.58 | 31.7 |
| Total N° of inputs: 11 | 7 | 0.56 | 30.5 |
| hidden neurons:5 | 8 | 0.53 | 36.8 |
| **Average error (%)** | | **5.2** | **9.2** |
| exogenous input | 1 | 0.51 | 30.9 |
| delay:3 | 2 | 0.56 | 31.1 |
| | 3 | 0.59 | 37.2 |
| Recurrent input | 4 | 0.48 | 29.8 |
| delay:1 | 5 | 0.52 | 34.8 |
| | 6 | 0.51 | 32.4 |
| Total N° of inputs: 17 | 7 | 0.59 | 30.6 |
| hidden neurons:5 | 8 | 0.50 | 33.5 |
| **Average error (%)** | | **3.6** | **6.9** |

**Table 3** Final CSD – hybrid model predictions (eqs. 31+TLNN)

| TLNN | batch No. | AM[mm] | CV[%] |
|---|---|---|---|
| Tap delay: 1 | 1 | 0.49 | 30.8 |
| | 2 | 0.51 | 37.1 |
| | 3 | 0.62 | 31.5 |
| Total N° of inputs: 8 | 4 | 0.60 | 35.5 |
| hidden neurons:5 | 5 | 0.57 | 36.2 |
| | 6 | 0.52 | 28.7 |
| | 7 | 0.55 | 38.6 |
| | 8 | 0.54 | 32.4 |
| **Average error (%)** | | **6.02** | **11.0** |
| Tap delay: 2 | 1 | 0.51 | 37.5 |
| | 2 | 0.49 | 31.6 |
| | 3 | 0.59 | 34.6 |
| | 4 | 0.53 | 40.3 |
| Total N° of inputs: 12 | 5 | 0.60 | 35.2 |
| hidden neurons:5 | 6 | 0.49 | 31.5 |
| | 7 | 0.51 | 29.6 |
| | 8 | 0.54 | 30.3 |
| **Average error (%)** | | **5.9** | **10.8** |
| Tap delay: 3 | 1 | 0.479 | 30.3 |
| | 2 | 0.559 | 41.2 |
| | 3 | 0.680 | 39.4 |
| | 4 | 0.494 | 35.7 |
| Total N° of inputs: 16 | 5 | 0.537 | 35.4 |
| hidden neurons:5 | 6 | 0.556 | 30.3 |
| | 7 | 0.560 | 29.9 |
| | 8 | 0.530 | 28.3 |
| **Average error (%)** | | **5.8** | **10.3** |

**Table 4** Final CSD – hybrid model predictions (eqs. 31+RCN)

| RCN | batch No. | AM[mm] | CV[%] |
|---|---|---|---|
| Reservoir | 1 | 0.53 | 31.2 |
| dimension: 100 nodes | 2 | 0.49 | 28.1 |
| | 3 | 0.57 | 43.6 |
| Total N° of inputs: 4 | 4 | 0.61 | 41.7 |
| | 5 | 0.59 | 39.6 |
| | 6 | 0.60 | 36.1 |
| | 7 | 0.51 | 30.4 |
| | 8 | 0.54 | 40.2 |
| **Average error (%)** | | **6.8** | **12.0** |
| Reservoir | 1 | 0.56 | 40.1 |
| dimension: 200 nodes | 2 | 0.51 | 37.4 |
| | 3 | 0.61 | 36.2 |
| | 4 | 0.56 | 38.6 |
| Total N° of inputs: 4 | 5 | 0.49 | 28.9 |
| | 6 | 0.59 | 34.7 |
| | 7 | 0.61 | 30.4 |
| | 8 | 0.54 | 39.2 |
| **Average error (%)** | | **5.9** | **10.2** |
| Reservoir | 1 | 0.59 | 33.9 |
| dimension: 300 nodes | 2 | 0.48 | 28.8 |
| | 3 | 0.57 | 39.7 |
| | 4 | 0.51 | 29.6 |
| Total N° of inputs: 4 | 5 | 0.53 | 31.8 |
| | 6 | 0.51 | 33.9 |
| | 7 | 0.49 | 30.7 |
| | 8 | 0.57 | 36.9 |
| **Average error (%)** | | **5.9** | **9.8** |

## 6 Conclusions

This work is focused on presenting a more efficient computational scheme for estimation of process reaction rates based on temporal artificial neural network (ANN) architectures. It is assumed that the kinetics coefficients are all known and do not change over the process run, while the process states are not all measured and therefore need to be estimated. It is a very common scenario in reaction systems with low or medium complexity.

The concepts developed here concern two aspects. On one side we formulate a hybrid ( temporal ANN+ analytical) model that outperforms the traditional reaction rate estimation approaches. On the other side a procedure for ANN supervised training is introduced when target (reference) outputs are not available. The network is embedded in the framework of a first principle process model and the error signal for updating the network weights is determined analytically. According to the procedure, first the unmeasured states are estimated independently of the reaction rates and then the ANN is trained with the estimated and the measured data. Ongoing research is related with the integration of the hybrid models proposed in this work in the framework of a model based predictive control.

**Acknowledgements.** This work was financed by the Portuguese Foundation for Science and Technology within the activity of the Research Unit IEETA-Aveiro, which is gratefully acknowledged.

## References

1. Antonelo, E.A., Schrauwen, B., Campenhout, J.V.: Generative modeling of autonomous robots and their environments using reservoir computing. Neural Processing Letters 26(3), 233–249 (2007)
2. Bastin, G., Dochain, D.: On-line estimation and adaptive control of bioreactors. Elsevier Science Publishers, Amsterdam (1990)
3. Bishop, C.M.: Pattern Recognition and Machine Learning. Springer, Heidelberg (2006)
4. Chen, L., Bastin, G.: Structural identifiability of the yeals coefficients in bioprocess models when the reaction rates are unknown. Mathematical Biosciences 132, 35–67 (1996)
5. Georgieva, P., Meireles, M.J., Feyo de Azevedo, S.: Knowledge Based Hybrid Modeling of a Batch Crystallization When Accounting for Nucleation, Growth and Agglomeration Phenomena. Chem. Eng. Science 58, 3699–3707 (2003)
6. Hagan, M.T., Demuth, H.B., Beale, M.H.: Neural Network Design. PWS Publishing, Boston (1996)
7. Haykin, S.: Neural Networks: A Comprehensive Foundation. Prentice-Hall, NJ (1999)
8. Jaeger, H.: The "echo state" approach to analysing and training recurrent neural networks. Technical Report GMD Report 148, German National Research Center for Information Technology (2001a)
9. Maass, W., Natschlager, T., Markram, H.: Real-time computing without stable states: A new framework for neural computation based on perturbations. Neural Computation 14(11), 2531–2560 (2002)
10. Mandic, D.P., Chambers, J.A.: Recurrent Neural Networks for Prediction: Learning Algorithms, Architectures and Stability (Adaptive & Learning Systems for Signal Processing, Communications & Control). Wiley, Chichester (2001)
11. Noykove, N., Muller, T.G., Gylenberg, M., Timmer, J.: Quantitative analysis of anaerobic wastewater treatment processes: identifiably and parameter estimation. Biotechnology and bioengineering 78(1), 91–103 (2002)
12. Oliveira, C., Georgieva, P., Rocha, F., Feyo de Azevedo, S.: Artificial Neural Networks for Modeling in Reaction Process Systems. In: Neural Computing & Applications, vol. 18, pp. 15–24. Springer, Heidelberg (2009)
13. Principe, J.C., Euliano, N.R., Lefebvre, W.C.: Neural and adaptive systems: Fundamentals through simulations, New York (2000)
14. Steil, J.J.: Backpropagation-Decorrelation: Online recurrent learning with O(N) complexity. In: Proc. Int. Joint Conf. on Neural Networks (IJCNN), vol. 1, pp. 843–848 (2004)
15. Walter, E., Pronzato, L.: Identification of parametric models from experimental data. Springer, UK (1997)

# Decentralized Adaptive Soft Computing Control of Distributed Parameter Bioprocess Plant

Ieroham S. Baruch and Rosalba Galvan-Guerra

**Abstract.** The paper proposed to use recurrent Fuzzy-Neural Multi-Model (FNMM) identifier for decentralized identification of a distributed parameter anaerobic wastewater treatment digestion bioprocess, carried out in a fixed bed and a recirculation tank. The distributed parameter analytical model of the digestion bioprocess is reduced to a lumped system using the orthogonal collocation method, applied in three collocation points (plus the recirculation tank), which are used as centers of the membership functions of the fuzzyfied plant output variables with respect to the space variable. The local and global weight parameters and states of the proposed FNMM identifier are implemented by hierarchical fuzzy-neural direct and indirect multi-model controllers. The comparative graphical simulation results of the digestion wastewater treatment system identification and control, obtained via learning, exhibited a good convergence, and precise reference tracking very closed to that of the optimal control.

## 1 Introduction

In the last decade, the Computational Intelligence tools (CI), including Artificial Neural Networks (ANN) and Fuzzy Systems (FS), applying soft computing, became universal means for many applications. Because of their approximation and learning capabilities, [1], the ANNs have been widely employed to dynamic process modeling, identification, prediction and control, [1]-[9]. Many applications have been done for identification and control of biotechnological plants too, [8]. Among several possible neural network architectures the ones most widely used are the Feedforward NN (FFNN) and the Recurrent NN (RNN), [1]. The main NN property namely the ability to approximate complex non-linear relationships without prior knowledge of the model structure makes them a very attractive alternative to the classical modeling and control techniques. This property has been proved for both types of NNs by the universal approximation theorem [1]. The preference

Ieroham S. Baruch · Rosalba Galvan-Guerra
CINVESTAV-IPN, Department of Automatic Control,
Ave. IPN No 2508, A.P. 14-470, Mexico D.F., C.P. 07360, Mexico
e-mail: {baruch,rgalvan}@ctrl.cinvestav.mx

V. Sgurev et al. (Eds.): Intelligent Systems: From Theory to Practice, SCI 299, pp. 201–228.
springerlink.com 

given to NN identification with respect to the classical methods of process identification is clearly demonstrated in the solution of the "bias-variance dilemma" [1]. The FFNN and the RNN have been applied for Distributed Parameter Systems (DPS) identification and control too. In [10], a RNN is used for system identification and process prediction of a DPS dynamics - an adsorption column for wastewater treatment of water contaminated with toxic chemicals. In [11], [12], a spectral-approximation-based intelligent modeling approach is proposed for the distributed thermal processing of the snap curing oven DPS that is used in semiconductor packaging industry. After finding a proper approximation of the complex boundary conditions of the system, the spectral methods can be applied to time–space separation and model reduction, and NNs are used for state estimation and system identification. Then, a neural observer has been designed to estimate the states of the Ordinary Differential Equation (ODE) model from measurements taken at specified locations in the field. In [13], it is presented a new methodology for the identification of DPS, based on NN architectures, motivated by standard numerical discretization techniques used for the solution of Partial Differential Equation (PDE). In [14], an attempt is made to use the philosophy of the NN adaptive-critic design to the optimal control of distributed parameter systems. In [15] the concept of proper orthogonal decomposition is used for the model reduction of DPS to form a reduced order lumped parameter problem. The optimal control problem is then solved in the time domain, in a state feedback sense, following the philosophy of adaptive critic NNs. The control solution is then mapped back to the spatial domain using the same basis functions. In [16], measurement data of an industrial process are generated by solving the PDE numerically using the finite differences method. Both centralized and decentralized NN models are introduced and constructed based on this data. The models are implemented on FFNN using Backpropagation (BP) and Levenberg-Marquardt learning algorithms.

Similarly to the static ANNs, the fuzzy models could approximate static nonlinear plants where structural plant information is needed to extract the fuzzy rules, [17], [18]. The difference between them is that the ANN models are global models where training is performed on the entire pattern range and the FS models perform a fuzzy blending of local models space based on the partition of the input space. So the aim of the neuro-fuzzy (fuzzy-neural) models is to merge both ANN and FS approaches so to obtain fast adaptive models possessing learning, [17]. The fuzzy-neural networks are capable of incorporating both numerical data (quantitative information) and expert's knowledge (qualitative information), and describe them in the form of linguistic IF-THEN rules. During the last decade considerable research has been devoted towards developing recurrent neuro-fuzzy models, summarized in [19]. To reduce the number of IF-THAN rules, the hierarchical approach could be used [19]. A promising approach of recurrent neuro-fuzzy systems with internal dynamics is the application of the Takagi-Sugeno (T-S) fuzzy rules with a static premise and a dynamic function consequent part, [20]-[23]. The papers of Mastorocostas and Theocharis, [22], [23], Baruch et al [19], proposed as a dynamic function in the consequent part of the T-S rules to use a Recurrent Neural Network Model (RNNM). Some results of this RNNM approach for centralized and decentralized identification of dynamic plants with distributed parameters is given in [24]. The difference between the used in [22], [23] fuzzy neural model and the approach used in [25], [26], [19] is that the first one uses the Frasconi, Gori and Soda RNN model, which is sequential one, and the second one uses the RTNN model , which is completely parallel one. But it is not still enough because the neural nonlinear dynamic function ought to be learned, and the Backpropagation learning algorithm is not introduced in the T-S fuzzy rule. For this reason in [27], [28], the RTNN BP learning algorithm [29] has been introduced in the antecedent part of the IF-THAN rule so to complete the learning procedure and a second hierarchical defuzzyfication BP learning level has been formed so to improve the adaptation and approximation ability of the fuzzy-neural system, [19]. This system has been successfully applied for identification and control of complex nonlinear plants, [19]. The aim of this paper is to describe the results obtained by this system for decentralized identification and control of wastewater treatment anaerobic digestion bioprocess representing a Distributed Parameter System (DPS).

The analytical anaerobic bioprocess plant model [30], used as an input/output plant data generator, is described by PDE/ODE, and simplified using the orthogonal collocation technique, [31], in three collocation points and a recirculation tank. This measurement points are used as centres of the membership functions of the fuzzyfied space variable of the plant.

## 2 Analytical Model of the Anaerobic Digestion Bioprocess Plant

The anaerobic digestion systems block diagram is depicted on Fig.1. It is conformed by a fixed bed reactor and a recirculation tank. The physical meaning of all variables and constants (also its values), are summarized in Table 1. The complete analytical model of wastewater treatment anaerobic bioprocess, taken from [30], could be described by the following system of PDE and ODE (for the recirculation tank):

$$\frac{\partial X_1}{\partial t} = (\mu_1 - \varepsilon D) X_1 , \quad \mu_1 = \mu_{1\max} \frac{S_1}{K_{s_1}^{'} X_1 + S_1} , \tag{1}$$

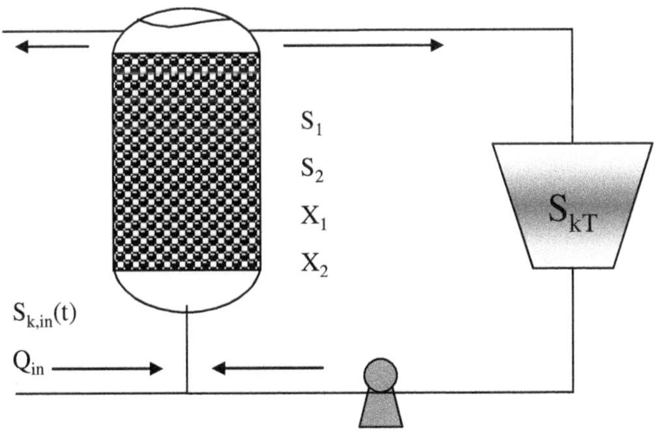


**Fig. 1** Block-diagram of anaerobic digestion bioreactor

**Table 1** Summary of the variables in the plant model

| *Variable* | *Units* | *Name* | *Value* |
|---|---|---|---|
| z | z∈[0,1] | *Space variable* | |
| t | D | *Time variable* | |
| $E_z$ | $m^2/d$ | *Axial dispersion coefficient* | 1 |
| D | 1/d | *Dilution rate* | 0.55 |
| H | m | *Fixed bed length* | 3.5 |
| $X_1$ | g/L | *Concentration of acidogenic bacteria* | |
| $X_2$ | g/L | *Concentration of methanogenic bacteria* | |
| $S_1$ | g/L | *Chemical Oxygen Demand* | |
| $S_2$ | mmol/L | *Volatile Fatty Acids* | |
| ε | | *Bacteria fraction in the liquid phase* | 0.5 |
| $k_1$ | g/g | *Yield coefficients* | 42.14 |
| $k_2$ | mmol/g | *Yield coefficients* | 250 |
| $k_3$ | mmol/g | *Yield coefficients* | 134 |
| $\mu_1$ | 1/d | *Acidogenesis growth rate* | |
| $\mu_2$ | 1/d | *Methanogenesis growth rate* | |
| $\mu_{1max}$ | 1/d | *Maximum acidogenesis growth rate* | 1.2 |
| $\mu_{2s,}$ | 1/d | *Maximum methanogenesis growth rate* | 0.74 |
| $K_{1s,}$ | g/g | *Kinetic parameter* | 50.5 |
| $K_{2s,}$ | mmol/g | *Kinetic parameter* | 16.6 |
| $K_{I2}$ | mmol/g | *Kinetic parameter* | 256 |
| $Q_T$ | $m^3/d$ | *Recycle flow rate* | 0.24 |
| $V_T$ | $m^3$ | *Volume of the recirculation tank* | 0.2 |
| $S_{1T}$ | g/L | *Concentration of Chemical Oxygen Demand in the recirculation tank* | |
| $S_{2T}$ | mmol/L | *Concentration of Volatile Fatty Acids in the recirculation tank* | |
| $Q_{in}$ | $m^3/d$ | *Inlet flow rate* | 0.31 |
| $V_B$ | $m^3$ | *Volume of the fixed bed* | 1 |
| $V_{eff}$ | $m^3$ | *Effective volume tank* | 0.95 |
| $S_{1,in}$ | g/l | *Inlet substr. concentration* | |
| $S_{2,in}$ | mmol/L | *Inlet substr. concentration* | |

$$\frac{\partial X_2}{\partial t} = (\mu_2 - \varepsilon D) X_2 , \quad \mu_2 = \mu_{2s} \frac{S_2}{K_{s_2}^{'} X_2 + S_2 + \frac{S_2^2}{K_{I_2}}} , \tag{2}$$

$$\frac{\partial S_1}{\partial t} = \frac{E_z}{H^2} \frac{\partial^2 S_1}{\partial z^2} - D \frac{\partial S_1}{\partial t} - k_1 \mu_1 X_1 , \tag{3}$$

$$\frac{\partial S_2}{\partial t}=\frac{E_z}{H^2}\frac{\partial^2 S_2}{\partial z^2}-D\frac{\partial S_2}{\partial t}+k_2\mu_1 X_1\ , \tag{4}$$

$$S_1(0,t)=\frac{S_{1,in}(t)+RS_{1T}}{R+1},\quad S_2(0,t)=\frac{S_{2,in}(t)+RS_{2T}}{R+1},\quad R=\frac{Q_T}{DV_{eff}}, \tag{5}$$

$$\frac{\partial S_1}{\partial z}(1,t)=0\ ,\quad \frac{\partial S_2}{\partial z}(1,t)=0\ . \tag{6}$$

For practical purpose, the full PDE anaerobic digestion process model (1)-(6), taken from [30], could be reduced to an ODE system using an early lumping technique and the Orthogonal Collocation Method (OCM), [31], in three points (0.25H, 0.5H, 0.75H) obtaining the following system of OD equations:

$$\frac{dX_{1,i}}{dt}=(\mu_{1,i}-\varepsilon D)X_{1,i}\ ,\quad \frac{dX_{2,i}}{dt}=(\mu_{2,i}-\varepsilon D)X_{2,i}\ , \tag{7}$$

$$\frac{dS_{1,i}}{dx}=\frac{E_z}{H^2}\sum_{j=1}^{N+2}B_{i,j}S_{1,j}-D\sum_{j=1}^{N+2}A_{i,j}S_{1,j}-k_1\mu_{1,i}X_{1,i}\ , \tag{8}$$

$$\frac{dS_{1T}}{dt}=\frac{Q_T}{V_T}\left(S_1(1,t)-S_{1T}\right),\quad \frac{dS_{2T}}{dt}=\frac{Q_T}{V_T}\left(S_2(1,t)-S_{2T}\right). \tag{9}$$

$$\frac{dS_{2,i}}{dx}=\frac{E_z}{H^2}\sum_{j=1}^{N+2}B_{i,j}S_{1,j}-D\sum_{j=1}^{N+2}A_{i,j}S_{2,j}+k_2\mu_{1,i}X_{2,i}-k_3\mu_{2,i}X_{2,i}\ , \tag{10}$$

$$\frac{dS_{1T}}{dt}=\frac{Q_T}{V_T}\left(S_{1,N+2}-S_{1T}\right),\quad \frac{dS_{2T}}{dt}=\frac{Q_T}{V_T}\left(S_{2,N+2}-S_{2T}\right), \tag{11}$$

$$S_{k,1}=\frac{1}{R+1}S_{k,in}(t)+\frac{R}{R+1}S_{kT}\ ,\quad S_{k,N+2}=\frac{K_1}{R+1}S_{k,in}(t)+\frac{K_1R}{R+1}S_{kT}+\sum_{i=2}^{N+1}K_iS_{k,i} \tag{12}$$

$$K_1=-\frac{A_{N+2,1}}{A_{N+2,N+2}},\quad K_i=-\frac{A_{N+2,i}}{A_{N+2,N+2}}, \tag{13}$$

$$A=\Lambda\phi^{-1}\ ,\quad \Lambda=\left[\varpi_{m,l}\right],\quad \varpi_{m,l}=(l-1)z_m^{l-2}\ , \tag{14}$$

$$B=\Gamma\phi^{-1}\ ,\quad \Gamma=\left[\tau_{m,l}\right],\quad \tau_{m,l}=(l-1)(l-2)z_m^{l-3}\ ,\phi_{m,l}=z_m^{l-1} \tag{15}$$

$$i=2,\ldots,N+2,\quad m,l=1,\ldots,N+2\ . \tag{16}$$

The reduced plant model (7)-(16) (here (9) represented the OD equations of the recirculation tank), could be used as unknown plant model which generate input/output process data for decentralized adaptive FNMM control system design, based on the concepts, given in [16], [25], [26], [19], [30]. The mentioned concepts could be applied for this DPS fuzzyfying the space variable z, which represented the height of the fixed bed. Here the centers of the membership functions with respect to z corresponded to the collocation points of the simplified plant model which are in fact the three measurement points of the fixed bed, adding one more point for the recirculation tank.

# 3 Description of the Direct Fuzzy-Neural Control System

The block-diagrams of the complete direct Fuzzy-Neural Multi-Model (FNMM) control system and its identification and control parts are schematically depicted in Fig. 2, Fig. 3 and Fig. 4. The structure of the entire control system, [19], [26] contained Fuzzyfier, Fuzzy Rule-Based Inference System (FRBIS), containing four identification, four feedback control and four feedforward control T-S rules ($RI_i$, $RCfb_i$, $RCff_i$) and a defuzzyfier.

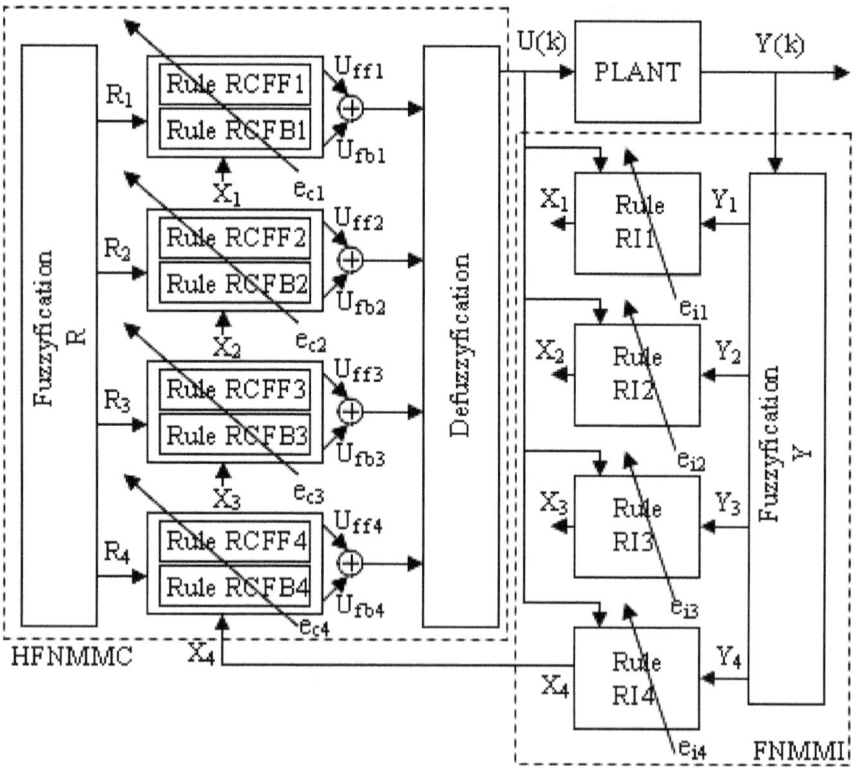


**Fig. 2** Block-Diagram of the FNMM Control System

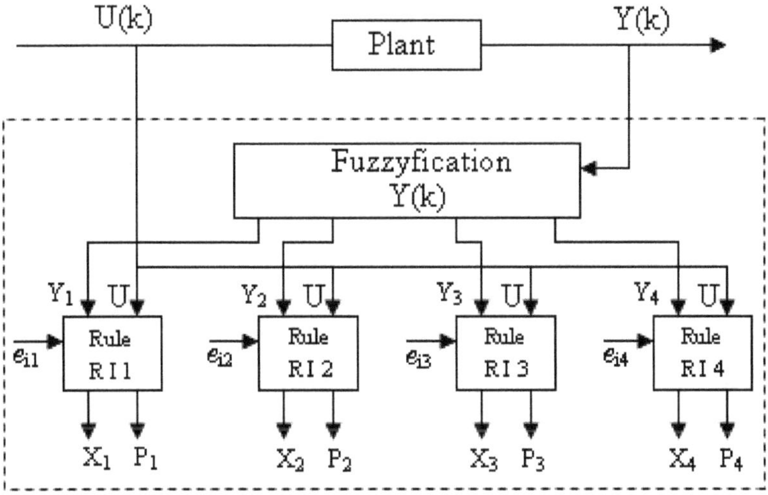


**Fig. 3** Detailed block-diagram of the FNMM identifier

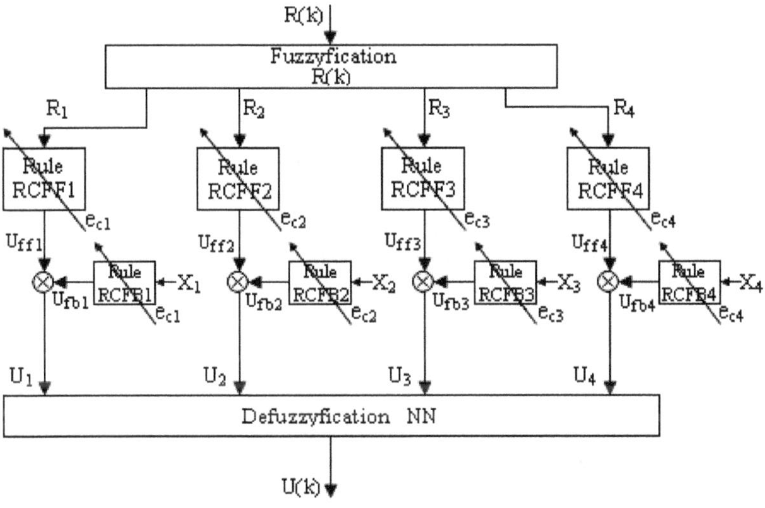


**Fig. 4** Detailed block-diagram of the HFNMM controller

## *3.1 Direct Adaptive FNMM Control System Design*

The plant output variable and its correspondent reference variable depended on space and time, and they are fuzzyfied on space and represented by four membership functions which centers are the four collocation points of the plant (three points for the fixed bed and one point for the recirculation tank). The main objective of the Fuzzy-Neural Multi-Model Identifier (FNMMI), containing four rules, is to issue states for the direct adaptive Fuzzy-Neural Multi-Model Feedback Controller (FNMMFBC) when the FNMMI outputs follows the outputs of the plant in the four measurement (collocation) points with minimum error of approximation. The direct fuzzy neural controller has also a direct adaptive Fuzzy-Neural Multi-Model Controller (FNMMC). The objective of the direct adaptive FNMM controller, containing four Feedback (FB) and four Feedforward (FF) T-S control rules is to reduce the error of control, so that the plant outputs in the four measurement points tracked the corresponding reference variables with minimum error of tracking.

The upper hierarchical level of the FNMM control system is one- layer- perceptron which represented the defuzzyfier, [19]. The hierarchical FNMM controller has two levels – Lower Level of Control (LLC), and Upper Level of Control (ULC). It is composed of three parts: 1) Fuzzyfication, where the normalized reference vector signal contained reference components of four measurement points; 2) Lower Level Inference Engine, which contains twelve T-S fuzzy rules (four rules for identification and eight rules for control- four in the feedback part and four in the feedforward part), operating in the corresponding measurement points; 3) Upper Hierarchical Level of neural defuzzification. The detailed block-diagram of the FNMMI, contained a space plant output fuzzyfier and four identification T-S fuzzy rules, labeled as $RI_i$, which consequent parts are RTNN learning procedures, [19]. The identification T-S fuzzy rules have the form:

$$RI_i\text{: } \textbf{\textit{If}}\ x(k)\ \textbf{\textit{is}}\ A_i\ \textbf{\textit{and}}\ u(k)\ \textbf{\textit{is}}\ B_i\ \textbf{\textit{then}}\ Y_i = \Pi_i\,(L,M,N_i,Y_{di},U,X_i,A_i,B_i,C_i,E_i),\ i=1\text{-}4 \quad (17)$$

The detailed block-diagram of the FNMMC, given on Fig. 4, contained a spaced plant reference fuzzyfier and eight control T-S fuzzy rules (four FB and four FF), which consequent parts are also RTNN learning procedures, [19], using the state information, issued by the corresponding identification rules. The consequent part of each feedforward control rule (the consequent learning procedure) has thc M, L, $N_i$ RTNN model dimensions, $R_i$, $Y_{di}$, $E_{ci}$ inputs and $U_{ffi}$, outputs used to form the total control. The T-S fuzzy rule has the form:

$$RCFF_i\text{: } \textbf{\textit{If}}\ R(k)\ \textbf{\textit{is}}\ B_i\ \textbf{\textit{then}}\ Uff_i = \Pi_i\,(M, L, N_i, R_i, Y_{di}, X_i, J_i, B_i, C_i, E_{ci}),\ i=1\text{-}4 \quad (18)$$

The consequent part of each feedback control rule (the consequent learning procedure) has the M, L, $N_i$ RTNN model dimensions, $Y_{di}$, $X_i$, $E_{ci}$ inputs and $U_{fbi}$, outputs used to form the total control. The T-S fuzzy rule has the form:

$$RCFB_i\text{: } \textbf{\textit{If}}\ Y_{di}\ \textbf{\textit{is}}\ A_i\ \textbf{\textit{then}}\ Ufb_i = \Pi_i\,(M, L, N_i, Y_{di}, X_i, X_{ci}, J_i, B_i, C_i, E_{ci}),\ i=1\text{-}4 \quad (19)$$

The total control corresponding to each of the four measurement points is a sum of its corresponding feedforward and feedback parts:

$$U_i\,(k) = -U_{ffi}\,(k) + U_{fbi}\,(k) \quad (20)$$

The defuzzyfication learning procedure, which correspond to the single layer perceptron learning is described by:

$$U = \Pi\ (M, L, N, Y_d, Uo, X, A, B, C, E) \qquad (21)$$

The T-S rule and the defuzzification of the plant output of the fixed bed with respect to the space variable z ($\lambda i,z$ is the correspondent membership function), [19], [20], are given by:

$RO_i$: ***If*** $Y_{i,t}$ ***is*** $A_i$ ***then*** $Y_{i,t} = a_i^T Y_t + b_i$, i=1,2,3

$$Y_z = [\Sigma_i\ \gamma_{i,z}\ a_i^T]\ Y_t + \Sigma_i\ \gamma_{i,z}\ b_i\ ;\ \gamma_{i,z} = \lambda_{i,z} / (\Sigma_j\ \lambda_{j,z})$$

The direct adaptive neural control algorithm, which is in the consequent part of the local fuzzy control rule RCFBi, (19) is a feedback control, using the states issued by the correspondent identification local fuzzy rule RIi (17).

## 3.2 *Description of the RTNN Topology and Learning*

The block-diagrams of the RTNN topology and its adjoint, are given on Fig. 5, and Fig. 6. Following Fig. 5, and Fig. 6, we could derive the dynamic BP algorithm of its learning based on the RTNN topology using the diagrammatic method of [32]. The RTNN topology and learning are described in vector-matrix form as:

$$X(k+1) = AX(k) + BU(k);\ B = [B_1\ ;\ B_0];\ U^T = [U_1\ ;\ U_2]; \qquad (22)$$

$$Z_1(k) = G[X(k)]; \qquad (23)$$

$$V(k) = CZ(k);\ C = [C_1\ ;\ C_0];\ Z^T = [Z_1\ ;\ Z_2]; \qquad (24)$$

$$Y(k) = F[V(k)]; \qquad (25)$$

$$A = \text{block-diag}\ (Ai),\ |Ai\,| < 1; \qquad (26)$$

$$W(k+1) = W(k) + \eta\ \Delta W(k) + \alpha\ \Delta W_{ij}(k-1); \qquad (27)$$

$$E(k) = T(k)-Y(k); \qquad (28)$$

$$E_1(k) = F'[Y(k)]\ E(k);\ F'[Y(k)] = [1-Y^2(k)]; \qquad (29)$$

$$\Delta C(k) = E_1(k)\ Z^T(k); \qquad (30)$$

$$E_3(k) = G'[Z(k)]\ E_2(k);\ E_2(k) = C^T(k)\ E_1(k);\ G'[Z(k)] = [1-Z^2(k)]; \qquad (31)$$

$$\Delta B(k) = E_3(k)\ U^T(k); \qquad (32)$$

$$\Delta A(k) = E_3(k)\ X^T(k); \qquad (33)$$

$$Vec(\Delta A(k)) = E_3(k)\square X(k); \qquad (34)$$

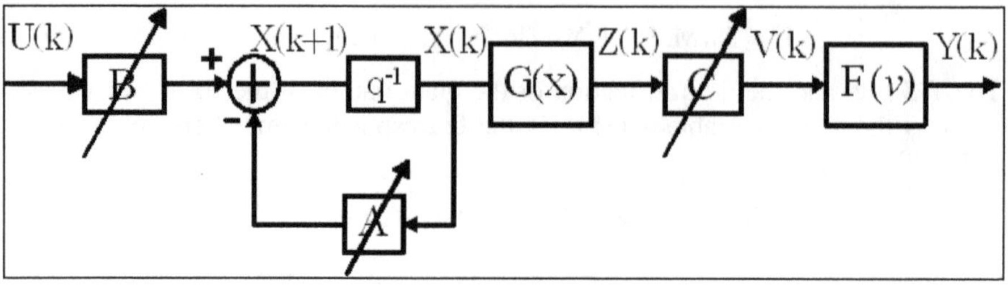


**Fig. 5** Block diagram of the RTNN model

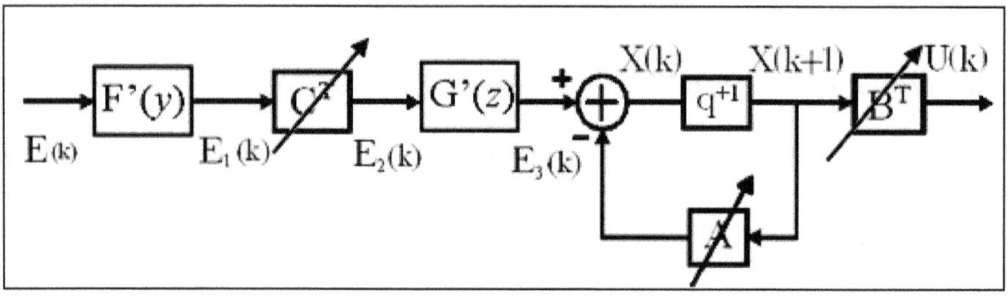


**Fig. 6** Block diagram of the adjoint RTNN model

Where: X, Y, U are state, augmented output, and input vectors with dimensions n, (l+1), (m+1), respectively, where $Z_1$ and $U_1$ are the (nx1) output and (mx1) input of the hidden layer; the constant scalar threshold entries are $Z_2 = -1$, $U_2 = -1$, respectively; V is a (lx1) pre-synaptic activity of the output layer; T is the (lx1) plant output vector, considered as a RNN reference; A is (nxn) block-diagonal weight matrix; B and C are [nx(m+1)] and [lx(n+1)]- augmented weight matrices; $B_0$ and $C_0$ are (nx1) and (lx1) threshold weights of the hidden and output layers; F[.], G[.] are vector-valued tanh(.)-activation functions with corresponding dimensions; F'[.], G'[.] are the derivatives of these tanh(.) functions; W is a general weight, denoting each weight matrix (C, A, B) in the RTNN model, to be updated; ΔW (ΔC, ΔA, ΔB), is the weight correction of W; $\eta$, $\alpha$ are learning rate parameters; ΔC is an weight correction of the learned matrix C; ΔB is an weight correction of the learned matrix B; ΔA is an weight correction of the learned matrix A; the diagonal of the matrix A is denoted by Vec(.) and equation (34) represents its learning as an element-by-element vector products; E, $E_1$, $E_2$, $E_3$, are error vectors with appropriate dimensions, predicted by the adjoint RTNN model, given on Fig. 6. The stability of the RTNN model is assured by the activation functions (-1, 1) bounds and by the local stability weight bound condition, given by (26). Below a theorem of RTNN stability which represented an extended version of Nava's theorem, [27], [28], [29] is given.

**Theorem of stability of the RTNN [29]:** Let the RTNN with Jordan Canonical Structure is given by equations (1)-(5) (see Fig.1) and the nonlinear plant model, is as follows:

$$X_{d.}(k+1) = G[\, X_d(k), U(k)\, ]$$

$$Y_d(k) = F[\, X_d(k)\, ]$$

Where: $\{Y_d(.), X_d(.), U(.)\}$ are output, state and input variables with dimensions l, $n_d$, m, respectively; F(.), G(.) are vector valued nonlinear functions with respective dimensions. Under the assumption of RTNN identifiability made, the application of the BP learning algorithm for A(.), B(.), C(.), in general matricial form, described by equation (27)-(34), and the learning rates $\eta$ (k), $\alpha$ (k) (here they are considered as time-dependent and normalized with respect to the error) are derived using the following Lyapunov function:

$$L(k) = L_1(k)+L_2(k)$$

Where: $L_1(k)$ and $L_2(k)$ are given by:

$$L_1(k) = \tfrac{1}{2}e^2(k)$$

$$L_2(k) = tr\left(\tilde{W}_A(k)\tilde{W}_A^T(k)\right)+tr\left(\tilde{W}_B(k)\tilde{W}_B^T(k)\right)+tr\left(\tilde{W}_C(k)\tilde{W}_C^T(k)\right)$$

Where: $$\tilde{W}_A(k) = \hat{A}(k)-A^*, \tilde{W}_B(k) = \hat{B}(k)-B^*, \tilde{W}_C(k) = \hat{C}(k)-C^*$$

are vectors of the estimation error and $(A^*,B^*,C^*)$, $(\hat{A}(k),\hat{B}(k),\hat{C}(k))$ denote the ideal neural weight and the estimate of the neural weight at the k-th step, respectively, for each case. Then the identification error is bounded, i.e.:

$$L(k+1) = L_1(k+1)+L_2(k+1) < 0$$
$$\Delta L(k+1) = L(k+1) - L(k)$$

Where the condition for $L_1(k+1)<0$ is that:

$$\frac{\left(1-\frac{1}{\sqrt{2}}\right)}{\psi_{max}} < \eta_{max} < \frac{\left(1+\frac{1}{\sqrt{2}}\right)}{\psi_{max}}$$

And for $L_2(k+1)<0$ we have:

$$\Delta L_2(k+1) < -\eta_{max}\left|e(k+1)\right|^2 - \alpha_{max}\left|e(k)\right|^2 + d(k+1)$$

Note that $\eta_{max}$ changes adaptively during the RTNN learning and:

$$\eta_{max} = \max_{i=1}^{3}\{\eta_i\}$$

Where all: the unmodelled dynamics, the approximation errors and the perturbations, are represented by the d-term. The Rate of Convergence Lemma used, [28], is given below. The complete proof of that Theorem of stability is given in [29].

**Rate of Convergence Lemma [28]:** Let $\Delta L_k$ is defined. Then, applying the limit's definition, the identification error bound condition is obtained as:

$$\overline{\lim_{k\to\infty}}\frac{1}{k}\sum_{t=1}^{k}\left(\left|E(t)\right|^{2}+\left|E(t-1)\right|^{2}\right)\leq d$$

**Proof:** Starting from the final result of the theorem of RTNN stability:

$$\Delta L(k)\leq-\eta(k)\left|E(k)\right|^{2}-\alpha(k)\left|E(k-1)\right|^{2}+d$$

and iterating from k=0, we get:

$$L(k+1)-L(0)\leq-\sum_{t=1}^{k}\left|E(t)\right|^{2}-\sum_{t=1}^{k}\left|E(t-1)\right|^{2}+dk$$

$$\sum_{t=1}^{k}\left(\left|E(t)\right|^{2}+\left|E(t-1)\right|^{2}\right)\leq dk-L(k+1)+L(0)\leq dk+L(0)$$

From here, we could see that $d$ must be bounded by weight matrices and learning parameters, in order to obtain: $\Delta L(k)\in L(\infty)$.

As a consequence: $A(k)\in L(\infty), B(k)\in L(\infty), C(k)\in L(\infty)$.

The stability of the HFNMMI could be proved via linearization of the activation functions of the RTNN models and application of the methodology, given in [33].

## 4 Description of the Indirect Fuzzy-Neural Control System

The block-diagram of the FNMM control system is given on Fig.7. The structure of the entire control system, [19], [25] consisted of Fuzzyfier, Fuzzy Rule-Based Inference System (FRBIS), containing four identification and four control T-S rules ($RI_i$, $RC_i$), and a Defuzzyfier. Due to the learning abilities of the defuzzifier, the exact form of the control membership functions is not need to be known.

### *4.1 Indirect Adaptive FNMM Control System Design*

The plant output variable and its correspondent reference variable depended on space and time, and they are fuzzyfied on space. The membership functions of the fixed-bed output variables are triangular or trapezoidal ones and that - belonging to the output variables of the recirculation tank are singletons. The centers of the membership functions are the respective collocation points of the plant. The main objective of the FNMM Identifier (FNMMI) (see Fig. 3), containing four T-S rules, is to issue states and parameters for the indirect adaptive FNMM Controller (FNMMC) when the FNMMI outputs follows the outputs of the plant in the four measurement (collocation) points with minimum Means Squared Error (MSE%) of approximation. The objective of the indirect adaptive FNMM controller,

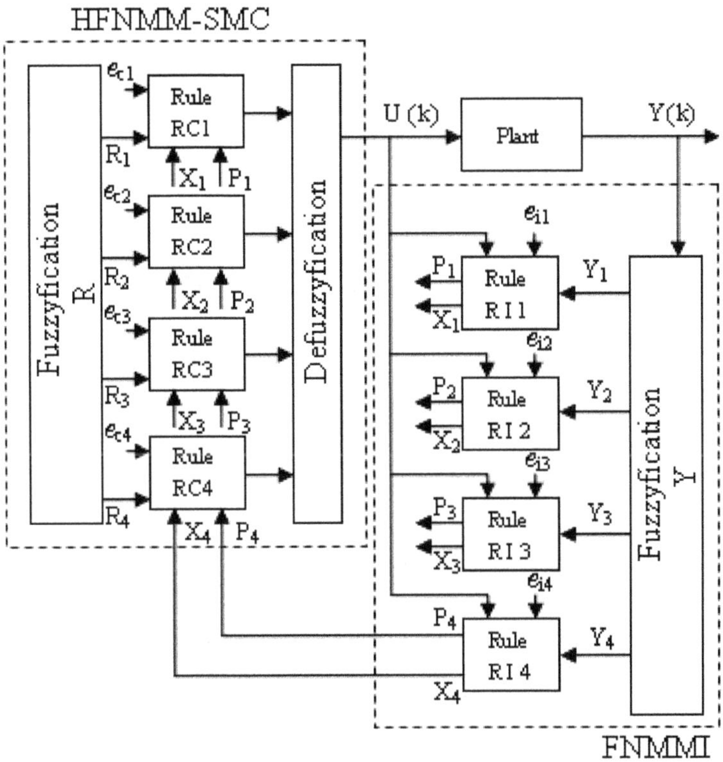


**Fig. 7** Block-diagram of the FNMM control system

containing four Sliding Mode Control (SMC) rules is to reduce the error of control, so that the plant outputs of the four measurement points tracked the corresponding reference variables with minimum MSE%. The hierarchical FNMM controller (see Fig. 8) has two levels – Lower Level of Control (LLC), and Upper Level of Control (ULC). It is composed of three parts: 1) Fuzzyfication, where the normalized reference vector signal contained reference components of four measurement points; 2) Lower Level Inference Engine, which contains eight T-S fuzzy rules (four rules for identification and four rules for control), operating in the corresponding measurement points; 3) Upper Hierarchical Level of neural defuzzification, represented by one layer perceptron, [19]. The detailed block-diagram of the FNMMI, given on Fig. 3, contained a space plant output fuzzyfier and four identification T-S fuzzy rules, labeled as $RI_i$, which consequent parts are learning procedures, [19], given by (17). The block-diagram of the FNMMC, given on Fig. 4, contained a spaced plant reference fuzzyfier and four sliding mode control T-S fuzzy rules, which consequent parts are SMC procedures, [19], using the state, and parameter information, issued by the corresponding identification rules. The control T-S fuzzy rules have the form:

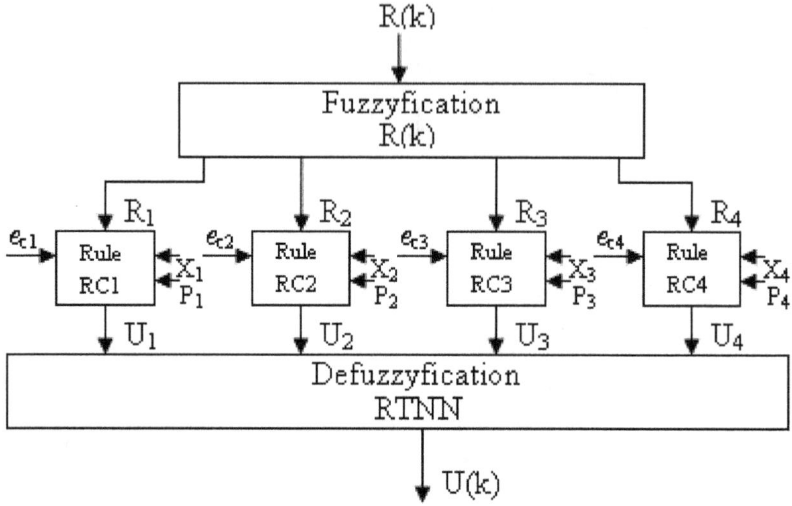


**Fig. 8** Detailed block-diagram of the HFNMM controller

$$RC_i: \textbf{\textit{If}}\ R(k)\ \textbf{\textit{is}}\ C_i\ \textbf{\textit{then}}\ U_i = \Pi_i\ (M, L, N_i, R_i, Y_{di}, X_i, A_i, B_i, C_i, E_{ci}),\ i=1\text{-}4 \tag{35}$$

The defuzzyfication of the control variable is a learning procedure, which correspond to the single layer perceptron learning, given by (21).

The T-S rule and the defuzzification of the plant output of the fixed bed with respect to the space variable z (here λi,z is the correspondent membership function), are given by:

$$RO_i: \textbf{\textit{If}}\ Y_{i,t}\ \textbf{\textit{is}}\ A_i\ \textbf{\textit{then}}\ Y_{i,t} = a_i^T Y_t + b_i,\ i=1,2,3 \tag{36}$$

$$Y_z=[\Sigma_i\ \gamma_{i,z}\ a_i^T]\ Y_t + \Sigma_i\ \gamma_{i,z}\ b_i\ ;\ \gamma_{i,z} = \lambda_{i,z} / (\Sigma_j\ \lambda_{j,z}) \tag{37}$$

Next the indirect SMC procedure will be briefly described.

## *4.2 Sliding Mode Control System Design*

Here the indirect adaptive neural control algorithm, which appeared in the consequent part of the local fuzzy control rule RCi (35) is viewed as a Sliding Mode Control (SMC), [19], [29], designed using the parameters and states issued by the correspondent identification local fuzzy rule RIi (17). Let us suppose that the studied nonlinear plant possess the following structure:

$$X_p(k+1) = F[X_p(k), U(k)]; Y_p(k) = G[X_p(k)], \tag{38}$$

Where: $X_p(k)$, $Y_p(k)$, U(k) are plant state, output and input vector variables with dimensions Np, L and M, where L>M is supposed; F and G are smooth, odd, bounded nonlinear functions. The linearization of the activation functions of the learned identification RTNN model, which approximates the plant (see equations. (22) to (26)), leads to the following linear local plant model:

$$X(k+1) = AX(k) + BU(k); Y(k) = CX(k) \tag{39}$$

Where L > M, is supposed. Let us define the following sliding surface with respect to the output tracking error:

$$S(k+1) = E(k+1) + \sum_{i=1}^{P} \gamma_i E(k-i+1)\ ; \quad |\gamma_i| < 1; \tag{40}$$

Where: S(.) is the sliding surface error function; E(.) is the systems output tracking error; $\gamma_i$ are parameters of the desired error function; P is the order of the error function. The additional inequality in (40) is a stability condition, required for the sliding surface error function. The tracking error is defined as:

$$E(k) = R(k) - Y(k); \tag{41}$$

Where R(k) is a L-dimensional reference vector and Y(k) is an output vector with the same dimension. The objective of the sliding mode control systems design is to find a control action which maintains the systems error on the sliding surface assuring that the output tracking error reached zero in P steps, where P<N, which is fulfilled if:

$$S(k+1) = 0. \tag{42}$$

As the local approximation plant model (39), is controllable, observable and stable,[29], the matrix A is block-diagonal, and L>M, the matrix product (CB) is nonsingular with rank M, and the plant states X(k) are smooth non- increasing functions. Now, from (39)-(42), it is easy to obtain the equivalent control capable to lead the system to the sliding surface which yields:

$$U_{eq}(k) = (CB)^{+}\left[-CAX(k) + R(k+1) + \sum_{i=1}^{P} \gamma_i E(k-i+1)\right] + Of, \tag{43}$$

$$(CB)^{+} = [(CB)^{T}(CB)]^{-1}(CB)^{T}$$

Here the added offset Of is a learnable M-dimensional constant vector which is learnt using a simple delta rule (see [1] for more details), where the error of the plant input is obtained backpropagating the output error through the adjoint RTNN model.

The SMC avoiding chattering is taken using a saturation function inside a bounded control level Uo, taking into account plant uncertainties. So the SMC has the form:

$$U(k) = \begin{cases} U_{eq}(k), & if \ \|U_{eq}(k)\| < U_0 \\ -U_0 U_{eq}(k) / \|U_{eq}(k)\|, & if \ \|U_{eq}(k)\| \geq U_0 \end{cases} \tag{44}$$

The proposed SMC cope with the characteristics of the wide class of plant model reduction neural control with reference model, and represents an indirect adaptive neural control, given by Baruch, [29]. Next simulation results of an indirect fuzzy-neural control of the DPS anaerobic wastewater treatment bioprocess are given.

# 5 Simulation Results

In this paragraph, graphical and numerical simulation results of system identification, direct, indirect and optimal control, will be given.

## *5.1 Simulation Results of the System Identification*

The decentralized FNMM identifier used a set of four T-S fuzzy rules containing in its consequent part RTNN learning procedures (17), (22)-(34). The topology of the first three RTNNs is (2-6-4) (2 inputs, 6 neurons in the hidden layer, 4 outputs) and the last one has topology (2-6-2) corresponding to the fixed bed plant behavior in each collocation point and the recirculation tank. The RTNNs identified the following fixed bed variables: $X_1$ (acidogenic bacteria), $X_2$ (methanogenic bacteria), $S_1$ (chemical oxygen demand) and $S_2$ (volatile fatty acids), in the following collocation points, z=0.25H, z=0.5H, z=0.75H, and the following variables in the recirculation tank: $S_{1T}$ (chemical oxygen demand) and $S_{2T}$ (volatile fatty acids). The graphical simulation results of RTNNs learning are obtained on-line during 100 days with a step of 0.1 day (To=0.1 sec.; Nt=1000 iterations). The learning rate parameters of RTNN have small values which are different for the different measurement point variables. The Figs. 9-13 showed graphical simulation results of open loop decentralized plant identification. The Figs. 9a, 10a, 11a, 12a, 13a, gives the 3d space-time T-S fuzzy approximation of the correspondent variables. The Figs. 9b,c,d, 10b,c,d, 11b,c,d, 12b,c,d, 13b,c,d, compared the time-dependent graphics of the correspondent plant output variables with the correspondent RTNNs outputs only at the collocation points. The input signals applied are:

$$S_{1in} = 0.55+0.2\cos\left(\frac{\pi t}{80}\right)+0.1\sin\left(\frac{3\pi t}{80}\right)+0.1\sin\left(\frac{\pi t}{80}\right),$$
$$S_{2in} = 0.55+0.1\sin\left(\frac{\pi t}{36}\right)+0.2\cos\left(\frac{3\pi t}{36}\right)+0.1\cos\left(\frac{8\pi t}{36}\right) \quad (45)$$

The MSE% of the decentralized FNMM approximation of plant variables are shown in Table 2.

**Table 2** MSE% of the decentralized FNMM approximation of the bioprocess output variables

| *Collocation point* | $X_1$ | $X_2$ | $S_1/S_{1T}$ | $S_2/S_{2T}$ |
|---|---|---|---|---|
| z=0.25H | 1.2524e-8 | 6.5791e-8 | 2.9615e-5 | 4.3302e-4 |
| z=0.5H | 5.0180e-9 | 2.9067e-8 | 1.1840e-8 | 2.7851e-6 |
| z=0.75H | 1.0487e-9 | 2.7977e-9 | 9.3562e-5 | 2.8941e-4 |
| *Recirculation. Tank* | - | - | 8.6967e-7 | 2.0205e-6 |

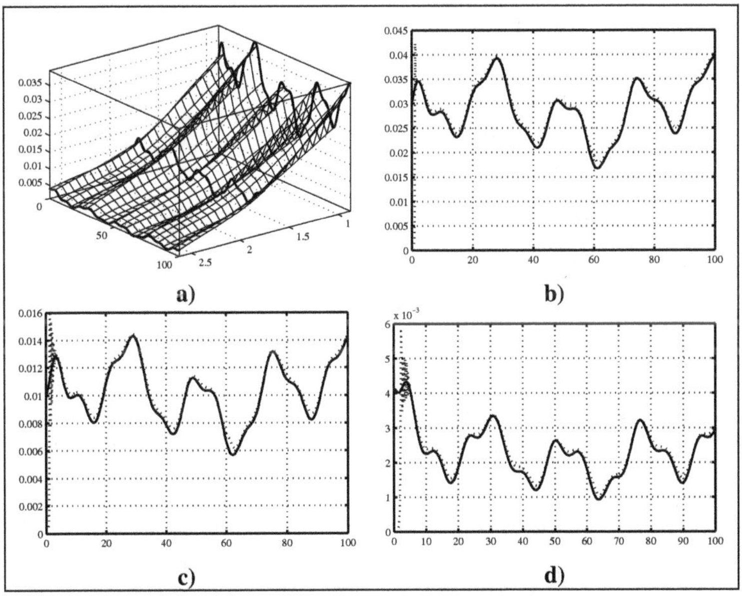

**Fig. 9** Graphical simulation results of the FNMM identification of X1 (acidogenic bacteria in the fixed bed) by three fuzzy rules RTNNs (dotted line-RTNN output, continuous line-plant output); a) 3d view of X1; b) X1 in z=0.25H; c) X1 in z=0.5H; d) X1 in z=0.75H

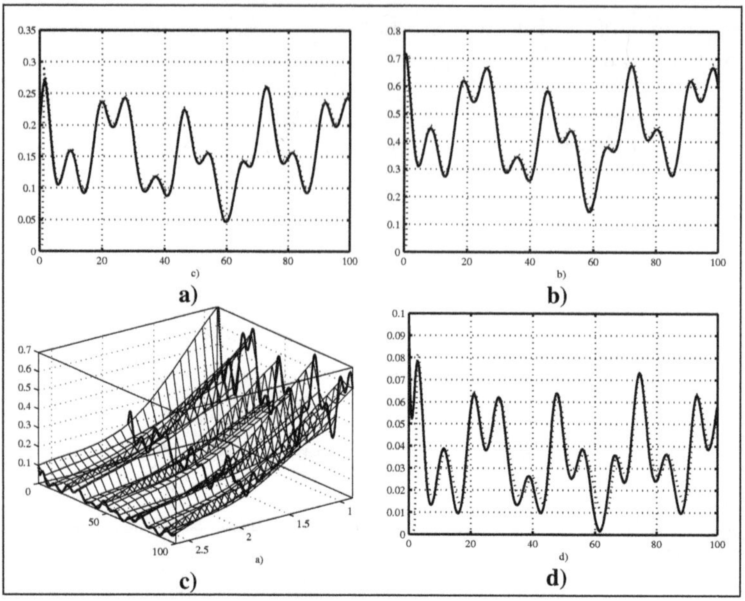


**Fig. 10** Graphical simulation results of the FNMM identification of S1 (chemical oxygen demand in the fixed bed) by three fuzzy rules RTNNs (dotted line-RTNN output, continuous line-plant output); a) 3d view of S1; b) S1 in z=0.25H; c) S1 in z=0.5H; d) S1 in z=0.75H

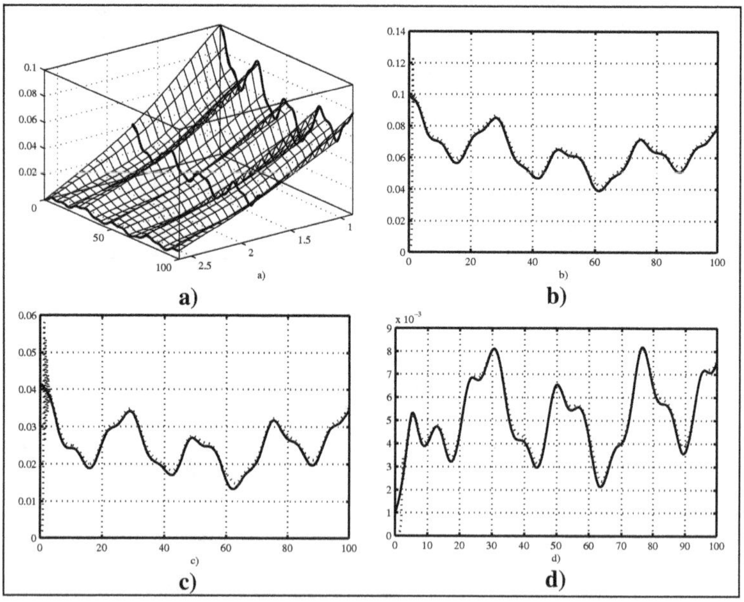


**Fig. 11** Graphical simulation results of the FNMM identification of $X_2$ (methanogenic bacteria in the fixed bed) by three fuzzy rules RTNNs (dotted line-RTNN output, continuous line-plant output); a) 3d view of $X_2$; b) $X_2$ in z=0.25H; c) $X_2$ in z=0.5H; d) $X_2$ in z=0.75H

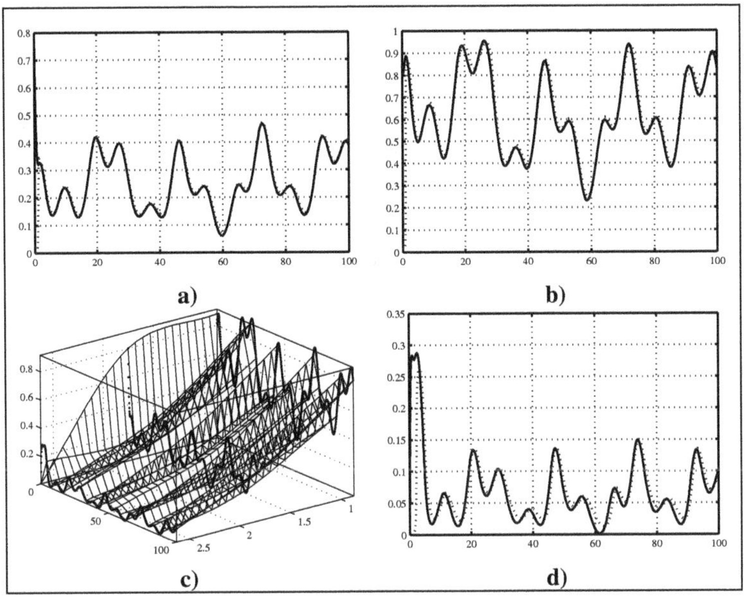

**Fig. 12** Graphical simulation results of the FNMM identification of $S_2$ (volatile fatty acids in the fixed bed) by three fuzzy rules RTNNs (dotted line-RTNN output, continuous line-plant output); a) 3d view of $S_2$; b) $S_2$ in z=0.25H; c) $S_2$ in z=0.5H; d) $S_2$ in z=0.75H

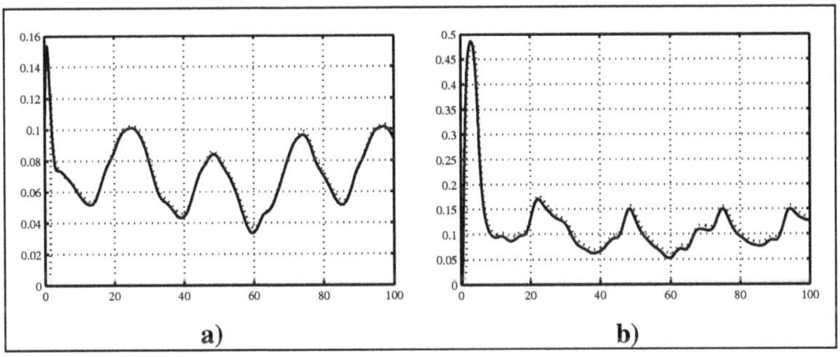

**Fig. 13** a) Graphical simulation results of the FNMM identification of $S_{1T}$ (chemical oxygen demand in the recirculation tank) (dotted line-RTNN output, continuous line-plant output); b) Graphical simulation results of the FNMM identification of $S_{2T}$ (volatile fatty acids in the recirculation tank)-the same as for $S_{1T}$

The graphical y numerical results of decentralized FNMM identification (see Fig. 9-13, and Table 2) showed a good HFNMMI convergence and precise plant output tracking (MSE% is 4.3302e-4 in the worse case).

Next some results of direct and indirect decentralized hierarchical fuzzy-neural multi-model control will be given.

## 5.2 Simulation Results of the Direct HFNMM Control

The Figs. 14-18 showed graphical simulation results of the direct decentralized HFNMM control, where the outputs of the plant are compared with the reference signals. The reference signals are train of pulses with uniform duration and random amplitude. The MSE% of control for each output signal and each measurement point are given on Table 3.

**Table 3** MSE% of the direct decentralized HFNMM control of the bioprocess plant

| *Collocation point* | $X_1$ | $X_2$ | $S_1/S_{1T}$ | $S_2/S_{2T}$ |
|---|---|---|---|---|
| z=0.25H | 5.5985e-7 | 1.1744e-6 | 1.1840e-4 | 9.6735e-5 |
| z=0.5H | 7.0402e-8 | 4.3666e-7 | 1.4949e-5 | 3.8093e-5 |
| z=0.75H | 3.0264e-9 | 2.2637e-8 | 6.3987e-7 | 2.0047e-6 |
| *Recirc.Tank* | | | 3.0627e-6 | 2.7773e-6 |

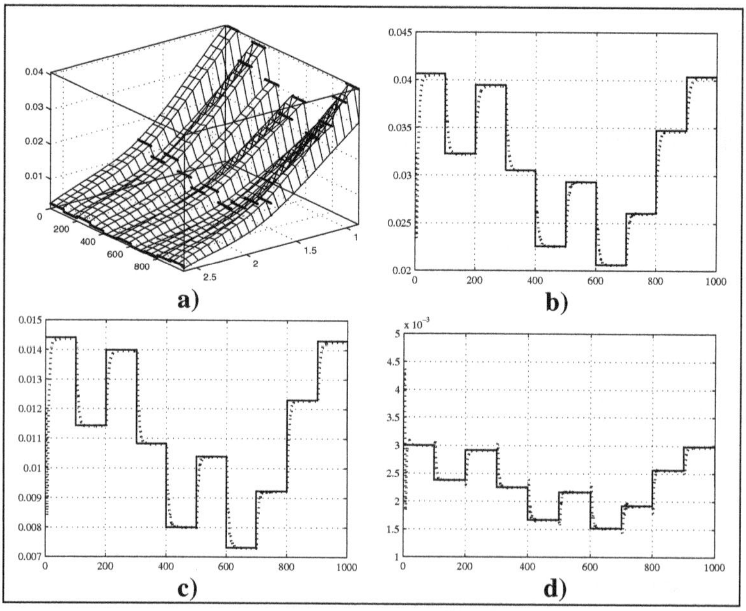

**Fig. 14** Results of the direct decentralized FNMM control of $X_1$ (acidogenic bacteria in the fixed bed) (dotted line-plant output, continuous-reference); a) 3d view of $X_1$ ; b) R1 vs $X_1$ in z=0.25H; c) R2 vs $X_1$ in z=0.5H; d) R3 vs $X_1$ in z=0.75H

The graphical y numerical results (see Fig. 14-18, and Table 3) of the direct decentralized control showed a good HFNMMC convergence and precise reference tracking (MSE% is about 1.184e-4 in the worse case).

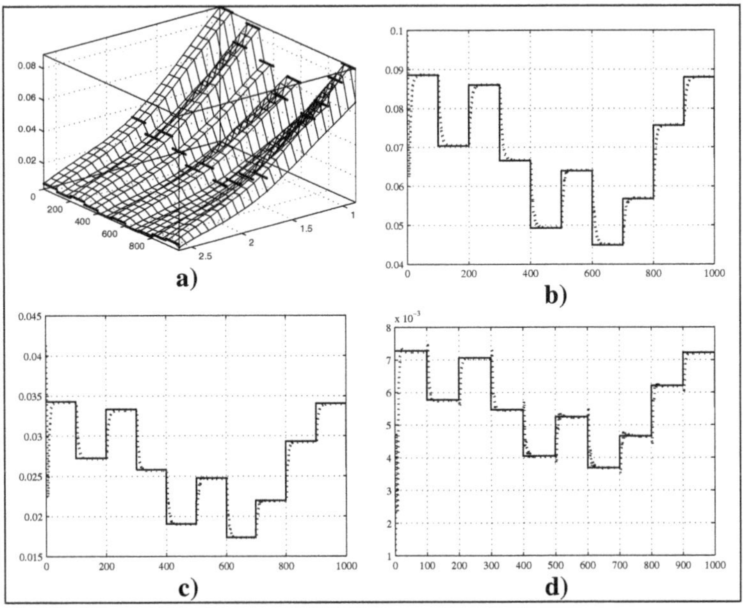


**Fig. 15** Results of the direct decentralized FNMM control of $X_2$ (methanogenic bacteria in the fixed bed) (dotted line-plant output, continuous-reference); a) 3d view of $X_2$ ; b) R1 vs $X_2$ in z=0.25H; c) R2 vs of $X_2$ in z=0.5H; d) R3 vs $X_2$ in z=0.75H

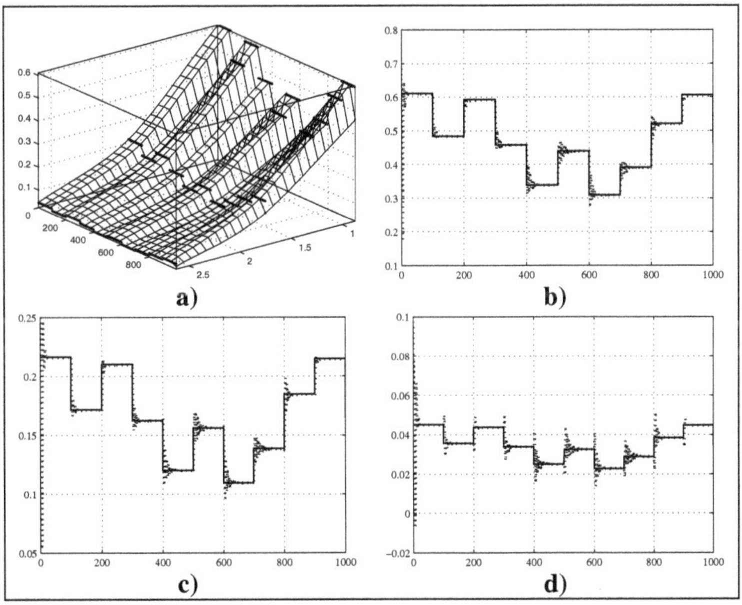


**Fig. 16** Results of the direct decentralized FNMM control of S1 (chemical oxygen demand in the fixed bed) (dotted line-plant output, continuous-reference); a) 3d view of $S_1$ ; b) R1 vs $S_1$ in z=0.25H; c) R2 vs $S_1$ in z=0.5H; d) R3 vs $S_1$ in z=0.75H

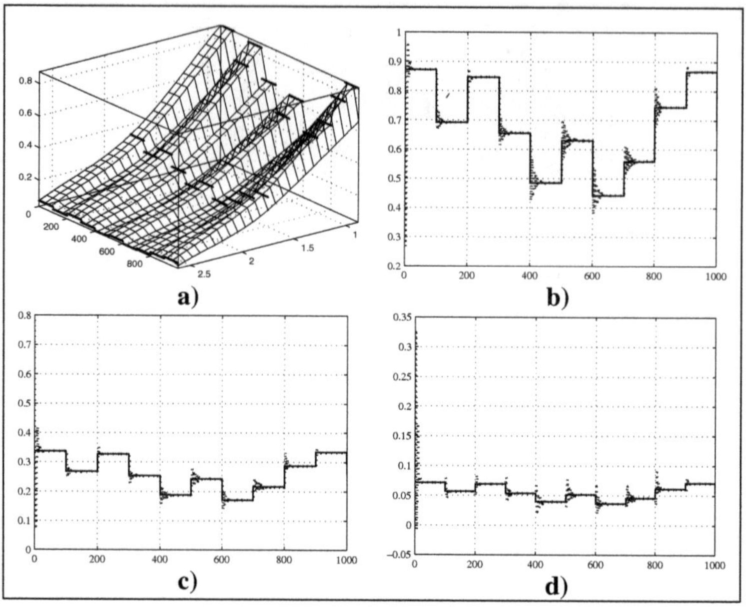

**Fig. 17** Results of the direct decentralized FNMM control of S2 (volatile fatty acids in the fixed bed) (dotted line-plant output, continuous-reference); a) 3d view of $S_2$ ; b) R1 vs $S_2$ in z=0.25H; c) R2 vs $S_2$ in z=0.5H; d) R3 vs $S_2$ in z=0.75H

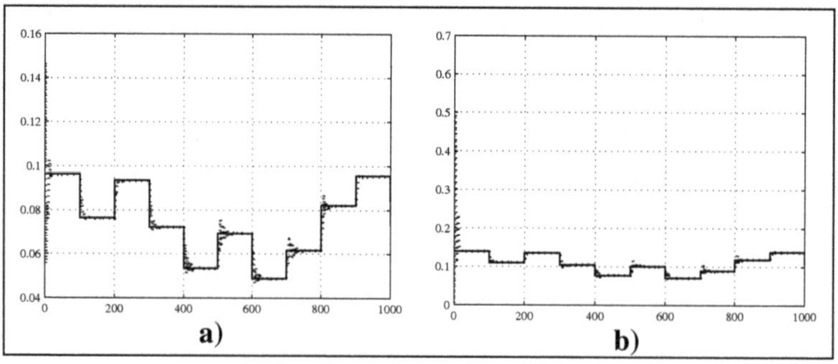

**Fig. 18** a) Results of the direct decentralized FNMM control of S1T (chemical oxygen demand in the recirculation tank) (dotted line-RTNN output, continuous line-plant output); b) Results of the direct decentralized FNMM control of S2T (volatile fatty acids in the recirculation tank)-the same as for S1T

## 5.3 Simulation Results of the Indirect HFNMM SMC

The Figs. 19-23 showed graphical simulation results of the indirect (sliding mode) decentralized HFNMM control. The MSE% of control for each output signal and each measurement point are given on Table 4.

**Table 4** MSE% of the indirect decentralized HFNMM control of the bioprocess plant

| *Collocation point* | $X_1$ | $X_2$ | $S_1/S_{1T}$ | $S_2/S_{2T}$ |
|---|---|---|---|---|
| z=0.25H | 4.6827e-7 | 9.9984e-7 | 9.8315e-5 | 8.0969e-5 |
| z=0.5H | 5.8922e-8 | 3.6415e-7 | 1.2420e-5 | 3.1382e-5 |
| z=0.75H | 2.5340e-9 | 1.8844e-8 | 5.3106e-7 | 1.6497e-6 |
| *Recir.. Tank* | | | 2.5533e-6 | 2.3211e-6 |

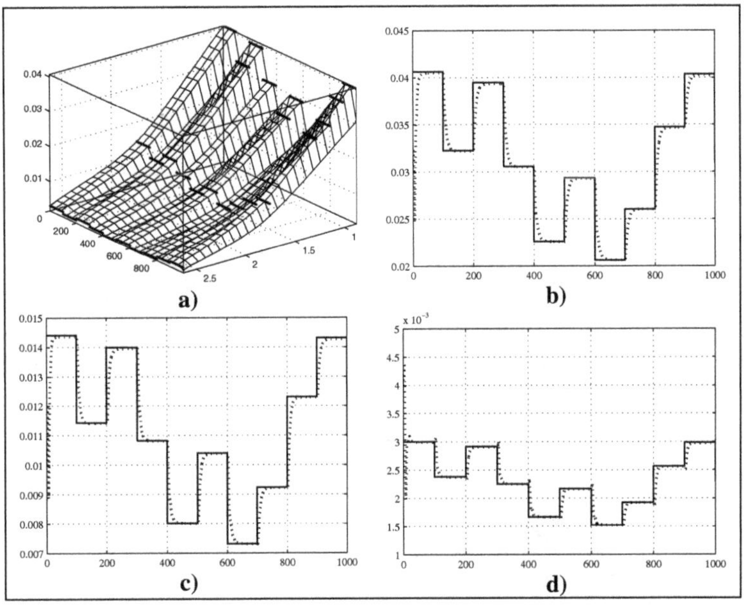


**Fig. 19** Results of the decentralized FNMM-SMC of X1 (acidogenic bacteria in the fixed bed) (dotted line-plant output, continuous-reference); a) 3d view of X1; b) SMC of X1 in z=0.25H; c) SMC of X1 in z=0.5H; d) SMC of X1 in z=0.75H

The reference signals are train of pulses with uniform duration and random amplitude and the outputs of the plant are compared with the reference signals.

The graphical y numerical results (see Fig. 19-23, and Table 4) of the indirect (sliding mode) decentralized control showed a good identification and precise reference tracking (MSE% is about 9.8315e-5 in the worse case).

The comparison of the indirect and direct decentralized control showed a good results for both control methods (see Table 3 and Table 4) with slight priority for the indirect control (9.8315e-5 vs. 1.184e-4) due to its better plant dynamics compensation ability and adaptation.

Next some simulation results of the linear optimal control will be given.

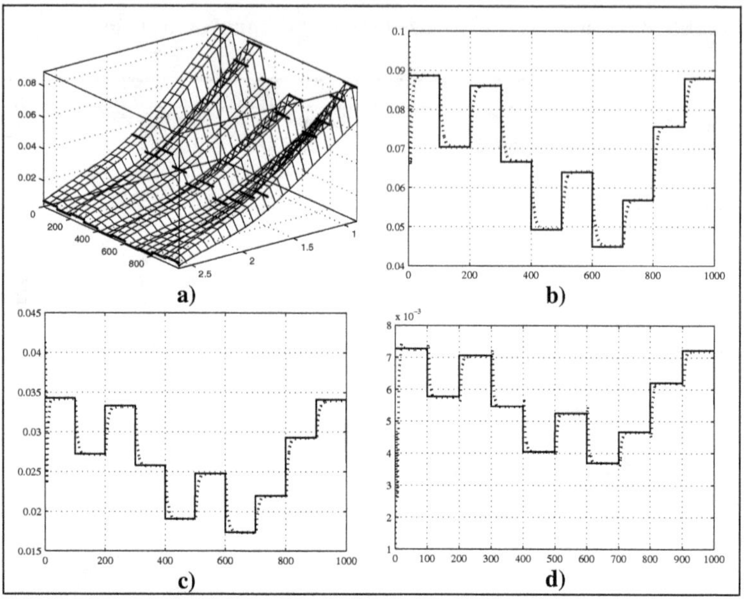

**Fig. 20** Results of the decentralized FNMM-SMC of $X_2$ (methanogenic bacteria in the fixed bed) (dotted line-plant output, continuous-reference); a) 3d view of $X_2$ ; b) SMC of $X_2$ in z=0.25H; c) SMC of $X_2$ in z=0.5H; d) SMC of $X_2$ in z=0.75H

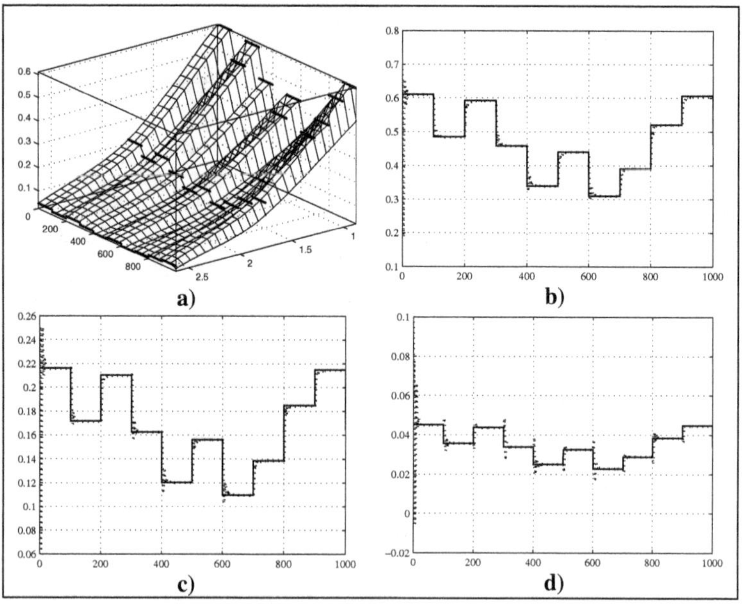

**Fig. 21** Results of the decentralized FNMM-SMC of S1 (chemical oxygen demand in the fixed bed) (dotted line-plant output, continuous-reference); a) 3d view of S1; b) SMC of S1 in z=0.25H; c) SMC of S1 in z=0.5H; d) SMC of S1 in z=0.75H

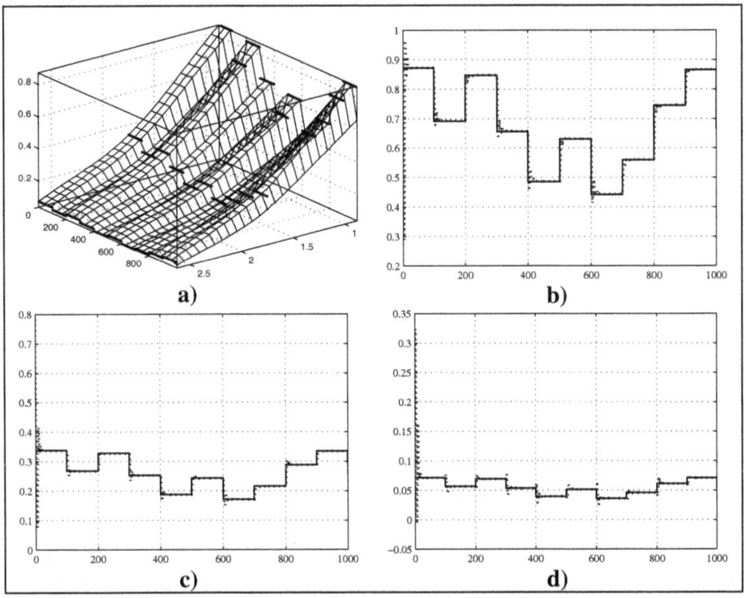


**Fig. 22** Results of the decentralized FNMM-SMC of S2 (volatile fatty acids in the fixed bed) (dotted line-plant output, continuous-reference); a) 3d view of $S_2$; b) SMC of $S_2$ in z=0.25H; c) SMC of $S_2$ in z=0.5H; d) SMC of $S_2$ in z=0.75H

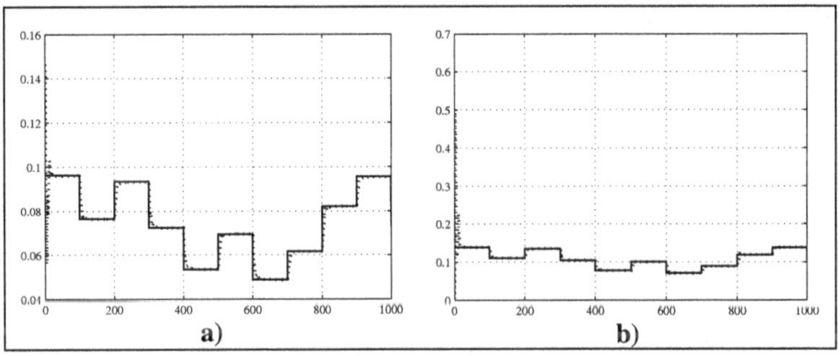


**Fig. 23** a) Results of the decentralized FNMM-SMC of $S_{1T}$ (chemical oxygen demand in the recirculation tank) (dotted line-RTNN output, continuous line-plant output); b) Results of the FNMM-SMC of $S_{2T}$ (volatile fatty acids in the recirculation tank)-the same as for $S_{1T}$

## *5.4 Simulation Results of the Linear Optimal Control*

For sake of comparison, in Table 5 are given MSE% results, obtained with optimal control of the linearized plant model (see equations (7)-(16)). The comparison of that MSE% results showed slight priority of the linearized optimal control over the direct and indirect decentralized FNMM control due to the exact plant model equations used in the control synthesis (2.1078e-5 vs. 1.184e-4 vs. 9.8315e-5 in the worse case).

**Table 5** MSE% of the Proportional optimal control of the bioprocess output variables

| *Col. point* | $X_1$ | $X_2$ | $S_1/S_{1T}$ | $S_2/S_{2T}$ |
|---|---|---|---|---|
| z=0.25H | 5.3057e-8 | 1.7632e-7 | 1.1978e-5 | 2.1078e-5 |
| z=0.5H | 6.6925e-9 | 4.2626e-8 | 1.4922e-6 | 4.4276e-6 |
| z=0.75H | 3.0440e-10 | 2.0501e-9 | 6.8737e-8 | 2.0178e-7 |
| *Recirc Tank* | | | 2.7323e-7 | 6.0146e-7 |

## 6 Conclusions

The paper performed decentralized recurrent fuzzy-neural identification, direct and indirect control of an anaerobic digestion wastewater treatment bioprocess, composed by a fixed bed and a recirculation tank, represented a DPS. The simplification of the PDE process model by ODE is realized using the orthogonal collocation method in three collocation points (plus the recirculation tank) represented centers of membership functions of the space fuzzyfied output variables. The obtained from the FNMMI state and parameter information is used by a HFNMM direct and indirect (sliding mode) control. The applied fuzzy-neural approach to that DPS decentralized direct and indirect identification and control exhibited a good convergence and precise reference tracking reaching the results obtained by the optimal control, which could be observed in the MSE% numerical results given on Tables 3, 4 and 5 (2.107e-5 vs. 1.184e-4 vs. 9.8315e-5 in the worse case).

**Acknowledgements.** The Ph.D. student Rosalba Galvan-Guerra is thankful to CONACYT for the scholarship received during her studies at the Department of Automatic Control, CINVESTAV-IPN, Mexico City, MEXICO.

## References

1. Haykin, S.: Neural Networks, a Comprehensive Foundation, 2nd edn., Section 2.13, pp. 84–89, Section 4.13, pp. 208–213. Prentice-Hall, Upper Saddle River (1999)
2. Narendra, K.S., Parthasarathy, K.: Identification and Control of Dynamical Systems Using Neural Networks. IEEE Transactions on Neural Networks 1(1), 4–27 (1990)
3. Chen, S., Billings, S.A.: Neural Networks for Nonlinear Dynamics System Modelling and Identification. International Journal of Control 56(2), 319–346 (1992)
4. Hunt, K.J., Sbarbaro, D., Zbikowski, R., Gawthrop, P.J.: Neural Network for Control Systems (A survey). Automatica 28, 1083–1112 (1992)
5. Miller III, W.T., Sutton, R.S., Werbos, P.J.: Neural Networks for Control. MIT Press, London (1992)
6. Pao, S.A., Phillips, S.M., Sobajic, D.J.: Neural Net Computing and Intelligent Control Systems. International Journal of Control, Special Issue on Intelligent Control 56(3), 263–289 (1992)
7. Su, H.-T., McAvoy, T.J., Werbos, P.: Long-Term Predictions of Chemical Processes Using Recurrent Neural Networks: A Parallel Training Approach. Industrial Engineering Chemical Research 31(5), 1338–1352 (1992)

8. Boskovic, J.D., Narendra, K.S.: Comparison of Linear, Nonlinear and Neural-Network-Based Adaptive Controllers for a Class of Fed-Batch Fermentation Processes. Automatica 31, 817–840 (1995)
9. Omatu, S., Khalil, M., Yusof, R.: Neuro-Control and its Applications. Springer, London (1995)
10. Bulsari, A., Palosaari, S.: Application of Neural Networks for System Identification of an Adsorption Column. Neural Computing and Applications 1, 160–165 (1993)
11. Deng, H., Li, H.X.: Hybrid Intelligence Based Modelling for Nonlinear Distributed Parameter Process with Applications to the Curing Process. IEEE Transactions on Systems, Man and Cybernetics 4, 3506–3511 (2003)
12. Deng, H., Li, H.X.: Spectral-Approximation-Based Intelligent Modelling for Distributed Thermal Processes. IEEE Transactions on Control Systems Technology 13, 686–700 (2005)
13. Gonzalez-Garcia, R., Rico-Martinez, R., Kevrekidis, I.: Identification of Distributed Parameter Systems: A Neural Net Based Approach. Computers and Chemical Engineering 22(4-supl. 1), 965–968 (1998)
14. Padhi, R., Balakrishnan, S., Randolph, T.: Adaptive Critic Based Optimal Neuro- Control Synthesis for Distributed Parameter Systems. Automatica 37, 1223–1234 (2001)
15. Padhi, R., Balakrishnan, S.: Proper Orthogonal Decomposition Based Optimal Neuro-Control Synthesis of a Chemical Reactor Process Using Approximate Dynamic Programming. Neural Networks 16, 719–728 (2003)
16. Pietil, S., Koivo, H.N.: Centralized and Decentralized Neural Network Models for Distributed Parameter Systems. In: Proc. of the Symposium on Control, Optimization and Supervision, CESA 1996, IMACS Multiconference on Computational Engineering in Systems Applications, Lille, France, pp. 1043–1048 (1996)
17. Lin, C.T., Lee, C.S.G.: Neural Fuzzy Systems: A Neuro - Fuzzy Synergism to Intelligent Systems. Prentice Hall, Englewood Cliffs (1996)
18. Babuska, R.: Fuzzy Modeling for Control. Kluwer, Norwell (1998)
19. Baruch, I., Beltran-Lopez, R., Olivares-Guzman, J.L., Flores, J.M.: A Fuzzy-Neural Multi-Model for Nonlinear Systems Identification and Control. Fuzzy-Sets and Systems 159, 2650–2667 (2008)
20. Takagi, T., Sugeno, M.: Fuzzy Identification of Systems and Its Applications to Modelling and Control. IEEE Transactions on Systems, Man, and Cybernetics 15, 116–132 (1985)
21. Teixeira, M., Zak, S.: Stabilizing Controller Design for Uncertain Nonlinear Systems, Using Fuzzy Models. IEEE Transactions on Systems, Man, and Cybernetics 7, 133–142 (1999)
22. Mastorocostas, P.A., Theocharis, J.B.: A Recurrent Fuzzy-Neural Model for Dynamic System Identification. IEEE Transactions on Systems, Man and Cybernetics, Part B: Cybernetics 32, 176–190 (2002)
23. Mastorocostas, P.A., Theocharis, J.B.: An Orthogonal Least-Squares Method for Recurrent Fuzzy-Neural Modeling. Fuzzy Sets and Systems 140(2), 285–300 (2003)
24. Galvan-Guerra, R., Baruch, I.S.: Anaerobic Digestion Process Identification Using Recurrent Neural Multi-Model. In: Gelbukh, A., Kuri-Morales, A.F. (eds.) Sixth Mexican International Conference on Artificial Intelligence, Aguascalientes, Mexico, Special Session, Revised Papers, CPS, November 4-10, 2007, pp. 319–329. IEEE Computer Society, Los Alamos (2008) ISBN 978-0-7695-3124-3

25. Baruch, I., Olivares-Guzman, J.L., Mariaca-Gaspar, C.R., Galvan-Guerra, R.: A Sliding Mode Control Using Fuzzy-Neural Hierarchical Multi-Model Identifier. In: Castillo, O., Melin, P., Ross, O.M., Cruz, R.S., Pedrycz, W., Kacprzyk, J. (eds.) Theoretical Advances and Applications of Fuzzy Logic and Soft Computing. ASC, vol. 42, pp. 762–771. Springer, Heidelberg (2007)
26. Baruch, I., Olivares-Guzman, J.L., Mariaca-Gaspar, C.R., Galvan-Guerra, R.: A Fuzzy-Neural Hierarchical Multi-Model for Systems Identification and Direct Adaptive Control. In: Melin, P., Castillo, O., Ramirez, E.G., Kacprzyk, J., Pedrycz, W. (eds.) Analysis and Design of Intelligent Systems Using Soft Computing Techniques. ASC, vol. 41, pp. 163–172. Springer, Heidelberg (2007)
27. Baruch, I.S., Flores, J.M., Nava, F., Ramirez, I.R., Nenkova, B.: An Advanced Neural Network Topology and Learning, Applied for Identification and Control of a D.C. Motor. In: Proc. 1st Int. IEEE Symp. Intelligent Systems, IS 2002, Varna, Bulgaria, vol. 1, pp. 289–295 (2002)
28. Nava, F., Baruch, I.S., Poznyak, A., Nenkova, B.: Stability Proofs of Advanced Recurrent Neural Networks Topology and Learning. Comptes Rendus (Proceedings of the Bulgarian Academy of Sciences) 57(1), 27–32 (2004)
29. Baruch, I.S., Mariaca-Gaspar, C.R., Barrera-Cortes, J.: Recurrent Neural Network Identification and Adaptive Neural Control of Hydrocarbon Biodegradation Processes. In: Hu, X., Balasubramaniam, P. (eds.) Recurrent Neural Networks, ch. 4, pp. 61–88. I-Tech Education and Publishing KG, Vienna (2008)
30. Aguilar-Garnica, F., Alcaraz-Gonzalez, V., Gonzalez-Alvarez, V.: Interval Observer Design for an Anaerobic Digestion Process Described by a Distributed Parameter Model. In: Proc. of the 2-nd International Meeting on Environmental Biotechnology and Engineering (2IMEBE), CINVESTAV-IPN, Mexico City, paper 117, pp. 1–16 (2006)
31. Bialecki, B., Fairweather, G.: Orthogonal Spline Collocation Methods for Partial Differential Equations. Journal of Computational and Applied Mathematics 128, 55–82 (2001)
32. Wan, E., Beaufays, F.: Diagrammatic Method for Deriving and Relating Temporal Neural Network Algorithms. Neural Computations 8, 182–201 (1996)
33. Guerra, T.M., Kruszewski, A., Vermeiren, L., Tirmant, H.: Conditions of Output Stabilization for Nonlinear Models in the Takagi-Sugeno's Form. Fuzzy Sets and Systems 157, 1248–1259 (2006)

# Effective Mutation Operator and Parallel Processing for Nurse Scheduling

Makoto Ohki, Shin-ya Uneme, and Hikaru Kawano

**Abstract.** This paper proposes an effective mutation operator and an effective parallel processing algorithm for cooperative genetic algorithm (CGA) to solve a nurse scheduling problem. The nurse scheduling is very complex task for a clinical director in a general hospital. Even veteran director needs one or two weeks to create the schedule. Besides, we extend the nurse schedule to permit the change of the schedule. This permission explosively increases computation time for the nurse scheduling. We propose the effective mutation operator for the CGA. This mutation operator does not lose consistency of the nurse schedule. Furthermore, we propose the parallel processing algorithm of the CGA. The parallel CGA always brings good results.

## 1 Introduction

A general hospital consists of several sections such as the internal medicine department and the pediatrics department. About fifteen to thirty nurses work in each section. A clinical director of the section arranges a shift schedule of all nurses in her/his section every month. The director has to consider more than fifteen requirements for arranging the shift schedule. Arrangement of the shift schedule, or the nurse scheduling, is very complex task. In our investigation, even a veteran director needs one or two weeks for the nurse scheduling. This means a loss of great work force and time. Therefore, computer software for the nurse scheduling has recently come to be required in the general hospital.

In this paper, we discuss a technique to generate and optimize the nurse schedule by using the cooperative genetic algorithm (CGA). The conventional CGA [2-5] optimize the nurse schedule by using only crossover operator, because the crossover is considered as the only one method which keeps consistency of relation between chromosomes in the CGA. We have proposed an effective mutation

Makoto Ohki · Shin-ya Uneme · Hikaru Kawano
Department of Information and Electronics, Graduate School of Engineering, Tottori University, 101, 4 Koyama-Minami, Tottori, Tottori 680-8552 Japan
Tel.: +81 857 31 5231
e-mail: mohki@ele.tottori-u.ac.jp

V. Sgurev et al. (Eds.): Intelligent Systems: From Theory to Practice, SCI 299, pp. 229–242.
springerlink.com 

operator [6, 7] for the CGA. The mutation operator brings small change into the population. However if some parts of the population are randomly changed, the population would be meaningless thing. These changes are very hard to recover by genetic operations. Therefore the mutation must preserve the consistency of the population. In this paper, we propose the mutation operator that does not lose the consistency of the population. Furthermore, we investigate a condition in which the mutation operator effectively works.

We discuss a case when the nurse schedule has been changed. At the beginning of a month, the schedule is made for coming four weeks. We accept that the schedule is changed in the past two weeks when two weeks have passed. By means of these changes, inconvenience occurs on the shift schedule, for example, defections of duty days and holidays. When there are few change points, it is not a problem that much. However, serious problems about duty management and the salary system occur when many points are changed on the schedule. We explain a technique to cancel problems by such the unfavorable changes by optimizing the schedule for coming four weeks in this paper.

The optimization of such the nurse schedule requires ten times the computational time in the case of the conventional nurse scheduling. Furthermore, in our investigation, we have to execute ten or more times of the optimization in order to acquire satisfactory results enough. Accordingly we require enormous computational time for the nurse scheduling with the changes. Now, we propose a technique to execute the nurse scheduling in parallel processing by using MPI™ technology. We call this technique parallel CGA for the nurse scheduling. Assigning the scheduling tasks to several computers with small computational load in night time, the nurse scheduling can be quickly executed. Besides, leisure computers are effectively utilized. The parallel CGA always gives better nurse schedule than the conventional technique.

## 2 Cooperative GA for Nurse Scheduling

### *2.1 Encoding the Nurse Schedule*

An individual and a population in the CGA for the nurse scheduling are defined as shown in Fig.1. The individual consists of the sequence of the duty symbols. The duty sequence consists of thirty fields, since one month includes thirty days in the practical example. Each individual expresses one-month schedule of the nurse $i$. There are not two or more individuals including the same nurse's schedule in the population. In the CGA, the population denotes the whole schedule.

### *2.2 Evaluation of the Nurse Schedule*

For arranging the nurse schedule, the clinical director must consider many requirements. For example, meeting, training, requested holiday, these must be accepted, where we assume that all the requested holidays have been confirmed by the director. Semi-night and midnight duty should be fairly arranged to all nurses.

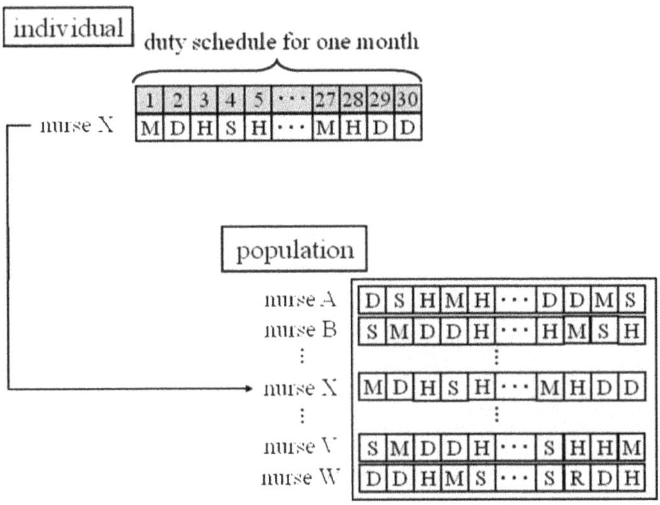


**Fig. 1** An individual coded in chromosome giving shift schedule of the nurse, X, for one month and a population including one-month schedules of all nurses. D, S, M and H denotes a day time duty, a semi-night duty, midnight duty and holiday respectively.

**Table 1** Duty Patterns for Eq. (2)

| Penalty | Duty pattern |
|---|---|
| 0 | DDD DDH DDM DHD DHH DHM DMS HDD HDM HHD HHH HMS SHH |
| 1 | DDS DSH DMH HDH HDS HHS HSH SHD MHD MHH MSH |
| 2 | DHS DSS HHM HSS HMH SHS SSH MDH MDS MHS MMH |
| 5 | DSD DSM DMD DMM HSD HSM HMD HMM SDD SDH SDS SDM SHM SSD SSS SSM SMD SMD SMS SMM MDD MDM MHM MSD MSS MSM MMD MMS MMM |

And it is prohibited to make nurses work for more than six consecutive days. We have summarized all the requirements into the twelve penalties. These penalties are classified into four penalty groups.

We define a penalty function on shift pattern as the following equation,

$$H_1 = \sum_{i=1}^{M} (h_{11}F_{1i} + h_{12}F_{2i} + h_{13}F_{3i}), \tag{1}$$

where $F_{1i}$, $F_{2i}$ and $F_{3i}$ denote the following penalty functions about the shift pattern.

We classify consecutive duty patterns for three days into four categories as shown in Table 1. In this table, Meeting and training are handled as day time duty, and requested holiday is handled as holiday. The first category denotes a top priority pattern, and its penalty value is defined to zero. The second category denotes a priority pattern, and its penalty value is defined to one. The third category denotes a compromised pattern, and its value is defined to two. The final category denotes a prohibited pattern, and its penalty value is defined to five. Comparing whole shift schedule of the nurse $i$ with Table 1, the penalty is given by the following equation.

$$F_{1i} = \sum_{j=2}^{D-1} p_{ij} , \tag{2}$$

where $p_{ij}$ denotes the penalty value given by Table 1.

It is not preferable for a night duty to be assigned for some nurse intensively. To suppress this situation, we define the following penalty function to prohibit four or more night duties to be assigned for consecutive six days.

$$F_{2i} = \sum_{j=1}^{D-5} \max\left(N_{night/6}(i, j) - 3, 0\right) , \tag{3}$$

where $N_{night/6}(i, j)$ denotes the number of night duties assigned for consecutive six days starting from date $j$ in the shift schedule of nurse $i$.

Depending on a hospital, there are some cases to prohibit a specific duty pattern. We define a penalty function $F_{3i}$ to implement such the prohibition.

We define a penalty function on the number of duty days as the following equation,

$$H_2 = \sum_{i=1}^{M} \left(h_{21}F_{4i} + h_{22}F_{5i} + h_{23}F_{6i}\right), \tag{4}$$

where $F_{4i}$, $F_{5i}$ and $F_{6i}$ denote the following penalty functions about the number of duty days.

The duty days should be fairly assigned among nurses. There is a threat that a total nursing level falls when much duty days were assigned to some nurses. We define the following penalty functions to suppress unevenness of the assigned duty days among nurses.

$$F_{4i} = \left|N_i^{\text{hom}} - N_{\text{hom}}\right|, \tag{5}$$

$$F_{5i} = \max\left(N_i^{sem} - N_{sem}, 0\right) + \max\left(N_i^{mid} - N_{smid}, 0\right), \tag{6}$$

where $N_i^{\text{hom}}$ , $N_i^{sem}$ and $N_i^{mid}$ denote the numbers of holidays, semi-night and midnight duties respectively assigned to nurse $i$ for one month. $N_{\text{hom}}$ denotes the number of Saturdays and Sundays on the current month. $N_{sem}$ and $N_{mid}$ denotes the limited numbers of semi-night and midnight duties, defined to four respectively in this paper.

If some nurses work many consecutive days, total nursing level falls. We define the following penalty function to restrain assigning more than six consecutive duty days.

$$F_{6i} = \sum_{j=1}^{D-5} \max\left(N_{serial}(i, j) - 5, 0\right), \tag{7}$$

where $N_{serial}(i, j)$ denotes the number of consecutive duty days starting from date $j$ in the shift schedule of nurse $i$.

We define a penalty function on nursing level as the following equation,

$$H_3 = \sum_{j=1}^{D} \left( h_{31} F_{7j} + h_{32} F_{8j} + h_{33} F_{9j} \right), \tag{8}$$

where $F_{7j}$, $F_{8j}$ and $F_{9j}$ denote the following penalty functions about the nursing level.

In our algorithm, the number of the nurses in each duty time is secured by all means. However, the nursing level deteriorates if new face nurses are only assigned. The expert or more skilled nurses should be assigned to keep nursing level. The nursing level of each nurse is given by ten phases as shown in Table 2. We assume that the manager evaluates the nursing level in ten phases. We define the following penalty functions to perform the nursing level of each duty time.

$$F_{7j} = \max\left\{ L_j^{day} - \sum_i L(n_i), 0 \right\}, n_i \in M_j^{day}, \tag{9}$$

$$F_{8j} = \max\left\{ L_j^{sem} - \sum_i L(n_i), 0 \right\}, n_i \in M_j^{sem}, \tag{10}$$

$$F_{9j} = \max\left\{ L_j^{mid} - \sum_i L(n_i), 0 \right\}, n_i \in M_j^{mid}, \tag{11}$$

where $L_j^{day}$, $L_j^{sem}$ and $L_j^{mid}$ denote the lowest nursing level at date $j$, and $M_j^{day}$, $M_j^{sem}$ and $M_j^{mid}$ denote sets of nurses assigned with day, semi night and mid night duty at date $j$. In this paper, we define $L_j^{day}$ to 54 on week day, 33 on Saturday and 28 on Sunday, and $L_j^{sem}$ and $L_j^{mid}$ both to 16.

We define a penalty function on nurse combination as the following equation,

$$H_4 = \sum_{j=1}^{D} \left( h_{41} F_{10j} + h_{42} F_{11j} + h_{43} F_{12j} \right), \tag{12}$$

where $F_{10j}$, $F_{11j}$ and $F_{12j}$ denote the following penalty functions about the nurse combination.

**Table 2** Nursing Level Defined for Twenty-Three Nurses

| nurse | $m_1$ | $m_2$ | $m_3$ | $m_4$ | $m_5$ | $M_6$ | $m_7$ | $m_8$ |
|---|---|---|---|---|---|---|---|---|
| level | 10 | 9 | 9 | 8 | 8 | 8 | 8 | 8 |
| nurse | $m_9$ | $m_{10}$ | $m_{11}$ | $m_{12}$ | $m_{13}$ | $m_{14}$ | $m_{15}$ | $m_{16}$ |
| level | 8 | 7 | 7 | 7 | 7 | 7 | 6 | 6 |
| nurse | $m_{17}$ | $m_{18}$ | $m_{19}$ | $m_{20}$ | $m_{21}$ | $m_{22}$ | $m_{23}$ | |
| level | 5 | 5 | 4 | 4 | 3 | 2 | 1 | |

The manager has to consider affinity between the nurses. Because of bad affinity between a certain nurses assigned to in the same time, there is the case that the nursing level deteriorates remarkably. We define a penalty function $F_{10j}$. When a pair of such the bad affinity is found in the shift schedule, penalty value 1 is added to the penalty function.

In the time of the midnight shift, only the nurse of few numbers of people is assigned to. The nursing level of the midnight shift deteriorates remarkably if the most of a nurse assigned to the time are new faces. To restrain such the unfavorable situation, we define the following penalty function,

$$F_{11j} = \begin{cases} 0 & , N_{j,new}^{mid} < 2 \\ \sum_{i=0}^{N_{j,new}^{mid}-2} \left(N_{j,new}^{mid} - i - 1\right) & , N_{j,new}^{mid} \geq 2 \end{cases} \quad (13)$$

where $N_{j,new}^{mid}$ denotes the number of new faces assigned to night duty on date $j$. In this paper, we define positions of nurse as shown by the Table 3. In this table, EX, BB and NF denote an expert, a backbone and a new face respectively.

In general, one or more expert or more skilled nurses must be assigned to day time and mid night duty. To restrain such an unfavorable situation, we define a penalty function, $F_{12j}$. If no expert or more skilled nurse is assigned to day time duty or mid night duty at date $j$, the function, $F_{12j}$, is increased with one point.

Finally, the whole shift schedule is performed by the following penalty function,

$$E = \sum_{k=1}^{4} H_k^{\ 2} \quad (14)$$

The smaller value the penalty function, $E$, gives, the better the shift schedule means.

**Table 3** Positions of Nurse

| $m_1$ | $m_2$ | $m_3$ | $m_4$ | $m_5$ | $m_6$ | $m_7$ | $m_8$ |
|---|---|---|---|---|---|---|---|
| chief | head | head | EX | EX | EX | EX | EX |
| $m_9$ | $m_{10}$ | $m_{11}$ | $m_{12}$ | $m_{13}$ | $m_{14}$ | $m_{15}$ | $m_{16}$ |
| EX | EX | BB | BB | BB | BB | BB | BB |
| $m_{17}$ | $m_{18}$ | $m_{19}$ | $m_{20}$ | $m_{21}$ | $m_{22}$ | $m_{23}$ | |
| BB | BB | NF | NF | NF | NF | NF | |

## 2.3 Basic Algorithm of CGA

The CGA basically searches for better nurse schedule by using the crossover operator [1] as shown in Fig. 2. The crossover operator brings ability for local search to the CGA. Only by using the crossover operator, the optimization stagnates because the CGA trapped in a local minimum area in the solution space. On the other hand, we have already proposed an effective mutation operator [6, 7]. By

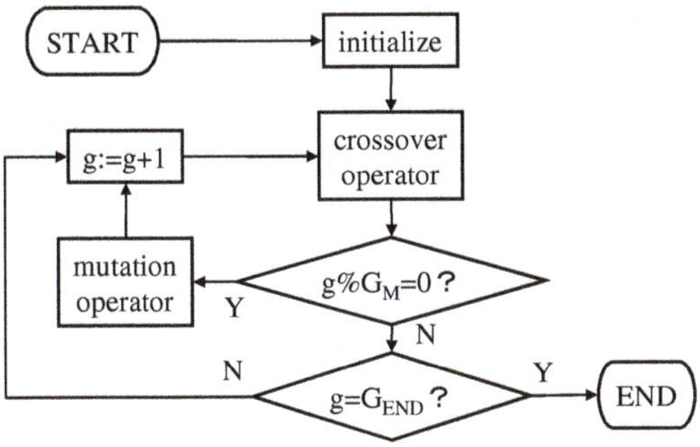


**Fig. 2** Basic algorithm of the CGA for the nurse scheduling. The CGA basically searches for the schedule by means of the crossover operator and periodically apply the mutation operator to find an area including better solution.

means of the mutation operator, the CGA can escape from the local minimum area when the optimization stagnates. In this paper, the mutation operator periodically applied to find a new solution area including better schedule.

## 3 Mutation Operator

The aim of mutation is to bring small change into the population. However if one of the duty fields of an individual is randomly changed, the schedule become meaningless thing, which is very hard to recover by a genetic operation. Therefore, the mutation operator must preserve the number of nurses in every duty at all date. The primitive operation of the mutation is shown in Fig.3. First, one of dates is randomly selected. Second, two nurses at the date are decided. Finally, the duties of these two positions are exchanged. If one of the selected duties is the fixed duty, another nurse is randomly selected again.

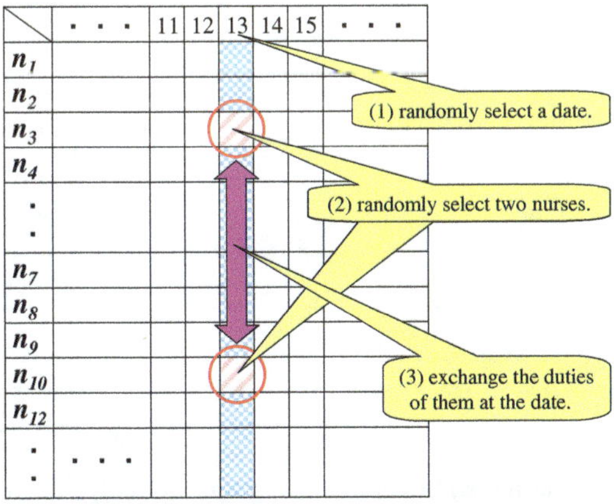


**Fig. 3** Primitive operation of the mutation operator

The primitive mutation operation is periodically applied as shown in Fig.4. Before the primitive operation is applied to the population, the penalty values of the current population and the population which given by the previous mutation operator are compared. Then the population giving smaller value of the population is selected as the best population. The primitive operation is periodically applied to the best population. The mutation period is defined as $G_M$.

We have tried several mutation periods as shown in Fig.5.Forty trials of optimization are performed under each condition. In the case of less than 1000, the optimization has brought a result of stable good average. Furthermore, in the case of less than 200, a good result has been provided generally. Therefore, we decide the mutation period with 200 in this paper.

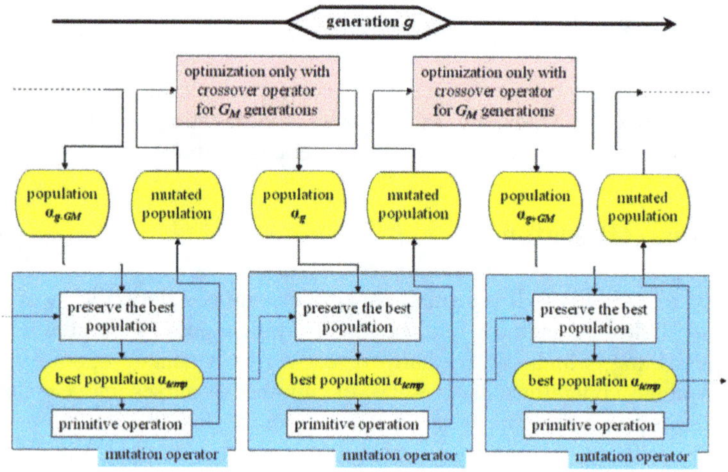


**Fig. 4** Mutation operator executed by a period for $G_M$ generations

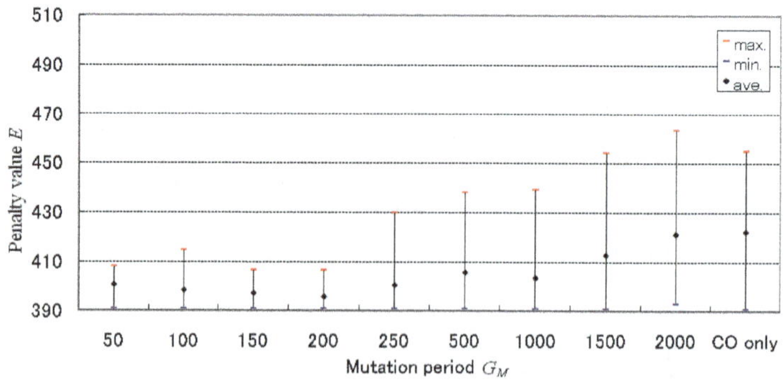


**Fig. 5** Comparison of the mutation period. The item “CO only” denotes the optimization only with the crossover operator

# 4 Parallel CGA for Nurse Scheduling

## *4.1 Expansion of the Nurse Scheduling Problem*

When we really apply the optimized nurse schedule, there is the case to change of duty contents by some kind of circumstances. Such a change of the shift schedule leads to the disproportion of the duty days. It causes the overwork of some nurses when such situation is ignored. Furthermore, there can be not only the fall of the nursing level but also the thing leading to a medical accident.

To restrain such an unfavorable situation, we propose a technique to accept and to relieve the changes in the past two weeks as shown in Fig.6. First, we have had the optimized shift schedule at the beginning of the *k*-th month. We suppose that two weeks passed and there are some changes during the two weeks now. The CGA optimizes the coming four weeks including third and fourth weeks of the *k*-th month and first and second weeks of the (*k*+*1*)-th month. When the coming four weeks are optimized, contents of the third and fourth weeks of the *k*-th month had better not be changed if possible. Then we define new penalty function, $F_{13}$, performing difference of the third and the fourth weeks between the original schedule and the newly optimized schedule. Whenever there is one difference, $F_{13}$ imposes one point of penalty. We define a total penalty function for evaluating the shift scheduling again as follows,

$$E = \sum_{k=1}^{4} H_k^{\ 2} + h_5 F_{13}. \tag{15}$$

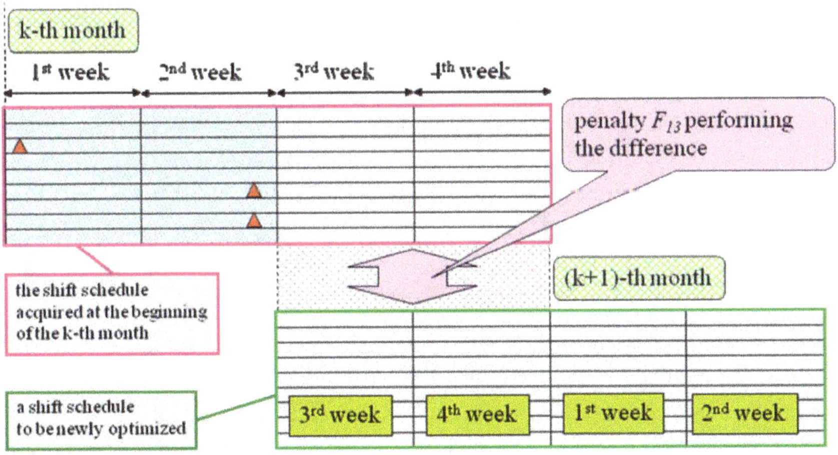


**Fig. 6** Expansion of the nurse scheduling to accept some changes in the past two weeks. When the two weeks have past, the coming four weeks are optimized to restrain inconvenience because of the changes

## 4.2 Parallel Processing

In the case optimizing the nurse schedule for four weeks without such the expansion, it takes a hundred thousand generations to acquire a good result, where this means about ten minutes. On the other hand, it takes million or more generations to acquire a good result of the expanded nurse scheduling. Furthermore, we have to try several times of the optimization to acquire a good result. We have tried five times of the optimization for a million generations as shown in Fig.7. If each optimization periodically restarts with the best schedule, more possibility is always searched for near the best solution.

Now, we propose to apply a parallel processing technique to the CGA for the nurse scheduling. Fig.8 shows an overview of the CGA in parallel processing. In this example, three processes are initially generated, where there is no limitation

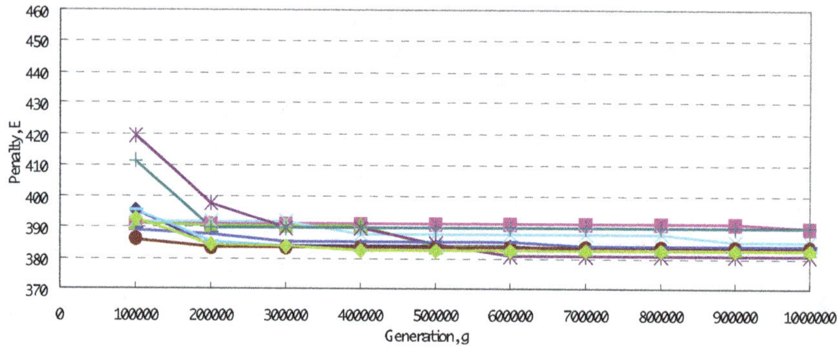


**Fig. 7** An example of five trials of the optimization for a million generations. We have to try several times of the optimization to acquire a good result.

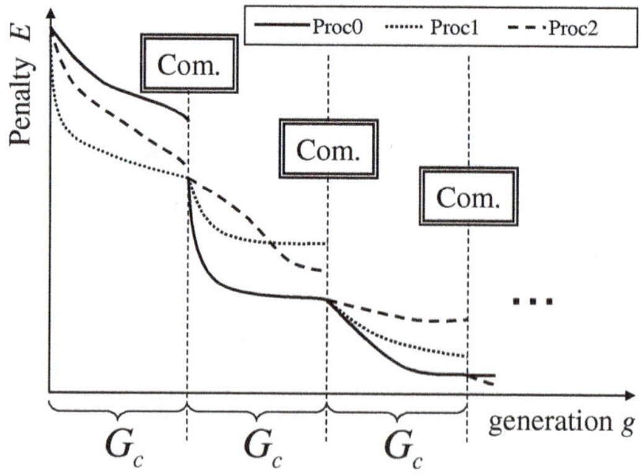


**Fig. 8** the CGA in parallel processing. In this example, three processes are initially generated and communicate every $G_C$ generation period. In the communication, those processes exchange the best schedule acquired by $G_C$ generations.

about the number of processes. The first process, Proc0, is generated on a computer that an operator starts the nurse scheduling. Those processes communicate every $G_C$ generation period. The value of $G_C$ should be defined to a multiple of $G_M$. The Proc0 manages the communication. In the communication, all the processes send the best schedule acquired by the optimization for $G_C$ generations to the process Proc0. The Proc0 selects the best schedule among the schedules sent from all the processes and sends it to all the processes. Each process starts their optimization again with the best schedule given by the communication. Therefore, more possibility is always searched for near the best solution.

This parallel algorithm is implemented by using MPI technology. Each process is composed of two threads. A main thread executes the CGA algorithm for $G_C$ generations. And one another thread treats the communication to other process. The communication thread works only when a message comes in from other process. Only the main thread always works for optimization. Actually, we can generate several processes on one CPU. When two or more processes work on one CPU, the performance of the CPU is lost remarkably. Therefore, only one process should be generated on each CPU.

We apply a simple parallel processing to the CGA. In our parallel algorithm, a process working on an expensive CPU of the computing power finish their thread early and have to wait for a slow process as shown in Fig. 9. The waiting time of the process finished early means the useless waste of computing resources. This problem must be improved in the future work.

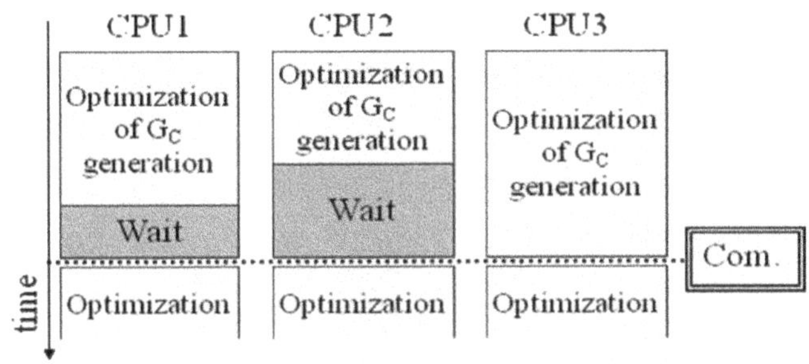


**Fig. 9** A simple parallel processing applied to the CGA. The optimization threads executed on the powerful CPU finish early and have to wait for a slow thread.

## 5 Practical Experiment

We have tried experiment of the nurse scheduling with practical data by the CGA in parallel processing. The number of the nurses is defined to twenty-three. One change is included in the past two weeks of shift schedule given at the beginning of the current month. The CGA in parallel processing is implemented on two PCs. Each PC has two CPUs. The specification of the PC1 is as follows: Intel ® core TM2 CPU 6600, 2.4GHz with 2GB RAM. The specification of the PC2 is as follows: AMD Athlon™ Dual Core Processor 4200+, 2.20GHz with 2GB RAM.

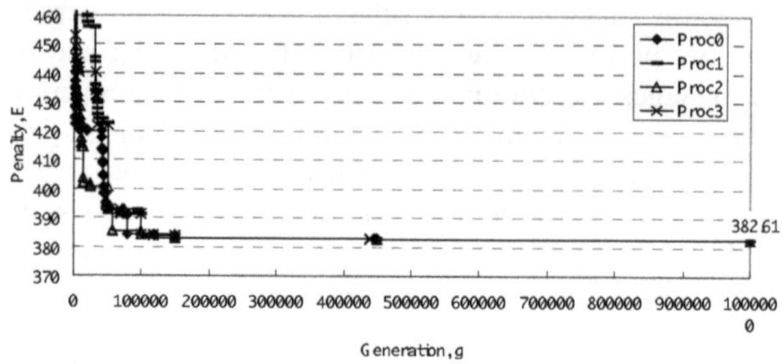


(a) trial 1

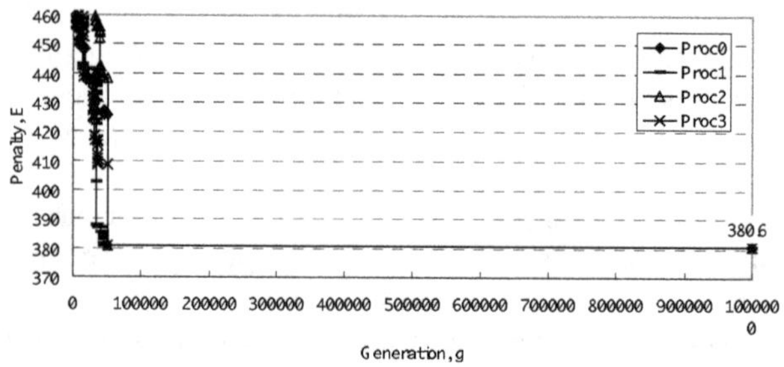


(b) trial 2

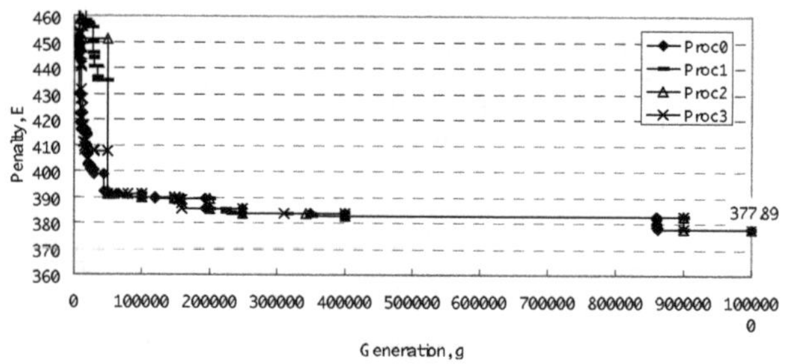


(c) trial 3

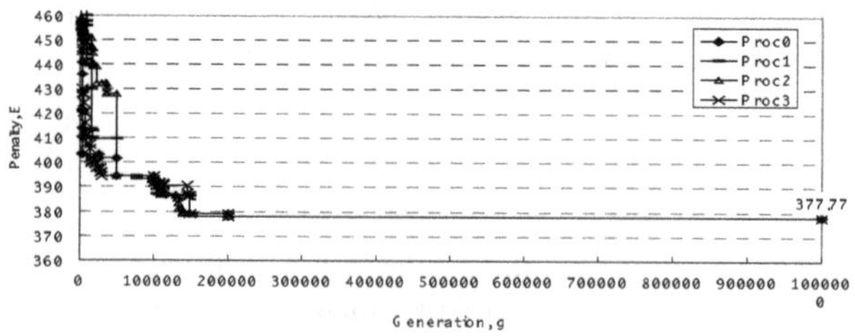


(d) trial 4

**Fig. 10** Result of the practical experiment. Each line denotes the value of the penalty, *E*, given at the generation. We have tried the parallel optimization with four processors four times.

Totally four CPUs are prepared for the experiment. We define four processes for the experiment. A million generations are executed on each process. The values of the mutation period, $G_M$ , and the communication period, $G_C$ , are defined to 200 and 50000 respectively. The algorithm of the CGA in parallel processing is implemented by using Visual Studio 2005.

Several results of the CGA in parallel processing are shown in Fig.10. We have tried the parallel optimization four times. Proc0 and Proc2 are carried in PC1. Proc1 and Proc3 are carried in PC2. In most cases, the CGA in parallel processing gives the most suitable solution through the optimization among 400,000 generations from 100,000 generations. This is equivalent to the speed from 2.5 times to 10 times than the speed when the conventional technique is carried out on the same computer resources. Furthermore, the CGA in parallel processing has given the good result that has not found by five times of experiments by using the conventional technique shown in Fig.7.

Finally, we have investigated total computing time and waiting time as shown in Fig.11. This result is given by the first trial. The processes Proc0 and Proc2 executed on the faster CPU take time to wait for the others executed on the slower CPU. In time used for wait idly, it is about 12% of the processing capacity of the PC1's CPU. This is a problem due to the technique of the parallelization of the CGA, and this must be settled in the future work.

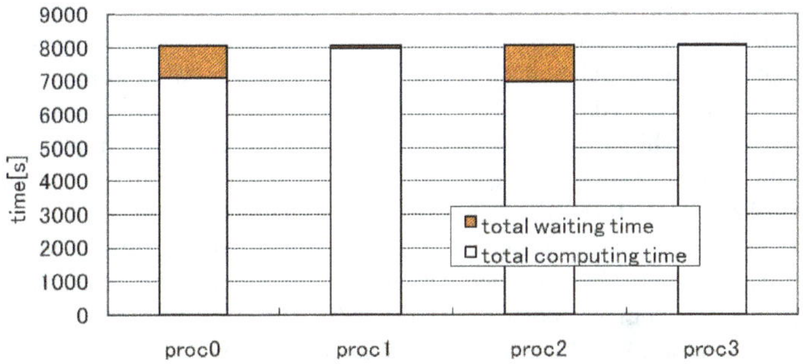


**Fig. 11** Total computing time and total waiting time at each CPU. This result is given by the first trial.

## 6 Conclusion

In this paper, we have handled the nurse scheduling by using the CGA. Several penalty functions are defined for evaluating the shift schedule. The nurse scheduling is expanded to accept the changes of the shift schedule. The new penalty function is defined for evaluating the difference of the part of the shift schedule between the original schedule given at the beginning of the current month and the schedule to be newly optimized. By means of the expansion of the nurse scheduling, the optimization of the schedule needs enormous time. We have proposed the mutation operator which does not lose the consistency of the schedule. We have investigated the mutation period in which the mutation operator effectively works.

When the conventional technique with the mutation operator is applied, in comparison with a case without problem expansion, we need about ten times the computing time. Furthermore, we have to execute ten or more times of the optimization in order to acquire satisfactory results enough. We have proposed the parallel CGA to restrain such a problem. By using the parallel CGA, the better nurse schedule is acquired in smaller number of generations than that by the conventional technique.

We have applied a simple parallel processing to the CGA. In our parallel algorithm, some processes to work with an expensive CPU of the computing power finish their thread early and have to wait for a slow process. In time used for wait idly, it is about 12% of the processing capacity of the faster CPU. This is a problem due to the technique of the parallelization of the CGA, and this must be settled in the future research work.

## References


1. Ikegami, A.: Algorithms for Nurse Scheduling. In: Proc. of 11th Intelligent System Symposium, pp. 477–480
2. Goto, T., Aze, H., Yamagishi, M., Hirota, M., Fujii, S.: Application of GA, Neural Network and AI to Planning Problems. NHK Technical report, No.144, pp. 78–85 (1993)
3. Kawanaka, S., Yamamoto, Y., Yoshikawa, D., Shinogi, T., Tsuruoka, N.: Automatic Generation of Nurse Scheduling Table Using Genetic Algorithm. Trans. on IEE Japan 122-C(6), 1023–1032 (2002)
4. Inoue, T., Furuhashi, T., Maeda, H., Takabane, M.: A Study on Interactive Nurse Scheduling Support System Using Bacterial Evolutionary Algorithm Engine. Trans. on IEE Japan 122-C(10), 1803–1811 (2002)
5. Itoga, T., Taniguchi, N., Hoshino, Y., Kamei, K.: An Improvement on Search Efficiency of Cooperative GA and Application on Nurse Scheduling Problem. In: Proc. of 12th Intelligent System Symposium, pp. 146–149
6. Ohki, M., Morimoto, A., Miyake, K.: Nurse Scheduling by Using Cooperative GA with Efficient Mutation and Mountain-Climbing Operators. In: 3rd International IEEE Conference Intelligent Systems, pp. 164–169 (2006)
7. Ohki, M., Uneme, S., Hayashi, S., Ohkita, M.: Effective Genetic Operators of Cooperative Genetic Algorithm for Nurse Scheduling. In: 4th International INSTICC Conference on Informatics in Control, Automation and Robotics, pp. 347–350 (2007)

# Case Studies for Genetic Algorithms in System Identification Tasks

Aki Sorsa, Riikka Peltokangas, and Kauko Leiviskä

**Abstract.** In this paper, genetic algorithms are used in system identification with reference to two case studies. The first case considers a structure identification problem of a black-box model. The identified black-box model is dedicated to the prediction of residual stress based on Barkhausen noise measurement. To find the most suitable model structure, a genetic algorithm with a cross-validation based objective function is utilized. The second case studies a parameter identification problem and uses a model of a biological reactor. The Chemostat model utilized in this work is nonlinear with two distinct operating areas and the model is identified separately for both operating areas. Optimization with genetic algorithms is repeated many times in both cases to guarantee the validity of results. The results from both cases are good, indicating that genetic algorithms can be used in system identification tasks.

## 1 Introduction

Process models are useful in many ways. They can be used in process design, planning and scheduling, process optimization and prediction and control [1]., There are different kinds of models, depending on the usage. For example, rather simple static models can be used in planning and scheduling, and dynamic models are needed in predicting future process states [1]. Models can be categorized into white-box and black-box models. White-box models are based on the laws of nature, while black-box models only try to map the input-output behaviour of the process. Somewhere between the white-box and black-box models are grey-box models, which typically incorporate black-box modelling tools inside a white-box model [1]. For example, in a white-box model of a chemical process, the reaction rate expression may be executed with black-box modelling tools.

Aki Sorsa · Riikka Peltokangas · Kauko Leiviskä
University of Oulu, Control Engineering Laboratory
P.O. Box 4300, FIN-90014
e-mail: {aki.sorsa,kauko.leiviska}@oulu.fi

V. Sgurev et al. (Eds.): Intelligent Systems: From Theory to Practice, SCI 299, pp. 243–260.
springerlink.com 

System identification problems can be divided into model structure and parameter identification problems. In the identification of a white-box model structure, process knowledge is beneficial especially while defining a valid reaction scheme and a kinetic model structure. If no process knowledge is available, a trial and error approach is used to choose a valid model structure [2]. Another possibility is to use black-box modelling tools. In black-box models, the structure identification problem easily turns into a variable selection problem [3]. In parameter identification, the model parameters are solved for the known or selected model structure. The solution is typically found by minimizing the squared error between the predicted and actual outputs [4].

Genetic algorithms have been used in system identification earlier. Model structure identification has been studied, for example, in [2] and [3]. [2] studied structure identification for three cases. They found genetic algorithms to be a useful tool for structure identification but still stated that some process knowledge is needed. In [3], the structure identification of a black-box model was studied. A variable selection problem was solved with genetic algorithms to find the optimal variable set for the prediction model. Model parameter identification with genetic algorithms has been studied, for example, in [5], [6], [7], [8] and [9]. [5] identified the parameters of the kinetic equation that describes the growth of cells in a biological reactor. [6] used genetic algorithms in identifying the model of a biological fed-batch reactor. A complex biological fed-batch process was also studied in [9]. In [8], a CSTR model was studied and its parameters were identified with genetic algorithms. They also studied the applicability of genetic algorithms to parameter identification with a couple of other nonlinear mathematical models. [7] studied the same models as [8].

Genetic algorithms have been shown to be suitable for parameter identification of large and complex systems [4]. It has been stated in the literature that conventional methods (such as gradient methods) have the drawback of getting trapped in local optima. The problem of overcoming the local optima does not exist with genetic algorithms because suitable solutions are sought from multiple dimensions and the whole search space is basically covered [7]. Thus, genetic algorithms are more likely to find the optimal (or at least a near-optimal) solution. One drawback of genetic algorithms is that the exact optimal solution may not be found but only a near-optimal one [7]. It has also been stated in the literature that optimization with genetic algorithms is time-consuming with complex systems [7].

This paper presents some results obtained by using genetic algorithms for system identification. The first results were obtained by using a binary coded genetic algorithm for determining a suitable input variable configuration and thus the model structure for a model used for predicting residual stress from Barkhausen noise measurement [3]. The second study uses a simple model of an ideally stirred Chemostat [10] to investigate the applicability of real-coded genetic algorithms to the parameter identification problem. A dynamic mass balance model of the process has been developed [11]. The model is the basis for the simulator, which is then used to generate data for the parameter identification [12].

This paper is organized as follows. First, the basics of genetic algorithms and a short review of available crossover and mutation operators are provided in Section 2. In Section 3, a case study concerning model structure identification is given, followed by a case study about model parameter identification in Section 4. Section 5 concludes the paper.

## 2 Genetic Algorithms

Genetic algorithm is an optimization technique that mimics evolution. The optimization is regulated by genetic operators: reproduction and mutation. These operators steer the population of chromosomes towards the optimal solution. Each chromosome holds a possible solution to the problem in encoded form. The information in chromosomes is decoded and passed to an objective function, which then determines the fitness of each chromosome. It can be said that the objective function forms the link between the optimization problem and the algorithm. In genetic algorithms, binary or real-valued coding can be used. The traditional genetic algorithm uses binary coding but it has some drawbacks. Maybe the most significant one is the loss of precision due to the change from real values to binary digits [4].

The genetic operators mentioned above regulate optimization. Probabilities are defined for both of them to obtain the desired behaviour of the population. The basic idea is that through reproduction the population evolves towards better solutions, while mutation adds random changes to chromosomes and thus prevents the algorithm from getting trapped into local optima. Reproduction has two steps: parent selection and crossover. In parent selection, two (or more) chromosomes are selected for crossover. Selection methods favour better chromosomes. Better chromosomes are more likely to reproduce and evolve the population towards better solutions. [13]

There are a lot of parent selection methods. Typical methods include roulette wheel and tournament selection. In roulette wheel selection, the cumulative sum curve of the fitness values for all chromosomes is drawn (Figure 1). A random number is then chosen and compared to the cumulative sum curve in order to determine which chromosome is selected. In the roulette wheel method, the probability of a chromosome to be selected is proportional to its fitness value [13]. In tournament selection, a predefined number of chromosomes are selected randomly as candidates and the fittest of them is selected to be a parent [14]. Usually, two candidate chromosomes are selected [14].

In recent years, new crossover and mutation operators have been developed for real-coded genetic algorithms. The following section consists of a short review of the operators found in the literature. A brief description is also given of the crossover and mutation operators typically used for binary coded algorithms.

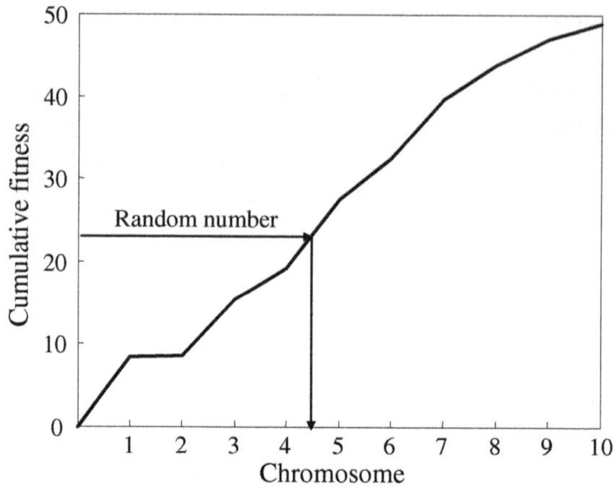


**Fig. 1** An example of roulette wheel parent selection. A random number is chosen between 0 and the sum of the fitness values. In this case, chromosome 5 is selected as a parent.

## 2.1 Crossover Operators

The selected parents are combined to create offspring which are then placed in the new population. There are a lot of operators used for crossover. In binary algorithms, a one-point crossover is typically used [15]. A random point is chosen and segments of the parent chromosomes after that point are switched. Another alternative is to use an n-point operator, in which the n points are selected randomly and the segments of parents defined by the random points are switched [16]. The third crossover operator used in binary coded algorithms is the uniform crossover, where each bit of the offspring is taken randomly with equal probabilities from the parents [15].

There are many more proposed crossover operators for real-coded genetic algorithms. In arithmetic crossover, two parents produce two offspring according to [17]

$$\begin{aligned} y_i^{(1)} &= \alpha_i x_i^{(1)} + (1-\alpha_i)x_i^{(2)} \\ y_i^{(2)} &= \alpha_i x_i^{(2)} + (1-\alpha_i)x_i^{(1)}. \end{aligned} \tag{1}$$

Above, $\alpha_i$ are uniformly distributed random numbers. $\alpha$ is constant in uniform arithmetical crossover and varies as a function of generations in non-uniform crossover [14]. In linear crossover, three offspring are generated with predefined values of $\alpha$ (0.5, 1.5 and -0.5). The parents are then substituted by two of the most promising offspring [16].

In flat crossover, one offspring is produced from two parents by taking it randomly from the interval [ $x_i^{(1)}, x_i^{(2)}$ ] [18]. Flat crossover is a special case ($\alpha$ = 0) of the BLX-$\alpha$ crossover. In BLX-$\alpha$ crossover, one offspring is taken from the interval [ $x_{i,min} - I\alpha, x_{i,max} + I\alpha$ ] ( $I = x_{i,max} - x_{i,min}$ ) [16].

Heuristic crossover produces one offspring according to [16]

$$y_i = r(x_i^{(2)} - x_i^{(1)}) + x_i^{(2)}. \tag{2}$$

Above, $r$ is taken from the interval [0 1]. It should also be noticed that parent $x^{(2)}$ is better than $x^{(1)}$. Parent-centric crossover operators create the offspring in the neighbourhood of the female parent using a probability distribution. The male parent is considered to define the range of the probability distribution [19]. For example, a parent-centric BLX-α crossover operator (PBX-α) can also be used [20]. The offspring is generated similarly as in the BLX-α crossover but the interval is defined slightly differently.

A multi-crossover operator with more than two parents is proposed in [21]. The offspring in multiple crossover are given by

$$\begin{aligned} y_i^{(1)} &= x_i^{(1)} + r\left(2x_i^{(1)} - x_i^{(2)} - x_i^{(3)}\right) \\ y_i^{(2)} &= x_i^{(2)} + r\left(2x_i^{(1)} - x_i^{(2)} - x_i^{(3)}\right) \\ y_i^{(3)} &= x_i^{(3)} + r\left(2x_i^{(1)} - x_i^{(2)} - x_i^{(3)}\right). \end{aligned} \tag{3}$$

Above, $x^{(1)}$ is the best parent and $r$ is taken from the interval [0 1]. The parents for multiple crossover are selected randomly. [21] Other proposed crossover operators are the Laplace crossover [22] and crossover based on fuzzy connectives [16].

## 2.2 *Mutation Operators*

Mutation creates new material in the population. In mutation, a random part of a chromosome is changed [23]. Mutation in binary coded algorithms is usually carried out by browsing through all the binary digits, choosing a random number for each digit, comparing the random number to mutation probability and changing the digit if the random number is smaller than the probability. [15]

The uniform mutation operator is essential in early stages of the evolution process. It simply takes a random element of a chromosome and replaces it with a feasible random number. The uniform mutation allows solutions to move freely within the search space [14]. A variation of the uniform mutation is boundary mutation. In boundary mutation, a random segment of a chromosome is replaced by the lower or upper boundary of the feasible area [14]. Evidently, the boundary mutation is applicable to optimization problems where the solution is near the boundary of the feasible area.

Non-uniform mutation is one of the most commonly used mutation operators in real-coded genetic algorithms [14]. The operator is designed for obtaining high precision. During the early generations, the non-uniform mutation searches the space uniformly but during the later generations it tends to search the space locally [17].

A new mutation operator based on the power distribution has been introduced recently [24]. This mutation operator is used to create a solution in the vicinity of a parent solution.

# 3 Model Structure Identification

Material properties can be measured destructively or non-destructively. Destructive methods are obviously not applicable to quality control and thus non-destructive methods are preferred. Some methods exist for measuring residual stress non-destructively, such as X-ray diffraction, but these methods have a rather slow response time. Barkhausen noise measurement is an intriguing non-destructive method for estimating residual stress due to its fast response time and low costs. [25] Barkhausen noise has been shown to be sensitive to different material properties but the results found in the literature are more or less qualitative. However, quantitative prediction of material properties is required for better utilization of the measurement. [26]

A typical Barkhausen noise signal is shown in Figure 2. Usually, a single feature from the Barkhausen noise signal is calculated and compared to studied material properties. The calculated feature could be, for example, the RMS value of the signal, peak height, peak position or peak width. In [26] and [27], calculational procedures for calculating more features from the Barkhausen noise signal have been introduced. This leads to a situation where the selection of features for the prediction model becomes crucial.

This case illustrates the use of genetic algorithms in the structure identification of the model used for predicting residual stress from the features obtained through Barkhausen noise measurements. A linear regression model is identified. The structure of the linear regression model is determined by the input variables used. Also, mathematical operations such as logarithms and powers can be used [2] but are omitted in this case.

The data used in this case includes about 50 features. These variables have high cross-correlations, which makes the modelling more complicated. The data sets used and data acquisition are described in [26]. In addition, a more thorough description of the procedures and results used are given in [3]. The material studied is case-hardened steel as described in [26].

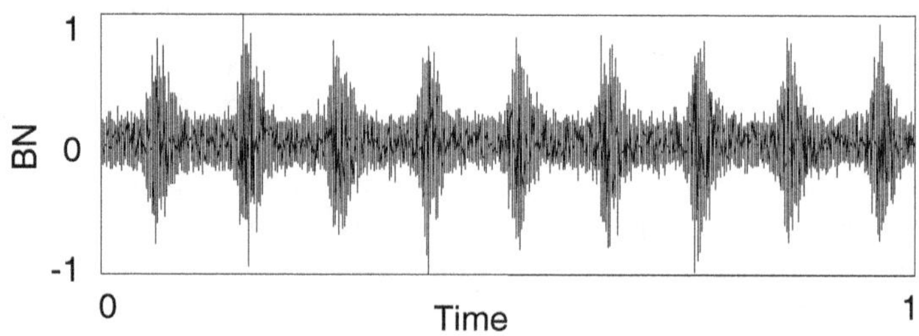


**Fig. 2** A typical Barkhausen noise signal in scaled units

## *3.1 Applied GA*

The model structure is identified with binary coded genetic algorithms. The coding is carried out in such a way that all the potential variable combinations are possible, even though it was expected that the suitable number of variables is rather low. The objective function utilizes a 10-fold cross-validation procedure due to the limited amount of data. Also, a penalty term is used for excess variables to guarantee that the model structure is simple enough. [3] The objective function used is [3]

$$J = SSEP \times \lambda p. \tag{4}$$

Above, $J$ is the value of the objective function, $SSEP$ is the sum of the squared error of prediction obtained through the 10-fold cross-validation, $\lambda$ is the penalty constant, and $p$ is the number of variables included in the model. In this case, $\lambda$ is set to one but other values can also be used.

### 3.2 Results

In this case, optimization with the genetic algorithm was repeated 50 times. The parameters for the algorithm were determined through a couple of experimental optimization runs. The results obtained showed some variations as the suitable number of variables ranged between 4 and 7. However, about 60 percent of the runs gave a four-variable set as the best solution, indicating that four variables is a suitable number. [3]

To validate the results, the models obtained from all the optimization runs were analyzed further. Each model was investigated by evaluating the significance of the model's input variables. The significance was evaluated by identifying the corresponding model without the variable. Doing this determined the relative change of SSEP and the significance was evaluated from that change. [3]

The results from the validation showed that the model with four variables obtained as the best result from about 60 percent of the optimizations was the most suitable for prediction purposes. This study showed that genetic algorithms can be used in model structure identification, as has been noticed earlier [2]. More specific details about this study can be found in [3].

## 4 Model Parameter Identification

In chemical and biochemical applications, macroscopic models are used to give insight to real-life processes. Typically, they are based on an assumed reaction scheme and flow conditions [28]. This paper discusses the parameter identification of a nonlinear model of a biological wastewater treatment process, namely the activated sludge process. Macroscopic models with different levels of detail can be derived for the activated sludge process [11], [29], [30]. In [30], a very detailed model of the activated sludge process is used while [11] and [29] utilize simpler models. The latter studies use a mass balance model of the ideally stirred Chemostat introduced in [10].

The Chemostat model studied in this case is potentially a multistable system because the reactions are inhibited by high substrate concentrations . In such a case, the otherwise linear Chemostat model becomes nonlinear. The Chemostat model is given by [11]:

$$\begin{aligned}\frac{dc_s}{dt} &= \frac{Q_{in}}{V}c_{s,in} - \frac{Q_{out}}{V}c_s - \mu(c_s)c_b \\ \frac{dc_b}{dt} &= \frac{Q_{in}}{V}c_{b,in} - \frac{Q_{out}}{V}c_b + \mu(c_s)c_b \\ \frac{dV}{dt} &= Q_{in} - Q_{out}.\end{aligned} \tag{5}$$

Above, $c_s$ and $c_b$ are the substrate and biomass concentrations, respectively, $Q$ is the flow rate, $V$ the volume of the reactor and $\mu$ the reaction rate. The reaction rate follows Haldane kinetics given by

$$\mu(c_s) = \frac{\mu_0 c_s}{K_S^{-1}c_s^2 + c_s + K_I}. \tag{6}$$

Above, $\mu_0$, $K_S$ and $K_I$ are experimental constants. The values for the constants are $\mu_0 = 0.74$, $K_S = 15$ and $K_I = 9.28$ and they are adopted from [31]. With the parameters given above, the Chemostat model becomes bistable. The bistability can be observed from the histograms of the simulated output variables given in Figure 3. The figure shows two-peaked histograms, indicating that the process has two distinct operating areas depending on the substrate concentration. The behaviour of the process in these operating areas varies. A more thorough description of the model behaviour is found in [11].

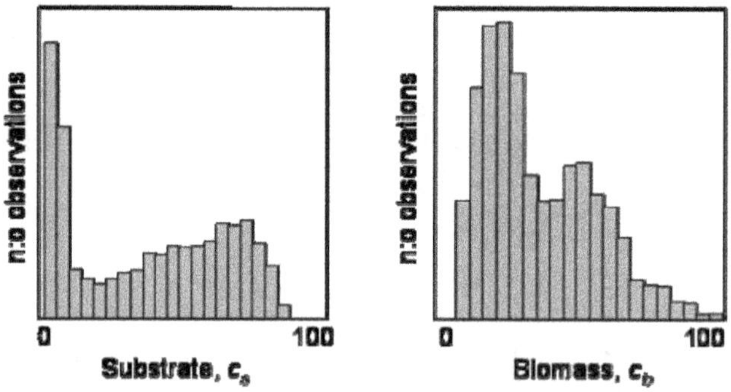


**Fig. 3** The simulated output variables. The figure shows that the process has two operating areas leading to two-peaked distributions [11].

## 4.1 Data Sets Used

Based on (5) and (6), a simulator was developed for the system, which is used in generating data for the parameter identification and validation of the results. The data sets are generated by applying step changes to the input concentration of substrate. Normally distributed disturbances are also added to the reactor volume and the concentrations of the substrate and biomass. Table 1 presents the means and variances of the applied disturbances. The output variables ($c_s$ and $c_b$) have a high cross-correlation and thus only one of them is used for parameter identification. In this case, the substrate concentration is used because it was stated in [11] that it is more significant where process control is concerned.

**Table 1** The means and variances of the applied disturbances [12]

| | $c_{s,in}$ | $c_{b,in}$ | $V$ |
|---|---|---|---|
| Mean | 0 | 1 | 300 |
| Variance | 1 | 0.4 | 1 |

It was stated above that the process behaviour varies depending on the operating area. Therefore, parameter identification is carried out separately for both operating areas. The simulated substrate concentrations for the 1$^{st}$ operating area (high conversion) are given in Figure 4. The initial input concentration of substrate in the data set given in Figure 4 is 40 and it undergoes step changes of sizes 5 (Figure 4a) and -5 (Figure 4b). The substrate concentrations for the 2$^{nd}$

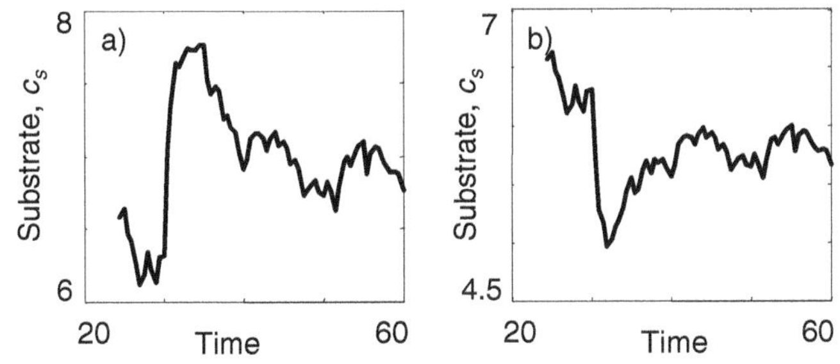


**Fig. 4** The simulated step responses of the substrate concentration for the 1st operating area. a) the positive step change and b) the negative step change [12].

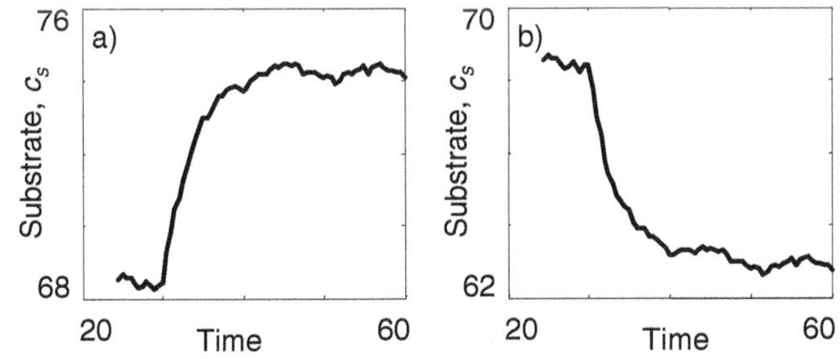


**Fig. 5** The simulated step responses of the substrate concentration for the 2nd operating area. a) the positive step change and b) the negative step change [12].

operating area (low conversion) are given in Figure 5. The initial substrate concentration is 75 and it undergoes step changes of sizes 5 (Figure 5a) and -5 (Figure 5b). Positive and negative step changes are applied to obtain data from three levels and thus to capture the nonlinearity of the system better [12].

## 4.2 Applied Genetic Algorithms

Real-coded genetic algorithms were used in this case. Probabilities were defined for crossing ($p_c$) and mutation ($p_m$) to achieve the desired behaviour of the population. Elitism is also used as the worst chromosome of the new population is replaced by the best chromosome of the previous population. This prevents the best chromosome from disappearing through genetic operations. Other tuneable parameters for regulating the optimizations are the population size ($n_{pop}$) and the number of generations ($n_{gen}$). For parent selection, the tournament method was used with the number of candidates set to 2. The non-uniform arithmetic crossover given in (1) and the uniform mutation were used [12].

The parameters for regulating the evolution of the population were determined through a couple of experimental runs. The desired population behaviour is given in Figure 6. As the figure shows, the aim was that initially the population would undergo a fast evolution towards the optimum, followed by a more gentle decrease in the mean of the fitness values. The parameter values that lead to the desired behaviour of the population are: $p_c = 0.9$, $p_m = 0.1$, $n_{pop} = 50$ and $n_{gen} = 30$ [12].

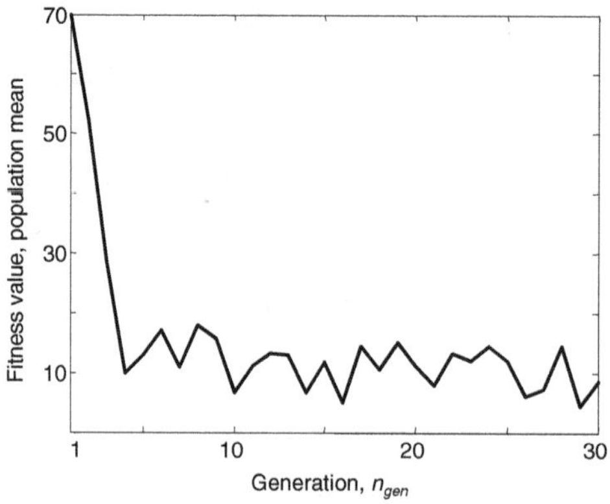


**Fig. 6** The mean of the fitness values as a function of the number of generations. Initially, the population undergoes a fast development towards the optimum. The initial decrease is followed by a more gentle decrease of the mean of the fitness values [12].

The initial population is taken randomly from the uniform distribution. To guarantee valid results, optimizations are repeated 200 times with different initial populations. The overall algorithm can be summarized as follows: [12]

1. Create a random initial population with $n_{pop}$ chromosomes.
2. Evaluate the fitness of the chromosomes through the objective function. Apply elitism.
3. If $n_{gen}$ generations are reached, go to step 5.
4. Apply reproduction and mutation. Go back to step 2.
5. Obtain the results.

The objective function uses the parameter values from each chromosome and predicts the output of the system. The predicted outputs are then compared to the actual (simulated in this case) outputs and the root mean squared value of the prediction error (RMSEP) is calculated. RMSEP is calculated for the data sets of positive and negative step changes. The objective function can be written as [12]

$$J_k = \sqrt{\frac{1}{n}\sum_{i=1}^{n}\left(c_{s,i} - \hat{c}_{s,i}\right)^2} + \sqrt{\frac{1}{m}\sum_{j=1}^{m}\left(c_{s,j} - \hat{c}_{s,j}\right)^2}. \tag{7}$$

Above, $n$ and $m$ are the number of data points in the step response data sets.

## *4.3 Results for the 1st Operating Area*

As mentioned earlier, the optimizations were repeated 200 times. The best set of parameters from the repetitions is given in Table 2. The best set of parameters is the one giving the lowest RMSEP. The parameters obtained were not the same as the actual ones, as can be seen from the relative error given in Table 3. Figure 7 presents the actual and the predicted substrate concentrations for the positive and negative step change data sets. The figure shows that the prediction accuracy is good even though the parameters are not the same. This is due to the fact that the data used for parameter identification is limited to low values of substrate input concentration. Therefore, the model is accurate only for low substrate concentrations and not necessarily for high concentrations. Figure 8 presents the reaction rate obtained with the true values and the reaction rate with the optimized values in Table 2. Figure 8 shows that the reaction rate obtained for the low substrate concentrations follows the actual reaction rate with the low substrate concentrations very closely. At higher concentrations, deviations are noticeable [12].

**Table 2** The best set of parameters and RMSEP for the 1st operating area

| Parameter | True value | Search space | Value | Relative error |
|---|---|---|---|---|
| $\mu_0$ | 0.74 | [0 4] | 0.88 | 0.18 |
| $K_S$ | 15 | [0 50] | 10.64 | 0.29 |
| $K_I$ | 9.28 | [0 15] | 11.51 | 0.24 |

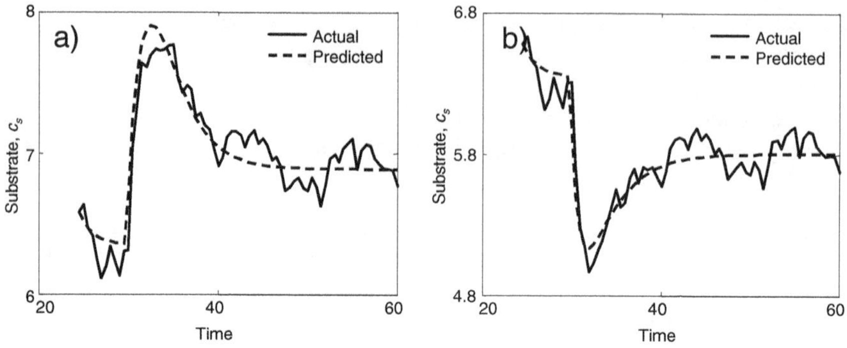


**Fig. 7** The actual and the predicted outputs of the system for the 1st operating area: a) the positive step change and b) the negative step change. The predicted output follows the actual output accurately [12].

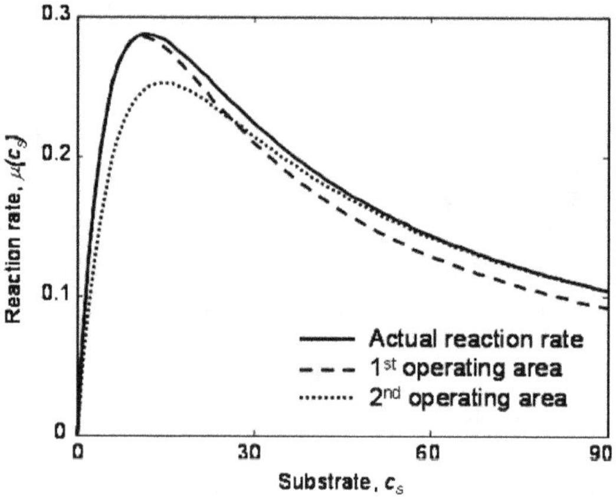


**Fig. 8** The actual reaction rate and the reaction rates identified for the 1st and the 2nd operating area. The reaction rate identified for the 1st operating area follows the actual reaction rate with low substrate concentrations very closely, while the reaction rate identified for the 2nd operating area is accurate with high substrate concentrations [12].

## 4.4 Results for the 2nd Operating Area

Table 3 shows the model parameters identified with the data sets from the 2nd operating area. The parameters apart from $K_I$ were very close to the true values. The prediction accuracy of the identified model was good, as shown in Figure 9. The influence of the difference between the actual and identified parameter values is given in Figure 8. It shows that with high substrate concentrations the reaction rate with the parameters in Table 3 follows the actual reaction rate very closely. Deviations occur at lower substrate concentrations [12].

**Table 3** The best set of parameters and RMSEP for the 2nd operating area

| Parameter | True value | Search space | Value | Relative error |
|---|---|---|---|---|
| $\mu_0$ | 0.74 | [0 4] | 0.73 | 0.01 |
| $K_S$ | 15 | [0 50] | 15.34 | 0.02 |
| $K_I$ | 9.28 | [0 15] | 13.70 | 0.48 |

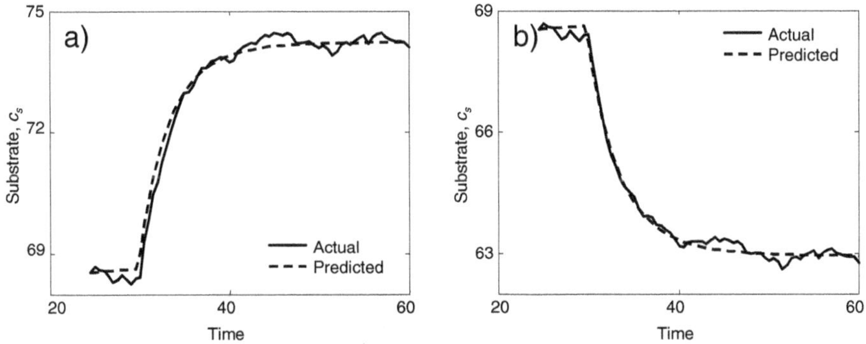


**Fig. 9** The actual and predicted outputs of the system for the 2nd operating area: a) the positive step change and b) the negative step change. The prediction accuracy is very good [12].

## *4.5 Validation of the Results*

The validity of the results was evaluated in many ways: visually from Figures 7 and 9, using the correlation coefficients and RMSEP values, by studying the variability of the parameter values, and by visually analyzing the histograms of the parameters. Table 4 shows the RMSEP values and the correlation coefficients for both data sets. Table 4 shows that the identified models in both operating areas are accurate. The correlations between the predicted and the actual outputs are high in both operating areas, especially in the second one. The accuracy can also be observed in Figures 7 and 9.

**Table 4** Statistical values for model validation

| | Step response 1 | | Step response 2 | |
|---|---|---|---|---|
| | RMSEP | Correlation | RMSEP | Correlation |
| 1st operating area | 0.14 | 0.94 | 0.12 | 0.93 |
| 2nd operating area | 0.22 | 0.99 | 0.15 | 0.99 |

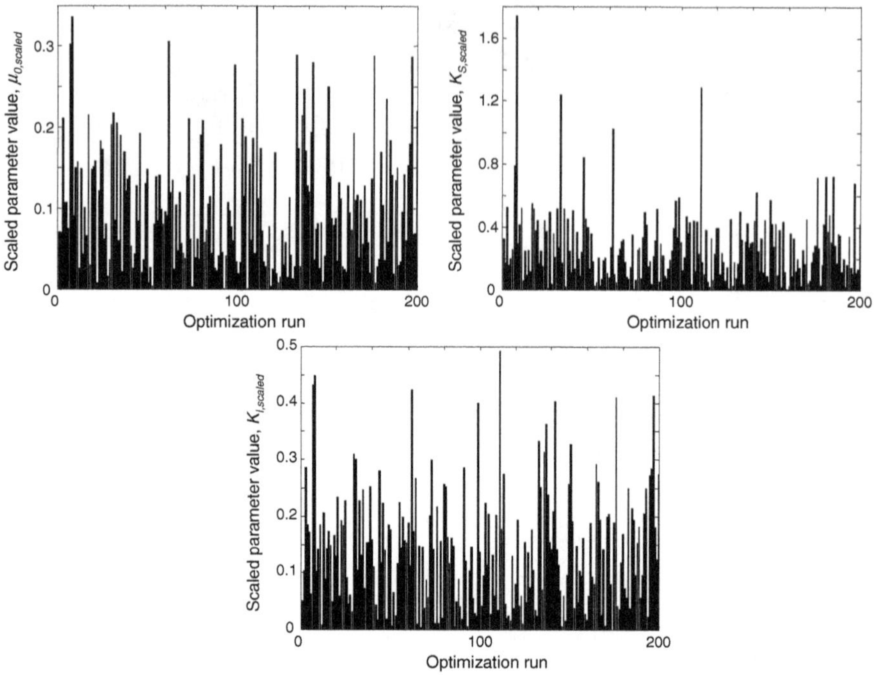


**Fig. 10** The scaled parameter values identified for the 1st operating area

The validity of the results obtained with genetic algorithms was evaluated in two ways:

1. The parameter values were scaled as given in (8) and the scaled values were plotted and analyzed [32] and
2. The histograms of the parameters were plotted and evaluated visually [33]. For the analysis of the histograms, the coefficient of variation was also calculated as given in (9).

The equations needed for validation are:

$$k_i = \frac{\left|k_i - \bar{k}_i\right|}{\bar{k}_i} \text{ and} \tag{8}$$

$$c_{v,i} = \frac{s_i}{m_i}. \tag{9}$$

Above, $k$ is the parameter value, $\bar{k}$ is the average parameter value, $c_v$ is the coefficient of variation, $s$ is the standard deviation, $m$ is the mean and subscript $i$ refers to an i:th parameter. The scaled parameter values are plotted in Figure 10, which shows that parameters $\mu_0$, $K_S$ and $K_I$ vary within about 35%, 175% and 50%, respectively. This indicates that the predictions are sensitive to the values of

$\mu_0$ and $K_I$ but insensitive to the value of $K_S$. This is also revealed by the histograms of the parameters given in Figure 11. The spread of the histograms of $\mu_0$ and $K_I$ are much lower than the spread of $K_S$. This is clearly indicated by the coefficient of variation, which is much higher for $K_S$. All the results indicate that the observations about the sensitivity of the prediction model given above are valid for the 1st operating area. The numerical results are collected in Table 5.

A similar analysis was also carried out for the model identified for the 2nd operating area. The results are tabulated in Table 5, which shows that the variability is highest for parameter $\mu_0$. Nevertheless, the coefficient of variation is lowest for that parameter. At the 2nd operating point, all the coefficients of variation are rather high. It means that the prediction accuracy is not as sensitive to changes in parameter values.

**Table 5** Statistical values for model validation (1st operating area)

| | 1st operating area | | 2nd operating area | |
|---|---|---|---|---|
| | Variability | $c_v$ | Variability | $c_v$ |
| $\mu_0$ | <40% | 0.13 | <150% | 0.28 |
| $K_S$ | <175% | 0.36 | <100% | 0.29 |
| $K_I$ | <50% | 0.18 | <100% | 0.34 |

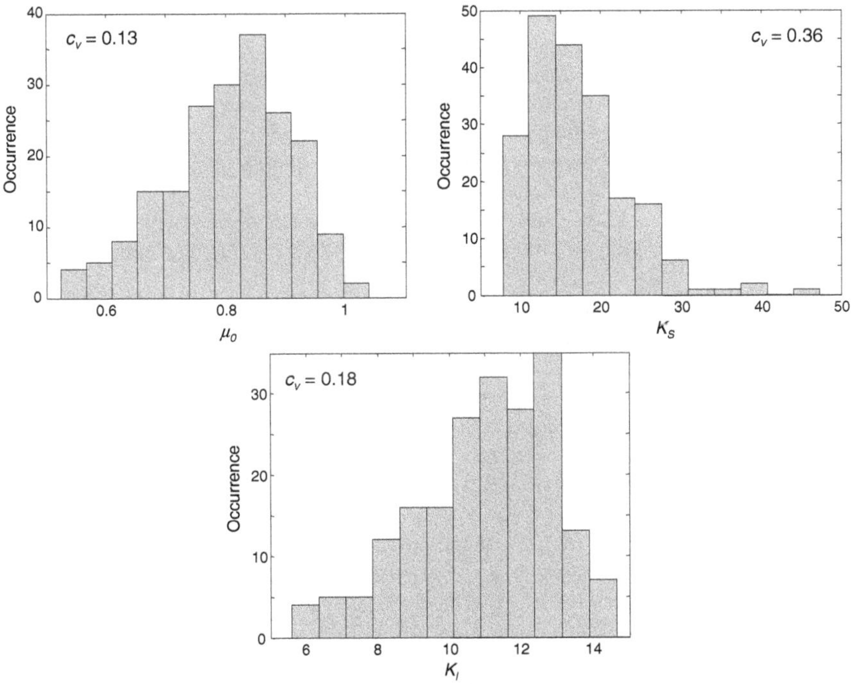


**Fig. 11** The histograms of the parameter values together with the coefficients of variation

## 5 Conclusions

This paper discussed system identification using genetic algorithms. The evolution of the population towards better solutions was regulated through certain parameters such as population size and probabilities for crossover and mutation. Suitable values for these parameters were obtained through a couple of experimental runs. The problem of model structure identification was studied by means of a case where a linear regression model was identified. The modelled phenomenon was known to be very complex. The model parameter identification was studied through a model of a theoretical bioreactor. The process model was known to be bistable and thus the parameters were identified for both operating areas separately. The results of the structure identification showed that a suitable model structure could be obtained with an appropriate objective function. The results with the Chemostat model were accurate. The results from both studies were validated and shown to be good. Based on this study, it can be concluded that genetic algorithms can be used in system identification tasks.

## References

[1] Roffel, B., Betlem, B.H.: Process dynamics and control: Modeling for control and prediction. John Wiley & Sons, Chichester (2006)
[2] Gray, G.J., Murray-Smith, D.J., Li, Y., Sharman, K.C., Weinbrenner, T.: Nonlinear model structure identification using genetic programming. Control Engineering Practice 6, 1341–1352 (1998)
[3] Sorsa, A., Leiviskä, K.: Feature selection from Barkhausen noise data using genetic algorithms with cross-validation. In: Proceedings of International Conference on Adaptive and Natural Computing Algorithms, p. 10 (2009)
[4] Chang, W.-D.: Nonlinear system identification and control using a real-coded genetic algorithm. Applied Mathematical Modelling 31, 541–550 (2006)
[5] Park, L.J., Park, C.H., Park, C., Lee, T.: Application of genetic algorithms to parameter estimation of bioprocesses. Medical and Biological Engineering and Computing 35, 47–49 (1997)
[6] Ranganath, M., Renganathan, S., Gokulnath, C.: Identification of bioprocesses using Genetic Algorithm. Bioprocess and Biosystems Engineering 21, 123–127 (1999)
[7] Nyarko, E.K., Scitovski, R.: Solving the parameter identification problem of mathematical models using genetic algorithms. Applied Mathematics and Computation 153, 651–658 (2004)
[8] Khalik, M.A., Sherif, M., Saraya, S., Areed, F.: Parameter identification problem: Real-coded GA approach. Applied Mathematics and Computation 187, 1495–1501 (2007)
[9] Wang, G., Feng, E., Xiu, Z.: Modeling and parameter identification of microbial bioconversion in fed-batch cultures. Journal of Process Control 18, 458–464 (2008)
[10] Smith, H.L., Waltman, P.: The Theory of the Chemostat. Cambridge University Press, Cambridge (1995)
[11] Sorsa, A., Leiviskä, K.: State detection of a wastewater treatment plant. In: Plesu, V., Agachi, P.S. (eds.) Proceedings of 17th European Symposium on Computer Aided Process Engineering, pp. 1337–1342 (2007)

[12] Sorsa, A., Peltokangas, R., Leiviskä, K.: Real-coded genetic algorithms and nonlinear parameter identification. In: Yager, R.R., Sgurev, V.S., Jotsov, V.S. (eds.) Proceedings of 4th International Conference on Intelligence Systems, vol. 10, pp. 42–47 (2008)
[13] Mitchell, M.: An introduction to genetic algorithms. MIT Press, Cambridge (1998)
[14] Michalewicz, Z.: Genetic Algorithms + Data Structures = Evolution Programs. Springer, New York (1996)
[15] Davis, L.: Handbook of genetic algorithms. Van Nostrand Reinhold, New York (1991)
[16] Herrera, F., Lozano, M., Verdegay, J.L.: Tackling real-coded genetic algorithms: Operators and tools for behavioural analysis. Artificial Intelligence Review 12, 265–319 (1998)
[17] Kaelo, P., Ali, M.M.: Integrated crossover rules in real coded genetic algorithms. European Journal of Operational Research 176, 60–76 (2007)
[18] Radcliffe, N.J.: Equivalence class analysis of genetic algorithms. Complex Systems 2, 183–205 (1991)
[19] García-Martínez, C., Lozano, M., Herrera, F., Molina, D., Sánchez, A.M.: Global and local real-coded genetic algorithms based on parent-centric crossover operators. European Journal of Operational Research 185, 1088–1113 (2008)
[20] Lozano, M., Herrera, F., Krasnogor, N., Molina, D.: Real-coded memetic algorithms with crossover hill-climbing. Evolutionary Computation Journal 12, 273–302 (2004)
[21] Chang, W.-D.: Coefficient estimation of IIR filter by a multiple crossover genetic algorithm. Computers and Mathematics with Applications 51, 1437–1444 (2006)
[22] Deep, K., Thakur, M.: A new crossover operator for real coded genetic algorithms. Applied Mathematics and Computation 188, 895–911 (2007)
[23] Arumugan, M.S., Rao, M.V.C., Palaniappan, R.: New hybrid genetic operators for real coded genetic algorithm to compute optimal control of a class of hybrid systems. Applied Soft Computing 6, 38–52 (2005)
[24] Deep, K., Thakur, M.: A new mutation operator for real coded genetic algorithms. Applied Mathematics and Computation 193, 211–230 (2007)
[25] Lindgren, M., Lepistö, T.: Application of Barkhausen noise to biaxial residual stress measurements in welded steel tubes. Materials Science and Technology 18, 1369–1376 (2001)
[26] Sorsa, A., Leiviskä, K., Santa-aho, S.: Prediction of residual stress from the Barkhausen noise signal. In: Proceedings NDT 2008, p. 10 (2008)
[27] Sorsa, A., Leiviskä, K.: An entropy-based approach for the analysis of the Barkhausen noise signal. In: Proceedings of 7th International Conference on Barkhausen Noise and Micromagnetic Testing, pp. 85–96 (2009)
[28] Grosfils, A., Vande Wouver, A., Bogaerts, P.: On a general model structure for macroscopic biological reaction rates. Journal of Biotechnology 130, 253–264 (2007)
[29] Holck, P., Sorsa, A., Leiviskä, K.: Parameter identification in the activated sludge process. Chemical Engineering Transactions 17, 1293–1298 (2009)
[30] Keskitalo, J., Sorsa, A., Heikkinen, T., Juuso, E.: Predicting COD concentration of activated sludge plant effluent using neural networks and genetic algorithms. In: Troch, I., Breitenecker, F. (eds.) Proceedings MATHMOD 2009, Vienna (2009) (Full Papers CD Volume)

[31] Vesterinen, T., Ritala, R.: Bioprocesses and other production processes with multistability for method testing and analysis. In: Puigjaner, L., Espuna, A. (eds.) 38th European Symposium of the Working Party on Computer Aided Process Engineering, pp. 859–864 (2005)
[32] Katare, S., Caruthers, J.M., Delgass, W.N., Venkatasubramanian, V.: A hybrid genetic algorithm for efficient parameter estimation of large kinetic models. Computers and Chemical Engineering 28, 2569–2581 (2004)
[33] Marseguerra, M., Zio, E., Podofillini, L.: Model parameters estimation and sensitivity by genetic algorithms. Annals of Nuclear Energy 30, 1437–1456 (2003)

# Range Statistics and Suppressing Snowflakes Detects for Laser Range Finders in Snowfall

Sven Rönnbäck and Åke Wernersson

**Abstract.** This paper presents statistics on registrations from laser range finders in snowfall. The sensors are standard laser range finders in robotics, the LMS200 and the URG-04LX. Three different working cases were identified for the pulsed laser range finder.

1) Normal operation with background objects present within the range of the sensor.
2) Close range objects where ranges to objects are shorter than the pulse length.
3) Free-space in the background.

The findings are summarized as:

- Two laser range finders have been used, one that sends out a pulsed wide beam and one with a modulated narrow laser beam. The narrow beam laser has better penetration between the snowflakes.
- Median filtering shows a substantial reduction in snowflake detects.
- The gamma distribution describes fairly well the range distribution of detected snowflakes.
- In an intense snowfall where about 24% of the ranges detected snowflakes.
- A time-polar median filter showed good results in suppressing snowflakes in range data.

## 1 Introduction

This paper presents statistics on range registrations in snowfall. It is based on registrations made with two different laser range finders and two snowfalls with different temperatures and snowfall intensity.

Sven Rönnbäck
Umeå University, 90187 Umeå, Sweden
e-mail: sven.ronnback@tfe.umu.se

Åke Wernersson
Luleå University of Technology, 97187 Luleå, Sweden
e-mail: ake.wernersson@ltu.se

V. Sgurev et al. (Eds.): Intelligent Systems: From Theory to Practice, SCI 299, pp. 261–277.
springerlink.com 

How will a pulsed laser beam be disturbed by falling snowflakes? A snowfall will attenuate a laser beam and reduce the visibility of the laser[2]. As a beam hits snowflakes a scattering effect will occur, some of the laser energy is reflected back to trigger the detector[5]. Blowing snow will also attenuate and scatter a pulse as it hits a snowflake[11][12].

How reliable is a laser range finder as navigation sensor in snowy conditions? We made this statistics to get some idea. This is especially important when they are used as the main sensors to navigate autonomous robots during winter conditions.

Snow changes its light reflectance properties over time and temperature, especially new fallen snow has a very high reflectance in IR-wavelength, above 80%, compared to icy snow that has below 20% reflectance[8]. Low reflectance materials need objects to be closer to the sensor to trigger the detector. To meet desired properties for a laser range finder in winter conditions a research group considered building special laser range finder especially for the winter conditions at Antarctic plateau[1].

Can we distinguish between false registrations and registrations on obstacles? For robust mobile robot navigation it is important with predictable properties. With this paper we try to find initial answers to the questions. Two different snowfalls were monitored and statistic analysis on range registrations were made.

## 2 Registrations

The first snowfall was monitored the 24th of February 2007. It had a precipitation of 0.9mm/hour (converted to rain), and the snowflakes were rather large compared to a standard snowfall in the region. The temperature was $-8^{o}$C with an average wind of 10m/s[10]. The location of the testbed was an inner yards, so the wind was turbulent and had vortexes.

The research testbed, a wheelchair equipped with two laser range finders, was placed outdoors to monitor it. The snowfall and the testbed is visible in Figure 1.

**Fig. 1** An electric wheelchair testbed was placed outdoors to monitor a snowfall. Two laser range finders were used for registrations. One of them, a LMS200 laser range finder, is visible on the wheelchair table.

The first snowfall was monitored for 10x2 minutes with two scanning range sensors, a LMS200[9] and a URG-04LX[4]. The second snowfall, in April 2007, was monitored with the LMS200 sensor only.

Some properties of the laser scanners are listed in Table 1. The sensors were placed horizontally, and the URG-04lx sensor was placed on the LMS200. Figure 2 shows a picture of the test environment and the first monitored snowfall, visible are also the reference objects.

**Table 1** Some properties for the LMS200 and URG-04LX scanning laser range finders

| | LMS200 | URG-04LX |
|---|---|---|
| Angular resolution | $0.5^o$ | $0.36^o$ |
| Max range | 80m | 4m |
| Power consumption | 20Watt | 2.5Watt |
| Scanning rate | configured to $\approx$ 4.7Hz | 10Hz |
| Wavelength | 905nm, Class 1 (eye-safe) | 785nm, Class 1 |
| Working principle | Pulsed | Modulated |
| Weight | 4.5kg | 0.14kg |
| Max range | 80m | 4.095m |
| Interface | RS232 | USB-serial |

**Fig. 2** The test scene with three reference objects visible; a snowpile in the centre and two walls, one to the left and one to the right. The wall behind the snowpile was not detected by the sensors. The test yard is located at latitude $65^o$37'1.01"N and longitude $22^o$8'13.01"E, Google Earth[3].

The LMS200 sensor is a pulsed laser scanner and was configured to scan a $180^o$ sector with $0.5^o$ in angular resolution. It sends out laser pulses that are reflected on objects. As a laser beam hits snowflakes scattering effects occur. Some of the light is reflected backwards to the detector, and if the reflections are strong enough the detector is triggered and a range can be calculated.

Figure 3 shows a typical LMS200 scan registered in snowfall, with the laser placed at (0,0) and reference walls visible to left and right. Detected snowflakes are visible in the first quadrant near (0,0), see Figure 3. Figure 4 shows four consecutive registered scans from the LMS200 in snowfall.

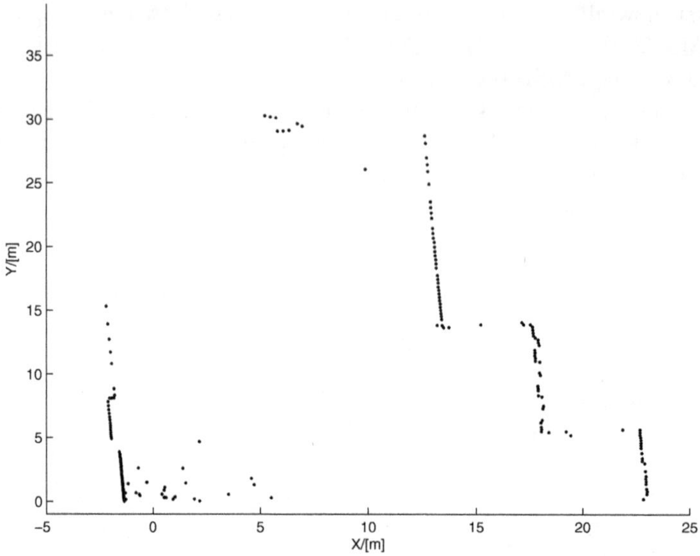


**Fig. 3** A single LMS200 scan from a snowfall registered with a LMS200. A snowpile is located 30m away is visible in the range data. Detected snowflakes are visible near the laser scanner at (0,0).

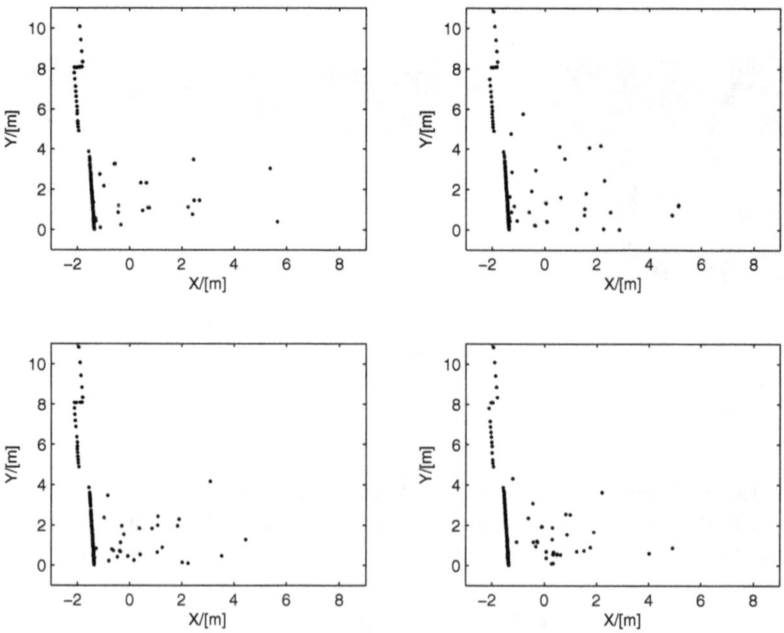


**Fig. 4** A sequence of four consecutive LMS200 laser scans in snowfall. A reference object, the left wall in Figure 2 and Figure 11, is visible in the data. The irregular detects, the differences between the scans, are detected snowflakes and essentially the same as in a flash photo of a snowfall in night time.

## 3 Snowfall Statistics

To get some more insight about snowflake detections a free-space sector area was identified and used to sort out only snowflakes.

The Cartesian coordinates to snowflakes were stacked on top on each other with the time axis along the vertical z-axis, see Figure 5. Of 38277 pulses were 9226 registered snowflakes, which is $\approx 24\%$.

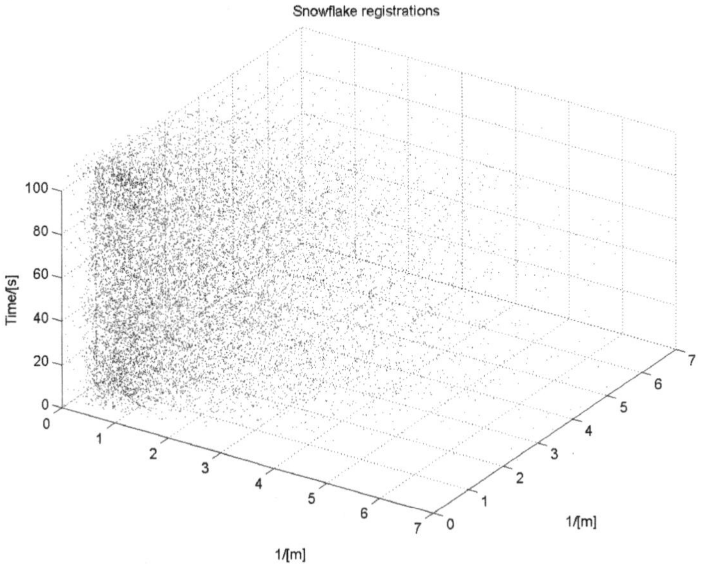


**Fig. 5** A snowflake cloud built of detected coordinates to snowflakes over a 100 sec. period. The LMS200 sensor is located at (0,0,t). Scans with snowflakes only were stacked on top of each other with time along vertical z-axis.

### *3.1 Bearing and Range Histograms to Snowflakes*

Next step was to get some statistics about the bearing angle to the detected snowflakes; a basic histogram was used for this, see Figure 6. There are some fairly large fluctuations but we consider it to be an uniform distribution.

Second we did a histogram on detected ranges, see Figure 7. The histogram gives us valuable information about suitable probability density functions. In the figure there is a significant peak close to 1m and almost no detected snowflakes beyond 8m. Note that there are some detects close to the sensor at (0,0).

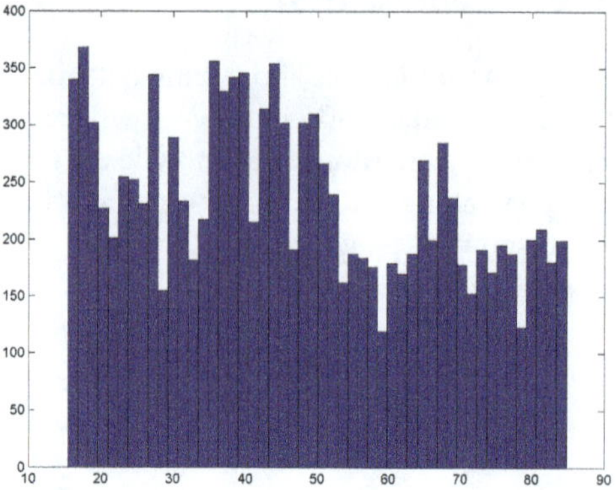


**Fig. 6** Histogram over bearing angles to detected snowflakes. Note the fairly large fluctuations.

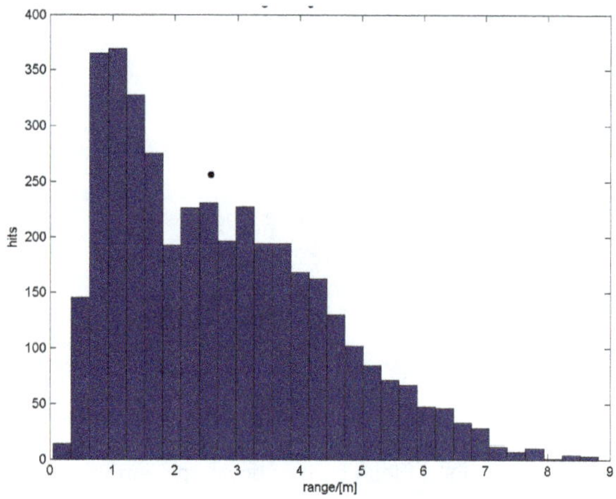


**Fig. 7** Histogram over registered ranges to snowflakes with the LMS200. Note that there are few detects near the sensor.

## 3.2 A Suitable Probability Density Function for the Range Distribution?

The gamma distribution can be used to model processes where the Poisson processes exists[13]. The probability density function (pdf) for a gamma distribution is given by

$$f(r,k,\theta) = r^{k-1}\frac{exp(-r/\theta)}{\theta^k \Gamma(k)}, r > 0. \tag{1}$$

Parameter $r$ is the range, $k > 0$ is called the shape parameter and $\theta > 0$ is the scale parameter. Variable $\Gamma$ is the gamma function[7].

Gamma distribution was fit to the histogram in Figure 7. The registered ranges were used compute the maximum likelihood estimates for parameters of the gamma distribution. This was done with the *gamfit* function provided in MATLAB[6]

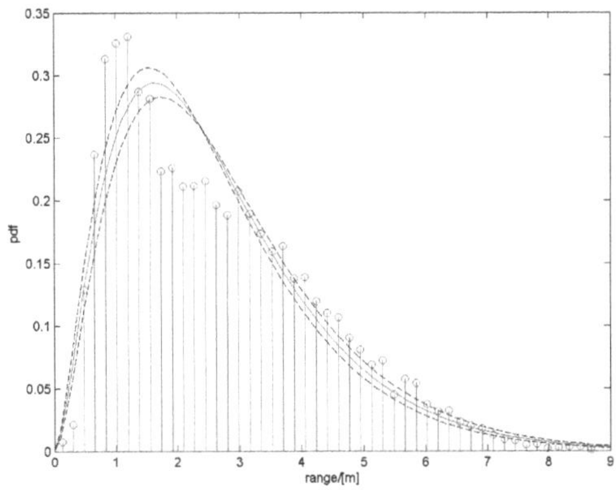


**Fig. 8** A probability density function, estimated from registered ranges to snowflakes, is plotted as the sold line. The sensor was LMS200. The estimated mean is 2.62m. The upper and lower confidence intervals are plotted dashed.

Statistics Toolbox, which also returns the confidence interval of the estimated parameters. The shape parameter was estimated to $k \approx 2.62$, and the scale parameter to $\theta \approx 1.01$. The 95% confidence intervals were [2.56, 2.67] and [0.99,1.04] respectively.

The estimated probability density function is plotted in Figure 8. The cumulative density function (cdf),

$$cdf = F(r,k.\theta) = \int_0^r f(r,k,\theta) = \frac{\gamma(r,k/\theta)}{\Gamma(k)}, \tag{2}$$

for the gamma distribution in Figure 8, is plotted as a dashed line in Figure 9. Where $\gamma(...)$ is called the lower incomplete gamma function,

$$\gamma(r,x) = \int_0^x t^{r-1} e^{-t} dt. \tag{3}$$

The upper incomplete gamma function is $\Gamma(...)$,

$$\Gamma(r,x) = \int_x^\infty t^{r-1} e^{-t} dt. \tag{4}$$

The cumulative probability function for the estimated parameters is plotted as the solid line in Figure 9. The dashed line is the accumulated probability based on the range histogram in Figure 7. The horizontal lines mark the 5% and 95% levels for the cumulative probability function.

In a set of over 11000 registered snowflakes were only seven registered beyond 10m. The most distant snowflake was detected at 13.2m. Since the beam of the LMS200 becomes wider on bigger distance, which means that more snowflakes will fall inside its volume. The spot diameter is about 18cm at 13m[1].

[1] SICK, Technical Description, LMS200 Manual, Fig 3-2, page 7.

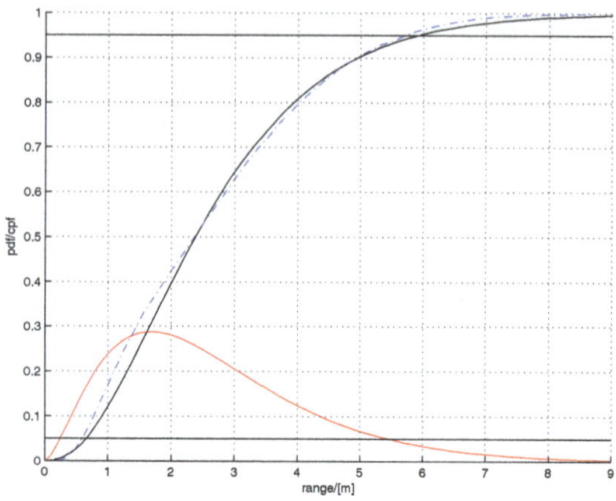


**Fig. 9** The probability density function and its cumulative probability function for the range distribution to detected snowflakes. The data was registered with a LMS200. The dashed line is the cumulative probability computed from the histogram in Figure 7. The horizontal lines mark 5% and 95% for the cumulative probability function.

# 4 The URG-04LX Scanning Range Finder

The previous section gave some information about the range distribution for the pulsed laser range finder, the LMS200. The URG04-LX is smaller, lighter, and sends out a narrow coherent laser beam with less energy than the LMS200. The max range for the URG-04LX was 4.095m. It is a modulated laser range finder [4].

Are the shapes of the range histograms different for the laser sensors? Figure 10 gives some information about this. It is based on data registered at the same time as the data used for the statistics presented in Section 3.2.

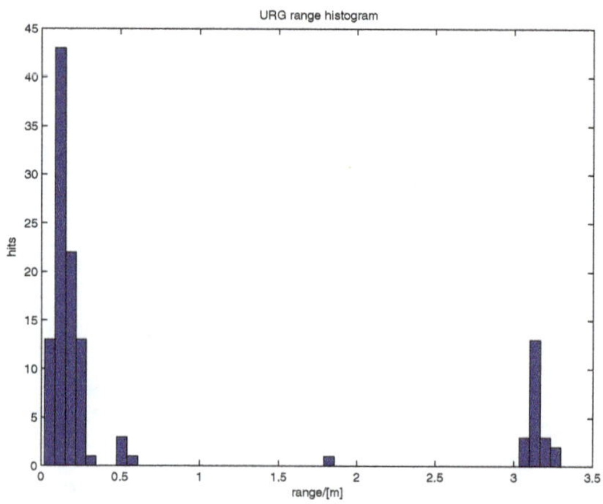


**Fig. 10** A histogram over detected ranges to snowflakes in February 2007 snowfall. The sensor was the URG-04LX scanning range finder. Only a few snowflakes were detected, compared to the LMS200 sensor.

The URG-04LX sensor did not detect as many snowflakes as the LMS200; 115 of 99556 ranges, about 0.12%. The histogram also looks different with the mean value closer to the sensor. The reason for may be that the sensor has a narrow coherent beam so it is has less probability to hit a snowflake.

## 5 Range Statistics in Different Sectors

Reference objects were identified to get more information about the data. The objects are; one reference wall to the left, one more distant wall to the right, and a snowpile. The reference objects are all visible in Figure 2. The right wall is also visible in Figure 10, note also the heavy snowfall.

- The left reference wall was divided into three sectors. The closest distance was approximately 1.1m.
- The right reference wall was one sector, this wall is more distant that the left wall. The distance to one of its corner was approximately 20m.
- The snowpile was one sector. The distance to it was 30m, see Figure 3 and Figure 4.

Range statistics for the left wall is found in Section 5.1. Range statistics for the right wall is found in Section 5.2, and range statistics for the snowpile in Section 5.3. Statistics for the different sectors are listed in Table 2.

**Table 2** Comparing table on snowflakes detects in the different regions/sectors. The columns are; No detects, detects on snowflakes, detects on the object in the background, and total number of laser pulses.

| AREA/SECTOR | NO DET. | DET. SNOWFLAKES | WALL DET. | PULSES |
|---|---|---|---|---|
| Sector 1 | 0.0% | 5.0% | 94.5% | 20400 |
| Sector 2 | 0.02% | 7.2% | 92.7% | 22805 |
| Sector 3 | 0.15% | 15.8% | 84.0% | 23437 |

### *5.1 Range Statistic from Detects towards Left Reference Wall*

The wall depicted in Figure 11 is the reference wall to the left. It was divided into three different sectors, see Figure 12. One sector close to the laser scanner, one with an angle of approximately $45^o$ to the wall, and sector 3 the remaining part of the wall. The resulting histograms for the three sectors are different, see Figure 13 to Figure 15.

The histogram for sector three follows the same shape as the histogram built on only snowflake registrations as presented in Section 3.2. The question is what happens in sector 1? Why is the shape of that histogram different?

### *5.2 Range Statistics from Detects towards Right Reference Wall*

A second wall was used for range statistics over longer distances, see Figure 16. Ranges from a sector of $13.5^o$ were used for the statistics.

**Fig. 11** The wall left of the laser range finder was selected as a reference object, a background object. In the picture it is also possible to see how the visibility is degraded by the snowfall.

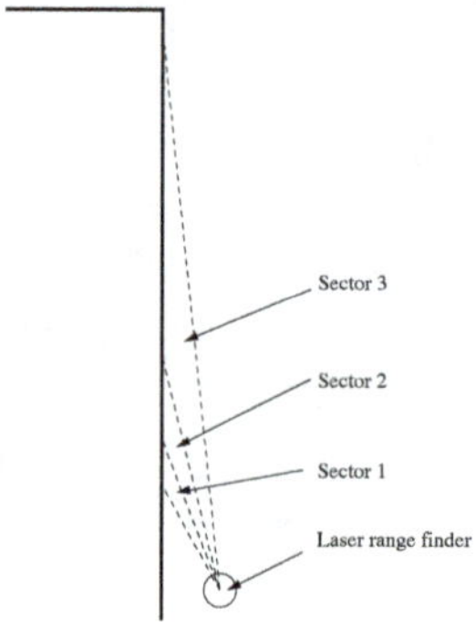


**Fig. 12** The laser shots towards the left wall were divided into three different sectors labelled from sector 1 to sector 3

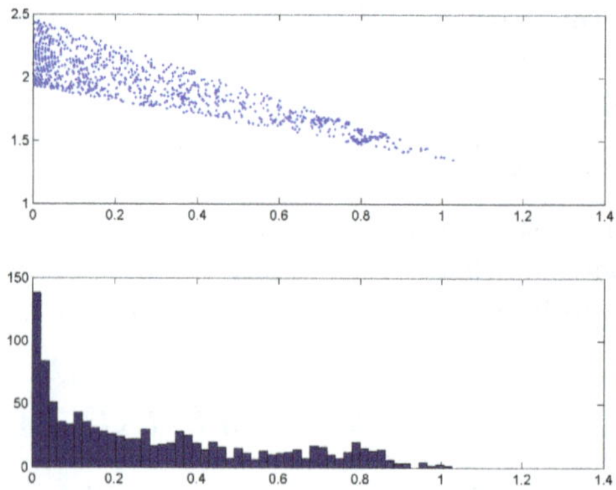


**Fig. 13** Registered snowflakes and a range histogram for sector 1. It is possible to see that this histogram has a different shape than in Figure 6 and Figure 15.

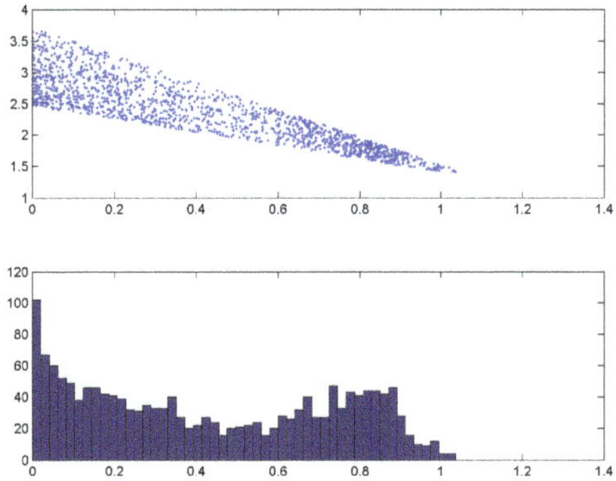


**Fig. 14** Registered snowflakes and a range histogram for sector 2

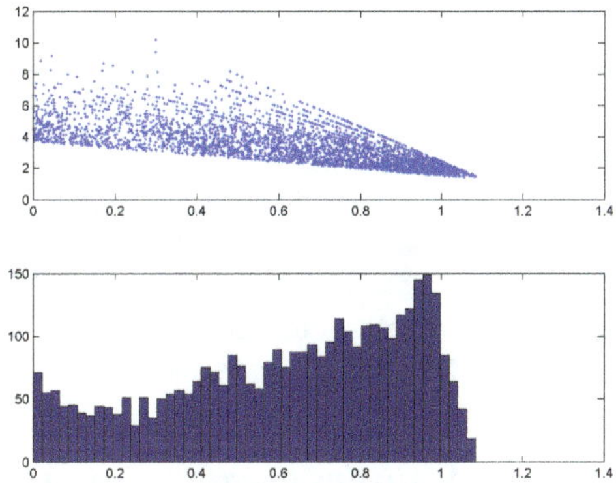


**Fig. 15** The dots in the upper sub figure are registered snowflakes for sector 3. In the lower subfigure we can see the shape of the gamma distribution.

**Fig. 16** A part of this wall was used as reference object to get statistics about percentage of snowflakes, wall detections, and the number of no detects. It is possible to see falling snowflakes in the picture.

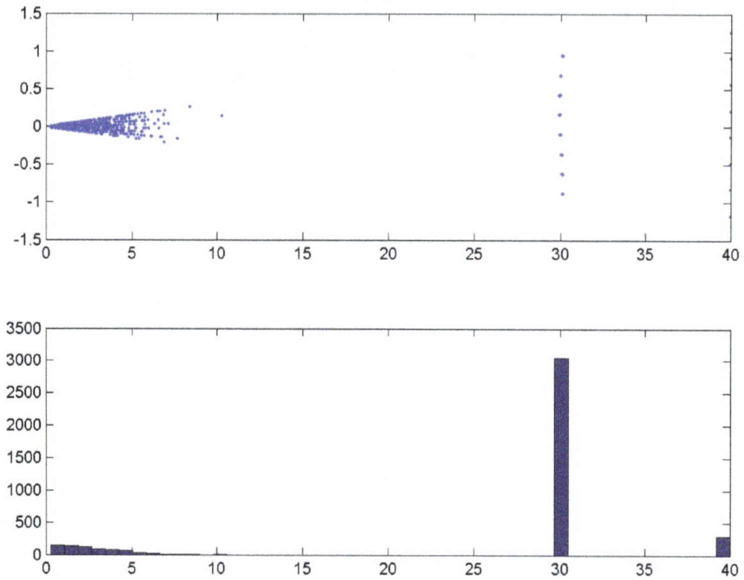


**Fig. 17** Range histogram made from range registrations on a snowpile. The bar at 30m represents pulses reflected from the snow on the snowpile. The histogram bar at 40m represents laser pulses that fell on the snowpile with no detected reflections.

From a set of close to 13700 registered ranges were approximately 19% snowflakes, 81% fell on the wall, and about 0.2% were absorbed laser pulses.

## *5.3 Range Statistics towards Snowpile*

A snowpile was also selected as a reference object. It is visible in Figure 2. The question was; how many of the laser pulses are absorbed by the snow on the snowpile?

A histogram over the ranges was made. As indicated in Figure 17 most of the pulses on the snowpile resulted registrations at 30m.

Some pulses, approximately 7.4%, resulted in no detects and shown as the peak at 40m. This means that some of the reflections from the snowpile snow were not strong enough to trigger the detector.

At shorter range the histogram in Figure 17 has the same shape as the one in Figure 7.

## *5.4 The Second Monitored Snowfall*

A second snowfall was monitored the 10th of April 2007. The temperature was just below zero, no wind, small snowflakes, and not as intense snowfall as in February. Only the LMS200 scanning range finder was used.

**Table 3** Comparing table over percentage of detected snowflakes in the second snowfall in April 2007. The snowflakes were smaller, and the outside temperature was higher; was just below zero.

| Area/Sector | No Detections | Det. Snowflakes | Wall Det. | Pulses |
|---|---|---|---|---|
| Sector 1 | 0 | 0.41% | 99.59% | 20740 |
| Sector 2 | 0 | 0.69% | 99.31% | 23179 |
| Sector 3 | 0 | 1.43% | 98.57% | 23174 |

The location and the pose of the experimental platform were close to the previous registration.

The statistics are based on 550 laser scans and are presented in Table 3.

A range histogram to detected snowflakes can be seen in Figure 18. The estimated probability density function for the range data is shown in Figure 19. The shape parameter is estimated to $k \approx 3.0$ and the scale parameter to $\theta \approx 0.48$. The cumulative probability function is plotted in Figure 20.

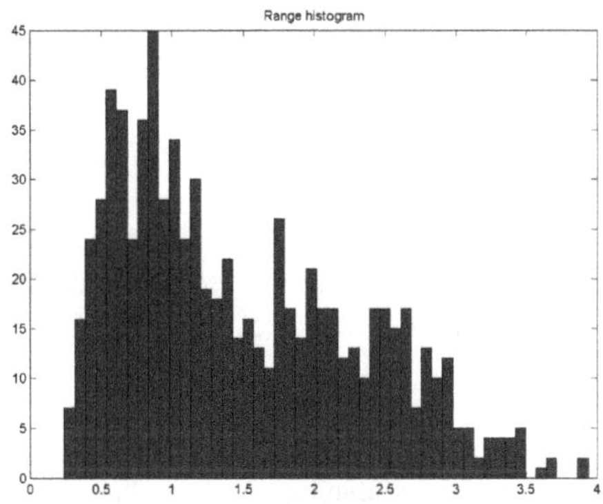


**Fig. 18** The second snowfall (April 2007): A histogram made from detected ranges to registered snowflakes by the URG-04LX sensor. This snowfall had a temperature just above zero and significant smaller snowflakes than the snowfall in February 2007.

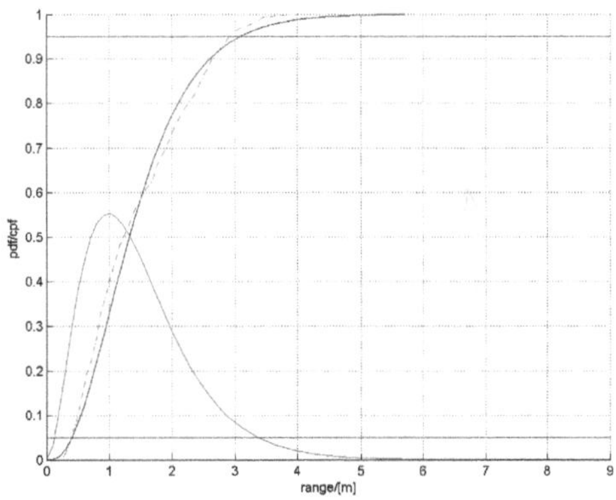


**Fig. 19** The second snowfall (April 2007): A gamma probability density function was fitted to the data in Fig 18. The result is the solid line. The dashed lines represent the upper and lower confidence intervals for the estimated parameters.

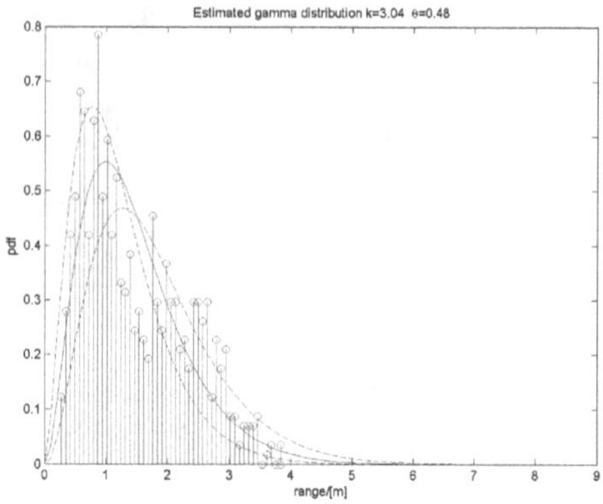


**Fig. 20** The cumulative probability function built from the histogram in Figure 18 is plotted as the dashed line. The solid line is the estimated cumulative probability function.

## 6 Median Filtering of Range Data

The median filter is well known from the area of image processing and was tested here to reduce the noise due to snowflakes. Therefore we implemented the filter both in polar space and in time-polar space.

### *6.1 Median Filter in Polar Space*

Two median filters, with window size 5 and 7, were used to process the range data.

The median filters show a reduction of flakes from about 19% to about 3.8% for a median window with size 5 and 2.2% for a window size of 7, see Table 4 for more information.

**Table 4** Comparing table on how snowflakes are suppressed in the different regions/sectors and after applying median filtering. The columns are; No detects, detects on snowflakes, detects on the object in the background, and total number of laser pulses.

| AREA/SECTOR | NO DET. | DET. SNOWFLAKES | WALL DET. | PULSES |
|---|---|---|---|---|
| Snowpile | 7.4% | 16.3% | 76.4% | 3996 |
| Right wall | 0.2% | 19% | 81.0% | 13692 |
| Sec.1 /Med.Filt.7 | 0 | 0.46% | 99.54% | 17340 |
| Sec.2/Med.Filt.7 | 0 | 0.32% | 99.68% | 1938 |
| Sec.3/Med.Filt.7 | 1 | 2.25% | 97.75% | 17340 |
| Sec.l/Med.Filt.5 | 0 | 0.55% | 99.45% | 17340 |
| Sec.2/Med.Filt.5 | 0 | 0.52% | 99.48% | 19380 |
| Sec.3/Med.Filt.5 | 1 | 3.84% | 96.16% | 19861 |
| Sector 1 | 0.0% | 5.0% | 94.5% | 20400 |
| Sector 2 | 0.02% | 7.2% | 92.7% | 22805 |
| Sector 3 | 0.15% | 15.8% | 84.0% | 23437 |

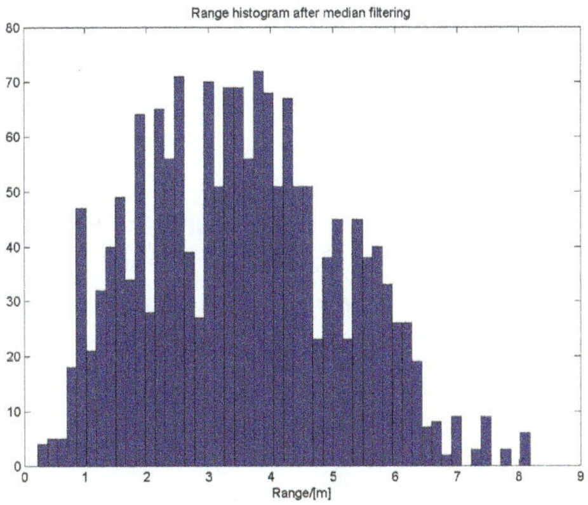


**Fig. 21** Range histogram after median filtering. We can clearly see a shift of the histogram, relative to Figure 6, and that the shape of it more reminds of a normal distribution. The mean value is estimated to be at approximately 3.7m.

Figure 21 shows a range histogram after median filtering, compare it with Figure 6. The mean value has shifted to the right from 2.6m to 3.7m and its shape has changed dramatically. The median filter in polar space eliminate s a big numbers of detected snowflakes, but still some remains, therefore we implemented a time-polar median filter.

## 6.2 *Median Filter in Time-Polar Space*

We also implemented a median filter that filters in time-polar space. It can almost be seen as median filtering in 2D. The median filtering was done in two steps; first in time space on three consecutive laser scans, and after that in polar space from right to left.

The last three collected laser scans where used for the median filter in time domain, the size of median window was three.

Table 5 shows an example of the output of the time median filter given a specific input where T is the sample time and $\alpha$ is the scan bearing angle, $\alpha \in [0, 180^o]$.

**Table 5** This table shows an example of the output from the time polar median filter. The angle is chosen arbitrary in the scanning sector.

| Time stamp | $\alpha$ | $\alpha + 0.5^o$ | $\alpha + 1^o$ |
|---|---|---|---|
| t-2T | 10 | 15 | 32 |
| t-T | 12 | 57 | 52 |
| t | 14 | 5 | 40 |
| Result, at t | 12 | 15 | 40 |

## 7 Results and Conclusions

We show statistics where laser scanners were used to monitor snowfalls. Two different types laser range finders were used.

Histograms indicate that the pulsed laser range finder is better to suited to detected snowflakes.

The URG-04LX scanning range finder was slightly affected by snowflakes, and only about 0.1% of the shots were registered snowflakes compared to the LMS200 which registered above 10% as snowflakes.

Results in form of statistic and histograms indicate that the gamma distribution may be suitable to describe the range distribution to detected snowflakes.

A median filter in time space combined with a median filter on range data resulted in almost total suppression of snowflakes in range data.

The range distribution to detected snowflakes is close to a gamma distribution. Statistics show that almost no snowflakes were registered beyond 10m by the LMS200; the maximum range to a detected snowflake was 13.2m.

The exact design of each detector is a secret for each sensor manufacturer.

A time-space median filter was implemented and showed good results in suppressing snowflake detects.

**Acknowledgements.** The authors thank Ove Steinwall at Swedish Defence Research Agency, FOI Linköping, for valuable references. First author also wants to thank Applied Physics and Electronics at Umeå University for supporting this work.

## References

1. Bartolini, L., Bordone, A., Fantoni, R., et al.: Development of a laser range finder for the antarctic plateu. In: Proceedings of EARSeL-SIG-Workshop LIDAR, Dresden/FRG, June 16-17 (2000)
2. O'Brien, H.W.: Visibility and Light Attenuation in Falling Snow. Journal of Applied Meteorology 9(4), 671–683 (1970)
3. Google, "Google Earth" (August 2009), http://www.earth.google.com
4. Hokuyo: Scanning range finder (SOKUIKI sensor) URG-04LX (2009), http://www.hokuyo-aut.jp/02sensor/07scanner/urg_04lx.html (Cited August 2009)
5. Hutt, D.L., Bissonnette, L.R., St. Germain, D., Oman, J.: Extinction of visible and infrared beams by falling snow. Applied Optics 31(24), 5121–5132 (1992)
6. MATLAB Statistic Toolbox, http://www.mathworks.com (Cited August 2009)
7. Råde, L., Westergren, B.: BETA and Mathematics Handbook, 2nd edn., Studentlitteratur (1990)
8. Salvatori, R., Casachia, R., et al.: Snow surface classification in the western Svalbard Island. In: 31st International Symposium on Remote Sensing of Environment (2005)
9. SICK, LMS200 laser measurement systems (2009), http://www.mysick.com, http://www.sick.de, https://www.mysick.com/saqqara/get.aspx?id=im0012759 (Cited August 2009)

10. SMHI, Swedish meteorological and hydrological institute (2007), http://www.smhi.se/en (Cited February 2007)
11. Vandapel, N., Moorehead, S.J., "Red" Whittaker, W., Chatila, R., Murrieta-Cid, R.: Preliminary Results on the use of Stereo, Color Cameras and Laser Sensors in Antarctica. Lecture Notes in Control and Information Sciences, vol. 250, pp. 59–68. Springer, Heidelberg (2000)
12. Vandapel, N., Moorehead, S.J., "Red" Whittaker, W., Chatila, R., Murrieta-Cid, R.: Preliminary Results on the use of Stereo, Color Cameras and Laser Sensors in Antarctica. In: Proceedings of International Symposium of Experimental Robotics, pp. 450–468 (1999)
13. Weisstein, E.W.: "Gamma distribution", From MathWorld (2009), http://mathworld.wolfram.com/GammaDistribution.html (Cited August 14, 2009)

# Situational Modelling for Structural Dynamics Control of Industry-Business Processes and Supply Chains

Boris Sokolov, Dmitry Ivanov, and Alexander Fridman

**Abstract.** The paper describes a flexible information technology for decision making support by situational modelling of possible alternative structures of industry-business processes and supply chains both in statics and dynamics. It provides a state analysis and prognosis of spatial hierarchical objects. The technology allows to use data coming from simulation modules of the object structural components, embedded geographic information system and expert system equally.

**Keywords:** intelligent information technology, structural dynamics control, auto-reconfigurable highly engineered system, situational analysis and synthesis, subject domain conceptual model.

## 1 Introduction

The term industry-business processes (IBP) will here stand for flexible enterprises and technologies, regional business organisations, etc., functioning at an essential dependency on spatial features of their components and time. Experiments upon

---

Boris Sokolov
St. Petersburg Institute for Informatics and Automation of RAS, 39, 14th line,
St.-Petersburg, 199178, Russia
e-mail: sokol@iias.spb.su

Dmitry Ivanov
Chemnitz University of Technology
D-09107 Chemnitz, Germany
e-mail: idm@hrz.tu-chemnitz.de

Alexander Fridman
Institute for Informatics and Mathematical Modelling of Technological Processes of RAS,
24A Fersman str., 184209, Apatity Murmansk reg., Russia
e-mail: fridman@iimm.kolasc.net.ru

V. Sgurev et al. (Eds.): Intelligent Systems: From Theory to Practice, SCI 299, pp. 279–308.
springerlink.com 

IBP are usually impossible or prohibited because of either lack of time, or high cost, or hazard of unrecoverable changes. That is why modelling is the main method of IBP behaviour studying and forecasting.

To acquire any practicable results, IBP have to be modelled as complex spatial dynamical systems with variable structure, and multiple inner and outer links. The modelling process must both consider different informational, financial, material, and power streams and stipulate analysis of consequences of the structural changes, possible emergencies, etc. Due to unavoidable incompleteness of knowledge regarding IBP it is reasonable not to confine to classic analytic and simulating models only, but to grant means for using experts' experience. So, any IBP modelling system ought to incorporate a geographical information system (GIS) and an expert system (ES), to admit dynamical simulation and to be open for operative changes resulting from previous modelling stages. With these goals, the modelling system must implement relevant procedures, techniques and algorithms for parametrical and structural adaptation of models to current and predicted dynamics both of IBP themselves and of their environment.
The main features of any IBP organization are as follows:

- network and enterprise collaboration are project-oriented;
- supply chains (SC) in IBPs are customer-oriented and do not have any pre-determined long-term suppliers' structures and product programs;
- each network participant specializes in certain functional competencies;
- network participants are independent, they act autonomously according to their own goals;
- each project has a coordinator, who is responsible for the project success (customer side) and for coordination of network participant activities (supply side) (Ivanov 2009).

The main advantages of IBP are a faster and more flexible reaction to market changes as well as in reducing time-to-market through operative distribution and coordination of resources (competencies) through all phases of product life cycle. However, the vertical and horizontal customer-oriented networking leads to new problems for the supply chain management (SCM) (Ivanov 2003). Traditionally, improvements in supply chain planning and scheduling have been algorithmic (Kreipl and Pinedo 2004). However, in recent years, the works on supply chain management have been broadened to cover the whole supply chain dynamics. The special feature of the SCM in IBP consists in flexibly configurable supply chains, conditioned by an enlargement of alternatives to search for partners suitable for the cooperation. Thus, the problem of comparative analysis of the alternatives becomes topical. In our approach, IBP planning and control are macro-functions comprised of synthesis of a virtual enterprise (VE) structure (building of a competence pool in accordance with the technological details of the product), synthesis of the supply chain structure (building of the supply chain through partner selection for this pool in accordance with the economical and administrative information, e.g. due dates, priorities, costs, soft-factors, etc.) and monitoring/reconfiguration of the supply chain.

VEs unite independent multi-business partners (real enterprises) within a temporal task-oriented technical-organizational structure through information technologies and telecommunications (Camarihna-Matos et al. 2004, Wang and Norrie 2001, Ivanov 2003). VEs are highly adaptive to consumer needs and benefit juridical and physical persons providing them with dynamic use of common resources during remote collaboration within a business project.

A VE is a typical example of a modern integrated transport, production and trading network performing intensive structural dynamics. This issue makes the structural synthesis of a VE more complicated. In particular, the structure dynamics have a complicate influence upon the following tasks: partner selection (producers and suppliers of components, retailers, etc); end products configuring; placement of orders; configuring transport network and information resources (Camarihna-Matos et al. 2004, Wang and Norrie 2001).

An IBP system must be configured according to the project goals and reconfigured in dynamics according to the current execution environment. In practice, IBP re-design decisions are centred on adapting and rationalizing the supply chains in response to permanent changes of network itself and its environment (Harrison et al. 2005).

Thus, our project aims at uniting all the tasks and forms of knowledge representation mentioned above into an integrated modelling system based upon an open subject domain model. Now we proceed with step-by-step description of modelling basics and techniques.

## 2 Problem Description

IBP model at a regional or state scale shall represent a multilevel system of sub-objects linked with different signals. These signals are modelled by data streams and treated as *resources* used and/or spent by the sub-objects during their life cycle. To avoid problems with small changes of variables and to provide a common logic-computational data processing, resources in our modelling system are treated as discrete variables and respectively attributed with lists of values. This also allows to detail the link control to each resource, thus raising description adequacy of links among the object components and decision control at branching links among the model elements, as well as providing processing of time data sets during simulation mode.

From the introduced point of view, we will describe IBP in the given project by using generalised logic-dynamical systems with a controlled structure. Such systems can function and develop under changing conditions for goals, tasks, situations as well as for criteria estimating efficiency and effectiveness of their performance. Catering for GIS integration into the developing Integrated Situational Modelling System (ISMS), we apply an additional restriction on the IBP model. Any IBP model in the ISMS must be discrete in area. That is, it must be possible to assign a finite set of standard GIS-elements to any IBP component.

The ISMS provides an objective (within the developed VE model) comparative analysis of situations and a simulation of consequences of the made decisions. Embedded GIS and expert systems allow to equally use analytical and numerical models of the VE components, heuristic rules and graphical information. The GIS is not used for object mapping only, but also for topology description, task setting, spatially dependent calculations and displaying of modelling results. Spatial-temporal ordering of different data attributes based on GIS technology provides an efficient ground for a reasonable information fusion regarding diverse aspects of VE, IBP and SC descriptions delivered by other kinds of models. Some of the models are briefly introduced in the section 7.

The opportunity to unify diverse sub-models differing in data representation, time increments, model ranks, and moreover, models that were developed by different researchers at different time and had different structural grounds (for instance, by means of logic-algebraic or logic-linguistic structures) is basically important for the proposed approach. The unification of different knowledge representation forms is inevitable within "poorly investigated" subject domains according to many authors (for example, Okhtilev et al. 2006). In this purpose, the ISMS will correspond to the modern paradigm of creation and usage of information systems with service-oriented architecture. In the field under consideration, this approach gives us an additional opportunity to implement the strategy of rapid software deployment by means of "rapid prototyping" that has already become conventional in service-oriented architectures. "Rapid prototypes" allow to substitute some absent or not ready sub-models of certain model components with a set of expert rules thus supporting an acceptable level of the whole system maintenance. To provide this feature, the ISMS will comprise data representation and transformation toolkits as well as intelligent interfaces in order to let non-programmers interact with models (create, modify and apply them) constructively in domain-oriented language.

Due to their complexity and changeability IBP and SC belong to complex technical organizational systems (CTOS). So, before giving a description of the ISMS itself, it looks reasonable to introduce a general survey of our approach to CTOS modelling.

## 3 CTOS Structural Dynamics Control

One of the main features of modern CTOSs is the variability of their parameters and structures caused by objective and subjective factors at different phases of the CTOS life cycle. In other words, we always come across the CTOS structural dynamics in practice. Under the existing conditions the CTOS potentialities increment (stabilization) or degradation reduction makes a CTOS structure control (including the control of structure reconfiguration) necessary. The general aim of our investigation is to develop principles, models, methods and algorithms for the CTOS structural dynamics control. There are many possible variants of such a control. For example, they are: *alteration of CTOS functioning means and objectives; alteration of the order solving for observation tasks and control tasks; redistribution of functions, problems and control algorithms among the CTOS levels; reserve resources' control; control of motion of CTOS elements and subsystems; reconfiguration of CTOS different structures.*

According to the contents of the structural dynamics control problems, they belong to the class of the structure–functional synthesis problems and problems of program construction, providing for CTOS development.

The major feature and the difficulty of this kind of problems result from the following opposition: programs optimal for the CTOS main elements and subsystems' control can be implemented only when the lists of functions as well as the control algorithms and information-processing algorithms for the elements and subsystems are known. In turn, proper distribution of functions and algorithms among the CTOS elements and subsystems depends upon the structure and parameters of the control rules that are actual for these elements and subsystems. The described contradictory situation becomes even more complicated due to the changes in CTOS parameters and structures that occur on different reasons during the CTOS life cycle.

At present, the class of problems under review has not been examined quite thoroughly. The new theoretical and practical results were obtained in the following lines of the investigation: *synthesis of the CTOS technical structure for known laws of CTOS functioning (the first direction)* (Zvirkun and Akinfiev 1993, Tsurkov 1989, Singh and Titli 1978)*; synthesis of the CTOS functional structure, in other words, synthesis of the control programs for the CTOS main elements and subsystems under the condition that the CTOS technical structure is known (the second direction)* (Moiseev 1974, Singh and Titli 1978, Athans and Falb 1963)*; synthesis of programs for CTOS construction and development without consideration of periods of parallel functioning of the actual and the new CTOS (the third direction )* (Zvirkun and Akinfiev 1993, Moiseev 1974, Tsurkov 1989)*; parallel synthesis of the CTOS technical structure and the functional one (the forth direction)* (Siliak 1990, Singh and Titli 1978).

Several iterative procedures for solving of the joint problems concerning the first and the second directions are known at present. Some particular results were obtained within the third and the forth directions of investigations. All the existing models and methods for the CTOS structure–functional synthesis and for the construction of the CTOS development programs can be applied during the period of the internal and external design when the time factor is not very essential.

The problem of CTOS structural dynamics control consists of the following groups of tasks: the tasks of structure dynamics analysis of CTOS;the tasks of evaluation (observation) of structural states and CTOS structure dynamics;the problems of optimal program synthesis for structure dynamics' control in different situations. Therefore, the development of new theoretical bases for CTOS structural dynamics control is very important now. From our point of view, the theory of structural dynamics control will be interdisciplinary and will combine the results of classical control theory, operations research, artificial intelligence, systems theory, and systems analysis. These ideas are summarized in Fig. 1.

The latter two fields of knowledge will provide a structured definition of the structural dynamics control problem instead of a poorly structured definition. Here, as the first step to the new theory, we introduce a conceptual and formal description of CTOS structural dynamics.

Since decision making on CTOS structure changes requires a lot of computations, a description of suitable software environment follows.

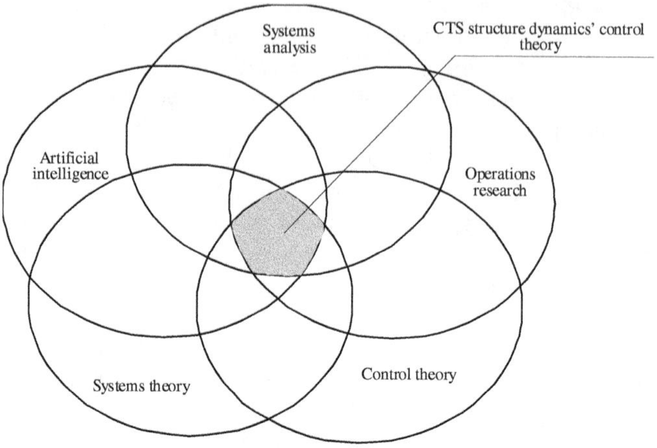


**Fig. 1** The theory of structure dynamics' control as a scope of interdisciplinary investigations

# 4 Grounds of Situational Modelling

An integrated subject domain conceptual model (SDCM) described in (Fridman et al. 2004) is the core of the ISMS. One more distinction of the ISMS from prototypes is that besides user-defined types of model elements, some automatically assigned *categories of elements* are brought into their description to provide a more detailed formal analysis of our model (we call it a "subject domain conceptual model" - SDCM).

To use the SDCM, one shall represent an IBP system as a hierarchy of *objects* (components of IBP) reflecting their organizational relations:

$$O = \{ o^{\gamma}_{\alpha \beta_\alpha} \} ::= \bigcup_{\alpha=1}^{N_L} O_\alpha, \tag{1}$$

where: $\alpha = \overline{1, N_L}$ is the level number in the object tree the given object belongs to ($L$ – the total number of decomposition levels);

$\beta_\alpha = \overline{1, N_\alpha}$ is the ordinal number of the object on its decomposition level;

$\gamma = \overline{1, N_{\alpha-1}}$ is the ordinal number of the super-object dominating the given object on the higher decomposition level;

$O_\alpha$ is the set of objects belonging to the level $\alpha$.

Here and further symbols ::= denote equality by definition. If necessary, other symbols of BNF-notation are used as well.

Depending on their position in the object tree and on the map, three object categories are distinguished. They are primitive objects (LEAF category) that have no details from the point of view of global modelling goal, elementary objects (GISC category) that are geographically linked to one GIS-element (either a polygon, or an arc, or a dot on a GIS-cover) and compound objects (COMP category) that comprise several objects. Thus, the set (1) can be split into non-overlapping subsets basing on categories of objects:

$$O ::= O^{LEAF} \cup O^{COMP} \cup O^{GISC}. \tag{2}$$

Adding the set of elementary GIS-objects to the set (2), we will get the complete set of the SDCM objects:

$$O' ::= O \cup O^{ELEM}, \tag{3}$$

where the complete set of standard GIS-objects is as follows:

$$O^{GIS} ::= O^{GISC} \cup O^{ELEM}. \tag{4}$$

Every object may be assigned with a number of *processes* simulating transformation of a set of input resources into a set of output ones. An *executor* is one of the main attributes of a process that determines its dynamical features and computer realization. Any executor may be specified either explicitly (as a function) or implicitly by referring to a name of a software module that is to implement this process.

To formalize expert knowledge, the ISMS includes a specific ES that allows a user to construct expert rules in an ordinary IF-THEN-ELSE format. The ES is used when there is no other way to model a part of the IBP under investigation. The ES may function as the executor of any resource or a process declared in the model. Then this resource or all the output resources of this process must be defined in right parts of some expert rules. In this way, we provide the opportunity to combine expert notions with mathematical models and graphic parameters.

We propose to consider uncertain factors caused by environmental impact upon modelling objects by means of analytic-simulating models developed within the bounds of a respective simulating system (Sokolov and Ivanov 2006). Some details are given below in the sections 7 and 8. Results of this modelling will serve for parametrical and structural adaptation of object models. The concept of service-oriented architectures mentioned above makes it possible to widely incorporate both conventional simulating toolkits (GPSS World, AnyLogic, PowerSim, etc.) and proprietary products into the developing ISMS.

The set of names of resources (data) is split into a set of names of variables that have numerical values (VAR category) and a set of names of parameters that have string values (PAR category):

$$D ::= \langle Var, Par \rangle, \tag{5}$$

$$Var ::= \{var_i\}, i=\overline{1,N_v}\ ;\ Par ::= \{par_j\}, j=\overline{1,N_p}\ , \tag{6}$$

where $N_v$ and $N_p$ stand for cardinality of the respective sets. Data may constitute resources (quantitative characteristics) of objects and processes, in which case they have the RES category. Variables may be also used as adjusting parameters of performance quality functions (criteria) for SDCM elements (ADJ category). Respectively, the set of names of variables can be split into a subset of names of resources and a subset of names of adjusting parameters:

$$Var::=\langle Res, Adj\rangle \tag{7}$$

A separate GIS category comprises graphic characteristics of SDCM objects directly calculated by a GIS.

One or several criteria (functions or functionals defined in some output resource sets) are ordinarily used to analyse IBP and compare possible alternatives of their implementation. We propose that the results of the comparison of these alternatives be used mostly in decision making on changing (or not changing) inner structure of processes or links among them depending on general state and revealed trends of the IBP system as a whole.

The *scheme of the SDCM* is as follows:

$$S_{ISMS} ::= \langle O, P, D^{CM}, H, OP, PO, U \rangle, \tag{8}$$

where: $O$ is the set of the SDCM objects defined in (2);

$P ::= \{p_n\},\ n = \overline{1, N_p}$ is the set of processes;

$D^{CM} \subseteq D$ (5) is the set of the SDCM data (resources);

$H$ is the object hierarchy relation that takes the following form considering (1):

$$H = \bigcup_{\alpha=1}^{N_{L-1}} H_\alpha, \tag{9}$$

where: $H_\alpha \subseteq O_{\alpha-1} \times B'(O_\alpha)$ are hierarchical relations for every level of the object tree and $B'(O_\alpha)$ means a partition of the set $O_\alpha$;

$OP \subseteq O \times B(P)$ is "an object – the processes generating its output data" relation and $B(P)$ means the power set (Boolean) of the set $P$;

$PO \subseteq P \times B(O)$ is the relation "a process – the objects generating its input data";

$U ::= U_p \cup U_o$ is the relation formalising the SDCM-based calculation process control. It has the following components:

$U_p \subseteq P \times B(Res)$ is "a process – the control datum" relation;

$U_o \subseteq O \times B(Res)$ is "an object – the control datum" relation.

The relations defined within the model (8) can be represented as functions partially defined upon sets $O$ and $P$ with ranges of values $B(P)$, $B(O)$, $B(Res)$ or $B'(O_\alpha)$. Names of the functions are denoted below with small letters corresponding to the names of relations. Ranges of values of these functions associated with cross-sections of the defined relations by an element of their ranges of definition are written in bold letters.

Cross-sections of the defined relations by a subset of their ranges of definition are defined as joins of all cross-sections on elements of this subset and are denoted in the same way. For instance, $\boldsymbol{h}_\alpha(O_i)$ where $O_i \subseteq O_{\alpha-1}$ means a set of objects from the level $\alpha$ dominated with the subset of objects $o_j \in O_i$ belonging to the level

$\alpha$ - 1. The formulas below also use the total dominated subset of an object $o_i$ (subordination set):

$$\underline{\underline{h}}^{\alpha}(o_i) ::= \bigcup_{\alpha \le k \le L} h_k^{\alpha}(o_i) \tag{10}$$

The hierarchy relations $H$ (9) allow to easily correlate any compound SDCM object with a subset of the set (4). The subordination set $\underline{\underline{h}}^{\alpha}(o_i)$ (10) of any object not belonging to the set of GIS-elements set (4) was proved to comprise the "main subset" of the dominated GIS-elements which are included into any alternative structure of this object, and a set of alternative GIS-elements that are revealed under a certain realization of the object:

$$(\forall o_i \in O \setminus O^{GIS}) \, (\underline{\underline{h}}^{\alpha}(o_i) = O_0(o_i) \cup Alt(o_i)), \tag{11}$$

where the symbol \ stands for the calculation of the set difference and

$$Alt(o_i) ::= \bigcup_{j=1}^{n} O_j(o_i), \tag{12}$$

while $O_j(o_i) \subset O^{GIS}$, $j = \overline{0,n}$ ; $O_0(o_i) \cap O_j(o_i) = \varnothing$, $j = \overline{1,n}$ .

The set $O_0$ comprises the GIS-elements that have no super-objects attributed by the OR-decomposition. The GIS representation of any alternative structure of the object $o_i$ is realized by the set $O_0$ and one of the sets $O_j$:

$$Alt_j(o_i) ::= O_0(o_i) \cup O_j(o_i \tag{13}$$

At forming the IBP conceptual model, some alternate structures of the object under investigation are input into the model by using OR-relations in object decomposition and resource generation. A specific extension of the situational approach (Pospelov 1986) on the problem of IBP investigation based on the SDCM was proposed to compare alternative object structures (Fridman et al. 2004). It is aimed at decision making support by choosing one of the possible structures to realize the object under investigation.

After constructing the object tree (1), the resource specification begins (Fridman et al. 2004). The resources providing object interaction must be specified first. To do so, one must define quality criteria, lists of input and output resources for every object. Forming of every input resource for every object is necessary for the model performance, so every element of lists of input resources for primitive objects must either be an external resource, or directly come from GIS, or belong to the lists of output resources of another primitive object.

External resources may come either from the ISMS database ("subject domain database" – SDDB) or from an explicitly defined time function, or from a software module with a single input, namely time. Particularly, it is a good way to form different scenarios of input resources' variations for the modelling object. By the end of link specification every input resource (excluding external data) of every model object must be assigned with objects generating this resource and every resource must be assigned with an executor.

Fig. 2 illustrates data subsets (ovals) used for description of the main ISMS subsystems (boxes).

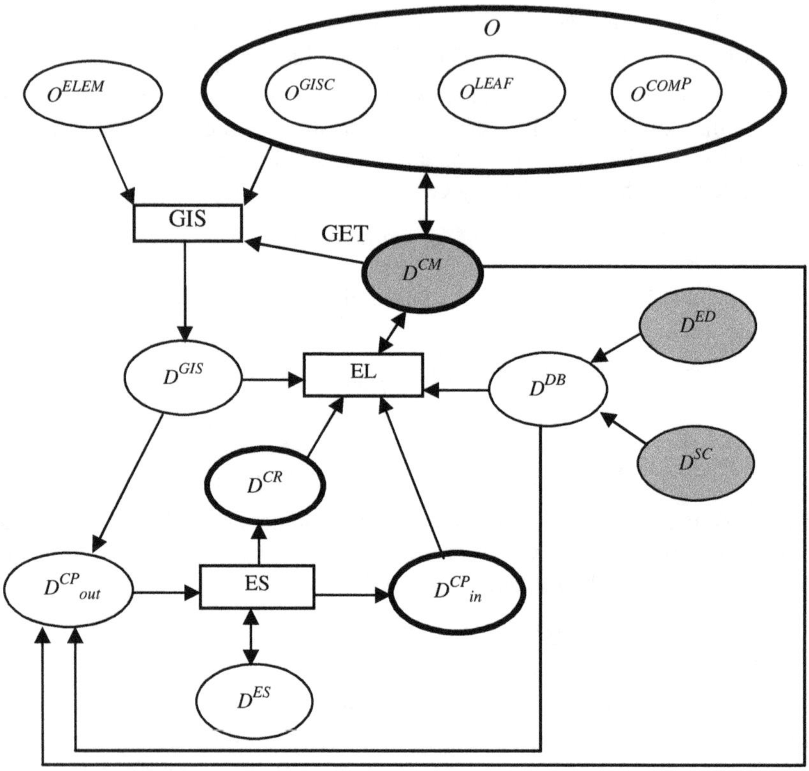


**Fig. 2** Data exchange among the ISMS subsystems

Shaded data sets belong to the SDDB, and ones outlined in bold belong to the SDCM. The SDDB and the SDCM share elements of the set $D^{CM}$ (8). EL means executors' library.

Considering (8), the data set (5) can be partitioned into the following subsets:

$$D ::= D^{CM} \cup D^{ES} \cup D^{GIS} \cup D^{DB} \cup D^{C}, \tag{14}$$

where: $D^{CM}$ - data used within the SDCM;

$D^{ES}$ - internal data of the embedded ES;

$D^{GIS}$ - graphical parameters of model objects measured by the GIS;

$D^{DB} ::= D^{ED} \cup D^{SC}$ – external data comprising experimental data and scenarios' data;

$D^{C}$ – common data used by both the SDCM and the ES.

All elements of the set $D^{GIS}$ belong to the set Res (7) by definition; elements of the rest sets may have both VAR and PAR categories (6).

The set $D^{C}$ can be partitioned into two disjoint subsets, namely the set of resources executed by the ES ($D^{CR}$) and the set of input and output resources of the processes executed by the ES ($D^{CP}$). The last set comprises two possibly overlapping subsets of input and output resources:

$$D^{C} ::= D^{CR} \cup D^{CP}, \;\; D^{CR} \cap D^{CP} = \varnothing, \tag{15}$$

$$D^{CP} ::= D^{CP}_{in} \cup D^{CP}_{out}. \tag{16}$$

The SDDB maintains all elements of $D^{DB}$ as well as the read-only elements of $D^{CM}$ calculated during simulations and intended for comparing with other results. Particularly, the SDDB stores parameters of *GIS-elements' transformations* (GET in the Fig. 2) resulted from simulations. Relationships (10) – (13) provide the basis for establishing (during the SDCM constructing) and supporting (at any admissible modifications of the model) a unique correspondence between conceptual and graphic representation of the IBP under investigation. This way we provide automation of mapping and graphic comparison of modelling results as well as allow for prompt changes of the model proceeded from previous modelling steps. Classification (14) - (16) is needed for model analysis.

As soon as the resource specification is over, one must specify processes generating output resources of objects.

The SDCM constructing ends with description of internal processes assigned to model objects when the SDCM connectedness shall be achieved for every resource.

## 5 Processing of Situations

The completed SDCM provides comparative analysis of IBP alternative structures. A specific extension of the situational approach (Pospelov 1986) on the problem of investigation of spatial dynamical complexes based on the SDCM was proposed to compare alternative object structures (Fridman et al. 1998). It is aimed at decision making support by choosing one of possible structures to realize the object under investigation. The modelling process steps as follows (Fridman A. and Fridman O. 2007).

The atomic information in the ISMS is a *fact* containing any actual or desirable values of a certain resource. For IBP it can include, for instance, features of a product for sale or a utility to buy. A finite set of facts inputted by a user constitutes an *initial situation*. It is interpreted as a task to be achieved at a certain IBP structure.

So, as soon as a user has defined an initial situation by either choosing a list of the SDCM resources or a map area of interest to him/her, the automatically found *decisive object* with its dominated connected fragment of the model will determine the *complete situation* for the given problem. This fragment must comprise all information necessary to investigate the problem, and the level of the decisive object in the object tree defines the organisational level of decision making. The fragment can be specified by a root object ${}_{\alpha_0}O$ and a set of leaf objects:

$$O^L ::= \{o_i^L\} ::= \bigcup_{\alpha=\alpha'}^{\alpha_0+1} O_\alpha^L . \tag{17}$$

To calculate parameters of the fragment (17), we need few more definitions.

*Definition 1.* Process generating relation $PP \subseteq P \times B(P)$ assigns every SDCM process with all other processes generating resources, which are immediately used as input resources of the given process.

*Theorem 1.*

$$\boldsymbol{pp}\ (p_i) = \boldsymbol{op}\ (\boldsymbol{po}\ (p_i) \setminus \{oa^{-1}\ (p_i)\}) \tag{18}$$

The scheme of a fragment can be defined similarly to (8):

$$S_{Fr} ::= <O_{Fr}, P_{Fr}, H_{Fr}, OP_{Fr}, PO_{Fr}, U_{Fr}>. \tag{19}$$

*Theorem 2.* A fragment includes the following set of objects and processes:

$$O_{Fr} ::= O_{Fr}^{pt} \cup O_{Fr}^{add} = \boldsymbol{oa}^{-1}(P_{Fr}^{add}) \cup O_{Fr}^{pt}; \tag{20}$$

$$\begin{aligned} P_{Fr} &::= P_{Fr}^{pt} \cup P_{Fr}^{add} \cup P_{Fr}^{srv} = \\ &= P_{Fr}^{pt} \cup (\boldsymbol{pp}^*(P_{Fr}^{pt}) \cap (\boldsymbol{pp}^*)^{-1}(P_{Fr}^{pt})) \cup \boldsymbol{oa}(O_{Fr}^{add}), \end{aligned} \tag{21}$$

where: $O_{Fr}^{pt} ::= \bigcup_{\alpha=\alpha'}^{\alpha_0} O'_\alpha$ is the set of participating objects,

$O'_{\alpha_0} ::= \{{}_{\alpha_0}o\}$, $O'_{\alpha'} ::= O^L_{\alpha'}$, $O'_{\alpha-1} = h_\alpha^{-1}(O'_\alpha) \cup O^L_{\alpha-1}$,

$\alpha = \alpha_0 + 1,\ \alpha_0 + 2,\ \ldots,\ \alpha'$;

$P_{Fr}^{pt} = \boldsymbol{oa}\ (O_{Fr}^{pt})$ is the set of participating processes;

$P_{Fr}^{add} = (\boldsymbol{pp}^*\ (P_{Fr}^{pt}) \cap (\boldsymbol{pp}^*)^{-1}(P_{Fr}^{add})) \setminus P_{Fr}^{pt}$ is the set of additional processes;

$O_{Fr}^{add} = \boldsymbol{oa}^{-1}(P_{Fr}^{add}) \setminus O_{Fr}^{pt}$ is the set of additional objects;

$P_{Fr}^{srv} = \boldsymbol{oa}(O_{Fr}^{add}) \setminus P_{Fr}^{add}$ is the set of service processes.

Here $PP^*$ stands for the transitive closure of the relation $PP$ (Def.1). The minimal fragment defined above has been proved to exist and be unique for every possible task setting in the ISMS. Relations (17) – (21) make the basis for procedures of initial situation extending and model consistency analysis.

Any complete situation possibly includes some structure alternatives as it may contain some objects with OR-decomposition or alternative sets of resources.

Each of the alternative structures is called a *sufficient situation*, it can be classified by comparing with other sufficient situations for the same decisive object in statics. To do so, we developed a hierarchical system of criteria (Fridman A. and Fridman O. 2007) allowing to choose a sufficient situation or few situations preferable for optimizing a quality criterion for the decisive object. To learn probable consequences of the chosen alternatives, a decision maker (DM) may proceed with a simulation of every sufficient situation. Then the modelling system will generate an appropriate invoking of executors (from EL, see Fig.1) for elements of this situation and provide them with all necessary and properly updated information during the simulation. This way it is easy to investigate different variants of the IBP future by modelling the whole IBP system or its any part. A *scenario* is formed by a series of sufficient situations for the same decisive object. Scenario simulation is possible both with and without automatic classification of situations.

The most important benefits of using the introduced conceptual modelling technology are as follows:

- it supports a hierarchical model of a spatial dynamic complex and the model is open for prompt editing;
- it allows to use the terms and expert notions on the subject domain familiar and clear to a DM while forming formal description of the problems to be investigated;
- it provides integration of the general expert knowledge on the subject domain into a single formal model;
- it contains automatic means for a unity and completeness control of the integrated conceptual model;
- it allows for automation of forming the computing experiment' executive environment.

The opportunity to apply the subject domain terminology for formal description is provided by a special conceptual model editor. While forming the model, a user is just to fill in the attribute description of the conceptual model elements. The "translation" of the user's description into the conceptual model units is realised automatically.

Integration of the considered problem descriptions, which are presented by different experts, is realised by "joining" the model fragments (sub-trees) formed by them into a general model using some appropriate hierarchy relations (9). In the course of conceptual model constructing, users are provided with reference directories of the subject domain elements (objects, processes and resources) already included into the model as well as with electronic maps to identify the geographical layout of the described object. These means used during model forming simplify the consequent co-ordination of the presented descriptions and ensure the model unity.

The conceptual model is characterized as a declarative one imposing no sufficient limits on the range of methods and implementation means for computing experiments. Thus, different modelling methods can be involved into data processing by the conceptual model, namely analytical, statistical, logical, simulating ones. The modelling system can easily integrate convenient and conventional reliable solutions. Above this, the conceptual modelling technology allows to identify the model components necessary to use for a certain problem solution and to specify a structure appropriate for modelling realization environment.

Selection of the conceptual description fragment relevant to a certain problem is delivered by the special algorithms intended for complete definition of the single solution structure. The editor marks some nodes (initial, goal, intermediate ones) on the model graphs. Then extra-definition procedures analogous to the solvability analysis algorithms follow to extend the set of model elements in order to provide the declarative solvability of the given problem considering the set up initial conditions.

The marked conceptual model fragment guides specification procedures to synthesise the executing environment for simulating of the problem presented.

The conceptual model structure as well as its software means enable one to form a common model of IBP by evoking some models of the ready subsystems. Similar in formal structure processes related to different problems or objects of the subject domain invoke one executor according to the reuse principle for standard solutions.

The access to the integrated SDCM enables a user to find and readily involve any earlier worked out fragments meeting his interests. Moreover, a DM is apt to develop his purpose models using models developed by experts in various subject domains. Thus, the DM can integrate knowledge about different features of any object declared in the conceptual model of the subject domain and considered by experts within their certain problems.

The described approach will bring the opportunity to significantly automate considering of spatial-depending restrictions and expenses caused by a IBP reconfiguration, particularly shipping and power-related expenses.

Having the SDCM described, we can proceed with introducing the algorithms "filling" the model with information.

## 6 Structural Dynamics Control for IBP and SC

The IBP model introduced above can be extended to the problem of comprehensive planning in a spatially distributed system of real enterprises in order to create VEs (Arkhipov et al. 2004). This problem is considered within a class of IBP structural dynamics control problems and requires a multi-criteria optimization of IBP performance as well as reallocation of IBP control functions among production network nodes. Basics of our approach to structural dynamics control are given it this section.

The general technology of supply chain tactical-operational planning and the technology of supply chain structural dynamics control include the following stages:

- structure-functional synthesis of the supply chain structures, plans, and schedules;

- structural and parametric adaptation of planning models, as well as algorithms for the past and present states of the supply chain and for the past and present states of the environment;
- supply chain scheduling, control program construction for supply chain structure dynamics, simulation of possible scenarios of supply chain functioning according to the schedule, structural and parametric adaptation of the schedule, models and algorithms for future supply chain states and environment states (predicted via simulation models).

The mathematical models presented in this section are an extended application of the models of complex technical systems (Kalinin 1981, Kalinin and Sokolov 1985, Skurikhin et al. 1989, Kalinin and Sokolov 1996, Sokolov and Yusupov 2004; Okhtilev et al. 2006) for the SCM domain.

One of the main supply chain features is the multiple structure design and changeability of structural parameters because of objective and subjective factors at different stages of the supply chain life cycle. At different stages of the supply chain evolution, the elements, parameters, and structural interrelations change. In these settings, a supply chain can be considered as a *multi-structural process*.

The main supply chain structures are as follows:

- product structure (bill-of-materials),
- functional (structure of management functions and business processes),
- organizational (structure of facilities, enterprises, managers and workers),
- technical-technological (structure of technological operations for product production and structure of machines, devices, etc.),
- topological (geographical) structure,
- informational (structure of information flows according to the coordination strategy), and
- financial (structure of supply chain costs and profit centres).

According to the theory of structural dynamics control (Okhtilev et al., 2006), the control is composed of both state and structure control. The proposed approach to the problem of supply chain control in terms of general context of supply chain structure dynamics control enables:

- general goals of supply chain functioning to be directly linked with those implemented (realized) in supply chain control process,
- reasonable deciding and selection (choice) of an adequate consequence of problems solved and operations fulfilled related to structure dynamics to be made (in other words, synthesis and developing of the supply chain control method),
- a compromise distribution (trade-off) of restricted resources appropriate for structural dynamics' control to be found voluntary.

The main idea of structural dynamics control is based on the functional-structural approach to describing objects of various nature. At the same time, problems of structural dynamics control are the generalization of structural-functional synthesis problems, which are traditionally formulated in the domain of automation of complicated technical-organizational complexes.

Let us introduce the main definitions.

*Supply chain macro-state* is a general supply chain state that can occur in a number of supply chain objects.

*Structural state* is a supply chain macro-state that reflects the current states of objects in a supply chain structure as well as interrelations among these objects.

*Multi-structural macro-state* is a supply chain macro-state that reflects the current states of objects and structures in supply chains as well as interrelations among them.

*Structure dynamics* is a process of supply chain structure transition from one planned macro-state to another.

*Structural dynamics control* is a process of producing control inputs and implementing the supply chain transition from the current macro-state to a given one.

The synthesis (selection) of technical-organizational structure (in our case, a supply chain structure/structures) is usually reduced to the following general optimization problem (Zvirkun and Akinfeev 1993), (Kreipl and Pinedo 2004):

$$\bar{S}\left\{\left[\bar{f} \subset \bar{F}(\bar{\pi})\right]\bar{R}\left[\bar{m} \subset \bar{M}\right]\right\} \rightarrow extr \tag{22}$$

$$\bar{\pi} \subset \bar{P} \tag{23}$$

$$\bar{f} \subset \bar{F}(\bar{\pi}) \tag{24}$$

$$\bar{m} \subset \bar{M} \tag{25}$$

where $\bar{P}$ is a set of feasible control principles (algorithms) and $\bar{F}$ is a set of interrelated functions (tasks, operations), which can be performed by the system. For each subset $\bar{\pi} \subset \bar{P}$ there exists a set $\bar{F}(\bar{\pi})$ from which the realizations sufficient for the given principles (algorithms) should be chosen, i.e., it is necessary to choose a $\bar{f} \subset \bar{F}(\bar{\pi})$; $\bar{M}$ is a set of supply chain possible elements such as information processing and transceiving facilities, control units, service terminals, etc; the map $\bar{R}$ takes $\bar{F}$ to $\bar{M}$.

It is stated that the optimal map $\bar{F}$ delivers an extremum to some objective function (functions) $\bar{S}$ under given constraints. The modifications of the considered problem concern some aspects of uncertainty and multi-criteria decision-making. The complexity of the synthesis problem (22) – (25) is mainly caused by its high dimensionality due to the number of variables and constraints in the detailed problem statement. That is why the methods of decomposition, aggregation and sub-problem coordination are widely used. Another peculiarity complicating the problem is the integer-valued variables.

In this section, we will present a complex of supply chain control models. These models provide a unified technology for an analysis and optimization of various processes concerning supply chain planning and execution. The main advantage of the constructed models is that structural and functional constraints for supply chain control are defined explicitly. The common conceptual basis facilitates constructing of a complex of unified dynamic models for supply chain control. The models describe functioning supply chains along with collaboration processes in them. The unified description of various control processes allows to synthesize different supply chain structures simultaneously. Moreover, it allows to establish dependence relation among control technology applied to supply chains and the supply chain management goals. This statement is exemplified by an analysis of supply chain functional abilities and goal abilities. It is important that the presented approach extends new scientific and practical results obtained in the modern control theory for the SCM domain. By now, the prototype programs realizing these models have been developed and used for evaluation of supply chain goal abilities and for planning of supply chain operations.

Now we introduce some notation for problem definition. Let $A = \{A_i, i \in N=\{1,...,n\}\}$ be a set of business processes (and corresponding control functions) to be implemented in a certain node of a SC at a given time interval $T = [t_0, t_f]$. To achieve the SC goals during the interval T, some certain IBP have to be fulfilled. We distinguish between functions of goal definition, planning (long-term and operational planning), real-time control, SC state analysis, external situation analysis and coordination. The set $A$ is related to sets of informational-technological operations $D^{(i)} = \{D^{(i)}_{æ}, æ \in K = \{1,..., s_i\}\}$, that are necessary for implementation of an IBP $A_i$, $i = 1,...,n$. Let $B=\{B_j, j \in M=\{1,...,m\}\}$ be a set of a VE main elements and subsystems. Each element $B_j$ can include technical facilities $C^{(j)} = \{C^{(j)}_{\lambda}, \lambda \in L = \{1,...,l\}\}$ with appropriate computer equipment and software. Technical facilities are used for implementation of control functions.

Let $E(t) = \|e_{ij}(t)\|$ be a known matrix function, with $e_{ij}(t)=1$ in case if the subsystem $B_j$ is carrying out the function $A_i$ at the time $t$ in accordance with time-spatial, technical and technological constraints, $e_{ij}(t)=0$ otherwise.

Now a verbal description of a functions-distribution problem can be presented as follows. It is necessary to select the best variants of functions' distribution among the nodes of the SC for each structural state $R_1, R_2, ..., R_k$ (under known time, spatial, technical and technological constraints) and to find the best variants of the functions' implementation. The structural states of the SC should be sorted according to their preference. The preference relation can be expressed through quality functions characterizing efficiency of the SC and its structural and technologic characteristics. Most of these quality functions can be calculated by means of the embedded GIS described above in the section 4. The GIS also provides a natural basis for sorting and ordering of all SC components thus forming the topological (geographical) structure of the SC.

The described problem belongs to the class of multi-criteria choice problems with finite sets of alternatives (structural states of the SC).

## 7 Multi-Criteria Planning of SC Operations

The general algorithm for the problem solving includes the following steps.

*Step 1.* Models (analytical, simulation and combined ones) describing structural states $R_1, R_2, ..., R_k$ are used for optimal distribution of IBP and control functions among subsystems of the SC, for planning of technological operations and for evaluation of the SC efficiency. The following characteristics of SC efficiency can be used: the total number of functions implemented in subsystems during the interval T, the total number of IBP in given macro-states, the total number of technological operations executed over the time interval T, the total time of operations over the time period T. The above-mentioned characteristics can have stochastic or fuzzy interpretation if uncertainty factors are present (Okhtilev et al 2006, Orlovski. 1981).

The following dynamic model of functions' distribution can be used for evaluation of VE efficiency (Okhtilev et al 2006, Kalinin and Sokolov 1995, Zimin and Ivanilov 1971).

$$\dot{x}_i^{(\phi)} = \sum_{j=1}^{m} \varepsilon_{i\,j}(t) u_{i\,j}^{(\phi)};\ \dot{x}_{iæ\,j}^{(0)} = \sum_{\lambda=1}^{l} b_{i\,æ\,j\lambda} u_{i\,æ\,j\lambda}^{(0)};\quad \dot{y}_{i\,j}^{(\phi)} = v_{i\,j}^{(\phi)}; \tag{26}$$

$$\sum_{j=1}^{m} u_{i\,j}^{(\phi)} \left[ \sum_{\alpha \in \Gamma_{i1}} \left(a_\alpha^{(\phi)} - x_\alpha^{(\phi)}\right) + \prod_{\beta \in \Gamma_{i2}} \left(a_\gamma^{(\phi)} - x_\gamma^{(\phi)}\right) \right] = 0; \tag{27}$$

$$\sum_{\lambda=1}^{l} u_{i\,æ\,j\lambda}^{(0)} \left[ \sum_{\nu \in \Gamma_{i\,æ\,1}} \left(a_{i\,\nu\,j}^{(0)} - x_{i\,\nu\,j}^{(0)}\right) + \prod_{\mu \in \Gamma_{i\,æ\,2}} \left(a_{i\,\mu\,j}^{(0)} - x_{i\,\mu\,j}^{(0)}\right) \right] = 0; \tag{28}$$

$$\sum_{i=1}^{n} u_{i\,j}^{(\phi)}(t) \le 1;\ \forall j;\ \sum_{j=1}^{m} u_{i\,j}^{(\phi)}(t) \le 1; \forall i;\quad u_{i\,j}^{(\phi)}(t) \in \{0,1\}; \tag{29}$$

$$\sum_{j=1}^{m}\sum_{\lambda=1}^{l} u_{iæ\,j\lambda}^{(0)}(t) \le 1, \forall i, \forall æ;\ \sum_{i=1}^{n}\sum_{æ=1}^{s} u_{iæ\,j\lambda}^{(0)}(t) \le 1,\ \ \forall i, \forall æ;\ u_{iæ\,j\lambda}^{(0)}(t) \in \{0, u_{i\,j}^{(\phi)}\}; \tag{30}$$

$$v_{i\,j}^{(\phi)}\left(a_{i\,s\,j}^{(0)} - x_{i\,s\,j}^{(0)}\right) = 0;\ v_{i\,j}^{(\phi)}(t) \in \{0,1\}; \tag{31}$$

$$\begin{gathered} x_i^{(\phi)}(t_0) = x_{i\,æ\,j}^{(0)}(t_0) = y_{i\,j}^{(\phi)}(t_0) = 0; \\ x_i^{(\phi)}(t_f) = a_i^{(\phi)};\ \left(a_{iæ\,j}^{(0)} - x_{i\,æ\,j}(t_f)\right) y_{i\,j}^{(\phi)}(t_f) = 0; \end{gathered} \tag{32}$$

$$x_i^{(\phi)}(t_f) = a_i^{(\phi)};\ \left(a_{i æ j}^{(0)} - x_{i æ j}(t_f)\right) y_{i j}^{(\phi)}(t_f) = 0; \tag{33}$$

$$J_0 = \sum_{i=1}^{n} \sum_{j=1}^{m} v_{i j}^{(\phi)}(t_f);\ J_1^{(n)} = \sum_{i=1}^{m} v_{n j}^{(\phi)}(t_f);$$

$$J_0 = \sum_{i=1}^{n} \sum_{j=1}^{m} v_{i j}^{(\phi)}(t_f);\ J_1^{(n)} = \sum_{i=1}^{m} v_{n j}^{(\phi)}(t_f); \tag{34}$$

where $x_i^{(\phi)}(t)$ is equal to total duration of the business process $A_i$ fulfillment in the subsystem $B_j$ as $u_{i j}^{(\phi)}(t) = 1$; the variable $x_{i æ j}^{(0)}$ expresses the current state of the technological operation $D_{æ}^{(i)}$; $y_{i j}^{(\phi)}$ is equal to the time that passed after $A_i$ completion in $B_j$ until the time $t = t_f$ ; $a_{\alpha}^{(\phi)}, a_{\alpha}^{(0)}, a_{\gamma}^{(0)}, a_{i \nu j}^{(0)}, a_{i \mu j}^{(0)}$ are given values setting end conditions for $x_i^{(\phi)}(t), x_{\alpha}^{(\phi)}(t), x_{\gamma}^{(\phi)}(t), x_{i \nu j}^{(0)}(t), x_{i \mu j}^{(0)}(t)$ at the time point $t = t_f$; $u_{i j}^{(\phi)}, u_{i æ j \lambda}^{(0)}, v_{i j}^{(\phi)}$ are control inputs. Here $u_{i j}^{(\phi)}(t)=1$ if an IBP $A_i$ is being executed in the subsystem $B_j$ at the time $t$, $u_{i j}^{(\phi)}(t)=0$ otherwise; $u_{i æ j \lambda}^{(0)}(t)=1$ if the technological operation $D_{æ}^{(i)}$ is executed in the technical facility $C_{\lambda}^{(j)}$, $u_{i æ j \lambda}^{(0)}(t)=0$ otherwise; $v_{i j}^{(\phi)}=1$ if an IBP $A_i$ was implemented in the subsystem $B_j$, $v_{i j}^{(\phi)}=0$ otherwise. Here the sets $\Gamma_{i1}$, $\Gamma_{i2}$ include the number of functions that are direct predecessors of the control function $A_i$. The set $\Gamma_{i1}$ indicates predecessors connected by logical "and", the set $\Gamma_{i2}$ indicates predecessors connected by logical "or". The sets $\Gamma_{i æ 1}$, $\Gamma_{i æ 2}$ include the number of technological operations $D_{\nu}^{(i)}$ and $D_{\mu}^{(i)}$ that are direct predecessors of the operation $D_{æ}^{(i)}$. The subscripts 1 and 2 express the type of logical connection as stated above.

Therefore, constraints (27) and (28) define allowable sequences of control functions and technological operations. Constraints (29) and (30) specify that each IBP at each time point can be carried out only in one subsystem $B_j$ ($i$=1,...,$n$; $j$=1,...,$m$) and conversely, each subsystem $B_j$ can carry out only one IBP $A_i$ at the same time. Similar constraints are used for technological operations $D_{æ}^{(i)}$ that are executed at the technical facility $C_{\lambda}^{(j)}$.

Expression (31) states switching-on conditions for the auxiliary control input $v_{i j}^{(\phi)}(t)$. Expressions (32) and (33) specify end conditions for the state variables at the time $t = t_0$, $t = t_f$, $R^1$ is a set of positive real numbers. The functionals $J_0$, $J_1$, $J_2$

are quality measures for distribution of IBP in the SC. Here $J_0$ is equal to total number of functions by the time $t = t_f$, $J_1$ is equal to the number of subsystems the function $A_i$ is implemented in, $J_2$ expresses the elapsed time for implementation of all necessary functions.

A simulation model of real-time control can be used together with expressions (26)-(34) for taking uncertainty factors into account. In this case, special procedures of inter-model coordination can be used (Okhtilev et al 2006, Kalinin and Sokolov 1995).

Extreme values of functionals characterizing SC efficiency can be determined via solution of optimal control problem for finite-dimensional differential system with mixed conditions. The solution algorithms and different aspects of their programming are considered in (Okhtilev et al 2006, Kalinin and Sokolov 1995).

*Step 2.* Structure-topological characteristics of the SC are being evaluated (Zimin and Ivanilov 1971) including: the coefficient of attainability $J_4$, different measures of structure compactness (radius $J_5$ of the structure, diameter $J_6$ of the structure, integral measure $J_7$ of structural compactness), measures $J_8$ of structure centralization (decentralization).

The formulas for computation of these measures are proposed in (Zimin and Ivanilov 1971).

*Step 3.* The pairwise-comparison matrix $K_c$ is completed for measuring SC efficiency. Expert appraisal is used for completion of the matrix.

*Step 4.* The weights of measures (significance coefficients) are evaluated according to the matrix $K_c$. The algorithm proposed in (Orlovski 1981) is used here. The vector of coefficients is equal to the normalized eigenvector $\vec{\omega}_c$ corresponding to the maximal eigenvalue $L_{max}$ of the matrix $K_c$. Thus the following equation has to be solved:

$$(K_c - L_{max} * I)\,\vec{\omega}_c = 0, \tag{35}$$

where $I$ is a unitary matrix.

Then the weight of each structural state $(R_1, R_2, \dots, R_k)$ of the SC for each measure taken separately is evaluated. These weights complete the matrix $K_r$. Each column of the matrix $K_r$ includes relative weights of the states in respect of some measure. A weighted sum of measures is received for each alternative $R_1, R_2, \dots, R_k$. In other words, total sets of weights are determined for each structural state via the formula:

$$K_r\,\vec{\omega}_c = \vec{\omega}^* . \tag{36}$$

*Step 5.* The structural states are sorted according to their preference. The best one is characterized with the maximal element of the vector $\vec{\omega}^*$ (36). Each element of this vector can be interpreted as the total weight of some structural state.

## 8 Risk Assessment in SC

In the presented approach, *supply chains are selected and scheduled taking risk factors into account* (Sokolov and Fridman 2008). Here we briefly describe the problem of risk management in SCs. Risk assessment is done in terms of system measures of planned *reliability and stability*. These measures are introduced in order to form and constrict the set of Pareto-optimal supply chains. Risk factors are considered as an additional criterion of supply chain selection that allows re(configuration) of supply chains considering the probability of deviations from customer's order parameters.

Let a number of possible SC configurations at a moment $t$ be $S = \{(S_i), i \in N\}$. Let a system of coordinates be ($X$, 0, $C$). A parameter set for an $i$-SC in this system is the point ($x(t)$, $c(t)$, see Fig. 3). To solve a problem of scales in this system, it is necessary to substitute absolute indicators of time and costs with relative ones. Then we mark a number of alternative SCs (macro-states) selected with a planning algorithm on the graph. After that we mark the "ideal point" (ideal state). This point reflects client's order parameters (with coordinates (1;1)). Taking into account maximal acceptable deviations $\delta$ from these parameters (defined by a client and a network manager), we mark the area of acceptable SC states by the restrictive lines. Let it be the "attainable area" (AA). Every SC within this area may be chosen as a final one.

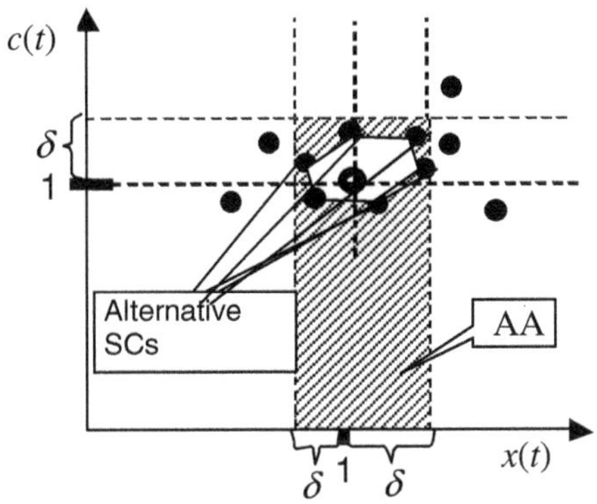


**Fig. 3** Attainable area and alternative SCs

Each alternative SC is characterized (above parameters $<x(t), c(t)>$) by a risk level $q$. Let it be the concept of the "economically accepted risk" $q$, which is determined by the network manager. All the points (SCs) above this level are rejected.

We introduce a concept of the average risk level $Q(t)$ for the AA, which is defined as the arithmetical mean of risks for selected alternative SCs. Then we determine a SC, which has a risk level closest to the average one, i.e. $|q_i(t) - Q\ (t)| \rightarrow \min$. This SC is to be selected as a plan for the VE. Its reliability $R_i$ is determined as $R_i(t) = 1 - q_i(t)$.

## 9 Modelling of Safety and Reliability

Our main hypothesis is that modelling of non-reliable and dangerous situations ought to be based on a model of normal functioning of an IBP since dangers arise from ordinary modes either when some parameters of a system exceed their limits or when a forbidden combination of some parameters occurs. So, when modelling dangers and risks, we suppose the following:

- the list of values (mentioned in the section 4) for every resource contains admissible values only and hence may be called a *safety range* (SR);
- one of the admissible values corresponds to the best performance of the process generating the given resource, this value is called the nominal one;
- if a resource can cause any danger, we can model the extent of this danger by expanding its SR with additional list of dangerous and critical values;
- dangerous values do not result in an accident, they only warn about abnormal or hazardous functioning whereas critical values correspond to an accident (there may be no more than two critical values for each resource);
- dangerous and critical values can be indexed (objectively or by experts) according to the extent of their danger: the farther a value is from the SR, the more dangerous it is (see Fig. 4).

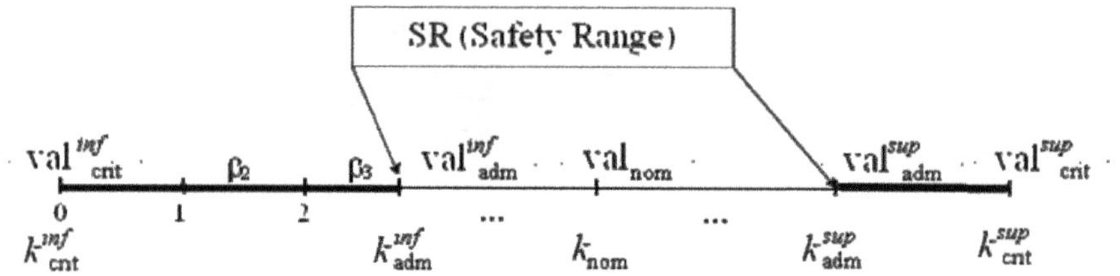


**Fig. 4** List of values of a resource

In this case, the total danger of an object or a situation may be estimated as a (weighted) sum of indices for dangerous values of the resources this object or situation comprises. If experts can weigh the extents of danger for every dangerous value of every considered resource, we can calculate the total danger using the algorithm of specific expenditures' calculating based on the same criterion we use for classification of situations (Fridman A. and Fridman O. 2004).

The criterion looks as follows:

$$\Phi_{ISMS}^{(s)} ::= \left( \frac{1}{m} \sum_{i=1}^{m} \left( \frac{a_i - a_{i0}}{\Delta a_i} \right)^s \right)^{1/s} ::= \left( \frac{1}{m} \sum_{i=1}^{m} \delta a_i^s \right)^{1/s} \qquad (37)$$

where: $s$ is an even positive integer;

$a_i$ –resources from the output list of a model element;

$a_{i0}$ and $\Delta a_i > 0$ – adjusting parameters reflecting requirements of a super-object to the nominal value $a_i$ of a resource and to the acceptable deviation from the value;

$\delta a_i ::= \dfrac{a_i - a_{i0}}{\Delta a_i}$ is the relative deviation of an actual value of a resource $a_i$ from its nominal value $a_{i0}$.

The criterion (37) has two advantages essential for this paper, namely:

- it explicitly reflects requirements of the super-element to a certain model element;
- it provides a natural index for estimation of any state or situation that occurred within IBP since it equals unity when all output resources of a given IBP element have their margin values:

$$\Phi^{(s)}_{ISMS} = 1, \text{ if } \left|a_i - a_{i0}\right| = \Delta a_i,\ i = \overline{1, m}, \text{ that is } \left|\delta a_i\right| = 1. \tag{38}$$

Conditions (38) allow to easily locate a source of a malfunctioning. This source is in performance of an element that is the lowest in the hierarchy of objects (1) and has the value of its criterion (37) far above unity.

To estimate quality of a certain resource producing, we proposed to use generalized expenditures (39) for every argument of the criterion (37):

$$\eta_i ::= \Phi^{(2)} \delta a_i + \frac{1}{m} \sum_{j=1}^{n} \eta_j , \tag{39}$$

where $\eta_j$ are expenditures for producing input resources of the given model element calculated according to (39) as well.

Generally, the ISMS distinguishes three groups of breakages and dangers (*initiating events* - IE), namely *functionally determined* (IE1), *spatially determined* (IE2), *temporally determined* (IE3) *and combined* IE (IE4 - spatial-temporal IE, compositions of IE2 and IE3). Within every group, IEs differ in danger levels assigned to them depending on the calculated total weight index for every object related to the given situation.

IE1 can be split into two subgroups, namely:

IE11 – IE1 is dangerous for the object consuming the resource, emerging condition is

$$(\forall res_m \in list\_out(o_i))\ (res_m \notin SR(res_m)); \tag{40}$$

IE12 – IE1 is dangerous for the object producing the resource, emerging condition is

$$(\forall res_m \in list\_in\ (o_i))\ (res_m \notin SR(res_m)). \tag{41}$$

In (40), (41) and below *list_in*(*) and *list_out*(*) stand for lists of input and output resources of a model element with the name *.

Safety logical model (SLM) for every IBP object considers a possibility of emerging of every type of initiating events. Denoting indicators of their emergence, according to (Ryabinin 2000), as $z$ with corresponding subscripts, we construct the Dangerous State Function (DSF) for every object $o_i$ as

$$y(o_i) = z_1(o_i) \vee z_2(o_i) \vee z_3(o_i) \vee z_4(o_i), \tag{42}$$

where:

$$z_1(o_i) = \left( \bigvee_{\text{res}_l \in \text{list\_out}(o_i)} z(\text{res}_l) \right) \vee \left( \bigvee_{\text{res}_l \in \text{list\_in}(o_i)} z(\text{res}_l) \right) \tag{43}$$

describes IE11 and IE12 respectively, and disjunctions in (43) include only the input and output resources of the object $o_i$ impacting on its safety and the safety of neighboring objects. Emerging of those IEs is declared by checking conditions (40), (41);

$$z_2(o_i) = z_2(\underline{\underline{h}}^{\alpha}(o_i)) \tag{44}$$

describes IE2 as a function from graphic attributes of the GIS-representation of the object $o_i$ depending in its turn upon the subordination set (10) of this object.

To model IE3 and IE4, it is necessary to consider a fairly wide class of spatial-temporal interrelations among characteristics of model objects. This can require extending the SDCM by adding some special auxiliary processes describing those interrelations to form $z_3(o_i) \vee z_4(o_i)$. It is also possible to use logical spatial-temporal functions (Fridman A. and Fridman O. 2007) embedded into the ES of the ISMS. Anyway, creating of the SLM cannot be automated completely though the above specified categories of IEs make the procedure easier.

As for reliability modelling, we base it on logic-probabilistical methods (for instance, Lisniansky and Levitin 2003). So, we assign every resource $\text{res}_m$ with a logical variable $x(\text{res}_m)$ reflecting the *structural reliability* of its producing. This variable depends on both the reliability of the model element producing this resource and the reliability of providing this element with its input resources:

$$(\forall \text{res}_m \in \text{list\_out}(o_i))\ (x(\text{res}_m) = x(o_i) \wedge \left( \bigwedge_{\text{res}_l \in \text{list\_in}(o_i)} x(\text{res}_l) \right)); \tag{45}$$

$$(\forall \text{res}_m \in \text{list\_out}(p_j))\ (x(\text{res}_m) = x(p_j) \wedge \left( \bigwedge_{\text{res}_l \in \text{list\_in}(p_j)} x(\text{res}_l) \right)), \tag{46}$$

In (45), (46) and below logical sums and products include only resources needed for a model element.

Then we form an algebraic logical function (ALF) (Ryabinin 2000) describing the total reliability of a model object as follows (Fridman and Yakovlev 2003).

$$x(o_i) = x_{\text{in}}(o_i) \wedge \left( \bigwedge_{\text{res}_l \in \text{list\_out}(o_i)} x(\text{res}_l) \right), \tag{47}$$

where the first component formalizes failures, which cannot be assigned to any process belonging to the object. Reliability of any connected fragment of the IBP SDCM can be represented similarly to (47).

If necessary, it is easy to consider the possibility of failures during transferring of resources among objects since they have a geographic siting within the SDCM:

$$(\forall((\text{res}_m \in \text{list\_in}(o_i) \wedge (\text{res}_m \in \text{list\_out}(o_k))) (x_{\text{in}}(\text{res}_m) = x_{\text{out}}(\text{res}_m) \wedge x_{\text{tr}}(\text{res}_m)), \tag{48}$$

where reliability of transfer for a resource depends upon subordination sets of the respective objects alike (44):

$$x_{\text{tr}}(\text{res}_m) = x_{\text{tr}}(\underline{\underline{\text{h}}}^{\alpha}(o_i) \cup \underline{\underline{\text{h}}}^{\alpha}(o_k)). \tag{49}$$

The set of the embedded operations with GIS-elements (Fridman A. and Fridman O. 2007) supports automatic measuring of necessary graphic parameters (coordinates, distances, areas, etc.) and their input into software modules.

The ALF for estimating reliability for every fragment of IBP can be generated automatically according to (45) – (49) and the following evident rules:

- if there is an alternative implementation of an object (an oi is decomposed into sub-objects oij, $j = \overline{1, n}$ by means of OR-operation, we get from (45):

$$(\forall \text{res}_m \in \text{list\_out}(o_i)) (x(\text{res}_m) = \bigvee_{j=1}^{n} \left[ x(o_{ij}) \wedge \left( \bigwedge_{\text{res}_l \in \text{list\_in}(o_{ij})} x(\text{res}_l) \right) \right]); \tag{50}$$

- if there is an alternative for producing of a resource (it is formalized by specification of an alternative set of resources $\text{set\_alt}(p_j) \subset \text{list\_in}(p_j)$ on the input of a process $p_j$), the reliability of producing for resources from this set by means of (46) considers all possible alternatives (let their number be $m$):

$$(\forall \text{res}_m \in \text{set\_alt}(p_j)) (x(\text{res}_m) = \bigvee_{k=1}^{m} x(\text{res}_{mk})). \tag{51}$$

- for aggregation of some resource sets by their summation on the input of a process $p_j$ (a set of resources $\text{set\_agr}(p_j)$ to be summed must have been specified before), we have, similarly to (51):

$$(\forall \text{res}_m \in \text{set\_agr}(p_j))\,(x(\text{res}_m) = \bigwedge_{k=1}^{m} x(\text{res}_{mk})) \quad (52)$$

Recursive implementation of (45) - (52) starting from leaf resources allows to synthesize the function of the up state (Ryabinin 2000) for the whole IBP system or any its part that relates to a connected fragment of the SDCM. If the model includes cycles of resources, these cycles can be detected by special algorithms (Fridman et al. 2004) preventing malfunctioning from ringing while generating the function of the up state. If a cycle (we call it *elementary cycle*, see Fig.5) comprises a sequence of $n$ processes $p_j$ connected with $n$ internal (with respect to the cycle) resources $\text{res}_j$, which are assigned to $s$ objects $o_i$, and the processes receive $m$ internal resources $\text{res}_l^{\text{in}}$ while generating $p$ output resources $\text{res}_k^{\text{out}}$, then the structural reliability of the cycle can be estimated as follows.

$$x(cycle1) = \left(\bigwedge_{l=1}^{m} x_{\text{in}}(\text{res}_l^{\text{in}})\right) \wedge \left(\bigwedge_{j=1}^{n} x(p_j)\right) \wedge \left(\bigwedge_{i=1}^{s} x_{\text{in}}(o_i)\right) \wedge \left(\bigwedge_{i=1}^{s} x_{\text{tr}}(\text{res}_j)\right) \quad (53)$$

where the last conjunction includes only resources transferred among the objects.

Reliability for every output resource of the cycle we calculate as follows: $x_{\text{out}}(\text{res}_k^{\text{out}}) = x(\text{cycle1})$.

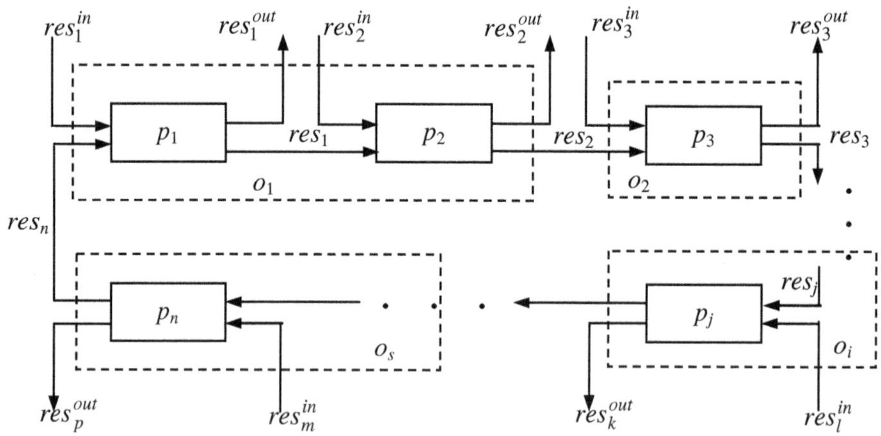


**Fig. 5** The structure of an elementary cycle

There are no fundamental difficulties to generalize (53) for more complicated structures admissible in the SDCM and exemplified in the Fig. 6. To do so, it is necessary to consider the joining order of certain SDCM elements (objects, processes, resources) into a given structure and modify conjunctions in (53) properly. We do not deduce explicit relations here since they are cumbersome.

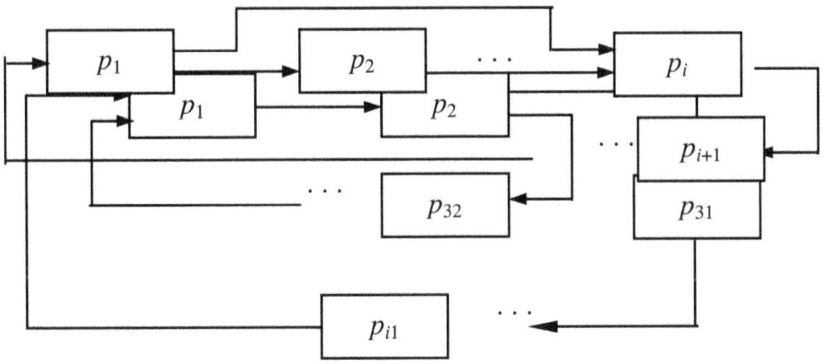


**Fig. 6** Structures of multiple cycles

Risk assessment for an accident initiation and development within integral technical-natural systems that include certain dangerous objects comprises the following main steps:

- revealing most probable reasons of accident and factors contributing to accident initiation and development;
- scenario description for possible accidents;
- quantitative evaluation of hazardous agents responsible for accidents;
- definition of affecting factors for the described accidents;
- probability (frequency) estimation of the described accidents;
- evaluation of possible losses;
- elaboration of action plans aimed at avoiding or eliminating of risks.

To prevent initiation and eliminate after-effects of technogeneous accidents, we propose an approved information technology for decision making support featured with automated synthesis of accident scenarios and risk assessment, which also considers local nature-climatic parameters (Fridman A. and Fridman O. 2007).

Thus, modelling of dangerous and critical situations is accomplished as an "extension" of normal performance modelling and uses the same software environment. This technique gives the opportunity to detail the danger analysis down to every resource, thus revealing the most dangerous situations and scenarios that resulted from combined and multiple failures within IBP. It also allows accumulation of diverse information about IBP for its complex study and usage.

## 10 Conclusions

As for software features, the developing situational modelling system is intended for automation of every modelling stage, wide usage of expert knowledge, employment of the embedded GIS not for object mapping only, but also for task setting, spatial-dependent calculations and displaying modelling results as well. The ISMS provides a DM with a support to ground altering (or not altering) the structure of his/her subordinated object and substitutes an expert council in this sense.

The hierarchical situational conceptual model of IBP is the core of the modelling system. It guides formalization and integration of the general expert knowledge on the subject domain and dynamics of its development.

The most important accomplishments of the conceptual modelling technology and its software means are as follows:

- usage terms and notions of the subject domain that are familiar and clear to an expert while he/she is forming a formal description of the problems to be investigated;
- integration of general expert knowledge on the subject domain into a single formal model; using automatic means for unity and completeness control of the integrated conceptual model;
- automatic forming of the computing experiments executive environment;
- providing comprehensive multi-model and multi-criteria simulation of IBP and SC involving a variety of techniques and methods for different degrees of uncertainty.

The developed modelling system allows using different methods for investigation of insufficiently formalized complex non-stationary spatial objects, ensures complex use of expert knowledge to form criteria and choose alternatives for more detail studying in the simulation mode.

Technologically, the ISMS is to narrow the gap in methods of organizational systems' modelling within the structural approach to the open subject domain model implementation. We have proven that the ISMS allows to model every main hierarchy type considered by the general theory of hierarchical systems (stratified, multilevel and organizational hierarchies).

Application of the ISMS to the subject domain of IBP and SC modelling by means of integrated multi-criteria operations planning in the context of VE structural dynamics control results in the following advantages.

Dynamic interpretation of operation planning in VEs allows to thoroughly describe and investigate interrelation and interaction of business processes and the processes of information processing, storing and interchange.

The goals of VE planning can be directly interrelated with the goals of business processes. Structure dynamics' operations (VE control technology) can be reasonably selected and substantiated.

Efficient compromise solutions can be found for allocation of control functions among the elements of SCs and for general programs (plans) of VE operation. The preliminary ordering of VE structural states allow to rapidly reconfigure the VE structure in case of failures (Okhtilev et al 2006).

Integration of different knowledge representation forms for the purpose of applying modern methods of state investigation and prognosis to complicated non-stationary nature-technical complexes treated as multilevel multi-component spatial objects constitutes the novelty of the approach described above.

Several prototype versions of software have been produced for structural dynamics control of VE in different application domains (cosmonautics, power industry, management, etc, see http://www.spiiras-grom.ru). Experiments with the software confirmed efficiency of the models applied.

**Acknowledgments.** The authors would like to thank the Russian Fund of Basic Researches (grants 09-07-00066, 08-08-00403 and 07-07-00169), Russian Foundation for Humanities (grants 09-02-00636, 09-02-43203a/C and 09-01-12305), the Department for Nanotechnologies and Information Technologies of RAS (project 2.3 of the current Programme of Basic Scientific Researches), and the Chair of RAS (project 4.3 of the Program #3) for their aid in partial funding of this research.

## References

Arkhipov, A., Ivanov, D., Sokolov, B.: Intelligent supply chain planning in ‘virtual organization’. In: Proceedings of 5th IFIP Working Conference on Virtual Enterprises (PRO-VE 2004), Toulouse, France, August 22-27, vol. 8, Part 8, pp. 215–224 (2004)

Athans, M., Falb, L.: Optimal control. An introduction to the theory and its applications. McGRAW-HILL BOOK COMPANY, New York (1963)

Camarihna-Matos, L., Kluwer, et al. (eds.): Virtual Enterprises and Collaborative Networks. Academic Publishers, Leiden (2004)

Fridman, A., Fridman, O.: Logic-analytical situative modelling in discretized state space. In: Management of natural-industrial systems safety, Apatity, vol. V, pp. 6–12 (2004) (in Russian)

Fridman, A., Yakovlev, S.: Situative approach to synthesis of logical models of safety and reliability for industry-natural complexes. In: Proceedings of International Sci. School Modelling and analysis of safety and risk in complex systems (MA SR 2003), St. Petersburg, Russia, August 20-23, pp. 375–381 (2003) (in Russian)

Fridman, A., Fridman, O.: Situative approach to modelling of performance and safety in nature-technical complexes. In: Lindfors, J. (ed.) Applied Information Technology Research – Articles by Cooperative Science Network, University of Lapland, Finland, pp. 44–59 (2007)

Fridman, A., Oleynik, A., Fridman, O.: Knowledge integrating in situative modelling system for nature-technical complexes. In: Proceedings of the European Simulation and Modelling Conference (ESMc 2004), Paris, France, October 25-27, pp. 25–29 (2004)

Fridman, A., Oleynik, A., Putilov, V.: GIS-based simulation system for state diagnostics of non-stationary spatial objects. In: Proceedings of 12th European Simulation Multiconference (ESM 1998), Manchester, UK, June 16-18, vol. 1, pp. 146–150 (1998)

Ivanov, D.A.: Virtual Enterprises and Logistics Chains: Integrated Approach to Organization and Control in New Forms of Production Cooperation. SPbSUEF (2003) (in Russian)

Kalinin, V.N., Sokolov, B.V.: Multiple-model approach to description of control processes in space systems. Control Theory and Systems # 1, 149–156 (1995)

Kreipl, S., Pinedo, M.: Planning and scheduling in supply chains: an overview of issues in practice. Prod. Oper. Manag. 13(1), 77–92 (2004)

Lisniansky, A., Levitin, G.: Multi-state system reliability. Assessment, optimization and applications. Word Scientific, Singapore (2003)

Moiseev, N.N.: Elements of the optimal systems theory. Nauka, Moscow (1974) (in Russian)

Okhtilev, M.Y, Sokolov, B.V., Yusupov, R.M.: Intellectual technology of complex-technical-object structural-dynamic monitoring and control, Nauka, Moscow (2006) (in Russian)

Ore, O.: Theory of graphs, vol. 38. AMS Colloquium Publications, AMS, RI, Providence (1962)

Orlovski, S.A.: Decision making under fuzzy information. Nauka, Moscow (1981) (in Russian)

Pospelov, D.: Situational control: theory and applications. Science, Moscow (1986) (in Russian)

Ryabinin, I.A.: Reliability and safety of structurally complex systems. Polytechnika, St.-Petersburg (2000) (in Russian)

Siliak, D.: Decentralized control of complex systems. Academic Press, New York (1990)

Singh, M., Titli, A.: Systems: decomposition, optimization and control. Pergamon Press, Oxford (1978)

Sokolov, B.V., Kalinin, V.N.: Multi-model approach to the description of the air-space facilities control process. Control theory and process # 1, 149–156 (1995) (in Russian)

Sokolov, B.V., Yusupov, R.M.: Complex simulation of automated control system of navigation spacecraft. Problems of informatics and control # 5, 103–117 (2002) (in Russian)

Sokolov, B.V., Ivanov, D.A., Zaychik, E.M.: The formalization and investigation of processes for structure-dynamics control models adaptation of complex business systems. In: Proceedings of 20th European Conference on Modelling and Simulation ECMS 2006, Bonn, Sankt Augustin, Germany, May 28-31, pp. 292–295 (2006)

Tsurkov, V.: Dynamic problems of large dimension. Nauka, Moscow (1989) (in Russian)

Wang, L., Norrie, D.H.: Process planning and control in a holonic manufacturing environment. Journal of Applied Systems Studies 2(1), 106–126 (2001)

Zimin, I.N., Ivanilov, Y. P.: Solving of network planning problems via a reduction to optimal control problems. Journal of Calculus Mathematics and Mathematical Physics 11(3), 632–641 (1971) (in Russian)

Zvirkun, A.D., Akinfiev, V.K., Filippov, V.A.: Simulation modelling in the problems of complex systems structure synthesis. Nauka, Moscow (1985) (in Russian)

Zvirkun, A.D., Akinfiev, V.K.: Structure of the multi-level systems. Nauka, Moscow (1993) (in Russian)

# Computational Study of Non-linear Great Deluge for University Course Timetabling

Joe Henry Obit and Dario Landa-Silva

**Abstract.** The great deluge algorithm explores neighbouring solutions which are accepted if they are better than the best solution so far or if the detriment in quality is no larger than the current water level. In the original great deluge method, the water level decreases steadily in a linear fashion. In this paper, we conduct a computational study of a modified version of the great deluge algorithm in which the decay rate of the water level is non-linear. For this study, we apply the non-linear great deluge algorithm to difficult instances of the university course timetabling problem. The results presented here show that this algorithm performs very well compared to other methods proposed in the literature for this problem. More importantly, this paper aims to better understant the role of the non-linear decay rate on the behaviour of the non-linear great deluge approach.

## 1 Introduction

The *great deluge* algorithm is a meta-heuristic approach proposed by Dueck [12] and is inspired by the behaviour that could arise when someone seeks higher ground to avoid the rising *water level* during constant rain. For a maximisation problem, the algorithm seeks to find the highest point on a certain surface with hills, valleys and plateaus (search space). Then, it starts to rain constantly and the algorithm walks around (explores the neighbourhood) but never makes a step into the increasing water level. As it continues raining, the algorithm can explore higher and lower ground

Joe Henry Obit
ASAP Research Group, School of Computer Science,
University of Nottingham, United Kingdom
e-mail: `jzh@cs.nott.ac.uk`

Dario Landa-Silva
ASAP Research Group, School of Computer Science,
University of Nottingham, United Kingdom
e-mail: `dario.landasilva@nottingham.ac.uk`

V. Sgurev et al. (Eds.): Intelligent Systems: From Theory to Practice, SCI 299, pp. 309–328.
springerlink.com 

(improving and non-improving positions) but is continually pushed to a high point (hopefully close to the optimum) until eventually it cannot escape the rising water level and it stops. The initial water level is set to a value below the fitness of the initial solution and then is increased in a linear fashion as the search progresses. Note that for a minimisation problem, the water level starts on a value above the fitness of the initial solution and decreases constantly. In this case, the algorithm seeks to find the lower point by exploring the surface and maintaining its head below the descreasing water level. One can see that great deluge is similar to simulated annealing (SA) [1] but while SA accepts non-improving solutions based on probability, great deluge does this in a more deterministic manner by controlling the water level. The original great deluge algorithm was applied to course timetabling problems by Burke, Bykov, Newall and Petrovic [6]. They observed good performance of great deluge on all the problem instances tackled.

In our previous work [15] we presented a simple but effective modification of the conventional great deluge algorithm. In that variant, the water level decreases in a non-linear fashion and it also rises from time to time in order to improve the explorative ability of the algorithm. In the present paper, our aim is to conduct a computational study of the *non-linear great deluge* (NLGD) algorithm in order to investigate the key mechanisms that make this algorithm very effective. For this study, we use a number of well-known and difficult instances of the university course timetabling problem. This problem is NP complete [11, 13] and real-world instances are very difficult mainly due to the associated constraints. The present study uses the 11 instances of the course timetabling problem proposed by Socha, Knowles and Samples [18] and the 20 instances of the 1st International Timetabling Competition. All these instances consist of a set of events that need to be assigned into timeslots and rooms ensuring the satisfaction of a number of constraints (e.g. events should not be timetabled at certain times). These instances have been proven to be very challenging for most of the methods proposed in the literature. In this problem, the quality of a solution is measured by the overall penalty due to the violation of soft constraints and the aim is to minimise such penalty.

The rest of the paper is organised as follows. Section 2 describes the non-linear great deluge algorithm. Section 3 describes the university course timetabling problem considered in this paper and the instances used in our experiments. Important algorithm implementation details are given in Section 4. Experiments and results are presented and discussed in Section 5 focusing on the overall performance of NLGD and the effect that the non-linear decay rate has on the overall performance of the algorithm. Conclusions and future work are the subject of Section 6.

## 2 The Non-linear Great Deluge Algorithm

Consider a problem in which the goal is to find the solution that minimises a given objective function. The distinctive feature of the conventional great deluge algorithm is that when the candidate solution $S^*$ is worse than the current solution $S$, then $S^*$ replaces $S$ depending on the current water level $B$. The water level is initially set according to the quality of the initial solution, that is, $B > f(S^0)$ where $f(S^0)$ denotes the objective function value of the initial solution $S^0$. The decay rate, i.e. the speed at which $B$ decreases, is determined by a linear function in the conventional great deluge algorithm:

$$B = B - \Delta B \text{ where } \Delta B \in \Re^+ \tag{1}$$

The non-linear great deluge algorithm uses a non-linear decay rate for decreasing the water level. The decay rate is given by the following expression:

$$B = B \times (\exp^{-\delta(rnd[min,max])}) + \beta \tag{2}$$

The various parameters in Eq. (2) control the speed and the shape of the water level decay rate. Parameter $\beta$ represents the minimum expected value corresponding to the optimal solution. In this paper, we set $\beta = 0$ because we want the water level to reach that value by the end of the search. This is because we know that an optimal value of zero is possible for the problem instances tackled in this paper (see Section 3). If for a given minimisation problem we knew that the minimum objective value that can be achieved is lets say 100, then we would set $\beta$ around that value. If there is no previous knowledge on the minimum objective value expected, then we suggest to tune $\beta$ through preliminary experimentation for the problem in hand. The role of the parameters $\delta$, *min* and *max* (more specifically the expression $\exp^{-\delta(rnd[min,max])}$) is to control the speed of the decay rate and hence the speed of the search process. By changing the value of these three parameters, the water level goes down faster or slower.

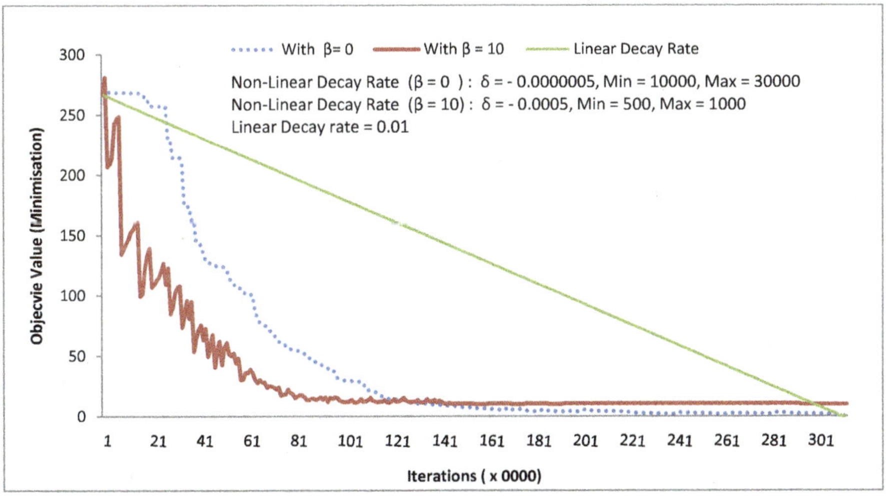


**Fig. 1** Comparison between linear (Eq. 1) and non-linear (Eq. 2) decay rates and illustration of the effect of parameters $\beta, \delta, min$ and $max$ on the shape of the non-linear decay rate

Figure 1 illustrates the difference between the linear and non-linear decay rates. The graph also illustrates the effect of parameters $\beta$, $\delta$, *min* and *max* on the non-linear decay rate. The straight line in Figure 1 corresponds to the linear decay rate (with $\Delta B = 0.01$) originally proposed by Dueck [12]. In this case, a non-improving candidate solution $S^*$ is accepted only if its objective value $f(S^*)$ is below the water level $B$. When $f(S^*)$ and $B$ converge the algorithm becomes greedy and it is more difficult for the search to escape from local optima. Figure 1 also illustrates the non-linear decay rate with different values for $\beta$, $\delta$, *min* and *max*.

---

**Algorithm 1**: Non-linear Great Deluge (NLGD) Algorithm

---

```
Construct initial feasible solution S
Set best solution so far S_best ← S
Set timeLimit according to problem size
Set initial water level B ← f(S)
while elapsedTime ≤ timeLimit do
   Select move at random from M1,M2,M3
   Define the neighbourhood N(S) of S
   Select candidate solution S* ∈ N(S) at random
   if ( f(S*) ≤ f(S) or f(S*) ≤ B ) then
      S ← S* {accept new solution}
      S_best ← S {update best solution}
   end if
   range = B − f(S*)
   if (range < 1) then
      if (Large or Small Problem) then
         B = B + rnd[B_min, B_max]
      else
         if ( f(S_best) < f_low ) then
            B = B + rnd[B_min, B_max]
         else
            B = B + 2
         end if
      end if
   else
      if f(S_best<=20) and Small then
         B = B × (exp^(−δ(rnd[min,max]))) + β (apply small instances parameters)
      else
         B = B × (exp^(−δ(rnd[min,max]))) + β
      end if
   end if
end while
```

---

Algorithm 1 corresponds to the Non-linear Great Deluge (NLGD) method and the use of the non-linear decay rate is shown in the last **else**. In addition to using a non-linear decay rate for the water level $B$, we also allow $B$ to go up when its value is about to converge with the penalty cost of the candidate solution $S^*$. This occurs when $range < 1$ in Algorithm 1. We increase the water level $B$ by a random number within the interval $[B_{min}, B_{max}]$. Full details of this strategy to control the non-linear decay rate are shown in Algorithm 1 and discussed in detail in [15].

# 3 The University Course Timetabling Problem

## *3.1 Benchmark Instances*

Educational timetabling refers to the allocation, subject to constraints on resources, of a set of timeslots and possibly rooms to events such as exams, lectures, lab sessions, etc. [20]. In general, educational timetabling problems can be classified into three types: school timetabling, course timetabling and examination timetabling [17]. Although these three timetabling problems share basic characteristics, significant differences among them still exist. In this paper, we are concerned with the *university course timetabling* problem which refers to the process of allocating, subject to constraints, a set of limited timeslots and rooms to events (courses), in such a way as to satisfy as nearly as possible a set of desirable objectives. In this problem, constraints can be distinguished into hard constraints and soft constraints. Hard constraints must be satisfied, i.e. a timetable is feasible only if no hard constraint is violated. Soft constrains might be violated but the number of violations has to be minimised in order to increase the quality of the timetable. Several formulations of the university course timetabling problem have been proposed in the literature. Next, we refer to the formulation by Socha, Knowles and Samples [18].

More formally, the university course timetabling problem consists of:

- $n$ events $E = \{e_1, e_2, \ldots, e_n\}$
- $k$ timeslots $T = \{t_1, t_2, \ldots, t_k\}$
- $m$ rooms $R = \{r_1, r_2, \ldots, r_m\}$ in which events can take place
- a set $F$ of room features satisfied by rooms and required by events
- a set $S$ of students

Each room has limited capacity. Each student attends a number of events (subset of $E$). The problem is to assign the $n$ events to the $k$ timeslots and $m$ rooms satisfying all hard constraints and minimising the violation of soft constraints.

There are four hard constraints in this problem:

- h1: A student cannot attend two events simultaneously, i.e. events with students in common must be timetabled in different timeslots.
- h2: Only one event can be assigned per timeslot in each room.

- h3: The room capacity must be equal to or greater than the number of students attending the event in each timeslot.
- h4: The room assigned to an event must satisfy the features required by the event.

There are three soft constraints in this problem:

- s1: Students should not have only one event timetabled on a day.
- s2: Students should not have to attend more that two consecutive events on a day.
- s3: Students should not have to attend an event in the last timeslot of a day.

We use two sets of benchmark instances for this problem. One is a set of 11 instances proposed by Socha, Knowles and Sampels [18].[1] The second set are the 20 instances used during the 1st International Timetabling Competition.[2] Details of these instances are given in Table 1 and Table 2.

**Table 1** There are 11 instances (5 small, 5 medium and 1 large) in the set by Socha, Knowles and Sampels [18]. The last four rows give some indication about the structure of the instances.

| | Small | Medium | Large |
|---|---|---|---|
| Number of events $n$ | 100 | 400 | 400 |
| Number of rooms $m$ | 5 | 10 | 10 |
| Number of room features $\|F\|$ | 5 | 5 | 10 |
| Number of students $\|S\|$ | 80 | 200 | 400 |
| Maximum events per student | 20 | 20 | 20 |
| Maximum students per event | 20 | 50 | 100 |
| Approximation features per room | 3 | 3 | 5 |
| Percent feature use | 70 | 80 | 90 |

### 3.2 *The Objective Function*

The objective is to find a feasible timetable that also minimises the violation of soft constraints. The problem can be formalised as follows. Let $X$ be the set of all possible solutions, $\otimes = \{h1, h2, h3, h4\}$ the set of hard constraints, $\oplus = \{s1, s2, s3\}$ the set of soft constraints and $\tilde{X} \subseteq X$ the set of all feasible solutions satisfying the hard constraints in $\otimes$. For each solution $x \in \tilde{X}$, $f(x)$ is the cost function measuring the violation of soft constraints in $\oplus$. The aim then is to find an optimal solution $x^* \in \tilde{X}$ such that $f(x^*) \leq f(x), \forall x \in \tilde{X}$. The cost function $f(x)$ is given by:

$$f(x) = \sum_{s \in S} (f_1(x,s) + f_2(x,s) + f_3(x,s)) \tag{3}$$

- $f_1(x,s)$: number of times a student $s$ in timetable $x$ has to attend a single event on a day (violation of s1). For example, $f_1(x,s) = 1$ if student $s$ has only 1 event in a day and if student $s$ has 2 days with only one event then $f_1(x,s) = 2$.

[1] These instances can be found at: http://iridia.ulb.ac.be/supp/IridiaSupp2002-001/index.html

[2] These instances can be found at: http://www.idsia.ch/Files/ttcomp2002/

**Table 2** There are 20 instances in the set for the 1st International Timetabling Competition. The last three columns give some indication about the structure of the instances.

| Instance | No. events $n$ | No. students $\|S\|$ | No. rooms $m$ | Rooms/event | Events/student | Students/event |
|---|---|---|---|---|---|---|
| com01 | 400 | 200 | 10 | 1.96 | 17.75 | 8.88 |
| com02 | 400 | 200 | 10 | 1.92 | 17.23 | 8.62 |
| com03 | 400 | 200 | 10 | 3.42 | 17.70 | 8.85 |
| com04 | 400 | 300 | 10 | 2.45 | 17.43 | 13.07 |
| com05 | 350 | 300 | 10 | 1.78 | 17.78 | 15.24 |
| com06 | 350 | 300 | 10 | 3.59 | 17.77 | 15.23 |
| com07 | 350 | 350 | 10 | 2.87 | 17.48 | 17.48 |
| com08 | 400 | 250 | 10 | 2.93 | 17.58 | 10.99 |
| com09 | 440 | 220 | 11 | 2.58 | 17.36 | 8.68 |
| com10 | 400 | 200 | 10 | 3.49 | 17.78 | 8.89 |
| com11 | 400 | 220 | 10 | 2.06 | 17.41 | 9.58 |
| com12 | 400 | 200 | 10 | 1.96 | 17.57 | 8.79 |
| com13 | 400 | 250 | 10 | 2.43 | 17.69 | 11.05 |
| com14 | 350 | 350 | 10 | 3.08 | 17.42 | 17.42 |
| com15 | 350 | 300 | 10 | 2.19 | 17.58 | 15.07 |
| com16 | 440 | 220 | 11 | 3.17 | 17.75 | 8.88 |
| com17 | 350 | 300 | 10 | 1.11 | 17.67 | 15.15 |
| com18 | 400 | 200 | 10 | 1.75 | 17.56 | 8.78 |
| com19 | 400 | 300 | 10 | 3.94 | 17.71 | 13.28 |
| com20 | 350 | 300 | 10 | 3.43 | 17.49 | 14.99 |

- $f_2(x,s)$: number of times a student $s$ in timetable $x$ has to attend more than two consecutive events (violation of s2). Every extra consecutive event receives 1 penalty point. For example $f_2(x,s) = 1$ if a student $s$ has three consecutive events and $f_2(x,s) = 2$ if the student $s$ has four consecutive events, and so on.
- $f_3(x,s)$: number of times a student $s$ in timetable $x$ has to attend an event in the last timeslot of the day (violation of s3).

# 4 Algorithm Implementation Details

## *4.1 Neighbourhood Structures*

We employ three neighbourhood moves in the overall approach from initialisation to improvement of solutions. Move M1 selects one event at random and assigns it to a feasible pair timeslot-room also chosen at random. Move M2 selects two events at random and swaps their timeslots and rooms while ensuring feasibility is maintained. Move M3 identifies an event that violates soft constraints and then it moves that event to another pair timeslot-room selected at random and also ensuring feasibility is maintained. Note that the three neighbourhood moves are based on random search but always seeking the satisfaction of hard constraints. Also note that the difference between moves M1 and M3 is whether the violation of soft constraints is taken into account or not when selecting the event to re-schedule. We use only these three simple neighbourhood moves (and not more sophisticated ones) to better assess the effectiveness of the non-linear decay rate in the NLGD algorithm.

## 4.2 Heuristic to Construct Feasible Timetables

To construct feasible timetables, we took the heuristic proposed by Chiarandini, Birattari, Socha and Rossi-Doria [10] and added the highest degree heuristic (a well-known graph colouring heuristic) to Step 1 as described next. This modification was necessary in our approach because otherwise, we were unable to generate feasible solutions for large problem instances. The resulting initialisation heuristic works as follows.

**Step 1 - Highest Degree Heuristic.** In each iteration, the unassigned event with the highest number of conflicts (other events with students in common) is assigned to a timeslot selected at random. Once all events have been assigned to a timeslot, the maximum matching algorithm for bipartite graph (see [10] for details) is used to assign each event to a room. At the end of this step, there is no guarantee for the timetable to be feasible.

**Step 2 - Local Search.** We use neighbourhood moves M1 and M2 to improve the timetable generated in Step 1. A move is only accepted if it improves the satisfaction of hard constraints (this is because the moves seek to achieve feasibility). This step terminates if after 10 iterations no move has produced a better (closer to feasibility) solution.

**Step 3 - Tabu Search.** We apply tabu search [14] using only move M1. The tabu list contains events that were assigned less than $tl$ iterations before calculated as $tl = rnd(10) + \alpha \times n_c$, where $rnd(10)$ is a random number from a uniform distribution U[0,10], $n_c$ is the number of events involved in hard constraint violations in the current timetable, and $\alpha = 0.6$. This step terminates if after 500 iterations no move has produced a better (closer to feasibility) solution.

In Steps 2 and 3 above, our initialisation heuristic uses simple local search and tabu search to achieve feasibility. The local search (Step 2) attempts to improve the solution but it also works as a disturbing operator, hence the reason for the maximum of 10 trials before switching to tabu search (Step 3). Note that in the tabu search, M1 selects only events that violate hard constraints. Then, Steps 2 and 3 are executed iteratively until a feasible solution is found. This three-step initialisation heuristic is capable of finding feasible timetables for most problem instances in reasonable computation times as shown in Tables 3 and 4. The exception is the large instance L1 from Table 1 which is the most difficult and it takes much longer time (a minimum of 300 seconds) to find a feasible timetable. The density matrix for this instance indicates a large number of conflicting events (with students in common).

**Table 3** Time range (in seconds) taken to construct an initial feasible timetable, for 10 runs of the initialisation heuristic on the instances by Socha, Knowles and Sampels [18] (see Table 1). Sx are small instances, Mx are medium instances and L1 is the large instance.

| | Minimum Time (s) | Maximum Time (s) |
|---|---|---|
| S1 | 0.07800 | 0.12500 |
| S2 | 0 .0790 | 0.10900 |
| S3 | 0.06800 | 0.11000 |
| S4 | 0.04700 | 0.11000 |
| S5 | 0.07800 | 0.11000 |
| M1 | 7.54600 | 9.3130 |
| M2 | 9.65600 | 10.9370 |
| M3 | 13.4370 | 21.7020 |
| M4 | 6.89100 | 7.76600 |
| M5 | 16.6700 | 143.560 |
| L1 | 300 | 3000 |

**Table 4** Time range (in seconds) taken to construct an initial feasible timetable, for 10 runs of the initialisation heuristic on the instances of the 1st International Timetabling Competition (see Table 2).

| | Minimum Time (s) | Maximum Time (s) |
|---|---|---|
| com01 | 1.93 | 5.492 |
| com02 | 1.36 | 2.644 |
| com03 | 1.34 | 2.22 |
| com04 | 4.464 | 28.98 |
| com05 | 2.112 | 11.028 |
| com06 | 1.33 | 3.272 |
| com07 | 2.644 | 42.402 |
| com08 | 1.82 | 11.086 |
| com09 | 1.496 | 8.088 |
| com10 | 4.644 | 29.045 |
| com11 | 3.14 | 13.75 |
| com12 | 3.016 | 12.632 |
| com13 | 2.26 | 6.976 |
| com14 | 5.816 | 50.675 |
| com15 | 1.564 | 8.956 |
| com16 | 1.092 | 3.884 |
| com17 | 2.136 | 13.048 |
| com18 | 1.292 | 2.948 |
| com19 | 3.228 | 20.753 |
| com20 | 1.804 | 0.085 |

# 5 Results with the NLGD Algorithm

## 5.1 Experimental Setting

We conducted several experiments using the two sets of benchmark instances described in Section 3. It is known that for each of those instances there is at least one assignment with an evaluation function value equal to zero, i.e. a feasible timetable satisfying all soft constraints too. For each type of instance (in terms of size) in Table 1, a fixed computation time (*timeLimit* in Algorithm 1) in seconds was set as the stopping condition: 3600 for small problems, 4700 for medium problems and 6700 for the large problem. This fixed computation time is only for the NLGD algorithm, i.e. starting from an already feasible solution. However, for every instance in Table 2, the *timeLimit* was set to 2500 seconds but including finding the initial feasible timetable. The reason for this is that the time taken by our initialisation heuristic (see subsection 4.2) on the instances of Table 2 is negible, but considerable for the large instance of Table 1. For each problem instance we executed the NLGD algorithm 10 times after generating an initial timetable.

The value of the parameters in Eq. (2) were determined by experimentation. We assigned $\delta$ the values of $5 \times 10^{-10}$, $5 \times 10^{-8}$ and $5 \times 10^{-9}$ for small, medium and large instances of Table 1 respectively. As said before, $\beta = 0$ for all problem instances. The values of *min* and *max* were set as follows: for medium and large problems we used $min = 100000$ and $max = 300000$ while for small problems we used $min = 10000$ and $max = 20000$. However, we should note that the parameter values given above for the small instance only apply when the penalty cost reach around 20. That is, the NLGD uses the same parameter values as for the medium instances and changes to the small instance parameter values the cost function reaches the value of 20. The interval $[B_{min}, B_{max}]$ (see Algorithm 1) was set as follows. For small instances it was [2,5] and for large instances it was [1,3]. For medium instances, we first check if the penalty of the best solution so far $f(S_{best})$ is lower than a parameter $f_{low}$. If this is the case, then we use [1,4]. Otherwise, we assume that the best solution so far seems to be stuck in local optima ($f(S_{best}) > f_{low}$) so we make $B = B + 2$ as shown in Algorithm 1.

## 5.2 The Computational Study

First, we evaluate how beneficial it is to have a non-linear decay rate and floating water level in the modified great deluge algorithm. In the first set of experiments, we compared the NLGD with other algorithms reported in the literature for the instances shown in Table 1. Results are reported in Table 5 where we can see the results obtained by the NLGD and by the original great deluge alongside other results reported in the literature. The table also shows the penalty of the initial solution provided to the great deluge approaches. The best results are shown in bold for each dataset. The main goal of this comparison is to assess whether great deluge with non-linear decay rate and floating water level performs better than or similar to other algorithms that have been reported in the literature. We also want to assess if

**Table 5** Comparison of results obtained by the non-linear great deluge (NLGD) against the best known results from the literature for the 11 instances of Table 1.

| Instance | Init. Sol. | GD | NLGD | Best Known |
|---|---|---|---|---|
| S1 | 198 | 17 | 3 | **0** (VNS-T) |
| S2 | 265 | 15 | 4 | **0** (VNS-T) |
| S3 | 214 | 24 | 6 | **0** (CFHH) |
| S4 | 196 | 21 | 6 | **0** (VNS-T) |
| S5 | 233 | 5 | 0 | **0** (MMAS) |
| M1 | 858 | 201 | **140** | 146 (CFHH) |
| M2 | 891 | 190 | **130** | 147 (HEA) |
| M3 | 806 | 229 | **189** | 246 (HEA) |
| M4 | 846 | 154 | **112** | 164.5 (MMAS) |
| M5 | 765 | 222 | 141 | **130** (HEA) |
| L1 | 1615 | 1066 | 876 | **529** (HEA) |

MMAS is the MAX-MIN Ant System in [18]
CFHH is the Choice Function Hyper-heuristic in [7]
VNS-T is the Hybrid of VNS with Tabu Search in [2]
HEA is the Hybrid Evolutionary Algorithm in [4]

the proposed modification to the water level decay rate produces better results than using the traditional linear and steady decay rate.

Table 5 shows that our algorithm outperforms some of the previous results and it is also competitive on the other instances. For the small problems, NLGD is able to solve instance S5 to optimality. For most of the medium problems, NLGD has shown significant improvement over other algorithms. However, for instance M5 the NLGD method is not able to improve the solution found by HEA. Still, NLGD is very competitive obtaining a solution quality of just around 8% worse than the best value for M5. Table 5 also shows that the NLGD algorithm obtained results that are much better than those produced with the conventional great deluge.

It must be said that adequate parameter tuning was required in our experiments, but the algorithm can definitely produce better results compared to the best results already published. But more importantly, the proposed algorithm can do that in short computation time, usually less than 700 seconds. We can also observe that in the small instances the algorithm is able to find solutions with low penalty cost but it cannot outperform those results reported previously. We need to further investigate this but we believe this is due to the ineffectiveness of the neighbourhood search for small instances, particularly when the penalty cost is too low. We plan to design a more effective strategy for exploring the neighbourhood of solutions and be sure to reach unexplored areas of the search space. We believe that the proposed non-linear great deluge algorithm has considerable potential to succeed in other timetabling and similar problems. This because the improvements achieved (4 new best results in the medium instances) are mainly due to the strategy used to control the water level decay rate. Remember that the neighbourhood moves and local search strategy implemented here are quite simple and general. That is, the local search is not dependant on the problem domain.

In the second set of experiments, we compared the NLGD with other algorithms reported in the literature for the instances shown in Table 2. These results are reported in Table 6 where we can see the best results obtained by different algorithms from the competition plus the results obtained by NLGD, best results are in bold. The table gives us an idea about the variability on the performance of different algorithm proposed in the competition. Results from Table 6 show that even though the NLGD did not obtain the best results, it is still very competitive particularly against the algorithms ranked fifth to ninth in the competition.

**Table 6** Comparison of results obtained by the non-linear great deluge (NLGD) against the best 9 ranked algorithms for the 20 instances of Table 2. Details of the competition algorithms are available at: http://www.idsia.ch/Files/ttcomp2002/results.htm.

| Instances | 1st | 2nd | 3rd | 4th | 5th | 6th | 7th | 8th | 9th | NLGD |
|---|---|---|---|---|---|---|---|---|---|---|
| com01 | **45** | 61 | 85 | 63 | 132 | 148 | 178 | 211 | 257 | 153 |
| com02 | **25** | 39 | 42 | 46 | 92 | 101 | 103 | 128 | 112 | 118 |
| com03 | **65** | 77 | 84 | 96 | 170 | 162 | 156 | 213 | 226 | 120 |
| com04 | **115** | 160 | 119 | 166 | 265 | 350 | 399 | 408 | 441 | 358 |
| com05 | 102 | 161 | **77** | 203 | 257 | 412 | 336 | 312 | 299 | 398 |
| com06 | 13 | 42 | **6** | 92 | 133 | 246 | 246 | 169 | 209 | 129 |
| com07 | 44 | 52 | **12** | 118 | 177 | 228 | 225 | 281 | 99 | 99 |
| com08 | **29** | 54 | 32 | 66 | 134 | 125 | 210 | 214 | 194 | 111 |
| com09 | **17** | 50 | 184 | 51 | 139 | 126 | 154 | 164 | 175 | 119 |
| com10 | **61** | 72 | 90 | 81 | 148 | 147 | 153 | 222 | 308 | 153 |
| com11 | 44 | 53 | 73 | 65 | **35** | 144 | 169 | 196 | 273 | 149 |
| com12 | 107 | 110 | **79** | 119 | 290 | 182 | 219 | 282 | 242 | 229 |
| com13 | **78** | 109 | 91 | 160 | 251 | 192 | 248 | 315 | 364 | 240 |
| com14 | 52 | 93 | **36** | 197 | 230 | 316 | 267 | 345 | 156 | 282 |
| com15 | **24** | 62 | 27 | 114 | 140 | 209 | 235 | 185 | 95 | 172 |
| com16 | **22** | 34 | 300 | 38 | 114 | 121 | 132 | 185 | 171 | 91 |
| com17 | 86 | 114 | **79** | 212 | 186 | 327 | 313 | 409 | 148 | 356 |
| com18 | **31** | 38 | 39 | 40 | 87 | 98 | 107 | 153 | 117 | 190 |
| com19 | **44** | 128 | 86 | 185 | 256 | 325 | 309 | 334 | 414 | 228 |
| com20 | **7** | 26 | **0** | 17 | 94 | 185 | 185 | 149 | 113 | 72 |

In more detail, Figures 2-7 summarise the performance of NLGD compared to other allgorithms. In these graphs, the x-axis represents the instance type while the y-axis represents the penalty cost. Figure 3 shows the strong performance of NLGD on the medium and large instances. Figures 4-7 show details of the results achieved by NLGD when compared to the algorithms from the competition.

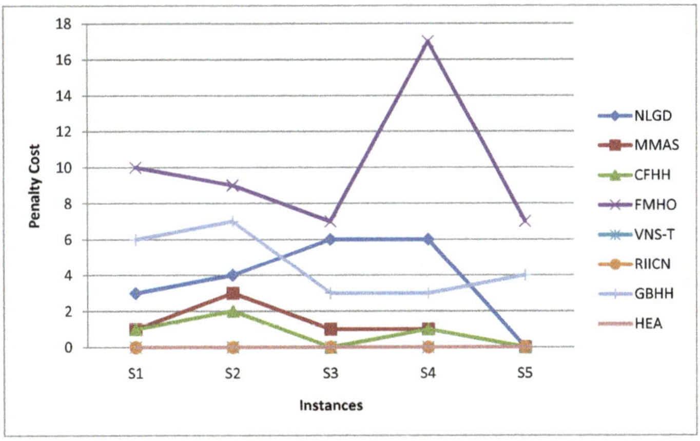


**Fig. 2** Detailed comparison of non-linear great deluge against other algorithms for small instances from Table 1

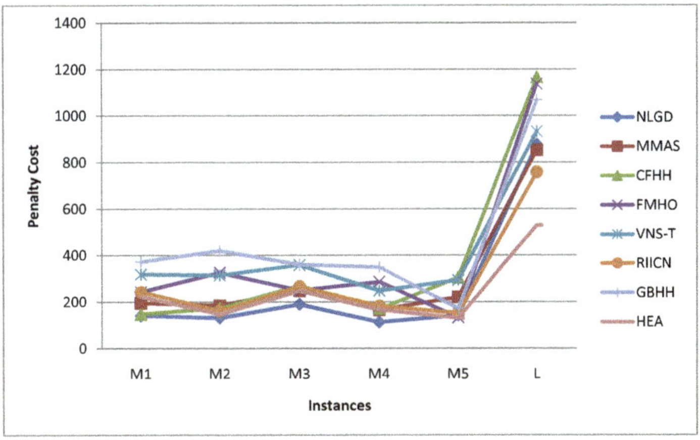


**Fig. 3** Detailed comparison of non-linear great deluge against other algorithms for medium and large instances from Table 1

## 5.3 *Effect of the Non-linear Decay Rate*

Here we present more results to illustrate the positive effect that the non-linear decay rate has on the performance of the NLGD algorithm. Figures 8-10 show the performance of linear great deluge (GD) across iterations for three problem instances while Figures 11-13 do the same but for the non-linear version of the algorithm (NLGD). Each graph in these Figures shows the search progress for one sample run of the corresponding algorithm. The dotted line corresponds to the water level and the solid line corresponds to the penalty of the best solution so far which should be

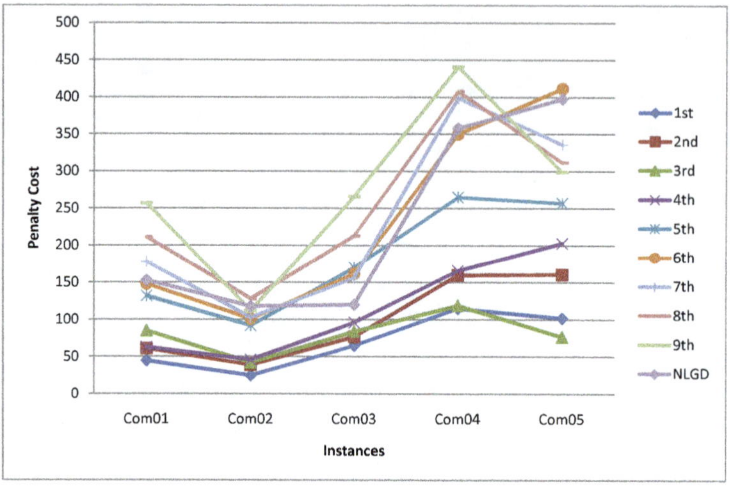


**Fig. 4** Detailed comparison of non-linear great deluge against other algorithms for com01-com05 instances from Table 2

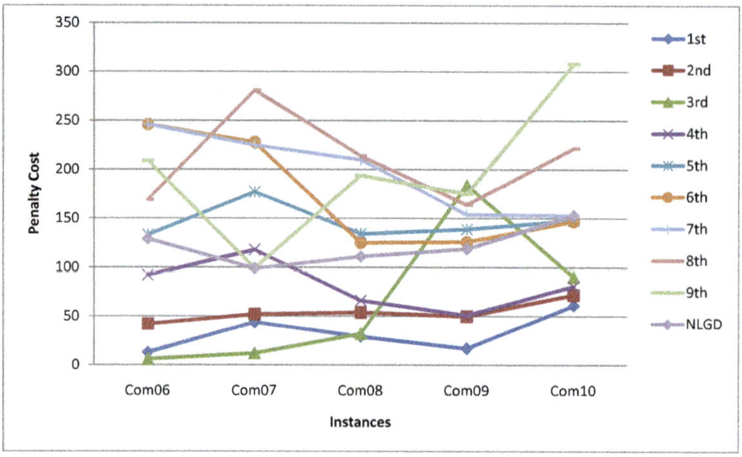


**Fig. 5** Detailed comparison of non-linear great deluge against other algorithms for com06-com10 instances from Table 2

minimised. The water level in the GD decreases at the same rate in every iteration while in the NLGD the water level decreases exponentially according to Eq. (2).

The first interesting observation is that the relation between the water level and the best solution varies for different instance sizes. The rigid and pre-determined linear decay rate appears to suit better the medium instance while for the small and large instances this decay rate seems to be less effective in driving the search for

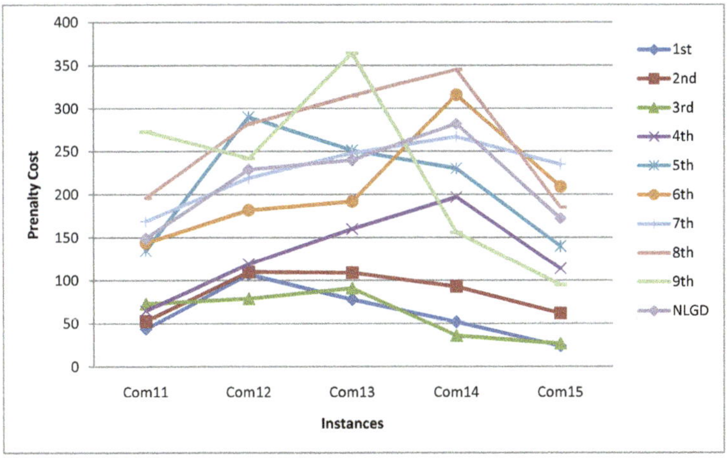


**Fig. 6** Detailed comparison of non-linear great deluge against other algorithms for com11-com15 instances from Table 2

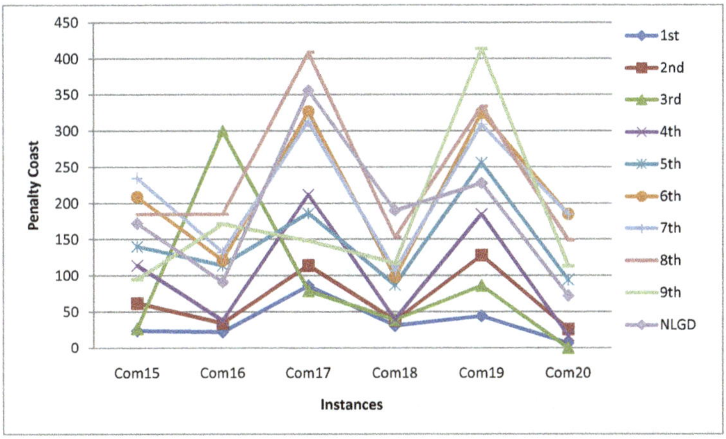


**Fig. 7** Detailed comparison of non-linear great deluge against other algorithms for com16-com20 instances from Table 2

the best solution. Figure 8 shows that in the small instance the water level is too high with respect to the best solution and this provokes that the best solution is not 'pushed down' for the first 60000 or so iterations, i.e. improvements to the best solution are rather slow. However, for the medium (Figure 9) and large (Figure 10) instances, the water level and the best solution are very close from the start of the search so the best solution is 'pushed down' as the water level decreases. We can also see that in the medium and large instances there is a point after which the water level continues decreasing but the best solution does not improve further, i.e. the

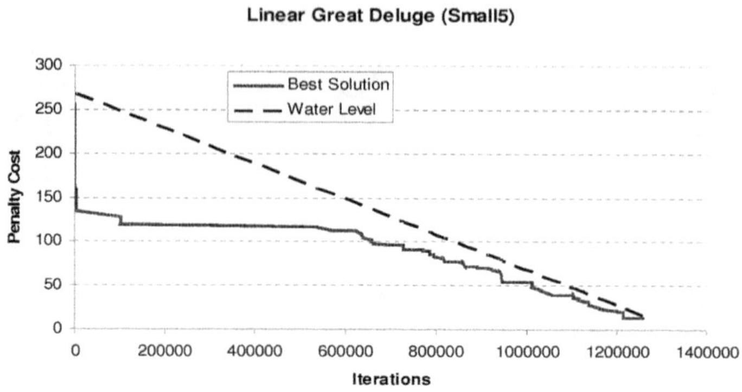


**Fig. 8** Sample of search progress behaviour of GD on small instance

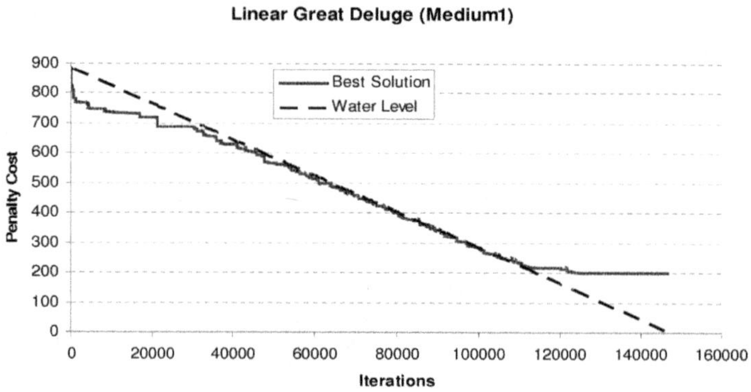


**Fig. 9** Sample of search progress behaviour of GD on medium instance

search stagnates. That is, when the water level and the best solution so far 'converge', the search becomes greedy and improvements are more difficult to achieve while the water level continues decreasing. This occurs around iteration 110000 in the medium instance and around iteration 8000 in the large instance. We argue that the simple linear water level decay rate in the original great deluge algorithm does not adapt easily to the quality of the best solution so far. This is precisely the shortcoming that we tackle with the non-linear great deluge algorithm.

Then, in the non-linear version of the algorithm, the decay rate is adjusted at every iteration and the size of the problem instance being solved is taken into account when setting the parameters of Eq.(2) as explained in Section 2. We can see in Figures 11-13 that this modification helps the algorithm to perform a more effective search regardless of the instance size. We can see that in the three sample runs of the non-linear great deluge algorithm, if drastic improvements are found then the water level also decreases more drastically. But when the improvement to the best solution so far becomes slower then the decay rate also slows in reaction to this. Moreover,

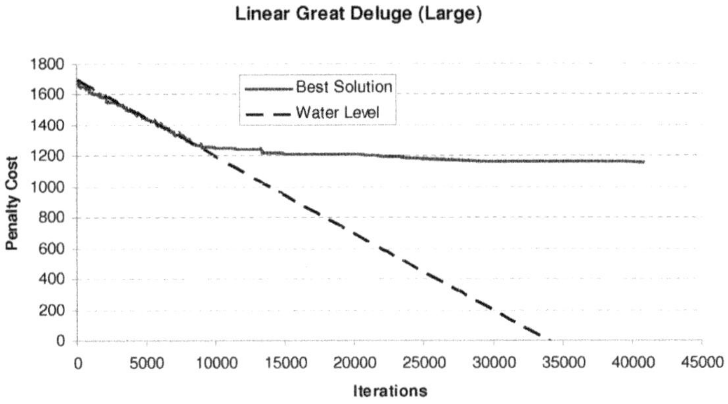


**Fig. 10** Sample of search progress behaviour of GD on large instance

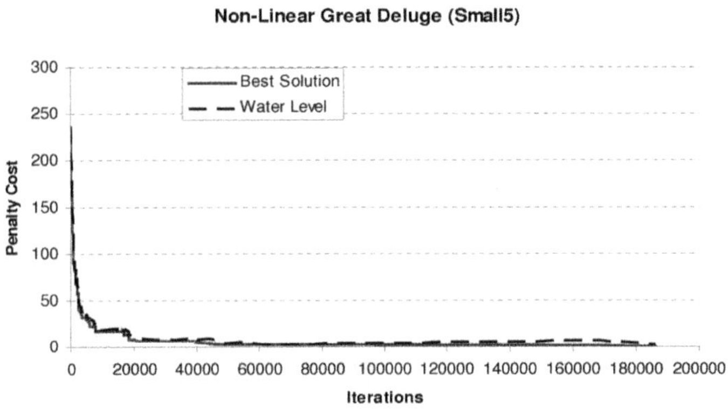


**Fig. 11** Sample of search progress behaviour of NLGD on small instance

to avoid (as much as possible) the convergence of the water level and the best solution, the water level is increased from time to time as explained in Section 2. This 'floating' feature of the water level explains the small increases on the best solution penalty observed in the graphs of Figures 11-13. As in many heuristics based on local search, the rationale for increasing the water level is to accept slightly worse solutions to explore different areas of the search space in the hope of finding better solutions.

The above observations help us to summarise the key differences between the linear (GD) and non-linear (NLGD) great deluge variants:

**Linear Great Deluge**

1. The decay rate is pre-determined and fixed
2. Mainly, the search is driven by the water level
3. When the best solution and water level converge the algorithm becomes greedy

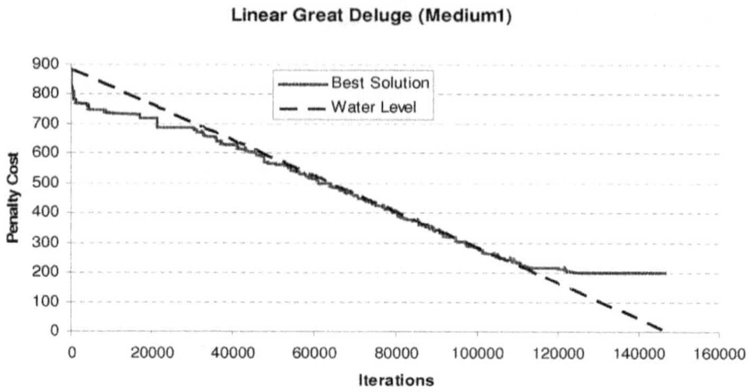


**Fig. 12** Sample of search progress behaviour of NLGD on medium instance

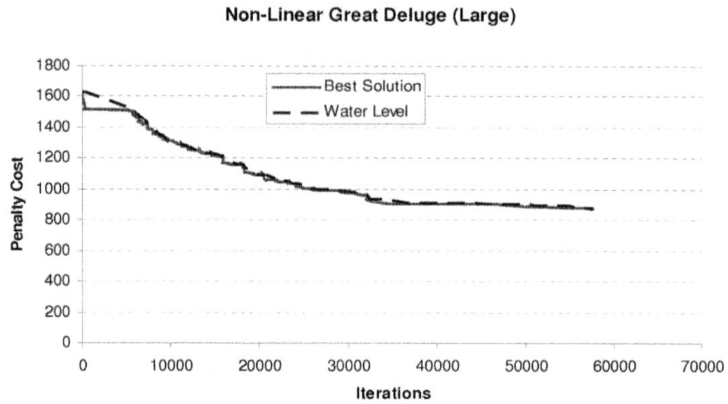


**Fig. 13** Sample of search progress behaviour of NLGD on large instance

**Non-Linear Great Deluge**

1. The decay rate changes every iteration based on Eq.(2)
2. Mainly, the water level is driven by the search
3. This algorithm never becomes greedy

## 6 Conclusions

This paper presented a computational study of the *non-linear great deluge* (NLGD) algorithm [15] which is an extension of the conventional great deluge method [12]. The NLGD approach incorporates a non-linear decay rate and floating water level. We applied this modified algorithm to well known benchmark instances of the university course timetabling problem: the 11 instances proposed by Socha, Knowles and Samples [18] and the 20 instances from the 1st International Timetabling Competition. The NLGD algorithm performs very well in both sets of instances and this study showed that the non-linear decay rate and floating water level are key components for the robust performance on this algorithm. In future work, we intend to investigate mechanisms to automatically adapt the non-linear decay rate to the size of the problem instance being tackled. Also, we want to investigate a population-based version of the non-linear great deluge algorithm taking into consideration the diversity among a set of timetables.

## References


1. Aarts, E., Korts, J.: Simulated Annealing and Boltzman Machines. Wiley, Chichester (1998)
2. Abdullah, S., Burke, E.K., McCollum, B.: An Investigation of Variable Neighbourhood Search for University Course Timetabling. In: Proceedings of MISTA 2005: The 2nd Multidisciplinary Conference on Scheduling: Theory and Applications, pp. 413–427 (2005)
3. Abdullah, S., Burke, E.K., McCollum, B.: A Hybrid Evolutionary Approach to the University Course Timetabling Problem. In: Proceedings of CEC 2007: The 2007 IEEE Congress on Evolutionary Computation, pp. 1764–1768 (2007)
4. Abdullah, S., Burke, E.K., McCollum, B.: Using a Randomised Iterative Improvement Algorithm with Composite Neighborhood Structures for University Course Timetabling. In: Metaheuristics - Progress in Complex Systems Optimization, pp. 153–172. Springer, Heidelberg (2007)
5. Asmuni, H., Burke, E.K., Garibaldi, J.: Fuzzy Multiple Heuristic Ordering for Course Timetabling. In: Proceedings of the 5th United Kingdom Workshop on Computational Intelligence (UKCI 2005), pp. 302–309 (2005)
6. Burke, E.K., Bykov, Y., Newall, J., Petrovic, S.: A Time-predefined Approach to Course Timetabling. Yugoslav Journal of Operations Research (YUJOR) 13(2), 139–151 (2003)
7. Burke, E.K., Kendall, G., Soubeiga, E.: A Tabu-search Hyperheuristic for Timetabling and Rostering. Journal of Heuristics 9, 451–470 (2003)
8. Burke, E.K., Eckersley, A., McCollum, B., Petrovic, S., Qu, R.: Hybrid Variable Neighbourhood Approaches to University Exam Timetabling. Technical Report NOTTCS-TR-2006-2, University of Nottingham, School of Computer Science (2006)
9. Burke, E.K., McCollum, B., Meisels, A., Petrovic, S., Qu, R.: A Graph Based Hyperheuristic for Educational Timetabling Problems. European Journal of Operational Research 176, 177–192 (2007)
10. Chiarandini, M., Birattari, M., Socha, K., Rossi-Doria, O.: An Effective Hybrid Algorithm for University Course Timetabling. Journal of Scheduling 9(5), 403–432 (2006)
11. Cooper, T., Kingston, H.: The Complexity of Timetable Construction Problems. In: Burke, E.K., Ross, P. (eds.) PATAT 1995. LNCS, vol. 1153, pp. 283–295. Springer, Heidelberg (1996)
12. Dueck, G.: New Optimization Heuristic: The Great Deluge Algorithm and the Record-to-record Travel. Journal of Computational Physics 104, 86–92 (1993)
13. Even, S., Itai, A., Shamir, A.: On the Complexity of Timetabling and Multicommodity Flow Problems. SIAM Journal of Computation 5, 691–703 (1976)
14. Glover, F., Taillard, E., De Werra, D.: A User's Guide to Tabu Search. Annals of Operations Research 41, 3–28 (1993)
15. Landa-Silva, D., Obit, J.-H.: Great Deluge with Nonlinear Decay Rate for Solving Course Timetabling Problems. In: Proceedings of the 2008 IEEE Conference on Intelligent Systems (IS 2008), pp. 8.11–8.18. IEEE Press, Los Alamitos (2008)

16. Rossi-Doria, O., Sampels, M., Birattari, M., Chiarandini, M., Dorigo, M., Gambardella, L., Knowles, J., Manfrin, M., Mastrolilli, M., Paechter, B., Paquete, L., Stuetzle, T.: A Comparion of the Performance of Different Metaheuristics on the Timetabling Problem. In: Burke, E.K., De Causmaecker, P. (eds.) PATAT 2002. LNCS, vol. 2740, pp. 333–352. Springer, Heidelberg (2003)
17. Schaerf, A.: A Survey of Automated Timetabling. Artificial Intelligence Review 13(2), 87–127 (1999)
18. Socha, K., Knowles, J., Sampels, M.: A Max-min Ant System for the University Course Timetabling Problem. In: Dorigo, M., Di Caro, G.A., Sampels, M. (eds.) Ant Algorithms 2002. LNCS, vol. 2463, pp. 1–13. Springer, Heidelberg (2002)
19. Socha, K., Sampels, M., Manfrin, M.: Ant Algorithms for the University Course Timetabling Problem with Regard to the State-of-the-Art. In: Raidl, G.R., Cagnoni, S., Cardalda, J.J.R., Corne, D.W., Gottlieb, J., Guillot, A., Hart, E., Johnson, C.G., Marchiori, E., Meyer, J.-A., Middendorf, M. (eds.) EvoIASP 2003, EvoWorkshops 2003, EvoSTIM 2003, EvoROB/EvoRobot 2003, EvoCOP 2003, EvoBIO 2003, and EvoMUSART 2003. LNCS, vol. 2611, pp. 334–345. Springer, Heidelberg (2003)
20. Wren, V.: Scheduling, Timetabling and Rostering A Specail Relationship? In: Burke, E.K., Ross, P. (eds.) PATAT 1995. LNCS, vol. 1153, pp. 46–75. Springer, Heidelberg (1996)

# Entropy Operator in Macrosystem Modeling

Yu S. Popkov

**Abstract.** Mathematical definition of a class macrosystem models with entropy operator is considered. The main properties of the entropy operator – existing, continuity, boundedness, differentiability are investigated. The stability conditions are obtained for the models with multiplicative and additive flows. The theoretical results are used for mathematical modeling of labor market and in dynamic procedure of the image restoration from projections.

**Keywords:** dynamic system, entropy, labor market modeling, Lipschitz constant, image restoration.

## 1 Introduction

The growing complexity of the objects under study required to regard the integration of branches of science not as the integration of tools of the investigation, but as a certain philosophical-world outlook problem of the interrelation between the "whole" and its "parts". The natural way of the human perception lies in the dissection of the object under study into parts and the study of these parts with the subsequent collection of derived knowledges. In this case, what is radically important is the understanding of integrity and systematization of the object involved, while any decomposition is merely a tool of the perception of its system properties.

The systems analysis as tools for the investigation of systems properties represents the aggregate of theories, methods, ways, and algorithmic and information resources. In this sense, it is a "roof" for many scientific trends that touched by one or another "side" the problems of discovering systems properties. Without claiming the chronology of the advent of these trends, the more so the ranking of their scientific value, we will mention some of them.

It would be valid to begin with the statistical physics and thermodynamics, where study was made of a system consisting of a large number of particles (molecules) with the random interactions and quite a determinate system state generated by them [1], [2]. Relations between the individual and the collective behavior comprise a section of the mathematical theory of behavior [3]. The order

Yu S. Popkov
Institute for Systems Analysis, Moscow

V. Sgurev et al. (Eds.): Intelligent Systems: From Theory to Practice, SCI 299, pp. 329–359.
springerlink.com 

parameters forming the basis of the synergetic concept are the real tool of mathematical modeling for the dynamics of shaping systems effects [4]. The concept of the spatial-time self-organization of systems and the transition from "chaos to order" explained many structural systems effects observed both in liquids and in social media [5], [6], [7].

A definite contribution to the study of systems effects was made by the theory of macrosystems that is set up on the generalized variational principle of the conditional maximization of entropy, in terms of which the notion of an equilibrium state is defined [8]. In the framework of this theory, it becomes possible to model the equilibrium prices in systems of the exchange of economic resources, the quasi-determinate distribution of material and information flows in stochastic networks (transport, pipeline, computer networks, etc.), the stationary migration of population, etc. The systems effects are taken to mean here the acquisition by appropriate systems in the equilibrium state of determinate properties (equilibrium prices, flow distribution, spatial structures of the migration), whereas their elements (producers-consumers, users of computer networks, migrants) have the stochastic type of behavior.

Certainly, as any other theory, the theory of equilibria in macrosystems has definite constraints. The main ones of them are the homogeneity of elements and the entropy hypothesis (the equilibrium state is the state with the conditionally maximum entropy). In the framework of this theory, a certain property of the equilibrium state is postulated, but here the process of its achievement falls out of consideration. It often proves to be rather important because the equilibrium is achieved in the infinite time prospect, while the real times of functioning or the lives of systems are finite.

Therefore, the study of dynamics of macrosystems with the aim to model appropriate processes acquired the definite scientific interest and pragmatic actuality. It is necessary to note at once that the problems of dynamics of macrosystems are quite complex and their essence consists in deriving suitable kinetic equations. All achievements that exist today are nominal, in particular, the kinetic equations of Leontovich [9], Stratonovich [10], Helbing [11], and Klimontovich [12]. These results rest to some degree or another on the theory of nonlinear Markov processes and Boltzmann equations.

The investigation of dynamics of macrosystems was performed by the scientific association not only in the direction of the development of the general theory, but also in the direction of the analysis and modeling of some classes of macrosystems. Thus, in 1980–1981, the book of A. G. Wilson [13] and the article of Yu. S. Popkov and A. N. Ryazantsev [14] appeared almost simultaneously, where processes of the reproduction and migration of the population were considered, which occur with appreciably various relaxation times and have a radically different nature: the quasi-determinate reproduction and stochastic migration. These features of the system made it possible to use the principle of local equilibria [15], which is known in nonequilibrium thermodynamics, and examine the spatial-time evolution of the system as a sequence of local-stationary states, each of which is specified by a conditional maximum of entropy [16].

Such an approach was found to be useful for the mathematical modeling of the processes of the exchange and distribution of resources in regional systems [17],

[18], [19], [20], [21], [22], the formation of spatial structures of biological communities, the processes of chemical kinetics with the common catalyst [24], the two-speed nonlinear Markov processes [25], and the dynamics of distribution of information flows in computer networks [26].

In all enumerated problems, use was made to some or another degree of one and the same mathematical structure of a nonlinear dynamic system, but which is different from the existing structures. The structure displays a specific nonlinearity, which is described by a parametric problem of mathematical programming with the entropy objective function. This structure is called the entropy operator (EO), that plays important role in dynamic macrosystem models.

## 2 Phenomenological Macrosystem Modeling

The phenomenological model is taken to mean a certain systems structure (generally speaking, an abstract one), with the aid of which the events (processes) that occur in it are described at the meaningful level. Then, on the basis of such a verbal description, a mathematical model is built up [16].

Macrosystem model consists of $n$ blocks (subsystems) linked to one another (Fig. 1).

Each block contains two types of elements: specific ones for a given block and nonspecific ones, i. e., identical elements for all blocks. For example, if the macrosystem is a reactor in which chemical reactions occur, whose blocks represent the blocks of the macrosystem, then the specific elements are substances that undergo chemical transformations in each reaction, while nonspecific elements can be molecules of the catalyst that is common to all reactions. In some cases, for nonspecific elements, it is convenient to consider portions of the universal product formed in each block.

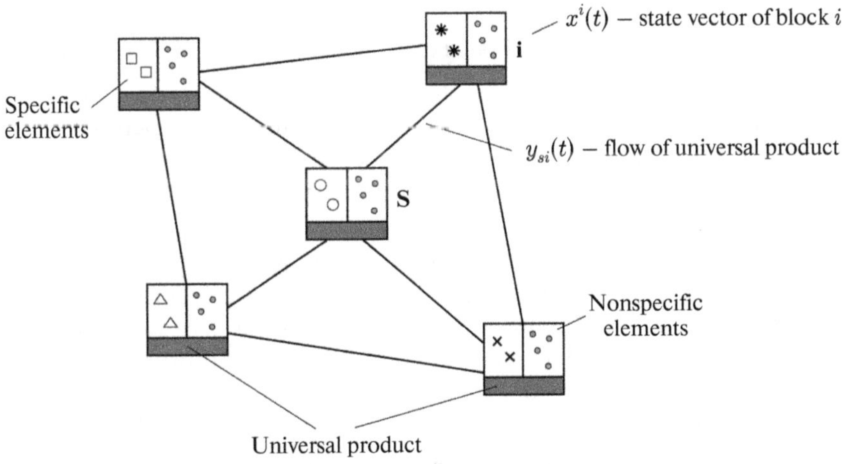


**Fig. 1**

The state of the $s^{\text{th}}$ block at the instant of time $t$ is defined by the vector $x^s(t) = \{x_1^s(t),\dots,x_l^s(t)\}$. To avoid the complication of the analysis by inessential generalizations, we will assume that the state vectors of blocks have the identical number of components, namely, one component $x^s(t) = x_s(t)$. Therefore, the vector $x(t) = \{x_1(t),\dots,x_n(t)\}$ specifies the state of blocks.

In each block, a universal resource $U_s(t) = u_s(x(t))$ is produced, where $u(\bullet)$ is a production function (it is not necessarily has the economic meaning), which generally depends on the state of all blocks. The universal product is redistiributed among blocks, so that flows $y_{si}(t)$ arise. The matrix $Y(t) = [y_{si}(t),\ s,i = 1,\dots,n]$ of flows defines the state of the distribution process.

Thus, the pair $\{x(t), Y(t)\}$ specifies the macrosystem state.

Basic components of the phenomenological model under consideration and assumptions laid in it are displayed in Table 1.

**Table 1** General properties of the macrosystems

| **components** | **in blocks** | **between blocks** |
|---|---|---|
| process | self-reproduction of $x(t)$ | distribution of $Y(t)$ |
| elements | specific | nonspecific |
| nature | determinate | stochastic |
| time scale | slow | fast |
| relaxation time | $\tau_{\text{slow}} \gg$ | $\tau_{\text{fast}}$ |

The model under study differs from the conventional one [2], [8] in the following:

(a) it includes the so-called discernible (specific) elements, whose evolution, on the one hand, affects the macrodistribution of indiscernible elements and, on the other, the model itself experiences the action of the latter;
(b) it lacks the systems variational principle postulating the equilibrium state.

A change in the *state of a block* occurs under the effect of the self-reproduction of specific elements with the participation of universal resources produced in other blocks. Assuming that the system is a Markov one, a change in the state of each $s$-block at the next instant of time will depend on the state of all blocks only at a given instant of time $x(t)$ and the matrix $Y(t)$ of flows of the universal product between blocks. In the situation of the general form, this dependence $\boldsymbol{L}(x(t), Y(t))$ is nonlinear. It defines a certain *material flow* that leads to a change in the state of blocks. Then, the evolution of the state of blocks can be described by nonlinear differential equations of the form

$$\frac{dx(t)}{dt} = \boldsymbol{L}\big(x(t), Y(t)\big), \tag{1}$$

where $\boldsymbol{L} = \{\boldsymbol{L}_1, \dots, \boldsymbol{L}_n\}$.

Here, it should be noted that the form of the right side of this differential equation depends on the properties of the macrosystem under study. As a result, equations of various classes may appear (equations with the retarded argument, integro-differential equations in convergences, etc.). In this work, we will consider ordinary differential equations (1) in which $\boldsymbol{L}(\cdot)$ is the continuously differentiable function.

According to Table 1, the *distribution of a universal product* between blocks is taken to be a random one with parameters depending on the state of blocks, i. e., the flows $y_{si}(t)$ represent a random quantity of the universal product in a unit of time. In this case, it is assumed that the prior probability $a_{si}(x)$ of the entry of a portion of the universal $s$-product into the $i$-block is known. The "receiver" of the universal product in each block has a limited capacity $G_{si}(x)$ and a cell structure.

In the distribution of portions of the universal product over blocks, the following situations are possible: each cell can be occupied by only one portion of the universal product (Fermi statistics), or any number of portions (Einstein statistics), or the mean number of portions of the block $s$ that fell into the block $i$ is small in comparison with the capacity $G_{si}(x)$ (Boltzmann statistics). A requisite generalized information entropy $H(Y,x)$ specifies each of these situations [8].

The distribution of the universal product between blocks commonly occurs with the expenditure of the resources of $r$ types, the amount of which is limited by values $q_1, \dots, q_k, q_{k+1}, \dots, q_r$. The dependence of the amount of the spent resources on the flows of the universal product and the state of blocks will be specified by the function

$$\Phi(Y,x) = \{\Phi_1(Y,x), \dots, \Phi_k(Y,x), \Phi_{k+1}(Y,x), \dots, \Phi_{k+p}(Y,x)\},$$

where $k + p = r$.

The existence of limited resources leads to the fact that the universal product distribution that can be carried out in the system must satisfy resource limits. These limits define the admissible set

$$\begin{aligned} D(x) = \{Y : \Phi_i(Y,x) = q_i, \quad & 0 \le \Phi_{k+j}(Y,x) \le q_{k+j}, \\ i = \overline{1,k}; \quad & j = \overline{1,p}\}, \\ q_i \ge 0, \quad & i = \overline{1,r}. \end{aligned} \tag{2}$$

The distribution of the universal product between blocks does not occur instantly. On the contrary, it takes a definite time, but in view of the adopted assumption, this time is substantially less than the relaxation time of the process of self-reproduction of the specific elements in blocks. Therefore, on the slow-time scale, over any sufficiently small time intervals, there is a local-stationary distribution $Y^*(t)$ of the universal product. For its definition, it is possible to use the generalization of the variational principle of static thermodynamics [2], according to which the realizable distribution of flows of the universal product is the distribution corresponding to a maximum of the entropy $H(Y,x)$ on the set $D(x)$ of bounded resources [8].

Because parameters of both the entropy and the admissible set depend on the states of blocks of the macrosystem in question, the realizable distribution is $Y^*(t) = Y^*(x(t)))$. Therefore, the material flow $\boldsymbol{L}$ that leads to a change in the state of blocks proves to be dependent on the local-stationary distribution $Y^*(x(t))$:

$$\boldsymbol{L}(x(t),Y(t))) = \boldsymbol{L}\left(x(t),Y^*(x(t)))\right). \tag{3}$$

The examined phenomenological model represents the systems illustration of the origin of dynamic systems with the entropy operator. It is precisely the systems illustration because other phenomenological models that are built up not on the basis of the systems paradigm are naturally possible, too.

## 3 Mathematical Model: Classification

For the mathematical description of a class of the dynamic systems, whose phenomenology is described in the preceding section, it is convenient to pass on to the vector description of the distribution of flows $Y^*(x)$. We will consider the vector $y^*(x)$ of flows instead of the matrix $Y^*(x)$. This transition can be performed by applying the following replacement of variables:

$$j = (s-1)r + i, \quad m = n \times r, \quad s,i = 1,\dots,n. \tag{4}$$

Then, the vector function $\boldsymbol{L}$ (3) can be represented in the form

$$\boldsymbol{L} = \boldsymbol{L}\left(x(t), y^*(x(t))\right). \tag{5}$$

It is possible to define in an appropriate way the vectors of prior probabilities $a(x)$ and capacities $G(x)$ of "receivers". Then, depending on the properties of cells of a receiver, we can write out the entropy of requisite distributions [8].

If each cell of the receiver can contain only one portion of the universal resource, then the distribution is specified by the generalized informational Fermi—Dirac entropy (F):

$$H_F(y,x) = -\sum_{j=1}^{m}\left( y_j \ln\frac{y_j}{a_j(x)} + (G_j(x) - y_j)\ln(G_j(x) - y_j) \right). \tag{6}$$

If the cells themselves have the unlimited capacity, then the forming distribution exhibits the generalized informational Bose-Einstein entropy (E):

$$H_E(y,x) = -\sum_{j=1}^{m}\left( y_j \ln\frac{y_j}{w_j(x)} - (G_j(x) + y_j)\ln(G_j(x) + y_j) \right) \tag{7}$$

If the fillability of the receiver is small in the mean, then the forming distribution features the generalized informational Boltzmann entropy (B):

$$H_B(y,x) = -\sum_{j=1}^{m} y_j \ln \frac{y_j}{e\, a_j(x)} \tag{8}$$

The admissible set of vectors y, according to (2), is given as

$$\begin{gathered} F(x) = \{ y : \Phi_i(y,x) = q_i, \quad 0 \le \Phi_{k+j}(y,x) \le q_{k+j}, \\ i = \overline{1,k}; \quad j = \overline{1,p}; \quad k+p=r; \quad y \in R_+^m, \quad x \in R^n \}. \\ q_i \ge 0, \qquad i = \overline{1,r}. \end{gathered} \tag{9}$$

**Definition 1.** *The operator* $y^*(x)$ *that maps the space* $R^n$ *into* $R_+^m$ *by the rule*

$$y^*(x) = \arg\max_{y} \{ H(y,x) \mid y \in F(x)),\, x \in R^n,\, y \in R_+^m \} > 0, \tag{10}$$

*will be called the entropy operator.*

Let us note that the entropy operator belongs to the class of *positive operators.*

**Definition 2.** *The mathematical model*

$$\frac{dx(t)}{dt} = \boldsymbol{L}\left( x(t), y^*(x(t)) \right), \tag{11}$$

*where* $y^*(x(t))$ *is the entropy operator (10), will be named the dynamic system with the entropy operator (DSEO).*

The appropriate classification is shown in Table 2.

The first row of the table displays the three types of entropy operators, which differ in models of the admissible set $F(x)$ : EQ identifies equalities, IEQ denotes inequalities, and MX stands for equalities + inequalities. The first column of the table indicates the two types of right sides in equation (11) with the separable flow (S) and the multiplicative flow (M), respectively, and the last column displays their mathematical expressions. The last row of the table shows the mathematical expressions for the requisite types of admissible sets.

**Table 2** Macrosystems classification

| | EQ | IEQ | MX | $\boldsymbol{L}$ |
|---|---|---|---|---|
| S | S-EQ | S-IEQ | S-MX | $f(x) + By^*(x)$ |
| M | M-EQ | M-IEQ | M-MX | $x \otimes (a - x \otimes Cy^*(x))$ |
| $F(x)$ | $\Phi_i(y,x) = q_i$ <br> $i = \overline{1,r}$ | $\Phi_i(y,x) \le q_i$ <br> $i = \overline{1,r}$ | $\Phi_i(y,x) = q_i$ <br> $\Phi_{k+j}(y,x) \le q_{k+j}$ <br> $i = \overline{1,k},\ j = \overline{1,p}$ | |

## 4 Entropy Operator

The entropy operator $y^*(x)$ (10) is the basic element of the DSEO, which defines most of its properties. The properties of the entropy operator in itself depend on the fact of how the admissible set $F(x)$ is arranged, i. e., on the subclass of operators (see the classification in Sec. 3). In this paper, we consider the entropy EQ-operators. For this subclass of entropy operators, the admissible set is described by the system of equalities

$$F(x) = \{y : \Phi_i(y,x) = q_i, \quad q_i \geq 0, \quad i = \overline{1,r}; \quad y \in R_+^m, \quad x \in R^n\}. \tag{12}$$

**Qualitative properties.** The general approach to their investigation relies on the Lagrange method. We will display it for the EQ-operators with the Boltzmann entropy (8) and make some remarks relative to the EQ-operators with the Fermi entropy function (6) and Einstein entropy function (7).

Consider a compact subset $X \in R^n$ and a point $x \in X$. Let us introduce the Lagrange function

$$L(y,x,\lambda) = H_B(y,x) - \sum_{i=1}^{r} \lambda_i \Phi_i(y,x). \tag{13}$$

For the given case, the optimality conditions can be represented in the form

$$\begin{aligned} \Psi_j(y,x,\lambda) = \ln \frac{a_j(x)}{y_j} - \sum_{i=1}^{r} \lambda_i \frac{\partial \Phi_i(y,x)}{\partial y_j} = 0, \quad j = 1,\ldots,m; \\ \Phi_i(y,x) = q_i, \quad i = 1,\ldots,r. \end{aligned} \tag{14}$$

**Theorem 4.1.** *Let the function*

$$\Phi(y,x) = \{\phi_1(y,x),\ldots,\phi_r(y,x)\}$$

*satisfy the following conditions for all points* $x \in X$ *:*

(a) *the functions* $\Phi_k(y,x), k = 1,\ldots,r$, *are twice continuously differentiable and monotonically increasing in y;*

(b) *the matrices* $\Phi_{yy}^k = \left[\frac{\partial^2 \Phi_k(y,x)}{\partial y_j \, \partial y_i}\right] > 0;$

(c) *there exists a subset* $M(x) \subset R_+^m$ *in which the matrix*

$$J_y = \left[\frac{\partial \Phi_i(y,x)}{\partial y_j}, \quad i,j \in \overline{1,r}\right]$$

*has the full rank equal to r for all* $y \in M(x)$ *;*

(d) *the following functions are given as*

$$\Phi_i(0,x) < q_i,$$
$$\Phi_i(y_1,\dots,y_{j-1},a_j(x),y_{j+1},\dots,y_m;x) > q_i,\ \ i=\overline{1,r}.$$

*Then, the Hessian* $L_{yy}$ *of the Lagrange function (13) is negatively defined.*

Similar results exist for the subclass of operators under examination with the Fermi and Einstein entropies if in addition to the theorem conditions, we will assume the separability of the functions $\Phi_i$

$$\Phi_i(y,x) = \sum_l \Phi_{il}(y,x), \quad i=1,\dots,r. \tag{15}$$

The general procedure of the proof relies on the investigation of the quadratic form of $K(y,x,\lambda) = \langle L_{yy}(y,x,\lambda)h,h\rangle$, where the vectors are $h \in R_+^m$. The key stage of this procedure is the proof of the nonnegativity of Lagrange multipliers under the conditions of Theorem 4.1.

**Theorem 4.2.** *Let the conditions of Theorem 4.1 be met and* $a_1(x),\dots,a_m(x)$ *be continuous. Then, for all* $x \in X$ *there exists a continuous EQ-operator* $y^*(x) \ge 0$ *(8, 10).*

**Theorem 4.3.** *Let the conditions of Theorem 4.1 be met and the functions* $a(x)$ *and* $\Phi(y,x)$ *have the smoothness order equal to p. Then, for all* $x \in X$ *, the EQ-operator (8, 10) has the smoothness order equal to p.*

**Theorem 4.4.** *Let the conditions of Theorem 4.1 be met. Then, the EQ-operator (8, 10) is bounded, i. e., there exists a positive constant C such that* $y(x) \le C$.

These assertions are valid for the EQ-operators with the Einstein entropy function (7) and the Fermi entropy function (6). For the Fermi operator, Theorem 4.4 not only does assert the existence of the constant $C$, but it also defines its value: $C = \max(G_1,\dots,G_m)$. The complete proof of these assertions is given in [27].

**Lipschitz constant.** In many problems of the investigation, the DSEO affords quite useful information on the existence and estimation of the local Lipschitz constant. The problem is rather complex. We will investigate it starting from EQ-operators of the form

$$y^*(x) = \arg\max_y \{H_B(y) \mid Ty = q(x),\ \ y \in R_+^m,\ \ x \in X\}, \tag{16}$$

where $T$ is the matrix of dimension $r \times m$, in which case $r < m$ with elements $t_{ij} \ge 0$.

Elements of the matrix $T$ satisfy the following conditions:

(a) the normalizations over columns,

$$\sum_{i=1}^r t_{ij} = 1, \quad j=\overline{1,m}. \tag{17}$$

(b) the principal diagonal,

$$\sum_{j=1}^{m} t_{ij}^2 - \sum_{l\neq i}^{r}\sum_{j=1}^{m} t_{ij}t_{lj} \geq \varrho > 0, \quad i=\overline{1,r}. \tag{18}$$

The parameters $a$ and $T$ of this operator are fixed and $q(x)$ is the strictly monotonically increasing vector function, for which

$$\left\| \left[ \frac{\partial q_i(x)}{\partial x_j} \,|\, i=\overline{1,r},\ j=\overline{1,m} \right] \right\| = J_{\max}^{q} = \max_{i,j} \frac{\partial q_i(x)}{\partial x_j}. \tag{19}$$

We will introduce an auxiliary EQ-operator of the form

$$y^*(q) = \arg\max_y \{H_B(y) \,|\, Ty = q,\ y \in R_+^m,\ q \in Q\}, \tag{20}$$

Consider an auxiliary matrix $C$ with the nonnegative elements

$$c_{kj} = \sum_{i=1}^{m} a_i t_{ki} t_{ji} \geq 0, \qquad k,j = 1,\ldots,r, \tag{21}$$

the determinant of which is

$$\det(C) \neq 0. \tag{22}$$

We will denote by $\det(C^k)$ the determinant of the matrix $C^k$ formed through the replacement of the $k^{\text{th}}$ column by the vector $q$ (see (16)). This determinant can be represented in the form

$$\det(C^k) = \sum_{i=1}^{r} A_{ki} q_i, \qquad A_{ki} = (-1)^{k+i} M_{ki}, \tag{23}$$

where $M_{ki}$ is the $(k,i)$-minor of the determinant of the matrix $C$ (21).

We will introduce for consideration the following polyhedral sets:

$$W_+ = \{q : \sum_{i=1}^{r} A_{ki} q_i \geq 0, \qquad k=1,\ldots,r\}, \tag{24}$$

$$W_- = \{q : \sum_{i=1}^{r} A_{ki} q_i \leq 0, \qquad k=1,\ldots,r\}, \tag{25}$$

$$G = G(\alpha^-, \alpha^+) = \{q : 0 < \alpha^- \leqq q \leqq \alpha^+\}, \tag{26}$$

where $\alpha^-$ and $\alpha^+$ are parameters of the faces of the parallelepiped $G$, in which case the vector $\alpha^-$ has rather small components; the sign $\leqq$ denotes component-wise inequalities.

Let us introduce the designations

$$\tilde{Q} = G \bigcap W_+, \quad \text{if} \ \det(C) > 0; \tag{27}$$

$$\tilde{Q} = G \bigcap W_{-}, \quad \text{if} \quad \det(C) < 0. \tag{28}$$

We define

$$\varepsilon_{-} = \min_{\tilde{Q}} \max_{i} q_i; \qquad \varepsilon_{+} = \max_{\tilde{Q}} \min_{i} q_i. \tag{29}$$

We will introduce vectors $\varepsilon$ и and $q^*$ with the components

$$\varepsilon_k = \varepsilon_{-}, \quad q_k^* = \varepsilon_{+}, \quad k = 1,\ldots,r. \tag{30}$$

Then, the set $Q$ in (16) will be defined in the form

$$Q = \{q : 0 < \varepsilon_k \le q_k \le q_k^*; \quad k = 1,\ldots,r\} \subseteq \tilde{Q}. \tag{31}$$

The local Lipschitz constant of the operator $y^*(q)$ (16) on the set $Q$ (31) is said to be the number $L_Q$, for which the following inequality is valid:

$$\left\| y(q^1) - y(q^2) \right\| \le L_Q \left\| q^1 - q^2 \right\|, \quad (q^1, q^2) \in Q. \tag{32}$$

**Theorem 4.5.** *Let there exist two positive numbers $\delta$ and $\Delta$ such that*

$$\delta = \min_{\xi \in Q} \min_{i} z_i(\xi), \quad \Delta = \max_{q \in Q} \max_{i} \tilde{z}_i(q),$$

*where*

(a) $z_i(\xi), i = \overline{1,r}$, *is the solution of the system of equations*

$$\sum_{j=1}^{m} a_i t_{ij} \prod_{l=1}^{r} z_l^{t_{li}} = \xi_i, \quad i = \overline{1,r};$$

(a) $\tilde{z}_i(q), \; i = \overline{1,r}$, *is the solution of the system of equations*

$$Cz = q, \quad z \underset{=}{>} 0$$

*with the matrix C (21).*

*Then, the upper bound of the local Lipschitz constant on the set Q has the form*

$$L_Q = M^{-1} J^q_{\max} t y^*_{\max},$$

*where $J^q_{\max}$ is defined by equality (19),*

$$M = \min_{j} a_j \min_{i} \left( \sum_{j=1}^{m} [t_{ij}^2 - \sum_{l \ne i}^{r} t_{ij} t_{lj}] \right) \sum_{j=1}^{m} \beta_j (-\ln \delta),$$

$$t = \left( \sum_{i=1}^{r} \sum_{j=1}^{m} t_{ij}^2 \right)^{1/2}, \qquad y^*_{\max} = \max_{j} \left( a_j \Delta^{\sum_{i=1}^{r} t_{ij}} \right).$$

*The proof of this theorem is given in [28].*

## 5 Macrosystem Model with the Separable Flow

We will examine the $S$ – EQ — DSEO (see Table 2):

$$\frac{dx}{dt} = f(x) + By^*(x), \tag{33}$$

where the entropy operator is

$$y^*(x) = \arg\max_y \{H_B(y) \mid Ty = q(x),\ \ y \in R_+^m,\ \ x \in R^n\}, \tag{34}$$

$$0 < \alpha^- \le q_i(x) \le \alpha^+, \qquad i = \overline{1,r}. \tag{35}$$

If the functions $q_i(x)$ are bounded below away from zero (35), then $0 < y^*(x) \leqq C(\alpha^+)$ for all $x \in R^n$. But if $q_i(x) = 0$, then $y^*(x) = 0$.

The matrix $B$ of dimension $(n \times m)$ has the full rank $n$, in which case $n < m$. The vector function $f(x)$ is continuous differentiable and $f(0) = 0$. The Jacobian $J_f(x)$ for all $x \in R^n$ has real numbers and various eigenvalues and at least one $\eta_{\max}$ for which Re $\eta_{\max} > 0$.

Let system (33) have a unique singular point $\hat{x} \in R^n$. This means that

$$f(\hat{x}) + By^*(\hat{x}) = 0. \tag{36}$$

We assume that $\hat{x} \neq 0$. Let us consider the behavior of system (33) in the neighborhood of this point, i. e., its stability "in the small". We will transform system (33), considering the properties of the entropy operator (34).

The domain of values of (34) depends on the optimality conditions of the Lagrange function (13):

$$L(y,x) = H_B(y) + \sum_{i=1}^{r} \lambda_i \left( q_i(x) - \sum_{j=1}^{n} t_{ij} y_j \right). \tag{37}$$

The general optimality conditions (14) are simplified and have the form

$$\Theta_i(x, \lambda(x)) = \sum_{j=1}^{n} a_j t_{ij} \exp\left( -\sum_{i=1}^{r} \lambda_i(x) t_{ij} \right) - q_i(x) = 0,\ \ i = \overline{1,r}, \tag{38}$$

$$y_j^*(\lambda(x)) = a_j \exp\left( -\sum_{i=1}^{r} \lambda_i(x) t_{ij} \right),\ \ j = \overline{1,m}, \tag{39}$$

Combining these expressions with the basic equation (33), we obtain

$$\begin{gathered} \frac{dx}{dt} = f(x) + By^*(\lambda(x)), \\ \Theta(x, \lambda(x)) = 0. \end{gathered} \tag{40}$$

Here, the vector functions $\Theta$ and $y^*$ have the components (37 and 39), respectively. The singular point of this system is $\hat{x}$. In this case, $\hat{\lambda} = \hat{\lambda}(\hat{x})$ is the solution of the nonlinear equation (38) for $x = \hat{x}$.

To investigate the stability of the singular point in the small, we will resort to the linear approximation of equations (40). For the first of equations (40), we will have

$$\frac{d\xi}{dt} = A|_{\hat{x}}\, \xi, \tag{41}$$

where $A|_{\hat{x}}$ is the matrix of the linear system, whose elements are estimated at the singular point, and the vector $\xi$ is the deviation from the singular point.

The matrix is given as

$$A|_{\hat{x}} = \hat{f}_x + B\, \hat{y}^{\lambda}\, \hat{\Lambda}_x. \tag{42}$$

In this equality,

$$\begin{aligned} \hat{f}_x &= J_f(\hat{x}) = \left[ \frac{\partial f_i}{\partial x_k} \,|\, i,k = \overline{1,n} \right]|_{\hat{x}}, \\ \hat{y}^{\lambda} &= \left[ \frac{\partial y_j^*}{\partial \lambda_s} \,|\, j = \overline{1,m};\ s = \overline{1,r} \right]|_{\hat{x}}, \\ \hat{\Lambda}_x &= \left[ \frac{\partial \lambda_s}{\partial x_k} \,|\, s = \overline{1,r};\ k = \overline{1,n} \right]|_{\hat{x}}. \end{aligned} \tag{43}$$

The elements of the matrix $\hat{y}^{\lambda}$, according to (39), have the form

$$\frac{\partial y_j^*}{\partial \lambda_s} = -a_j t_{sj} \exp\left( -\sum_{l=1}^{r} \lambda_l t_{lj} \right). \tag{44}$$

The matrix $\Lambda_x$ can be found from equations (38) by differentiating them with respect to $x$ in the neighborhood of the singular point. We derive the following matrix equation:

$$\Phi^{\lambda} \Lambda^x = Q^x, \tag{45}$$

where $\Phi^{\lambda}$ is the $(r \times r)$ matrix, $\Lambda^x$ is the $(r \times n)$ matrix, and $Q^x$ is the $(r \times n)$ matrix.

The elements of the matrices $\Phi^{\lambda}$ and $Q^x$ at the singular point $\hat{x}$, according to (38), have the form

$$\begin{aligned} \hat{\Phi}_{is}^{\lambda} &= -\sum_{j=1}^{m} t_{sj}\, \hat{y}_{\lambda,js}^*, \quad i,s = \overline{1,r}, \\ \hat{y}_{js}^{\lambda} &= -a_j t_{ij} \exp\left( -\sum_{l=1}^{r} \hat{\lambda}_l t_{lj} \right), \quad j = \overline{1,m};\ s = \overline{1,r}, \\ \hat{Q}_{ij}^x &= \frac{\partial q_i}{\partial x_j}|_{\hat{x}}, \quad i = \overline{1,r};\ j = \overline{1,m}. \end{aligned} \tag{46}$$

From equations (46), we obtain

$$\hat{\Lambda}^x = [\hat{\Phi}^\lambda]^{-1}\hat{Q}^x. \tag{47}$$

Thus, the matrix $A$ (42) of the linear system (41) has the form

$$\hat{A} = \hat{f}_x + B\,\hat{y}^\lambda\,[\hat{\Phi}^\lambda]^{-1}\,\hat{Q}^x. \tag{48}$$

To simplify the subsequent transformations, we will introduce the designations

$$\hat{A} = A = F + P,\;\; F = \hat{f}_x,\;\; P = B\,\hat{y}^\lambda[\hat{\Phi}^\lambda]^{-1}\hat{Q}^x. \tag{49}$$

With the use of the new designations, the linear system (41) takes the form

$$\frac{d\xi}{dt} = (F + P)\xi, \tag{50}$$

where the matrices $F$ and $P$ have constant elements. According to the initial assumptions, the matrix $F$ has real numbers and various eigenvalues, in which case

$$\eta_{\max} = \max_{1\le i\le n} \eta_i > 0. \tag{51}$$

$$\|F\| = \eta_{\max}. \tag{52}$$

Let the matrix $P$ also have real numbers and various eigenvalues $\varrho_i$ with the negative real part:

$$-\varrho_{max} = \max_{0\le i\le n}(-\varrho_i). \tag{53}$$

Consider the auxiliary lincar system

$$\frac{d\zeta}{dt} = P\zeta. \tag{54}$$

We will use the notion of a matriciant of the linear system (see [29]). For an auxiliary system, the matriciant is

$$\Omega_{t_0}^t(P) = \exp(P(t - t_0)). \tag{55}$$

It is known that there exists a number $C(P, -\varrho_{\max})$ such that

$$\|\exp(Pt)\| \le C(P, -\varrho_{\max})\exp(-\varrho_{\max}t),\;\; t > 0. \tag{56}$$

According to the properties of the matriciant,

$$\begin{gathered} \Omega_{t_0}^t(F + P) = \Omega_{t_0}^t(P)\Omega_{t_0}^t(S), \\ S = [\Omega_{t_0}^t(P)]^{-1} F\, \Omega_{t_0}^t(P). \end{gathered} \tag{57}$$

**Lemma 5.1.** Let F and P be matrices with constant elements. For any $0 \le s \le t$, the following estimate is valid:

$$\left\| \Omega_s^t (F+P) \right\| \le C \exp(-\varrho_{\max} (t-s)) \exp(C \| F \| (t-s)).$$

The proof of the lemma is given in the Appendix[1]. In view of (52), this estimate assumes the form

$$\left\| \Omega_s^t (F+P) \right\| \le C \exp((-\varrho_{\max} + \eta_{\max})(t-s)). \tag{58}$$

Because the matriciant is the matrix of the fundamental solutions of the linear system (50), there exists

**Theorem 5.1:** *The DSEO (40) has a stable singular point* $\hat{x}$ *in the small if the following conditions are met:*

(a) *the matrices P and F of the linearized system (50) have real numbers and various eigenvalues;*
(b) *the maximum eigenvalues* $-\varrho_{\max} < 0$ *and* $\eta_{\max} > 0$ *of the matrices P and F satisfy the condition*

$$-\varrho_{\max} + \eta_{\max} < 0.$$

Let us note that the singular point $\hat{x} = 0$ is unstable because $P = 0 \, (-\varrho_{\max} = 0)$, while $\eta_{\max} > 0$.

The instability of nonzero singular points can also take place. This instability depends on the parameters $\alpha^-$ and $\alpha^+$ of the entropy operator (34, 35) (at all other fixed parameters).

## 6 Macrosystem Model mwith the Multiplicative Flow

We will consider the DSEO equation with a multiplicative flow (see the third row of Table 2):

$$\frac{dx}{dt} = x \otimes (a - x \otimes P y^*(x)), \tag{59}$$

where $y^*(x)$ is the entropy operator and $P = [p_{ij} \ge 0]$ is the $(r \times m)$ matrix.

It is easy to note that under the initial conditions $x(0) > 0$, the solution of the system is $x(t) > 0$ for all $t > 0$ because at $x = 0$, the right side of the equation goes to zero. Thus, the DSEO with the multiplicative flow belongs to the class of positive dynamic systems.

The trajectories of the DSEOs under examination are not only positive, but they belong to the closed and bounded domain of the positive orthant $R_+^n$.

---

[1] The lemma is proved by B. S. Darkhovsky.

**Theorem 6.1.** *Let the following conditions be met:*

(a) *the conditions of Theorem 4.1;*
(b) *there exist compact subsets* $X_i^* \subset R_+^n, i = 1,\dots,n$ *, such that for all* $x \in X_i^*$

$$a_i - (P^i y(x))x_i \geq 0,$$

*and for all* $x \in X_i^*$

$$a_i - (P^i y(x))x_i < 0,$$

*where* $X_i^*$ *is an additional subset of* $X_i^*$ *and* $P^i$ *is the* $i^{th}$ *row of the matrix P.*

*Then, in* $R_+^n$ *there exists a bounded set* $Y \supset (\bigcup X_i^*)$ *such that for all initial points* $x(0) \in R_+^n$ *, the trajectory* $x(t) \in Y$ *for* $t \geq 0$ *.*

The proof of the theorem is given in [27].

## 7 Dynamic Procedures of the Entropy Restoration of Images by Projections

We will consider the orthogonal tomograph (Fig. 2) that illuminates a plane specimen by beams of photons in two perpendicular directions. The specimen is monochromatic and is represented by a collection of pixels with digital coordinates $(i, j)$, where $i = 1,\dots,n$ and $j = 1,\dots,m$. The distribution of the degree of darkening (gradations of the "gray" color) over the specimen surface, which accounts for the distribution of the absorbed photons in the specimen, is specified by the density function $\psi(i, j)$:

$$0 \leq \psi(i, j) \leq 1, \qquad i = \overline{1,n}; \; j = \overline{1,m}. \tag{60}$$

Two orthogonal projections are accessible for the observation:

(a) the horizontal projection,

$$\tilde{h}_i = \sum_{j=1}^{m} \psi(i, j), \qquad i = 1,\dots,n; \tag{61}$$

(b) the vertical projection,

$$\tilde{v}_j = \sum_{i=1}^{n} \psi(i, j), \qquad j = 1,\dots,m-1; \tag{62}$$

The measuring device of the tomograph (a detector) is not ideal, i. e., the measured projections contain errors $\xi$, which we will consider to be independent random quantities with M $\xi = 0$ and $\xi^2 = \sigma_\xi^2$.

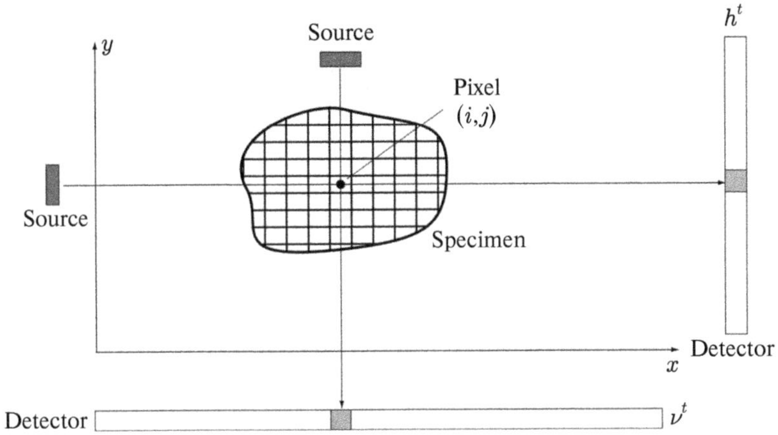


**Fig. 2**

Considering (61 and 62), equations for the measured horizontal and vertical projections can be represented in the form

$$\begin{aligned}&\sum_{j=1}^{m}\psi(i,j)=h_i=\tilde{h}_i+\xi_i,\quad i=1,\dots,n;\\&\sum_{i=1}^{n}\psi(i,j)=v_j=\tilde{v}_j+\xi_{j+n},\quad j=1,\dots,m-1.\end{aligned}\tag{63}$$

In order that we can specify the level of "noise" with respect to "useful signals", we will introduce the relative intensity of noise in the detector:

$$\beta=\frac{\sigma_\xi}{\max(\max_i\tilde{h}_i,\max_j\tilde{v}_j)}.\tag{64}$$

Thus, knowing the projections $h$ и and $v$ (63), it is necessary to restore the density function of the image $\psi$. In solving this problem, we will follow the variational principle of image generation [31], the basis for which is the stochastic model of the distribution of photons in the specimen. The natural functionals for this model are taken to be the generalized informational Fermi—Dirac entropy because there are lower and upper bounds on the level of the density function (60).

The stochastic mechanism presupposes the existence of some of its prior characteristics, namely, the prior probabilities of absorption of a photon in a pixel $(i,j)$. In the given context, such a characteristic is a *prior image* with the density function $E(i,j),\ i=\overline{1,n};\ j=\overline{1,m};\ 0\le E(i,j)\le 1$.

The generalized informational Fermi—Dirac entropy for the given case has the form

$$H(\psi)=-\sum_{i,j=1}^{n,m}\left(\psi(i,j)\ln\frac{\psi(i,j)}{E(i,j)}+(1-\psi(i,j))\ln(1-\psi(i,j))\right),\tag{65}$$

The standard "entropy" procedure of computer tomography consists in defining the complete set of projections with the aid of irradiation of a specimen and solving the entropy maximization problem [30], [31]:

$$H(\psi) \Rightarrow \max_{\psi}, \quad \psi \in D, \tag{66}$$

where the set $D$ is defined by equalities (63). As a result, we obtain the entropy-optimal restoration of the image. Bearing in mind that before the beginning of this procedure (irradiation and measurement of projections, the solution of the entropy maximization problem), we have the prior image $E$, it is natural to designate the function $\psi_*$ as a *posterior image*.

Such a procedure will be called *static*. It requires quite a high intensity of the flow of photons so as to afford a sufficient noise protectability of a recoverable image. However, for some classes of tomographic investigations, for example, medical ones, a high irradiation intensity is quite undesirable on account of side effects.

It becomes possible to avoid these shortcomings if one reaches the necessary noise protection not through an increase of the irradiation intensity, but through a sequential refinement of the image synthesized. This procedure, which is called *dynamic*, lies in the sequential irradiation in time at substantially small intensities and the multiple restoration of images with the aid of special algorithms for processing these images.

The basic idea of the dynamic procedure lies in the sequential receiving of information on the specimen under investigation in the form of its projections and in the solution of the sequence of appropriate entropy maximization problems. In other words, the dynamic procedure consists of the sequence of stages $t = 1,2,3,\ldots$, at the input of which there is a *t-prior image* $E^t$ and projections $h^t, v^t$, while at the output there is a *t-posterior image* $\psi_*^t$.

The prior image for the $(t+1)^{\text{th}}$ step is set up on the basis of $t, t-1, t-2, \ldots$ posterior images. The block diagram of the procedure is shown in Fig. 3. In block 1, the $t$-posterior image $\psi_*^t$ is estimated on the basis of the prior information on the projections $h^t, v^t$, and the $t$-prior image $E^t$. The transformation of the initial information to the $t$-posterior image is described by the entropy operator (66).

Block 2 implements the feedback principle in the given procedure. In this block, the calculations are made of the $t+1$-prior image $E^{t+1}$ on the basis of information on the $t$-prior image $E^t$ and $t, t-1, t-2, \ldots$ posterior images $\psi_*$, i. e.,

$$E^{t+1} = \boldsymbol{L}(E^t, \psi_*^t, \psi_*^{t-1}, \psi_*^{t-2}, \ldots), \tag{67}$$

where $\boldsymbol{L}$ is the operator of the given transformation.

It is seen from the presented block diagram that this procedure of the restoration of images represents a discrete DSEO, in which block 1 is described by the entropy operator (66) and block 2, by the dynamic nonlinear operator (67).

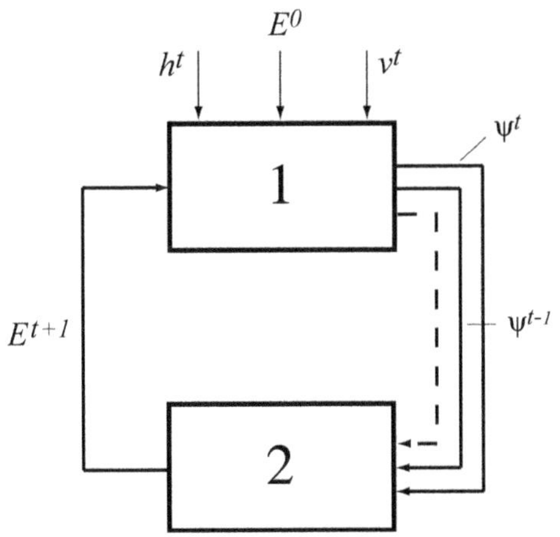


**Fig. 3**

The right side of equality (67) includes the common collection of information arrays, which, in principle, can be used for the formation of the $(t+1)$-prior image. Let us consider some special cases.

One of the cases relates to the consideration of the Markov version of the procedure when for the formation of $E^{t+1}$ use is made of the information on only the $t$-posterior image (Fig. 4*a*):

$$E^{t+1} = \mathbf{L}(E^t, \psi_*^t); \tag{68}$$

In the second case, information collections are used at $t, t-1, t-2, \ldots$ steps to define the current mean (CM) of the t-posterior image (Fig. 4*b*):

$$E^{t+1} = \mathbf{L}(E^t, \overline{\psi}_*^t); \tag{69}$$

Finally, in the third case use is also made of information collections at $t, t-1, t-2, \ldots$ steps to define the current mean and the current dispersion (CD) of the t-posterior image (Fig. 4*c*):

$$E^{t+1} = \mathbf{L}(E^t, \overline{\psi}_*^t, d_*^t); \tag{70}$$

Let us consider the following classes of operators $mathcalL$:

(a) the identical operator (*I-feedback*),

$$E^{t+1}(i,j) = \psi_*^t(i,j); \tag{71}$$

(b) the operator at the maximum residual with the constant weight $\alpha$ (*RCW-feedback*),

$$E^{t+1}(i,j) = \begin{cases} E^t(i,j), \text{ where } \{(i,j) : | E^t(i,j) - \overline{\psi}_*^t(i,j) | \leq \delta\}, \\ E^t(i,j) + \alpha[E^t(i,j) - \overline{\psi}_*^t(i,j)], \\ \qquad \text{where } \{(i,j) : | E^t(i,j) - \overline{\psi}_*^t(i,j) | > \delta\}; \end{cases} \tag{72}$$

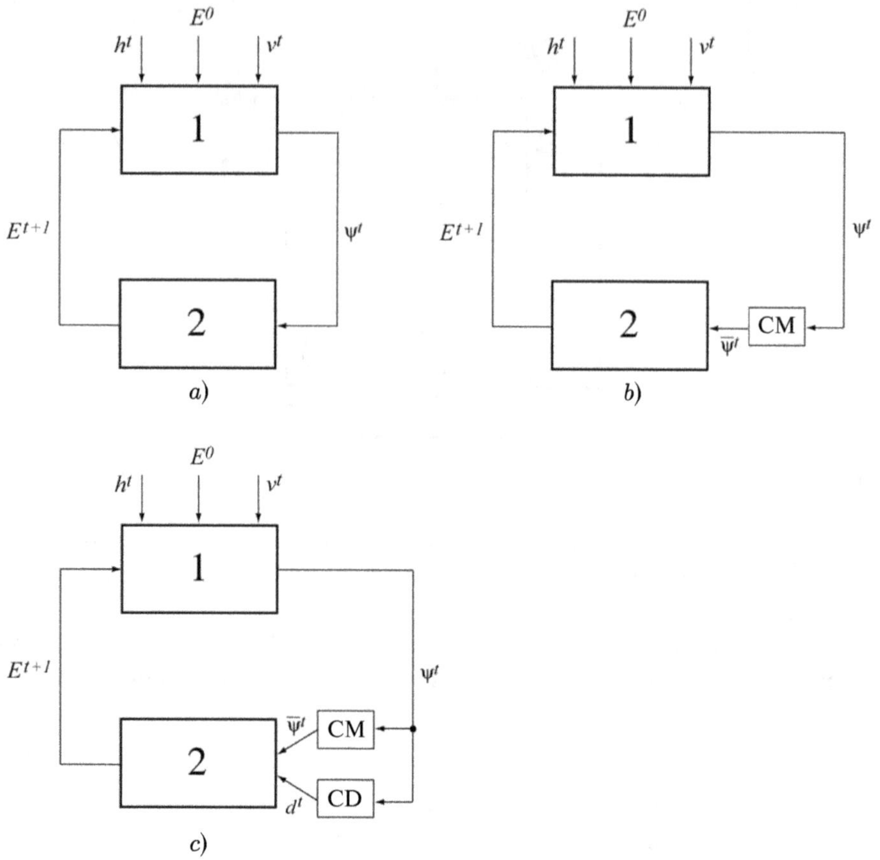


**Fig. 4**

(c) the operator at the maximum residual with the adaptation over the current dispersion $d_*^t$ (*RA-feedback*),

$$E^{t+1}(i,j)=\begin{cases}E^t(i,j), & \text{where } \{(i,j):|E^t(i,j)-\overline{\psi}_*^t(i,j)|\leq\delta\},\\ E^t(i,j)+\alpha(d_*^t)[E^t(i,j)-\overline{\psi}_*^t(i,j)], & \\ \qquad\text{where } \{(i,j):|E^t(i,j)-\overline{\psi}_*^t(i,j)|>\delta\}.\end{cases} \tag{73}$$

Thus, we have the following three classes of dynamic procedures of the entropy restoration of images over orthogonal projections:

*The dynamic procedure with the I-feedback:*

$$E^{t+1}=\alpha\arg\max_{\psi^t}\{H(\psi^t,E^t)\,|\,\psi^t\in D(h^t,v^t)\}. \tag{74}$$

*The dynamic procedure with the RCW-feedback:*

$$E^{t+1} = E^t + [1 - \lambda(E^t, \overline{\psi}_*^t)]\alpha[E^t - \overline{\psi}_*^t], \tag{75}$$

$$\lambda(E^t, \overline{\psi}_*^t) = \begin{cases} 1, & \text{if } |E^t(i,j) - \overline{\psi}_*^t(i,j)| \le \delta, \\ 0, & \text{if } |E^t(i,j) - \overline{\psi}_*^t(i,j)| > \delta. \end{cases} \tag{76}$$

*The dynamic procedure with the RA-feedback:*

$$E^{t+1} = E^t + [1 - \lambda(E^t, \overline{\psi}_*^t)]\alpha(d_*^t)[E^t - \overline{\psi}_*^t], \tag{77}$$

$$\lambda(E^t, \overline{\psi}_*^t) = \begin{cases} 1, & \text{if } |E^t(i,j) - \overline{\psi}_*^t(i,j)| \le \delta, \\ 0, & \text{if } |E^t(i,j) - \overline{\psi}_*^t(i,j)| > \delta. \end{cases} \tag{78}$$

In equations (71, 73), $i \in I$ and $j \in J$, the initial prior image $E^1$ and the value of $\delta$ are prescribed and the current mean $\overline{\psi}_*^t$ and the current dispersion $d_*^t$ are defined by the equalities

$$\begin{aligned} \psi_*^{t+1} &= \psi_*^t + \frac{1}{t+1}(\psi_*^{t+1} - \overline{\psi}_*^t), \\ d_*^{t+1} &= d_*^t + \frac{1}{t+1}(d_*^t + (\psi_*^{t+1} - \psi_*^t)^2). \end{aligned} \tag{79}$$

The posterior image $\psi_*^t$ is defined by the entropy operator (66).

The LENA experiment adopted in the IEEE Image Processing illustrates the effectiveness of the suggested dynamic procedures of the entropy restoration of images by projections. The initial information and the results are shown in Fig. 5.

The aim of the experiment is to restore the image on the test photograph using the information on two orthogonal projections measured with noise at a noise/signal ratio of $\beta = 0.3$.

Fig. 5 displays the screen to which the experiment results are brought out. In the upper left window, the test image (the spot on the eye) is located. To the right and below from it, the projections measured with noise are shown. The next window displays the 0-prior image (standard one). To the right upper window, the restored image is brought out with the aid of the static procedure. In the lower row in the left window, the result of the work of the dynamic procedure (10 iterations) with the *I*-feedback is shown. Finally, the result of 10 iterations of the dynamic procedure with the *RCW*-feedback is displayed in the lower right window.

## 8 Model of the Labor Market with the Entropy Operator

The problem of the incomplete employment (unemployment) remains to be one of the most pressing economic and social problems [32]. On the one hand, the market organization of the economic system presupposes the availability of the positive

**Fig. 5**

balance between the supply and demand of the labor force. This balance is a tool of the effect on the productivity and quality of the labor, but, on the other, it is found to be an additional load on the working population and is a source of social conflicts.

The age and cohort structures are built up in the labor market. This process is indeterminate. In the model presentation, it can be considered to be random. Therefore, we will specify the employment structure by the density functions of the distribution of the employed either in age $\tilde{a}$ (the age structure of employment (ASE) — $\chi(\tilde{a},\tilde{t})$) or in cohorts $\tilde{c}$ (the cohort structure of employment (CSE) — $w(\tilde{c},\tilde{t})$, which belong to productive (able-bodied) ages or cohorts.

Here, we will consider cohorts by the date of birth. Therefore, the cohort $\tilde{c}$, the age $\tilde{a}$, and the calendar time $\tilde{t}$ are related by the equality

$$\tilde{c} = \tilde{t} - \tilde{a}. \tag{80}$$

The availability of this relation enables us to use for the study of the labor market one of the characteristics of the employment structure, for example, the CSE function $w(\tilde{c},\tilde{t})$.

To model the dynamics of the CSE function, we will introduce the notion of the *quasi-elasticity*

$$\gamma(\tilde{c},\tilde{t}) = \frac{1}{w(\tilde{c},\tilde{t})}\frac{dw(\tilde{c},\tilde{t})}{d\tilde{t}}. \tag{81}$$

The evolution of the labor market state in time occurs under the effect of the social-economic system. In the given case, this system is the source of resources — the labor and economic ones (work places). The labor resources are specified by the supply of the labor force and the economic resources are specified by the demand for it. The internal mechanism of the labor market is the competition for work places. It is convenient to perform the quantitative description of this process in terms of the competitive capacity: the inherent one when the case in point involves the individuals who occupy the work places and strive to keep them and the comparative one for new participants of the labor market, who strive to push the former from their work places. The interaction of the resource factors and keep-push factors will be defined in terms of the *quasi-elasticity* $\gamma(\tilde{c},\tilde{t})$ (81).

The effect of the competition and the labor and economic resources on the quasi-elasticity will be described in terms of utility, in which case each cohort, as a member of the labor market, is defined by its own level of the utility. One of the components of the utility is the size of the cohort occupying the work places.

The distribution of productive cohorts over work places appears to be random, and in the time scale of the calendar time, it occurs substantially faster than the evolution of the competitive medium. This is the basis for the use of the principle of local equilibria in combination with the description of the equilibrium in terms of the conditional maximum of the entropy for modeling the distribution of productive cohorts.

Let us consider the population of the age $\tilde{a}$ from the interval of the capacity for work (age window) $\tilde{A}_w = [\tilde{a}_0, \tilde{a}_1]$, ($a \in \tilde{A}_w$).

We will examine the labor market dynamics in the final time interval $\tilde{T} = [\tilde{t}_0, \tilde{t}_1]$, where $\tilde{t}_1 = \tilde{t}_0 + \tilde{a}_0$. This consideration does not contracts the domain of application of the model, but enables us to represent the basic principles of the construction and the structure without technical details.

At the beginning of the time interval $\tilde{t}_0$, the elder cohort (of the age of $\tilde{a}_1$) has the birth date, according to (80), equal to

$$\tilde{c}_0 = \tilde{t}_0 - \tilde{a}_1. \tag{82}$$

The younger cohort, whose age at the beginning of the time interval is equal to $\tilde{a} = \tilde{a}_0$ and which enters into the age window $\tilde{A}_w$ at the end of the time interval $\tilde{t}_1$, has the date of birth equal to

$$\tilde{c}_1 = \tilde{t}_0. \tag{83}$$

Thus, the set of cohorts of interest to us is $\tilde{K} = [\tilde{c}_0, \tilde{c}_1]$.

We will introduce in parallel the standardized variables

$$\begin{aligned} &a = \tilde{a} - \tilde{a}_0, \quad a \in A_w = [0, a^*], \\ &t = \tilde{t} - \tilde{t}_0, \quad t \in T = [0, \tilde{a}_0], \\ &c = \tilde{c} - \tilde{c}_0, \quad c \in K = [0, \tilde{a}_1], \end{aligned} \tag{84}$$

In these equalities, $a^* = \tilde{a}_1 - \tilde{a}_0$. .

It should be noted that at each instant of time t, only a portion of the cohorts c from the set $K$ (84) belongs to the age window $A_w$. All these cohorts form the subset $C_t$ that takes the form

$$C_t = [t, t + a^*] \in K, \quad t \in T. \tag{85}$$

According to the definition of the quasi-elasticity $\gamma(c,t)$ (81), we obtain the following system of differential equations:

$$\frac{dw(c,t)}{dt} = w(c,t)\gamma(c,t), \quad t \in T, \quad c \in C_t. \tag{86}$$

Because all information that is necessary for the labor market investigation is related to the discrete time scale, it is expedient to pass on to the difference approximation of equations (86). Using the Euler diagram with the constant step equal to one year, we obtain

$$\begin{gathered} w(c,t+1) = w(c,t)[1 + \gamma(c,t)], \\ c \in C_t, \quad t \in T_1, \end{gathered} \tag{87}$$

where $C_t$ is defined by equality (85), and

$$T_1 = T \setminus \tilde{a}_0 = [0, \tilde{a}_0 - 1]. \tag{88}$$

The initial distribution w(c, 0) is preset by the conditions

$$\begin{gathered} w(c,0) = w^0(c), \quad 0 < w^0 < 1, \quad c \in C_0, \\ \sum_{c \in C_0} w^0(c) = 1. \end{gathered} \tag{89}$$

The boundary conditions $w^B(t+1)$ for equations (87) are

$$w(t + a^* + 1, t + 1) = w^B(t+1), \quad 0 < w^B(t+1) < 1, \; t \in T_1. \tag{90}$$

Here, it is pertinent to make some remarks on the properties of trajectories of equations (86) and (87).

The first of them concerns the fact that the differential equations (86) have positive solutions under the positive initial and boundary conditions. But in the transition to difference equations, this property is lost and it is necessary to control the positiveness of the trajectories of equations (87). The distribution $w(c,t+1)$ that results from equations (87) is not normalized. Therefore, at each step $t+1$, it is necessary to normalize the solution derived with due regard for the boundary conditions (90).

Summing up these remarks, we can represent the equations of the labor market model in the form

$$\hat{w}(c,t+1) = \begin{cases} w(c,t)[1 + \gamma(c,t)], \; c \in C_t, & \text{if } \hat{w}(c,t+1) \geq 0, \\ 0, & \text{if } \hat{w}(c,t+1) < 0. \end{cases} \tag{91}$$

$$w(c,t+1)=\frac{\hat{w}(c,t+1)}{N(t+1)},\quad c\in(C_t\setminus t), \tag{92}$$

where

$$N(t+1)=\sum_{c\in(C_t\setminus t)}\frac{\hat{w}(c,t+1)}{1-w^B(t+1)}. \tag{93}$$

In these equations, the condition $c\in(C_t\setminus t)$ permits us to exclude at each step the elder cohorts, which leave the labor market.

We will now refer to the quasi-elasticity $\gamma(c,t)$. Using the widespread presentation of the competitive medium in the labor market [32], we will represent the quasi-elasticity function in the form

$$\gamma(c,t)=\rho(c,t)+\kappa(c,t)+\sigma(c,t), \tag{94}$$

where

(a) $\rho(c,t)$ is the inherent competitive capacity of the cohort $c$;
(b) $\kappa(c,t)$ is the comparative competitive capacity of the cohort $c$ with respect to other cohorts $l\neq c$;
(c) $\sigma(c,t)$ is the interaction of the supply and demand of the labor force.

The inherent competitive capacity is a conservative factor. It keeps the cohort $c$ in the work places occupied by it. Therefore, the component $\rho(c,t)<0$ and it decreases with the growth of c. From this viewpoint, the suitable function is the function of the form

$$\rho(c)=\rho\exp(-\zeta c),\quad c\in C_t. \tag{95}$$

The comparative competitive capacity of the cohort $c$ is the sum of the comparative competitive capacities with respect to the cohorts $l\in(C_t\setminus c)$. These competitive capacities are proportional to $w(c,t)$ and the proportionality factor depends on the "utilities" of the cohorts $c$ and $l$. According to the theory of utility [7], the comparative competitive capacity depends on the "distance" between the cohorts $\Gamma(c,l)$ and the comparative utility $\Theta(c,l,t)$. They are given in the form

$$\Gamma(c,l)=\exp(-\alpha\,|\,c-l\,|), \tag{96}$$

$$\Theta(c,l,t)=\theta\exp[\eta_c(u(c,t)-u(l,t))], \tag{97}$$

where $\alpha$ is the parameter of the intensity of the effect of the distance between cohorts, $\theta$ and $\eta_c$ are the parameters of the scale and the intensity of the effect of the balance of utilities of the cohorts $c$ and $l$, respectively, and $u(c,t)$ is the utility function of the cohort $c$.

The choice of the class of utility functions with due regard for the features of a specific problem is quite not simple. Therefore, a common way is to invoke some general concepts, for example, the simplicity of such a class. From this viewpoint, it is possible to use a linear utility function. However, in this case it seems obvious that the utility of a cohort at its appreciable size $x(c,t)$ is not proportional to its dimension. Therefore, for the given case, a logarithmic utility function is more suitable:

$$u(c,t) = \ln x(c,t). \tag{98}$$

We will model the distribution of the size of working cohorts by applying the macrosystem approach developed in [8]. The factors that form the basis of this approach are two hypotheses with respect to the process of the distribution of workers over the cohorts. The first hypothesis presupposes that this process runs rather fast. Consequently, it is possible to examine its local-stationary state. The second hypothesis relates to a stochastic mechanism of the distributive process. According to this process, the workers are distributed over cohorts in a random way, independent of one another, and with prior probabilities, for which certain values of the CSE-function $w(c,t)$ are taken.

In this case, the distribution $x(c,t)$ is defined by the entropy operator of the form

$$\begin{gathered} H[X(t)] \Rightarrow \max_x, \\ \sum_{c\in C_t} x(c,t) = R^E(t), \\ 0 < x(c,t) < S(c,t), \quad c \in C_t, \end{gathered} \tag{99}$$

where the entropy is

$$\begin{aligned} H[X(t)] = -\sum_{c\in C_t} \Bigg( & x(c,t)\ln\frac{x(c,t)}{w(c,t)} + (S(c,t) - \\ & - x(c,t))\ln(S(c,t) - x(c,t))\Bigg), \end{aligned} \tag{100}$$

$X(t) = \{x(t,t),\dots,x(t+a^*,t)\}$, $R^E(t)$ is the common demand for the labor force, and $S(c,t)$ is the distribution of the supply of the labor force over cohorts.

In this expression, $x^*(c,t)$, $c \in C_t$, is the entropy-optimal distribution of occupied cohorts, which is defined by the solution of the problem (99, 100):

$$x^*(c,t) = \frac{w(c,t)S(c,t)}{w(c,t) + z^*(t)[1 - w(c,t)]}, \tag{101}$$

where the exponential Lagrange multiplier $z^*(t)$ is the solution of the equation

$$\psi(z,t) = \frac{1}{R^E(t)} \sum_{c\in C_t} \frac{w(c,t)S(c,t)}{w(c,t) + z^*(t)[1 - w(c,t)]} = 1. \tag{102}$$

Combining expressions (96), (97), and (98), we obtain an expression for the comparative competitive capacity in the form

$$\kappa(c,t) = \theta \sum_{l \in C_t \setminus c} \exp(-\alpha |c - l|) \left( \frac{x^*(c,t)}{x^*(l,t)} \right)^{\eta}. \tag{103}$$

The supply and demand of the labor force, which is the third component of the quasi-elasticity function (94), reflects the interaction of the economy (demand for the labor, work places $R^E(t)$ and the population (supply of the labor $S(t)$ and $S(c,t)$ ).

We will introduce the relative variables

$$r^E(t) = \frac{R^E(t)}{S(t)}, \qquad l(c,t) = \frac{S(c,t)}{S(t)}, \tag{104}$$

where $S(t)$ is the common supply of the labor at the instant of time *t*.

The component $\sigma(c,t)$ (94) is proportional to the relative supply $l(c,t)$ and is representable in the form

$$\sigma[r^E(t), l(c,t)] = \beta r^E(t) l(c,t), \tag{105}$$

where $\beta$ is the parameter of the scale.

Thus, equations (91, 92, 101, 102) describe the dynamic model of the labor market with the entropy operator.

The developed dynamic model was tested by means of real information on the state of labor markets in nine countries of the European Union (Belgium, Great Britain, Greece, Denmark, Ireland, Italy, Luxembourg, the Netherlands, and France). This information covers the period from 1983 to 1996. Using this information, the identification was performed of parameters of the model and its subsequent test with the defined optimal parameters.

Figures 6 and 7 display the plots of the functions $w(\tilde{c},1989)$ (real and model ones for optimal parameters), which show a sufficient adequacy of the model.

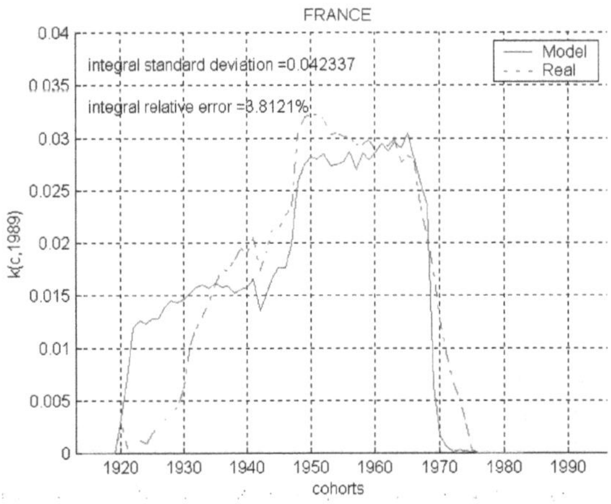


**Fig. 6**

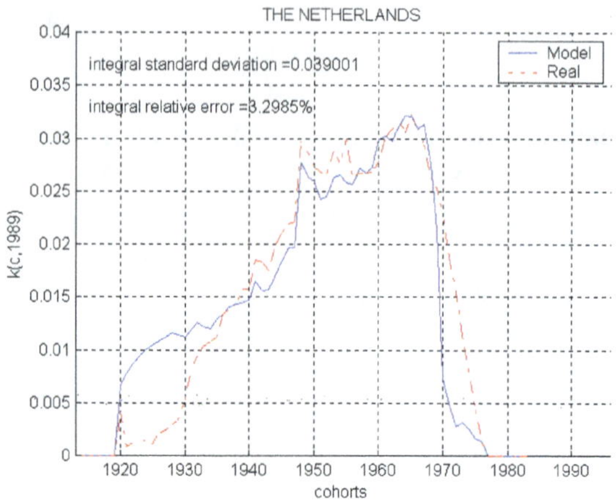


**Fig. 7**

## 9 Conclusions

Macrosystem models with the entropy operator form a new object of the mathematical theory of systems, the investigation of which is at the initial stage. They arose from applied problems in various branches of science, engineering, economics, demography, art, etc. This fact enables us to predict the growth of research interest in this object.

These models play a definite positive role from the viewpoint of the common understanding of the development of science. We will elucidate this thesis by the following example. Let us represent science in the form of a field “sown” by investigation objects that are studied during 10, 20, …, 50, … years. To support the fertility of this field, the lower inexhausted layers of the soil are upturned (“difficult” problems for conventional objects), while the upper layers, which noticeably lost their fertile properties, descend downward. As a result, a definite development of science occurs, but an extensive one in the time prospect, at least because the “thickness” of the fertile layer is finite. To prevent the occurrence of stagnant phenomena, it is necessary to try to go from time to time beyond this field to “virgin” soils. The macrosystem models with entropy operator implement this attempt to a definite degree.

## References

1. Boltzmann, L.: On the Link between the Second Beginning of Mechanical Calory Theory and Probability Theory in Theorems of Thermal Equilibrium. In: Collection of Works. Classics of Science, Nauka, Moscow, pp. 190–236 (1984)
2. Landau, L.D., Lifshitz, E.M.: Statistical Physics. Nauka, Moscow (1964)

3. Meerkov, S.M.: Mathematical Theory of Behavior — Individual and Collective Behavior of Retardable Elements. Math. Biosciences. 43, 41–106 (1979)
4. Haken, H.: Synergetics. Springer, Heidelberg (1974)
5. Prigogine, I., Stengers, I.: Order out of Chaos. Heinemann, Longon (1984)
6. Weidlich, W., Haag, G.: Interregional Migration: Dynamic Theory and Comparative Analysis. Springer, Berlin (1988)
7. Weidlich, W.: Sociodynamics. In: The Systems Approach to Modeling in Social Sciences. Harwood Academic Publishers, The Netherlands (2000)
8. Popkov Yu, S.: Theory of Macrosystems. Springer, Heidelberg (1995)
9. Leontovich, M.A.: The Basic Equation of Kinetic Theory of Gases from the Viewpoint of Theory of Random Processes. J. of Numerical Math. And Math. Physics 5(3-4), 211–231 (1935)
10. Stratonovich, R.L.: Nonlinear Nonequilibrium Thermodynamics. Nauka, Moscow (1985)
11. Helbing, D.: A Mathematical Model for the Behavior of Individuals in Social Field. J. Math. Sociology 19(3), 189–219 (1994)
12. Klimontovich Yu, L.: Statistical Theory of Open Systems. TOO "Yanus", Moscow (1995)
13. Wilson, A.G.: Catastrophy Theory and Bifurcations. Applications to Urban and Regional Systems. Croom Helm, London (1981)
14. Popkov Yu, S., Ryazantsev A. N.: Spatio-Functional Models of Demographic Processes, United Nations Fund for Population Activity, New York (1980) (Preprint)
15. De Groot, S.T., Mazur, P.: Nonequilibrium Thermodynamics. Mir, Moscow (1964)
16. Popkov Yu, S.: Locally Stationary Models of Nonequilibrium Dynamics of Macrosystems with Self-reproduction. Dokl. Akad. Nauk SSSR 303(3), 14–16 (1988)
17. Popkov Yu, S.: A New Class of Dynamic Macrosystem Models with Self-reproduction. Environment and Planning A (21), 739–751 (1989)
18. Shvetsov, V.I.: Stationary States and Bifurcations in Models of Regional Dynamics. Dynamics of Inhomogeneous Systems. In: Proc. VNIISI, Moscow, pp. 34–45 (1989)
19. Popkov Yu, S., Shvetsov, V.I.: The Principle of Local Equilibria in Regional Dynamic Models. Math. Modeling 2(5), 40–59 (1990)
20. Popkov Yu, S.: Dynamic Models of Macrosystems with Self-reproduction and their Applications to the Analysis of Regional Systems. Ann. Regional Sci. (27), 165–174 (1993)
21. Kitaev, O.V.: Macrosystem Models of Population Dynamics. Dynamics of Inhomogeneous Systems. In: Proc. VNIISI, Moscow, pp. 56–70 (1997)
22. Popkov Yu, S., Shvetsov, V.I., Weidlich, W.: Settlement Formation Models with Entropy Operator. Ann. Regional Sci. 32, 267–294 (1998)
23. Volterra, V.: Mathematical Theory of Struggle for Existence. Nauka, Moscow (1976)
24. Eigen, M., Schuster, P.: The Hypercycle. A Principle of Natural Self-organization. Springer, Berlin (1979)
25. Popkov Yu, S.: Macrodynamics of a Class of Nonlinear Markovian Processes (Comparison of Models of Geographical Systems). Geographical Syst. 3, 59–72 (1996)
26. Popkov Yu, S.: Macrosystem Models of Dynamic Stochastic Networks and GRID Technologies. Automation and Remote Control (12), 143–163 (2003)
27. Popkov Yu, S.: Positive Dynamic Systems with the Entropy Operator. Automation and Remote Control (3), 104–113 (2003)
28. Popkov Yu, S., Rublev, M.V.: Estimation of the Local Lipschitz Constant of the Bq-Operator. Automation and Remote Control (7), 54–65

29. Gantmacher, F.R.: The Theory of Matrices. Chelsea Publishing, New York (1959)
30. Byrne, C.L.: Iterative Image Reconstruction Algorithms Based on Cross-Entropy Minimization. IEEE Trans. Img. Proc. 2(1), 444–466 (1993)
31. Popkov Yu, S.: The Variational Principle of Restoration of Images by Projections. Automation and Remote Control (12), 131–139 (1997)
32. Wissen, L., Popkov, A.Y., Popkov, E.Y., Popkov Yu, S.: A Labor Market Model with the Entropy Operator. Ekon. And Mat. Metod. 40(2), 99–112 (2004)
33. Tsupkin Ya, Z., Popkov Yu, S.: Theory of Nonlinear Discrete Systems. Nauka, Moscow (1973)
34. Antipin, S.A.: Continuous and Iterative Processes with Operators of Design and Type of Design, Voprosy Kibern., Nauka, Moscow (1989)
35. Bobylov, N.A., Popkov, A.Y.: Periodic Oscillations in Nonlinear Systems with the Operator ArgMin. Automation and Remote Control (11), 13–23 (2002)

## Appendix

*Proof of Lemma 5.1.*

We will resort to the properties of the matriciant (57). Because P is a constant matrix, $\Omega_s^t(P) = \exp\big(P(t-s)\big)$. Therefore,

$$\begin{aligned} \Omega_s^t(F+P) &= e^{P(t-s)}\Omega_s^t\left(e^{-P(t-s)}Fe^{P(t-s)}\right) = \\ &= e^{P(t-s)}(E + \int_s^t e^{-P(u-s)}Fe^{P(u-s)}\,du + \\ &+ \int_s^t e^{-P(u-s)}Fe^{P(u-s)}\,du \int_s^u e^{-P(v-s)}Fe^{P(v-s)}dv + \cdots). \end{aligned} \tag{106}$$

The written row uniformly converges in any finite interval. For this reason, it can be multiplied termwise. Then, denoting the summands of the row by $a_i, i = 1,2,...$ we will derive for these summands the following expressions:

$$\begin{aligned} a_1 &= \exp(P(t-s)), \\ a_2 &= \int_s^t e^{P(t-u)}Fe^{P(u-s)}\,du, \\ a_3 &= \int_s^t e^{P(t-u)}Fe^{P(u-s)}\int_s^u e^{-P(v-s)}Fe^{P(v-s)}\,du\,dv, \\ &\ldots\ldots,\ldots. \end{aligned} \tag{107}$$

Therefore,

$$\begin{aligned} \|a_1\| &\le Ce^{-\eta_{\max}(t-s)}, \\ \|a_2\| &\le \int_s^t Ce^{-\eta_{\max}(t-u)}\|F\|Ce^{-\eta_{\max}(u-s)}\,du = \\ &= C^2\int_s^t e^{-\eta_{\max}(t-s)}\|F\|\,du, \end{aligned} \tag{108}$$

$$\begin{aligned} \|a_3\| &\le C^3e^{-\eta_{\max}(t-s)}\int_s^t \|F\|\,du\int_s^u \|F\|dv = \\ &= 1/2C^3e^{-\eta_{\max}(t-s)}\left[\int_s^t \|F\|\,du\right]^2, \end{aligned} \tag{109}$$

We put $N = \int_s^t \|F\|\,du$. Then,

$$\begin{aligned} &\left\|\Omega_s^t(F+P)\right\| \le \\ &\le Ce^{-\eta_{\max}(t-s)}\left(1 + CN + 1/2C^2N^2 + 1/6C^3N^3 + \ldots\right) = \\ &= C\exp(-\eta_{\max}(t-s))\exp\left(C\int_s^t \|F\|\,du\right). \end{aligned} \tag{110}$$

which is what we set out to establish.

# Generalized Net Model for Parallel Optimization of Feed-Forward Neural Network with Variable Learning Rate Backpropagation Algorithm with Time Limit

S. Sotirov, K. Atanassov, and M. Krawczak

**Abstract.** The used generalized net will give us a possibility for parallel optimization of a feed-forward neural network based on assigned training pairs with variable learning rate backpropagation algorithm with time limit.

**Index Terms:** Generalized Nets, Neural Networks.

## 1 Introduction

In a series of papers the process of functioning and the results of the work of different types of neural networks are described by Generalized Nets (GNs, see [1], [2]). Here, we shall discuss the possibility for training of feed-forward Neural Networks by backpropagation algorithm. The GN will optimize the NN-structure on the basis of a connections with a limit parameter.

The different types of NNs can be implemented in different ways [10], [17], [18] and can be learnt by different algorithms [7], [14], [16].

In the previous paper [3] we introduce a feed-forward neural network based on assigned training pairs with variable learning rate backpropagation algorithm.

S. Sotirov
Prof. Asen Zlatarov University, Bourgas-8000, Bulgaria
e-mail: ssotirov@btu.bg

K. Atanassov
CLBME, Bulgarian Academy of Sciences, Bl. 105, Sofia-1113, Bulgaria
e-mail: krat@bas.bg

M. Krawczak
Systems Research Institute – Polish Academy of Sciences,
Wyzsza Szkola Informatyki Stosowanej i Zarzadzania, Warsaw, Poland
e-mail: krawczak@ibs.pan.waw.pl

V. Sgurev et al. (Eds.): Intelligent Systems: From Theory to Practice, SCI 299, pp. 361–371.
springerlink.com 

The proposed generalized net model introduces parallel work in the training of two NNs with different structures. The difference between them is in the number of neurons in the hidden layer, which directly reflects on all of the network's properties. Through increasing their number the network is learnt within a smaller number of epoches to achieve its purpose. On the other hand, the great number of neurons complicates the implementation of the NN and makes it unusable in structures with elements' limits [5],[[6],[7],[11]. Sometimes learning requires too much time and all the processes get very long.

In the multilayer NNs, one layer exit become entries for the next one. The equations describing this operation are:

$$a^3=f^3(w^3f^2(w^2f^1(w^1p+b^1)+b^2)+b^3), \tag{1}$$

where:

- $a^m$ is the exit of the $m$-th layer of the NN for $m$ =1, 2, 3;
- $w$ is the matrix of the weight coefficients of each of the entries;
- $b$ is the neuron's entry bias;
- $f^m$ is the transfer function of the $m$-th layer.

The neuron in the first layer receives $p$ external entries.

The neurons' exits from the last layer determine the number $a$ of the network's exits.

A pair of numbers is submitted (an entry value and an achieving aim – on the network's exit) to the algorithm, since it

belongs to the training methods with teacher:

$$<p_1, t_1>, <p_2 , t_2>, ..., <p_Q , t_Q>, \tag{2}$$

where $Q \in \{1,...,n\}$, n – number of learning couple, where $p_Q$ is the entry value (on the network's entry), and $t_Q$ is the exit's value corresponding to the aim. Every network's entry is established in advance and constant, and the exit has to correspond to the aim. The difference between the entry values and the aim is the error: e = t-a.

The "back propagation" algorithm uses least-quarter error:

$$\hat{F} = (t-a)^2 = e^2. \tag{3}$$

In training the NN, the algorithm recalculates the network's parameters ($W$ and $b$) in order to achieve least-mean square error.

The "back propagation" algorithm [8],[12]for the $i$-th neuron, for $k$+1-st iteration uses equations:

$$w_i^m(k+1) = w_i^m(k) - \alpha \frac{\partial \hat{F}}{\partial w_i^m}, \tag{4}$$

$$b_i^m(k+1) = b_i^m(k) - \alpha \frac{\partial \hat{F}}{\partial b_i^m}, \tag{5}$$

where:

$\alpha$ is learning rate for neural network;

$\frac{\partial \hat{F}}{\partial w_i^m}$ is relation between the changes of mean square error and changes of the weights;

$\frac{\partial \hat{F}}{\partial b_i^m}$ is relation between the changes of mean square error and changes of the biases.

In standard steepest descent algorithm, the learning rate is held constant throughout the training. The performance of the algorithm is very sensitive to the proper adjustment of the learning rate. If the learning rate is set too high, the algorithm may oscillate and become unstable. If the learning rate is too low, the algorithm will take too long to converge. It is not practical to determine the optimal setting for the learning rate before training, and, in fact, the optimal learning rate changes during the training process, as the algorithm moves across the performance surface [9], [15], [12], [19], [13].

The rules of the variable learning rate backpropagation (VLBP) are:

1. If $e^2$ is increased more than some set percent $\xi$ (typically 0-5%) after a weight update, then the weight update is discarded, the learning rate is multiplied by some factor $0<\rho<1$.
2. If $e^2$ is decreased after a weight update, then the weight update is accepted and the learning rate is multiplied by some factor ($\eta>1$).
3. If $e^2$ is increased by less then $\xi$, then the weight update is accepted.

The network is trained when

$$e^2 < E_{max}, \tag{6}$$

where $E_{max}$ is the maximum mean square error.

For this case study a subject has been used as an example but there would be no essential algorithmic difference if the evaluation is related to a program form or a degree of education.

## 2 A GN-Model

All definitions related to the concept "GN" are taken from [1], [2]. The GN, describing the process of the NN learned by "backpropagation" algorithm [8], is shown on Fig. 1.

The GN constructed below, is a reduced GN. It does not have temporal components. The priorities of its transitions, places and tokens are equal, its place and arc capacities are equal to infinity.

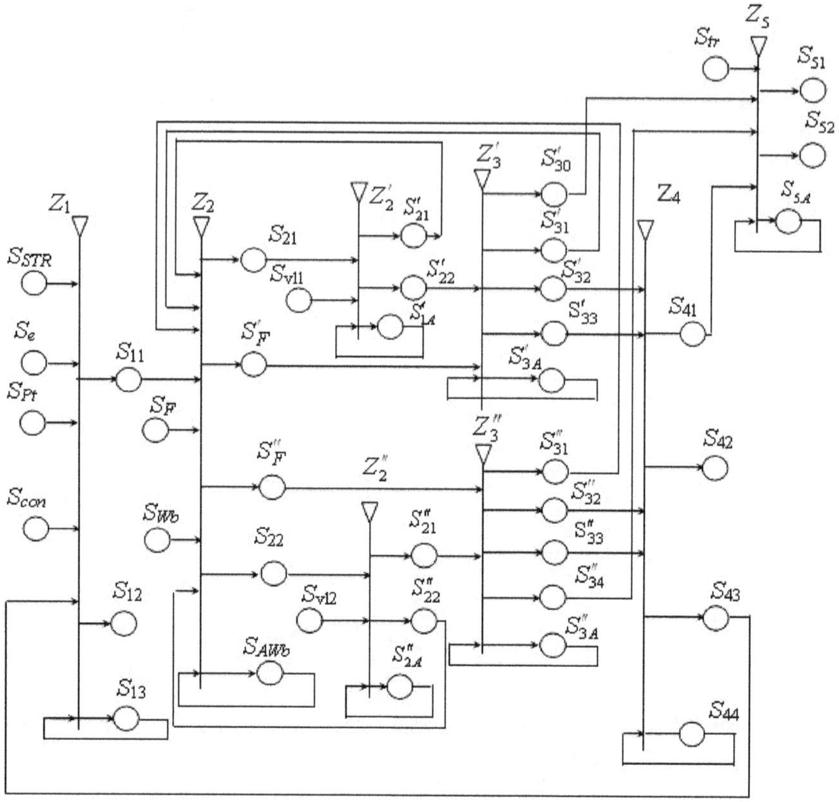


**Fig. 1** Generalized Net Model for Parallel Optimization of Feed-Forward Neural Network with time limit

Initially the following tokens enter in the GN:

- in place $S_{STR}$ – $\alpha$-token with initial characteristic

$$x_0^{\alpha} = \text{"number of neurons in the first layer, number of neurons in the output layer"};$$

- in place $S_e$ – $\beta$-token with initial characteristic

$$x_0^{\beta} = \text{"maximum error in neural network learning } E_{max}\text{"};$$

- in place $S_{Pt}$ – $\gamma$-token with initial characteristic

$$x_0^{\gamma} = \text{"}<p_1, t_1>, <p_2, t_2>, \ldots, <p_Q, t_Q>\text{"};$$

- in place $S_F$ – one $\delta$-token with initial characteristic

$$x_0^{\delta} = \text{"}f^1, f^2, f^3\text{"}.$$

- in places $S_{vl1}$ and $S_{vl2}$ – two $\psi$-tokens with initial characteristic

$$x_0^{\psi} = \text{“}\xi, \rho, \eta\text{”}.$$

The token splits into two tokens that enter respectively places $S'_F$ and $S''_F$;

- in place $S_{Wb}$ $\varepsilon$-token obtains characteristic

$$x_0^{\varepsilon} = \text{“}w, b\text{”};$$

- in place $S_{con}$ $\xi$-token obtains characteristics

$$x_0^{\xi} = \text{“maximum number of the neurons in the hiden layer of the neural network - } C_{max}\text{”};$$

- in place $S_{tr}$ $\tau$-token with characteristics

$$x_0^{\tau} = \text{“threshold value-}\tau\text{”}.$$

The GN is presented by a set of transitions:

$$A = \{ Z_1, Z_2, Z'_2, Z''_2, Z'_3, Z''_3, Z_4, Z_5\},$$

where transitions describe the following processes:

$Z_1$ – Formation of initial conditions and structure of the NNs;
$Z_2$ – Calculating of $a_i$ using (1);
$Z'_2$ – Calculation of the VL influence of the first NN;
$Z''_2$ – Calculation of the VL influence of the second NN;
$Z'_3$ – Calculation of the backward of the first NN using (3) and (4);
$Z''_3$ – Calculation of the backward of the secound NN using (3) and (4);
$Z_4$ – Checking for the end of the learning processes in the NNs;
$Z_5$ – Checking for the end of the processes in the NNs for time limits.

Let everywhere below:

$p$ be a vector of the inputs of the NN,
$a$ be a vector of outputs of NN,
$a_i$ be the output values of the $i$-th NN for $i = 1, 2$,
$e_i$ be the mean square error of the $i$-th NN, for $i = 1, 2$,
$E_{max}$ be the maximum error in the learning of the NN,
$t$ be the learnt target;
$w_{ik}$ be the weight coefficients of the $i$-th NN for $i = 1, 2$ for the $k$-th iteration;
$b_{ik}$ be the bias coefficients of the $i$-th NN for $i = 1, 2$ for the $k$-th iteration.

Transitions of the GN-model have the following forms.

$$Z_1 = \langle\{S_{STR}, S_e, S_{Pt}, S_{con}, S_{43}, S_{13}\}, \{S_{11}, S_{12}, S_{13}\}, R_1, \wedge(\vee(\wedge(S_e, S_{Pt}, S_{con}), S_{13}), \vee(S_{STR}, S_{43}))\rangle,$$

where:

$$R_1 = \begin{array}{c|ccc} & S_{11} & S_{12} & S_{13} \\ \hline S_{STR} & False & False & True \\ S_e & False & False & True \\ S_{Pt} & False & False & True \\ S_{con} & False & False & True \\ S_{43} & True & False & False \\ S_{13} & W_{13,11} & W_{13,12} & True \end{array},$$

and

$W_{13,11}$= "it is not possible to divide the current interval to subintervals";
$W_{13,12}$= "it is possible to divide the current interval to subintervals".

The token that enters place $S_{11}$ during the first activation of the transition $Z_1$ obtains characteristic

$$x_0^{\theta'} = "pr_1 x_0^{\alpha}, [1; x_0^{\xi}], pr_2\, x_0^{\alpha}, x_0^{\gamma}, x_0^{\beta}".$$

Next it obtains the characteristic

$$x_{cu}^{\theta'} = "pr_1 x_0^{\alpha}, [l_{\min}; l_{\max}], pr_2\, x_0^{\alpha}, x_0^{\gamma}, x_0^{\beta}",$$

where $[l_{\min}; l_{\max}]$ is the current characteristic of the token that enters place $S_{13}$ from place $S_{43}$.

The token that enters place $S_{12}$ obtains the characteristic $[l_{\min}; l_{\max}]$.

$$Z_2 = \langle \{ S'_{31}, S''_{31}, S_{11}, S_F, S_{Wb}, S_{AWb}, S'_{21}, S''_{22} \}, \{S_{21}, S'_F, S_{22}, S''_F, S_{AWb}\}, R_2, \vee(\wedge(S_F, S_{11}), \vee(S_{AWb}, S_{Wb}), (S'_{31}, S''_{31}, \vee(S'_{21}, S''_{22})) \rangle,$$

where

$$R_2 = \begin{array}{c|ccccc} & S_{21} & S'_F & S_{22} & S''_F & S_{AWb} \\ \hline S'_{31} & True & False & False & False & True \\ S''_{31} & False & False & True & False & True \\ S_{11} & True & False & True & False & False \\ S_F & True & True & True & True & False \\ S_{Wb} & True & False & True & False & False \\ S_{12} & True & False & True & False & False \\ S_{AWb} & True & False & True & False & False \\ S'_{21} & False & False & False & False & True \\ S''_{22} & False & False & False & False & True \end{array}.$$

The tokens that enter places $S_{21}$ and $S_{22}$ obtain the respective characteristics:

$$x_{cu}^{\eta'} = "x_{cu}^{\varepsilon'}, x_0^{\gamma}, x_0^{\beta''}, a_1, pr_1\ x_0^{\alpha}, [l_{\min}], pr_2 x_0^{\alpha}"$$

and

$$x_{cu}^{\eta''} = "x_{cu}^{\varepsilon'}, x_0^{\gamma}, x_0^{\beta''}, a_2, pr_1\ x_0^{\alpha}, [l_{\max}], pr_2 x_0^{\alpha}".$$

$$Z_2' = \langle\{ S_{21}, S_{vl1}\ S_{2A}', \}, \{ S_{21}', S_{22}', S_{2A}' \}, R_2', \wedge (S_{21}, S_{vl1}, S_{2A}')\rangle,$$

where

$R_2' =$

| | $S_{21}'$ | $S_{22}'$ | $S_{2A}'$ |
|---|---|---|---|
| $S_{21}$ | *False* | *False* | *True* |
| $S_{vl1}$ | *False* | *False* | *True* |
| $S_{2A}'$ | *True* | *True* | *True* |

The tokens that enter places $S_{21}'$ and $S_{22}'$ obtain the respective characteristics:

$$x_{cu}^{\hat{\eta}} = \begin{cases} 0, if & \Delta e^2 > \xi \ \&\ \Delta e^2 < 0 \\ 1, if & 0 < \Delta e^2 < \xi \end{cases}$$

$$x_{cu}^{\eta'} = \begin{cases} \alpha = \alpha * \rho, if & \Delta e^2 > \%\xi \\ \alpha = \alpha * \eta, if & \Delta e^2 < 0 \\ \alpha = \alpha, if & 0 < \Delta e^2 < \%\xi \end{cases}$$

$$Z_2'' = \langle\{ S_{22}, S_{vl2}, S_{2A}'' \}, \{ S_{21}'', S_{22}'', S_{2A}'' \}, R_2'', \wedge (S_{22}, S_{vl2}, S_{2A}'')\rangle,$$

where

$R_2'' =$

| | $S_{21}''$ | $S_{22}''$ | $S_{2A}''$ |
|---|---|---|---|
| $S_{22}$ | *False* | *False* | *True* |
| $S_{vl2}$ | *False* | *False* | *True* |
| $S_{2A}''$ | *True* | *True* | *True* |

The tokens that enter places $S_{21}''$, and $S_{22}''$ obtain the respective characteristics:

$$x_{cu}^{\hat{\eta}} = \begin{cases} 0, \text{ if} & \Delta e^2 > \xi \ \&\ \Delta e^2 < 0 \\ 1, \text{ if} & 0 < \Delta e^2 < \xi \end{cases}$$

$$x_{cu}^{\eta''} = \begin{cases} \alpha = \alpha * \rho, & \text{if} \quad \Delta e^2 > \%\xi \\ \alpha = \alpha * \eta, & \text{if} \quad \Delta e^2 < 0 \\ \alpha = \alpha, & \text{if} \quad 0 < \Delta e^2 < \%\xi \end{cases}$$

$$Z_3' = \langle\{ S_{22}', S_F', S_{3A}' \}, \{ S_{30}', S_{31}', S_{32}', S_{33}', S_{3A}' \}, R_3',$$
$$\wedge ( S_{22}', S_F', S_{3A}' )\rangle,$$

where

$$R'_3 = \begin{array}{c|ccccc} & S'_{30} & S'_{31} & S'_{32} & S'_{33} & S'_{A3} \\ \hline S'_{22} & False & False & False & False & True \\ S'_F & False & False & False & False & True \\ S'_{3A} & W'_{3A,30} & W'_{3A,31} & W'_{3A,32} & W'_{A3,33} & True \end{array}$$

and

$W'_{3A,31}$ = "$e_1 > E_{max}$";

$W'_{3A,32}$ = "$e_1 < E_{max}$";

$W'_{3A,33}$ = "$e_1 > E_{max}$ and $n_1 > m$";

$W'_{3A,33}$ = "$W'_{3A,33}$ or $W'_{3A,32}$";

where:

$n_1$ – current number of the first NN learning iteration,
$m$ – maximum number of the NN learning iteration.

The token that enters place $S'_{31}$ obtains the characteristic "first NN: $w(k+1)$, $b(k+1)$", according (4) and (5). The $\lambda'_1$ - and $\lambda'_2$-tokens that enter place $S'_{32}$ and $S'_{33}$ obtain the characteristic

$$x_0^{\lambda'_1} = x_0^{\lambda'_2} = "l_{\min}".$$

The token that enters place $S'_{30}$ obtains the characteristic "time for learning for the first NN – $t_1$".

$$Z''_3 = \langle \{ S''_{21}, S''_F, S''_{3A} \}, \{ S''_{31}, S''_{32}, S''_{33}, S''_{34}, S''_{3A} \}, R''_3, \wedge ( S''_{21}, S''_F, S''_{3A} ) \rangle,$$

where

$$R''_3 = \begin{array}{c|ccccc} & S''_{31} & S''_{32} & S''_{33} & S''_{34} & S''_{A3} \\ \hline S''_{21} & False & False & False & False & True \\ S''_F & False & False & False & False & True \\ S''_{3A} & W''_{3A,31} & W''_{3A,32} & W''_{A3,33} & W''_{A3,34} & True \end{array}.$$

and

$W''_{3A,31}$ = "$e_2 > E_{max}$",

$W''_{3A,32}$ = "$e_2 < E_{max}$",

$W''_{3A,33}$ = "$e_2 > E_{max}$ and $n_2 > m$",

$W''_{3A,34}$ = "$W''_{3A,32}$ or $W''_{3A,33}$",

where:

$n_2$ – current number of the second NN learning iteration,
$m$ – maximum number of the NN learning iteration.

The token that enters place $S''_{31}$ obtains the characteristic "second NN: $w(k+1)$, $b(k+1)$", according to (4) and (5). The $\lambda''_1$ - and $\lambda''_2$ -tokens that enter places $S''_{32}$ and $S''_{33}$ obtain characteristic

$$x_0^{\lambda''_1} = x_0^{\lambda''_2} = "l_{\max}".$$

The token that enters place $S''_{34}$ obtains the characteristic "time for learning for the last NN – $t_2$".

$$Z_4 = \langle \{ S'_{32}, S'_{33}, S''_{32}, S''_{33}, S_{44}\}, \{ S_{41}, S_{42}, S_{43}, S_{44}\}, R_4, \wedge(S_{44} \vee( S'_{32}, S'_{33}, S''_{32}, S''_{33}))\rangle,$$

where

$R_4=$

| | $S_{41}$ | $S_{42}$ | $S_{43}$ | $S_{44}$ |
|---|---|---|---|---|
| $S'_{32}$ | *False* | *False* | *False* | *True* |
| $S'_{33}$ | *False* | *False* | *False* | *True* |
| $S''_{32}$ | *False* | *False* | *False* | *True* |
| $S''_{33}$ | *False* | *False* | *False* | *True* |
| $S_{44}$ | $W_{44,41}$ | $W_{44,42}$ | $W_{44,43}$ | *True* |

and

$W_{44,41}$= "$e_1 < E_{max}$" & "$e_2 < E_{max}$";
$W_{44,42}$= "$e_1 > E_{max}$ and $n_1 > m$" & "$e_2 > E_{max}$ and $n_2 > m$";
$W_{44,43}$= "( $e_1 < \mathrm{E}_{\max}$ and ($e_2 > \mathrm{E}_{\max}$ and $n_2 > m$)) or ($e_2 < E_{max}$ and ($e_1 > E_{max}$ and $n_1 > m$))".

The token that enters place $S_{41}$ obtains the characteristic

"Both NN satisfy the conditions. The network that has smaller number of the neurons is used for the solution".

The token that enters place $S_{42}$ obtains the characteristic

"There is no solution (both NN do not satisfy the conditions)".

The token that enters place $S_{44}$ obtains the characteristic

"The solution is in the interval [$l_{\min}$; $l_{\max}$] – the interval is changed using the golden sections algorithm [4]".

$$Z_5 = \langle \{ S_{tr}, S'_{30}, S''_{34}, S_{41}, S_{5A} \}, \{ S_{51}, S_{52}, S_{5A}\}, R_5, \vee( S_{tr}, S'_{30}, S''_{34}, S_{41}, S_{5A} ))\rangle,$$

where

$$R_4 = \begin{array}{c|ccc} & S_{51} & S_{52} & S_{5A} \\ \hline S_{tr} & False & False & True \\ S_{30} & False & False & True \\ S_{34} & False & False & True \\ S_{41} & False & False & True \\ S_{5A} & W_{5A,51} & W_{5A,52} & True \end{array}$$

and

$W_{5A,51}$=" $\tau < t_1$ or $\tau < t_2$",
$W_{5A,52}$=" $\tau > t_1$ or $\tau > t_2$".

The token that enters place $S_{51}$ obtains the characteristic

"NN not satisfy the conditions for the time".

The token that enters place $S_{52}$ obtains the characteristic

"NN satisfy the conditions for the time".

## 3 Conclusion

The proposed GN-model introduces the parallel work in the training of two NNs with different structures. The difference between the nets is in the number of neurons in the hidden layer and that affects directly the properties of the whole network.

On the other hand, the great number of neurons complicates the implementation of the NN. The constructed GN-model allows simulation and optimization of the architecture of the NNs using the golden section rule and time limit for learning.

## References

1. Atanassov, K.: Generalized nets. World Scientific, Singapore (1991)
2. Atanassov, K.: On Generalized Nets Theory. Prof. M. Drinov. Academic Publishing House, Sofia (2007)
3. Atanassov, K., Krawczak, M., Sotirov, S.: Generalized Net Model for Parallel Optimization of Feed-Forward Neural Network with Variable Learning Rate Backpropagation Algorithm. In: 4th International IEEE Conference Intelligent Systems, Varna, pp. 16-16–16-19 (2008)
4. Atanassov, K., Sotirov, S., Antonov, A.: Generalized net model for parallel optimization of feed-forward neural network. Advanced studies in contemporary Mathematics (2007)

5. Bellis, S., Razeeb, K.M., Saha, C., Delaney, K., O'Mathuna, C., Pounds-Cornish, A., de Souza, G., Colley, M., Hagras, H., Clarke, G., Callaghan, V., Argyropoulos, C., Karistianos, C., Nikiforidis, G.: FPGA Implementation of Spiking Neural Networks - an Initial Step towards Building Tangible Collaborative Autonomous Agents. In: FPT 2004. International Conference on Field-Programmable Technology, The University of Queensland, Brisbane, Australia, December 6-8, pp. 449–452 (2004)
6. Gadea, R., Ballester, F., Mocholi, A., Cerda, J.: Artificial Neural Network Implementation on a Single FPGA of a Pipelined On-Line Backpropagation. In: 13th International Symposium on System Synthesis (ISSS 2000), pp. 225–229 (2000)
7. Hagan, M., Demuth, H., Beale, M.: Neural Network Design. PWS Publishing, Boston (1996)
8. Haykin, S.: Neural Networks: A Comprehensive Foundation. Macmillan, NY (1994)
9. Kostopoulos, A.E., Sotiropoulos, D.G., Grapsa, T.N.: A new efficient variable learning rate for Perry's spectral conjugate gradient training method. In: Proceedings of the 1st International Conference: From Scientific Computing to Computational Engineering (2004)
10. Krawczak, M.: Generalized Net Models of Systems, Bulletin of Polish Academy of Science (2003)
11. Maeda, Y., Tada, T.: FPGA Implementation of a Pulse Density Neural Network With Training Ability Using Simultaneous Perturbation. IEEE Transactions on Neural Networks 14(3) (2003)
12. Mandic, D., Chambers, J.: Towards the Optimal Learning Rate for Backpropagation. Neural Processing Letters 11(1), 1–5 (2000)
13. Plagianakos, V.P., Sotiropoulos, D.G., Vrahatis, M.N.: Automatic adaptation of learning rate for backpropagation neural networks. In: Mastorakis, N.E. (ed.) Recent Advances in Circuits and Systems, pp. 337–341. World Scientific Publishing Co. Pte. Ltd., Singapore (1998)
14. Rumelhart, D., Hinton, G., Williams, R.: Training representation by back-propagation errors. Nature 323, 533–536 (1986)
15. Sotiropoulos, D.G., Kostopoulos, A.E., Grapsa, T.N.: A spectral version of Perry's conjugate gradient method for neural network training. In: Proceedings of 4th GRACM Congress on Computational Mechanics, University of Patras, Greece, June 27-29 (2002)
16. Sotirov, S.: A method of accelerating neural network training. Neural Processing Letters 22(2), 163–169 (2005)
17. Sotirov, S.: Modeling the algorithm Backpropagation for training of neural networks with generalized nets – part 1. In: Proceedings of the Fourth International Workshop on Generalized Nets, Sofia, September 23, pp. 61–67 (2003)
18. Sotirov, S., Krawczak, M.: Modeling the algorithm Backpropagation for training of neural networks with generalized nets – part 2, Issue on Intuitionistic Fuzzy Sets and Generalized nets, Warsaw (2003)
19. Vogl, T.P., Mangis, J.K., Zigler, A.K., Zink, W.T., Alkon, D.L.: Accelerating the convergence of the back-propagation method. Biological Cybernetics 59, 257–263 (1988)

# Towards a Model of the Digital University: A Generalized Net Model for Producing Course Timetables and for Evaluating the Quality of Subjects

A. Shannon, D. Orozova, E. Sotirova, M. Hristova, K. Atanassov, M. Krawczak, P. Melo-Pinto, R. Nikolov, S. Sotirov, and T. Kim

A. Shannon
Warrane College, the University of New South Wales, Kensington, 1465, Australia
e-mail: t.shannon@warrane.unsw.edu.au

D. Orozova
Free University of Bourgas, Bourgas-8000, Bulgaria
e-mail: orozova@bfu.bg

E. Sotirova · S. Sotirov
Prof. Asen Zlatarov University, Bourgas-8000, Bulgaria
e-mails: esotirova@btu.bg, ssotirov@btu.bg

M. Hristova
Higher School of Transport "Todor Kableshkov", Sofia, Bulgaria
e-mail: mhristova@vtu.bg

K. Atanassov
CLBME, Bulgarian Academy of Sciences, Bl. 105, Sofia-1113, Bulgaria
e-mail: krat@bas.bg

M. Krawczak
Systems Research Institute – Polish Academy of Sciences,
Wyzsza Szkola Informatyki Stosowanej i Zarzadzania, Warsaw, Poland
e-mail: krawczak@ibs.pan.waw.pl

P. Melo-Pinto
CITAB - UTAD, Quinta de Prados, Apartado 1013, 5001-801 Vila Real, Portugal
e-mail: pmelo@utad.pt

R. Nikolov
Faculty of Mathematics and Informatics, University of Sofia "St. Kliment Ohridski"
Bulgaria
e-mail: roumen@fmi.uni-sofia.bg

T. Kim
Institute of Science Education, Kongju National University, Kongju 314-701, S. Korea
e-mails: taekyun64@hotmail.com, tkim@kongju.ac.kr

V. Sgurev et al. (Eds.): Intelligent Systems: From Theory to Practice, SCI 299, pp. 373–381.
springerlink.com 

**Abstract.** In a series of research papers, the authors study some of the most important processes of functioning of universities and construct their Generalized Net (GN) models. The main focus in this paper is to analyse the process of the production of course timetables in a digital university and to evaluate their quality. The opportunity of using GNs as a tool for modeling such processes is also analyzed.

**Index Terms:** Generalized Nets, Digital University, E-Learning, Teaching Quality.

## 1 Introduction

The wide penetration of Information and Communication Technologies (ICT) into society has catalyzed the need for a global educational reform which will break the monopoly of the print and paper based educational system. The ICT based distance education is considered as "*the most significant development in education in the past quarter century*" [3]. The pattern of growth in the use of ICT in higher education can be seen through [7]:

- increasing computing resources, including web-based technologies, encouraging supplemental instructional activities; a growth of academic resources online; and administrative services provided through networked resources;
- organisational changes in policies and approaches;
- an increasing emphasis on quality of teaching and the importance of staff development;
- changes in social practice, e.g. a growth in demand for life-long learning opportunities, which consequently affect the need to adapt technology into instructional delivery; and an increase in average age of students.

One of the main conclusions related to the ongoing educational reform is that it is based on designing and using different virtual learning environments which do not put clear boundary between physical and virtual worlds. A key factor for success is to integrate them, not to separate them, and to apply relevant instructional design strategy based on a current learning theory. A tool for implementing such learning environment is an integrated information system which provides services and supports all university activities.

Generalized Nets (GN) [1, 2] have been used for the construction of a model, describing the process of produce course timetable and subjects' quality estimation in digital university. The aim of the Generalized Net constructed in this paper is to model the process, aiming at its optimization. Since the modeled processes are very complex in, the GN presented here have not been described in much details, yet with sufficient information to apply the model in practice. This paper is based on [5, 8]. It is a continuation of a series of research papers [3, 9, 10, 11, 12, 13, 14, 15, 16].

The University produces a course based timetable for existing courses on offer. This is then rolled over into the following academic year. Any new courses for that forthcoming academic year are then fitted into the existing timetable.

If the timetable is verified, it is then published. If the timetable cannot be verified, it is reconsidered. The process loops back to timetable update. Once the students' course selection has been verified, two outcomes are possible. Firstly, if the student numbers for a particular course are too low, the University can cancel the course. Also, if the demand for the course is higher than expected, the timetable can be rearranged to accommodate this demand.

After academic members of staff have been allocated to courses, the next step is to allocate courses to rooms.

In the end of the academic year the quality of subjects is evaluated based on the Multifactor method from [5].

Quality of the subject

$$Q = f_Q(k_1, k_2, \ldots, k_n, K_1, K_2, \ldots, K_n) \tag{1}$$

is examined [4] as a complex multi-measurement value quantitatively dependant on different criteria that not cover each other $K_i$ ($k_1, k_2,\ldots, k_n$ are the weight coefficients of the criteria) that are correlatively connected with the quality indicators $P_1, P_2,\ldots, P_m$.

$$\begin{aligned} K_1 &= f_1(b_{1,1}, b_{1,2},\ldots, b_{1,m};\ P_1, P_2,\ldots, P_m) \\ K_2 &= f_2(b_{2,1}, b_{2,2},\ldots, b_{2,m};\ P_1, P_2,\ldots, P_m) \\ K_n &= f_n(b_{n,1}, b_{n,2},\ldots, b_{n,m};\ P_1, P_2,\ldots, P_m), \end{aligned} \tag{2}$$

where the coefficients of the indicator significance in the criteria value $b_{i,j}$ ($i = 1,\ldots, n, j = 1,\ldots, m$) are expert-defined.

With evaluating a subject in [4], $m = 9$ indicators and $l = 6$ experts who should estimate the values of the indicators have been recommended.

The following have been assumed as experts (evaluating staff):

1. Students' opinions on the subject quality
2. Opinions of colleagues (teachers)
3. Opinions of potential employers
4. Opinion of the Head of the department
5. Opinion of the faculty administrations
6. Self-assessment of the team teaching the subject.

Each expert gives its evaluation for each of them in the accepted rating system.

The criteria to estimate the quality of the subject have been defined in [4] as follows:

Criteria of subject evaluation
$K_1$ Subject aim and results expected
$K_2$ Subject teaching contents
$K_3$ Subject teaching quality
$K_4$ Teacher's assistance for students
$K_5$ Resources for teaching the subject
$K_6$ Evaluation of students' knowledge and results obtained

## 2 A GN-Model

The GN-model (see Fig. 1) contains 4 transitions and 25 places, collected in five groups and related to the five types of the tokens that will enter respective types of places:

$\alpha$ - tokens and *l*-places represent the input data necessary for producing of course timetable,

$\beta$ - tokens and *p*-places represent the timetable,

$\chi$ - tokens and *e*-places represent the experts who should estimate the subject quality,

$\delta$ - tokens and *t*-places represent the Indicators $P_j$,

$\varepsilon$ - tokens and *c*-places represent the criterions $K_i$ and summarized quality estimation.

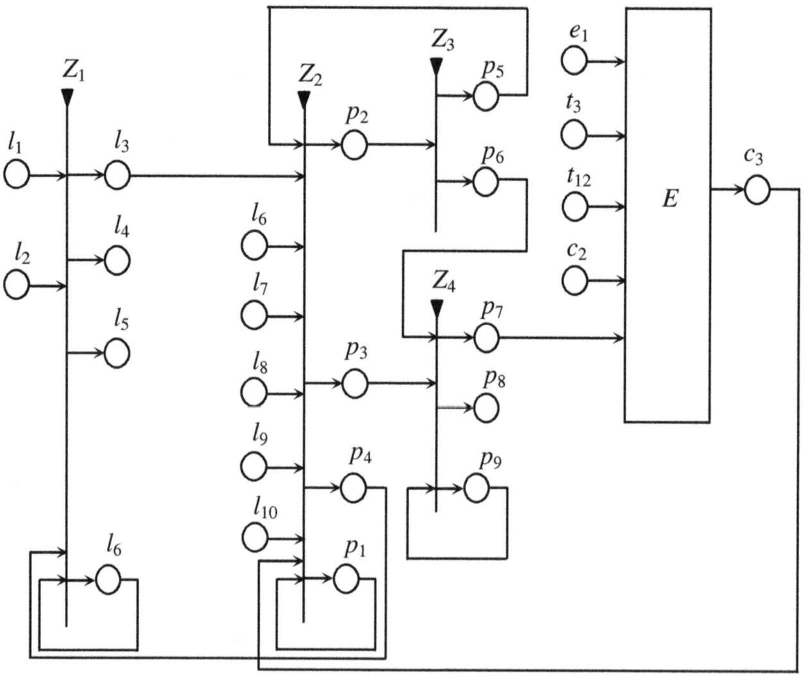


**Fig. 1** GN model of process of produce course timetable and subjects' quality estimation

For brevity, we shall use the notation $\alpha$- and $\beta$-tokens instead of $\alpha_i$- and $\beta_j$-tokens, where *i, j* are numerations of the respective tokens.

In the beginning $\beta$-tokens stay in place $p_1$ with initial characteristic:

"Initial (existing) timetable".

In the next time-moments this token is split into four. One of them, let it be the original $\beta$-token, will continue to stay in place $t_1$, while the other $\beta$-token will move to transitions $Z_1$, $Z_3$, $Z_4$, $Z_5$, passing via transition $Z_2$.

All tokens that enter transition $Z_2$ will unite with the original token. All information will be put as an initial characteristic of a token, generated by the original token.

The α-tokens with characteristics

"Cancelled course data"

and

"Live course data"

enter the net via places $l_1$ and $l_2$ respectively.

These data come from a variety of centrally and locally held systems with the University.

The $\alpha$-tokens with characteristics

"Course requirement",

"Student requirement",

"Non central rooms available",

"Teaching load model",

and

"Student number information"

enter the net via places $l_6$, $l_7$, $l_8$, $l_9$ and $l_{10}$, respectively.

The $v$ in number $\chi$-tokens with characteristics

"Expert $v$", $v = 1, 2, \ldots w$

enter the net via place $e_1$.

The $m$ in number $\delta_3$-tokens with characteristics

"Indicator $P_j$", $j = 1, 2, \ldots m$

enter the net via place $t_3$.

The $m$ in number $\varepsilon_{12}$-tokens with characteristics

"Weight $b_{i,j}$ of the indicators", $i = 1, 2, \ldots n; j = 1, 2, \ldots m$

enter the net via place $t_{12}$.

The $n$ in number $\varepsilon$-tokens with characteristics

"Weight $k_i$ of the criterions $K_i$"

enter the net via place $c_2$.

The forms of the transitions are the following.

$$Z_1 = <\{l_1, l_2, l_6, p_4\}, \{l_3, l_4, l_5, l_6\}, r_1, \vee(\wedge(l_1, l_2), p_4, l_6)>$$

where:

$$r_1 = \begin{array}{c|cccc} & l_3 & l_4 & l_5 & l_6 \\ \hline l_1 & false & false & false & true \\ l_2 & false & false & false & true \\ l_6 & W_{6,3} & W_{6,4} & W_{6,5} & false \\ p_4 & false & false & false & true \end{array},$$

$W_{6,3}$ = "The course delivery data is updated",

$W_{6,4}$ = "The first meeting date is sent to Timetab",

$W_{6,5}$ = "The information is fed into WebCT".

The $\alpha$-tokens obtains the characteristics:

"Concrete parameters of the updated course delivery data"

in place $l_3$,

"Concrete information for the first meeting date to Timetab"

in place $l_4$, and

"WebCT feed"

in place $l_5$.

$$Z_2 = <\{ l_3, l_7, l_8, l_9, l_{10}, l_{11}, p_1, p_5, c_3\}, \{p_1, p_2, p_3, p_4\}, r_2, \vee (\wedge(l_3, l_7, l_8, l_9, l_{10}, l_{11}), l_1, l_5, c_3)>,$$

where:

$$r_2 = \begin{array}{c|cccc} & p_1 & p_2 & p_3 & p_4 \\ \hline l_3 & true & false & false & false \\ l_7 & true & false & false & false \\ l_8 & true & false & false & false \\ l_9 & true & false & false & false \\ l_{10} & true & false & false & false \\ l_{11} & true & false & false & false \\ p_1 & false & true & true & true \\ p_5 & true & false & false & false \\ e_3 & true & false & false & false \end{array}.$$

The $\beta$-tokens that enter places $p_2$, $p_3$ and $p_4$ obtain characteristic

"The values of the completed timetable".

$$Z_3 = <\{p_2\}, \{p_5, p_6\}, r_3, \wedge(p_2)>,$$

where:

$$r_3 = \begin{array}{c|cc} & p_5 & p_6 \\ \hline p_2 & W_{2,5} & W_{2,6} \end{array}.$$

$W_{2,5}$ = "The timetable is not correct";

$W_{2,6}$ = "The timetable is correct".

The $\beta$-tokens obtain the characteristics:

"Revision query"

in place $p_5$ and

"Verified timetable"

in place $p_6$.

$$Z_4 = <\{p_6, p_3, p_9\}, \{p_7, p_8, p_9\}, r_4, \vee(\wedge(p_6, p_3), p_9)>$$

$$r_4 = \begin{array}{c|ccc} & p_7 & p_8 & p_9 \\ \hline p_3 & false & false & true \\ p_6 & false & false & true \\ p_9 & true & true & false \end{array}.$$

The $\beta$-tokens have the characteristics:

"Published final form of the course timetable"

in place $p_7$, and

"Concrete tutorial/Lab details, Initially allocated tutors/demonstrations"

in place $p_8$.

*E* the GN that represents the algorithmization of multifactor method for teaching quality estimation at universities and it is described in [5]. As a result of the work of the net *E* the quality is estimated on 9 steps:

Step 1. Election of the experts' staff;
Step 2. Determination of the experts' competence;
Step 3. Choose of the models of indicator evaluation;
Step 4. Choose of the teaching quality indicators according to experts and a check on the coordination of their opinions;
Step 5. Calculation of the averaged values of the indicators;
Step 6. Determination of the relative weight of the indicators for each criterion;
Step 7. Determination of the values of criteria;
Step 8. Determination of the criteria significance in the summarized quality estimation;
Step 9. Calculation of the summarized quality estimation.

The obtained estimations of the subjects' quality enter place $c_3$ and via transition $Z_2$ enter place $p_1$.

## 3 Conclusion

The research expounded in this paper is a continuation of previous investigations into the modelling of information flow within a typical university. The framework in which this is done is the theory of Generalized Nets (GNs) (and sub-GNs where appropriate). While the order of procedure might vary from one institution to another, the processes are almost invariant, so that the development of the GN in this paper can be readily adapted or amended to suit particular circumstances, since each transition is constructed in a transparent manner.

## References

1. Atanassov, K.: On Generalized Nets Theory. Prof. M. Drinov Academic Publishing House, Sofia (2007)
2. Atanassov, K.: Generalized Nets. World Scientific, Singapore (1991)
3. Dimitrakiev, D., Sotirov, S., Sotirova, E., Orozova, D., Shannon, A., Panayotov, H.: Generalized net model of process of the administration servicing in a digital university. In: Generalized Nets and Related Topics. Foundations. System Research Institute, vol. II, pp. 57–62. Polish Academy of Sciences, Warsaw (2008)
4. Hristova, M.: Quantitative methods for evaluation and control of university education quality. PhD Thesis, Sofia, Bulgaria (2007)
5. Hristova, M., Sotirova, E.: Generalized net model of the Multifactor method of teaching quality estimation at universities. IEEE Intelligent Systems, 16-20–16-24** (2008)
6. Moore, M.G.: Preface. In: Moore, M.G., Anderson, W. (eds.) Handbook of Distance Education, Lawrence Erlbaum Associates, Inc., Philadelphia (2003)
7. Price, S., et al.: Review of the impact of technology-enhanced learning on roles and practices in Higher Education, `http://www.lonklab.ac.uk/kscope/impact/dmdocuments/Reviewdocument.pdf`
8. Shannon, A., Orozova, D., Sotirova, E., Atanassov, K., Krawczak, M., Melo-Pinto, P., Nikolov, R., Sotirov, S., Kim, T.: Towards a Model of the Digital University: A Generalized Net Model for Producing Course Timetables. IEEE Intelligent Systems, 16-25–16-28* (2008)
9. Shannon, A., Langova-Orozova, D., Sotirova, E., Petrounias, I., Atanassov, K., Melo-Pinto, P., Kim, T.: Generalized net model of a university classes schedule. In: Advanced Studies in Contemporary Mathematics, S.Korea, vol. 8(1), pp. 23–34 (2004)
10. Shannon, A., Langova-Orozova, D., Sotirova, E., Petrounias, I., Atanassov, K., Melo-Pinto, P., Kim, T.: Generalized net model of the university electronic library, using intuitionistic fuzzy estimations. In: 18th Int. Conf. on Intuitionistic Fuzzy Sets, Sofia, August 2004, pp. 91–96 (2004)

11. Shannon, A., Langova-Orozova, D., Sotirova, E., Petrounias, I., Atanassov, K., Melo-Pinto, P., Kim, T.: Generalized net model of information flows in intranet in an abstract university. In: Advanced Studies in Contemporary Mathematics, S.Korea, vol. 8(2), pp. 183–192 (2004)
12. Shannon, A., Langova-Orozova, D., Sotirova, E., Atanassov, K., Melo-Pinto, P., Kim, T.: Generalized Net Model of a Training System. Advanced Studies in Contemporary Mathematics 10(2), 175–179 (2005)
13. Shannon, A., Orozova-Langova, D., Sasselov, D., Sotirova, E., Petrounias, I.: Generalized net model of the intranet in a university, using fuzzy estimations. In: Seventh Int. Conf. on Intuitionistic Fuzzy Sets, Sofia, August 23-24. NIFS, vol. 9(4), pp. 123–128 (2003)
14. Shannon, A., Riecan, B., Orozova, D., Sotirova, E., Atanassov, K., Krawczak, M., Georgiev, P., Nikolov, R., Sotirov, S., Kim, T.: A method for ordering of university subjects using intuitionistic fuzzy evaluation. In: Twelfth Int. Conf. on IFSs, Sofia, May 17-18. NIFS, vol. 14, pp. 84–87 (2008)
15. Shannon, A., Langova-Orozova, D., Sotirova, E., Petrounias, I., Atanassov, K., Krawszak, M., Melo-Pinto, P., Kim, T., Tasseva, V.: A Generalized Net Model of the Separate Information Flow Connections within a University, Computational Intelligence. IEEE Intelligent Systems, 755–759 (2006)
16. Shannon, A., Orozova, D., Sotirova, E., Atanassov, K., Krawczak, M., Chountas, P., Georgiev, P., Nikolov, R., Sotirov, S., Kim, T.: Towards a Model of the Digital University: A Generalized Net Model of update Existing Timetable. In: Proc. of the Nine International Workshop on Generalized Nets, Sofia, July 4, vol. 2, pp. 71–79 (2008)
17. Shannon, A., Atanassov, K., Sotirova, E., Langova-Orozova, D., Krawczak, M., Melo-Pinto, P., Petrounias, I., Kim, T.: Generalized Net Modelling of University Processes, Sydney. Monograph No. 7. KvB Visual Concepts Pty Ltd. (2005)

# Intuitionistic Fuzzy Data Quality Attribute Model and Aggregation of Data Quality Measurements

Diana Boyadzhieva and Boyan Kolev

**Abstract.** The model we suggest makes the data quality an intrinsic feature of an intuitionistic fuzzy relational database. The quality of the data is no more determined by the level of user complaints or ad hoc sql queries prior to the data load but it is stored explicitly in relational tables and could be monitored and measured regularly. The quality is stored on an attribute level basis in supplementary tables to the base user ones. The quality is measured along preferred quality dimensions and is represented by intuitionistic fuzzy degrees. To consider the preferences of the user with respect to the different quality dimensions and table attributes we create additional tables that contain the weight values. The user base tables are not intuitionistic fuzzy but we have to use an intuitionistic fuzzy RDBMS to represent and manipulate data quality measures.

**Index Terms:** data quality, quality model, intuitionistic fuzzy, relational database.

## 1 Introduction

Information systems map real-world objects into digital representation by storage of their qualifying characteristics, relationships and states. Usually the computerized object intentionally lacks many of the properties of its real-world counterpart as they are not considered interesting for analysis. The digital mapping of the important characteristics provides the fundamental set of data for the real object into the information system. However often the digital representation experiences

Diana Boyadzhieva
Faculty of Economics and Business Administration Sofia University
"St. Kliment Ohridski", blvd. Tzarigradsko shausee 125, bl. 3, Sofia-1113, Bulgaria
e-mail: dianaht@feb.uni-sofia.bg

Boyan Kolev
CLBME – Bulgarian Academy of Sciences, Bl. 105, Sofia-1113, Bulgaria
e-mail: bik@clbme.bas.bg

V. Sgurev et al. (Eds.): Intelligent Systems: From Theory to Practice, SCI 299, pp. 383–395.
springerlink.com 

some deficiencies that are the root for data quality problems. It is hard to define the exact essence of what data quality is and that's why a lot of definitions exist (R.Y. Wang, 1994), (Orr, 1998), (G.K. Tayi, 1998) that stress different aspects of the discipline. If we have to provide a short, informal and intuitive definition of the concept, we could say that *data quality gives information about the extent to which the data is missing or incorrect.* But we could also as (Jarke M., 1999) define the data quality with a focus on the process character of the task: *A high-quality data is one that is fit for its intended uses (in operations, decision-making, planning, production systems, science etc.) and data quality is the process that encompasses all the tasks involved in the assurance of these high-quality data.* Juran defines quality simply as "fitness for use" (Juran, 1974). The ISO 9000 revision IS9000:2005 defines quality as: "Degree to which a set of inherent characteristics fulfills requirements" (9000:2005, ISO, 2005).

## 2 The Model Justification

Data quality could be controlled across several different aspects of the existence and operation of an information system. The data quality could concern:

- The design of the database – i.e. the quality of the logical or physical database schema or
- Could refer the data values that are inserted, stored and updated during the entire data flow of the information.

Data anomalies could arise on every state of the data life cycle so, to have a high quality data it is fundamental to put multiple data quality checks in the system. The next efforts in a data quality initiative involve application of methodologies to deal with the data problem in a way that will just consider the lower data quality or will also make corrections. In (D. Boyadzhieva, 2008) is presented a framework for storage of quality level on an attribute-level basis. Correction methods could also be applied but the stress in the paper is that even if some data problem could not be corrected, the respective record should not be dismissed but stored with a respective designation of its lower quality.

Many approaches apply efforts to identify and clean the errors in data that arise during an integration process. The assertion is that upon their application only high quality data enter a database or a data warehouse. However the extent of this "high" quality is not exactly measured. Sometimes records are dropped when the application is not able to correct them or it makes corrections by assuming some propositions. These corrections could also introduce data quality issues. We have to note also that the quality of data usually degrades with the time of data existence in a system. As quality-enhancement initiatives are not always readily applied, we propose a framework to store data quality assessments during each state of the data movement in an information system.

A framework with four information quality categories is developed in (Huang K., 1999) – intrinsic, contextual, representational, accessibility. Each of the multiple data quality dimensions is related to one of these categories. The model presented in this paper is appropriate for storage of quality grades made along dimensions from the intrinsic or contextual categories as they could be assessed on an attribute or record-level basis with numerical values. Data quality is incorporated in the overall design of a database schema. The relational model is extended with supplementary tables where the exact quality level on an attribute level is explicitly saved. Such a model readily provides quality information at disposal. Attribute-based approach is presented also in (R.Y. Wang M. R., 1995) but we leverage intuitionistic fuzzy logic. We do not put requirements on the database to be an intuitionistic fuzzy one but we need to use an intuitionistic fuzzy RDBMS to represent and manipulate the data quality measures. We use the Intuitionistic Fuzzy PostgreSQL /IFPG/ (B., 2005), (Kolev B., 2005), giving the possibility to store and manage intuitionistic fuzzy relations.

## 3 The Intuitionistic Fuzzy Data Quality Attribute Model (IFDQAM)

Before the explanation of the model, we shortly describe the notion of quality dimensions. For many people data quality means just accuracy. However the quality of data is better represented if it is measured also along other - descriptive for the specific data - qualitative characteristics. Each of these descriptive qualitative characteristics is called a quality dimension. The choice of quality dimensions that will be measured depends on the user requirements and is the theoretical, empirical and intuitive approaches are described in (C. Batini, 2006).

In the intuitionistic fuzzy data quality attribute model, we store the quality on an attribute level basis – i.e. we store measures of the quality of the values in the user tables /tables 1 a)/. We keep these quality measures in supplementary table that we call quality table /tables 1 b)/. We propose to store and monitor data quality not for all attributes in a user table but only for some of them – those that bring critical values for the user. The user requirements, the potential type of tasks and requests to the data determine which these attributes of a special interest are. For each such attribute of a special interest we add in the quality table one record for each quality dimension that we want to measure. The table contains two attributes which represent $\mu$ and $\nu$ intuitionistic fuzzy degrees that measure the quality along the respective quality dimension.

Let we agree upon the following terminology. The attributes in the user tables (containing the source data) we will call ordinary attributes. The extent to which it is sure that a given characteristic of the data is present along a quality dimension we will call presence of quality or positive quality. The extent to which it is sure that a given characteristic of the data does not exist along a quality dimension we will call absence of quality or negative quality. The indefiniteness about the presence of quality we will call indefinable quality.

In the defined terminology, $\mu$ measures the degree of positive quality, $\nu$ measures the degree of negative quality and the indefinable quality is $1 - \mu - \nu$. If the user table contains a few attributes and if the tracked quality dimensions are not a lot, we could not create a separate quality table but keep the ordinary attributes and the quality attributes in a single table. However to keep the things clear we offer to follow an alternative approach – to create the attributes that will keep the quality measures in a separate table (we call it quality table) that refers the respective user table with the ordinary attributes /tables 1 a), b)/ The intuitionistic fuzzy degree μ is represented by the attribute MSHIP and the intuitionistic fuzzy degree ν is represented by the attribute NMSHIP.

The relative importance that the user assigns to each quality dimension of an ordinary attribute is modeled as a weight. This weight gives the share of the respective quality dimension in the calculation of the quality of a given value in the respective ordinary attribute. Actually these weights give the relative importance that the user assigns to each dimension. We assume the weights are normalized, i.e. for each ordinary attribute, the dimension weights sum up to 1. The weights are stored in a dimension-weights table /tables 1 c)/.

Furthermore, we expand the model with another metadata table which contains the weight of the quality of each ordinary attribute value in the calculation of the total quality of a tuple in a table /tables 1 d)/. These weights give the relative importance of an ordinary attribute for the calculation of the quality of a tuple. The table represents the attribute weights for the attributes of all tables in the database. We assume the weights are normalized, i.e. for each table, the attribute weights sum up to 1.

**Tables 1** a), b), c), d)

TableX

| Attr1_key | Attr2 | Attr3 | Attr4 |
|---|---|---|---|

a)

TableX_Quality

| Attr1_Key | Attribute_Name | Dimension_Name | MSHIP | NMSHIP |
|---|---|---|---|---|

b)

Dimension_Weights

| Table_Name | Attribute_Name | Dimension_Name | Weight |
|---|---|---|---|

c)

Attribute_Weights

| Table_Name | Attribute_Name | Weight |
|---|---|---|

d)

To calculate the quality measures, three methods could be utilized. In the first one the data editor introduces the measures based on user-defined criteria. In the second one, the system calculates the quality measures based on a set of user-defined logic or calculations (for instance a set of real-world categorical words like very weak, weak, strong, very strong, etc. could be automatically mapped to a number value). In the third one – the quality values could be result from the integration and data cleansing tool. In this case supplementary to the cleansed data, on the basis of the manipulations on the data the data cleansing tool should provide on its output also enough information for calculation of the intuitionistinc fuzzy degrees for the data quality along the respective quality dimensions. Principles that can help the users develop usable data quality metrics are described in (Leo L. Pipino, 2002).

**Tables 2** a), b), c), d)

Client

| ID | FName | LName | Address | Phone | Salary |
|---|---|---|---|---|---|
| 100001 | Peter | Ivanov | 18 Rakovski Str. | 844567 | 1000 |

a)

Client_Quality

| ID | Attribute_Name | Dimension | MSHIP | NMSHIP |
|---|---|---|---|---|
| 100001 | Address | Currency | 0.8 | 0.1 |
| 100001 | Phone | Currency | 0.7 | 0.1 |
| 100001 | Salary | Currency | 0.6 | 0.1 |
| 100001 | Salary | Believability | 0.8 | 0.1 |

b)

Dimension_Weights

| Table | Attribute_Name | Dimension | Weight |
|---|---|---|---|
| Client | Address | Currency | 1 |
| Client | Phone | Currency | 1 |
| Client | Salary | Currency | 0.4 |
| Client | Salary | Believability | 0.6 |

c)

Attribute_Weights

| Table | Attribute_Name | Weight |
|---|---|---|
| Client | Address | 0.4 |
| Client | Phone | 0.4 |
| Client | Salary | 0.2 |

d)

Let us consider an example where a company has to conduct a marketing campaign. We decide to keep track not only of the client data but also of the quality of data on an attribute-level basis. We extend the relational model with supplementary tables, which contain the quality measures for each attribute on one or more quality dimensions. In our example, this supplementary table for the table Client /tables 2 a)/ is the table Client_Quality /tables 2 b)/ presented only with records for a given Client ID. We can consider this table an intuitionistic fuzzy relation, where the degrees of membership and non-membership represent the extent to which the corresponding attribute value fulfils the quality requirements at a certain quality dimension. In the table Client_Quality we add one record for each quality dimension that has to be tracked for those client attributes that are of a special interest. Each record contains respectively the μ and ν measures of the quality along the respective dimension. For instance the Salary attribute has to be measured along two quality dimensions – currency and believability, thus for this attribute in the table Client_Quality we add two records / tables 2 b)/ In the record for client with ID 100001, the salary' currency MSHIP contains a measure showing the extent to which the Salary is current, NMSHIP contains a measure showing the extent to which the Salary is not current. The last row in our example measures the probability that the salary of the client with ID 100001 is the real one or the probability that the client lied about his salary. In other words, the intuitionistic fuzzy degrees of membership and non-membership answer the question (vague terms are highlighted) 'How *high* is the *believability* that the salary for client with ID 100001 is the one pointed in the database?'

We will use IFPG database engine to represent and manipulate data quality measures. An important feature of this intuitionistic fuzzy RDBMS is the processing of queries with intuitionistic fuzzy predicates, e.g. predicates which correspond to natural language vague terms like 'high', 'cheap', 'close', etc. These predicates are evaluated with intuitionistic fuzzy values, which reflect on the degrees of membership and non-membership of the rows in the query result, which is in fact an intuitionistic fuzzy relation.

## 4 Calculating the Quality for an Attribute Value at a Certain Dimension

We can create an intuitionistic fuzzy predicate which presents the quality of a certain attribute value at a certain dimension. Given this functionality the user is capable to filter the data on quality-measure basis.

```
CREATE PREDICATE
high_qualty_for_client_attribute_dimension
(integer, varchar, varchar)
AS `
SELECT MSHIP, NMSHIP
FROM Client_Quality
WHERE ID = $1
    AND Attribute_Name = $2
    AND Dimension = $3
' LANGUAGE sql;
```

The user can now make queries of the kind 'List all clients with *high believability* for the real value of their salaries' and even define threshold to filter those records with demanded minimal value of the quality measure:

```
SELECT ID, Address, Phone, Salary, 'Believability' as
Quality_Dim, MSHIP, NMSHIP
FROM Client
WHERE high_qualty_for_client_attribute_dimension (ID,
'Salary', 'Believability')
HAVING MSHIP > 0.6;
```

| ID | Address | Phone | Salary | Quality_Dim | MSHIP | NMSHIP |
|---|---|---|---|---|---|---|
| 100001 | 18 Rakovski Str. | 844567 | 1000 | Believability | 0.8 | 0.1 |

## 5 Calculating the Overall Quality for an Attribute Value

Since an attribute value may have more than one quality dimension, the overall quality of the attribute value has to be calculated considering the quality measures of all its dimensions. This may help the user make analyses on the basis of the total quality of a certain attribute value. For the purpose we introduce a metadata table *Dimension_Weights* /tables 2 c)/, containing weights of the quality dimensions, which participate in the calculation of the overall quality of each attribute value:

The calculation of the overall quality of attribute values in table *Client* is performed with the following SQL query:

```
SELECT Client_Quality.ID,
Client_Quality.Attribute_Name,
SUM(Client_Quality."mship" * Dimension_Weights.Weight),
SUM(Client_Quality."nmship" * Dimension_Weights.Weight)
FROM Client_Quality JOIN Dimension_Weights
    ON Client_Quality.Attribute_Name =
         Dimension_Weights.Attribute_Name
       AND Client_Quality.Dimension =
         Dimension_Weights.Dimension
WHERE Dimension_Weights.Table_Name = 'Client'
GROUP BY Client_Quality.ID,
           Clicnt_Quality.Attribute_Name;
```

Follows the result of the query applied on the table Client with the example data for just one client.

| ID | Attribute_Name | MSHIP | NMSHIP |
|---|---|---|---|
| 100001 | Address | 0.8 | 0.1 |
| 100001 | Phone | 0.7 | 0.1 |
| 100001 | Salary | 0.72 | 0.1 |

This intuitionistic fuzzy relation represents the overall quality of attribute values in table *Client*. For instance the third row of the table answers a question of the kind 'How *high* is the overall *possibility* that the salary of the client with ID 100001 is the one pointed in the database?'

Analogously we can create an intuitionistic fuzzy predicate which presents the overall quality of a certain attribute value. Thus the user is capable to filter the data based on the total attribute value quality.

```
CREATE PREDICATE
high_quality_for_client_attribute_value (integer,
varchar)
AS
'SELECT
SUM(Client_Quality."mship" * Dimension_Weights.Weight),
SUM(Client_Quality."nmship" * Dimension_Weights.Weight)
FROM Client_Quality JOIN Dimension_Weights
    ON Client_Quality.Attribute_Name =
         Dimension_Weights.Attribute_Name
       AND Client_Quality.Dimension =
         Dimension_Weights.Dimension
WHERE Dimension_Weights.Table_Name = ''Client''
    AND Client_Quality.Attribute_Name = $2
    AND Client_Quality.ID = $1 '
LANGUAGE sql;
```

The user can now make queries of the kind 'List all clients with *high* overall *possibility* for the real value of their salaries' and even define threshold to filter those records with demanded minimal value of the quality measure:

```
SELECT ID, Address, Phone, Salary, MSHIP, NMSHIP
FROM Client
WHERE high_quality_for_client_attribute_value (ID,
'Salary')
HAVING MSHIP > 0.6;
```

| ID | Address | Phone | Salary | MSHIP | NMSHIP |
|---|---|---|---|---|---|
| 100001 | 18 Rakovski Str. | 844567 | 1000 | 0.72 | 0.1 |

## 6 Calculating the Overall Quality of a Tuple

For some kind of analyses, the quality of data in a tuple as a whole may be of importance. For calculating the overall quality of a tuple we consider the overall qualities of each of the attribute values in the tuple. For the purpose we introduce another metadata table *Attribute_Weights* /tables 2 d)/, containing weights of the quality of attributes, which participate in the calculation of the overall quality of each tuple:

The calculation of the overall quality of tuples in the relation Client is performed with the following SQL query:

```
SELECT Client_Quality.ID,
SUM(Client_Quality."mship" * DW.Weight * AW.Weight),
SUM(Client_Quality."nmship" * DW.Weight * AW.Weight)
FROM Client_Quality
    JOIN Dimension_Weights DW ON
     Client_Quality.Attribute_Name = DW.Attribute_Name
     AND Client_Quality.Dimension = DW.Dimension
    JOIN Attribute_Weights AW ON
     Client_Quality.Attribute_Name = AW.Attribute_Name
WHERE DW.Table_Name = 'Client'
  AND AW.Table_Name = 'Client'
GROUP BY Client_Quality.ID;
```

The result intuitionistic fuzzy relation represents the overall quality of tuples in table *Client*, each row of which answers the question ‘How *high* is the overall *quality* of data about client with ID 100001 pointed in the database?’

| ID | MSHIP | NMSHIP |
|---|---|---|
| 100001 | 0.744 | 0.1 |

Analogously an intuitionistic fuzzy predicate *high_quality_tuple* may be created which can help the user make queries of the kind ‘List all the clients, the information about which is more than 60% *reliable*’:

```
CREATE PREDICATE high_quality_tuple (integer)
AS
'SELECT
SUM(Client_Quality."mship" * DW.Weight * AW.Weight),
SUM(Client_Quality."nmship" * DW.Weight * AW.Weight)
FROM Client_Quality JOIN Dimension_Weights DW
        ON Client_Quality.Attribute_Name =
             DW.Attribute_Name
            AND Client_Quality.Dimension = DW.Dimension
         JOIN Attribute_Weights AW
        ON Client_Quality.Attribute_Name =
             AW.Attribute_Name
WHERE DW.Table_Name = ''Client'' AND AW.Table_Name =
''Client'' AND Client_Quality.ID = $1
GROUP BY Client_Quality.ID '
LANGUAGE sql;
```

The following select uses the *high_quality_tuple* predicate and returns only those records that have positive quality grater than the specified threshold.

```
SELECT ID, Address, Phone, Salary, MSHIP, NMSHIP
FROM Client
WHERE high_quality_tuple (ID)
HAVING MSHIP > 0.6;
```

| ID | Address | Phone | Salary | MSHIP | NMSHIP |
|---|---|---|---|---|---|
| 100001 | 18 Rakovski Str. | 844567 | 1000 | 0.744 | 0.1 |

## 7 Calculating the Overall Quality of the Attributes

On the basis of the currently available values in a user table and their current quality, we could calculate the overall quality of the attributes in a user table. For a given attribute we consider the overall quality of an attribute value in a tuple and we average this quality along all the records. The following query performs these calculations for the table Client:

```
SELECT QS.Attribute_Name, avg(QS.sum_Quality_MSHIP) as
     Attr_Quality_MSHIP,
   avg(QS.sum_Quality_NMSHIP) as Attr_Quality_NMSHIP
FROM (SELECT ID, DW.Attribute_Name,
      sum (Client_Quality."mship" * DW.Weight) AS
            sum_Quality_MSHIP,
      sum (Client_Quality."nmship" * DW.Weight) AS
            sum_Quality_NMSHIP
      FROM Client_Quality
          JOIN Dimension_Weights DW
              ON Client_Quality.Attribute_Name =
                     DW.Attribute_Name
              AND Client_Quality.Dimension =
                     DW.Dimension
      WHERE DW.Table_Name = 'Client'
      GROUP BY ID, DW.Attribute_Name) AS QS
GROUP BY QS.Attribute_Name
```

The result is an intuitionistic fuzzy relation that contains as many rows as is the number of the attributes in *Client* whose quality we track. Each row represents the overall quality of the respective attribute on the basis of the current quality of the all the values in this attribute.

| Attribute_Name | Attr_Quality_MSHIP | Attr_Quality_NMSHIP |
|---|---|---|
| Address | 0.8 | 0.1 |
| Phone | 0.7 | 0.1 |
| Salary | 0.72 | 0.1 |

## 8 Attribute-Based Data Quality in a Data Warehouse

Data quality measures should be continuously updated during the life-cycle of data in an information system in order to reflect the actual quality of the attribute values which is not always a constant. For example prior to data load into a data warehouse, the source data sets are integrated and cleaned. If a data quality issue occurs and it could not be corrected (in short time or by the utilized data quality software), a readily workable decision could be not to reject the record but to store it with a diminished level of quality. Currently the widespread approach is to correct the data defects by overwriting the values in the source records that are considered wrong and loading into the data warehouse just a single value that is considered perfectly correct. However the correction itself could cause some data deficiencies as it could be based on wrong inference or outdated business rule. That's why sometimes it could be preferable to store the raw (uncorrected) data with a lower quality grades or to store multiple representations of the record. For example in tables 3 A) are represented the records for a given client. The second record has an update of the Salary field. The related table Client_Quality, shown on tables 3 B), stores each update of the data quality measures along the different dimensions for the records from table Client. The sample is for the Believability dimension. The records represent a case where the Believability for the Salary is tracked even for the outdated records. If some evidence is received that supports the old value of the Salary (i.e. 1000) then the respective intuitionistic fuzzy assessments are corrected and they could become even better then the data quality grades for the values of current client's record in table Client (as is the case in the sample). Furthermore the changes of data quality level could be analyzed on a historical basis.

**Tables 3** A), B), C)

Client

| SurrKey | NatKey | FName | LName | Address | Phone | Salary | FromDate | ToDate |
|---|---|---|---|---|---|---|---|---|
| 500226 | 100001 | Peter | Ivanov | 18 Rakovski Str. | 844567 | 1000 | 10.1.2009 | 14.06.209 |
| 500848 | 100001 | Peter | Ivanov | 18 Rakovski Str. | 844567 | 1500 | 15.6.2009 | |

a)

Client_Quality

| SurrKey | AttributeName | Dimension | FromDate | ToDate | MSHIP | NMSHIP |
|---|---|---|---|---|---|---|
| 500226 | Salary | Believability | 9.1.2009 | 11.5.2009 | 0.6 | 0.1 |
| 500226 | Salary | Believability | 12.5.2009 | 1.7.2009 | 0.4 | 0.2 |
| 500848 | Salary | Believability | 15.6.2009 | | 0.5 | 0.1 |
| 500226 | Salary | Believability | 2.7.2009 | | 0.8 | 0.1 |

b)

| SurrKey | LName | Salary | AttributeName | Dimension | MSHIP | NMSHIP |
|---|---|---|---|---|---|---|
| 500848 | Ivanov | 1500 | Salary | Believability | 0.5 | 0.1 |
| 500226 | Ivanov | 1000 | Salary | Believability | 0.8 | 0.1 |

c)

Such a design permits answering the question: "For a specific client, list the latest data quality grades for all values of his salary along the Believability dimension.". The following simple query provides the result:

```
SELECT C.SurrKey, C.LName, C.Salary, CQ.Dimension,
        CQ."MSHIP", CQ."NMSHIP"
FROM Client C JOIN Client_Quality CQ ON
      C.SurrKey=CQ.SurrKey
WHERE C.NatKey=100001 and Dimension='Believability' and
CQ.ToDate is NULL;
```

The result for the sample data is given on tables 3 C). We see that the intuitionistic fuzzy data quality grades for the value of the salary from the outdated record (i.e. Salary=1000) are better then the respective grades for the currently valid record. In such a case the analyst could decide to use the "outdated" value of the salary. If we want to have in the result just data for the currently valid customer record from `Client`, then we have to add in the where clause another simple requirement - that the field `C.ToDate` should also equal null.

## 9 Conclusion

The utility of this model could be in several directions. Whatever the application is, we could note the following main type of gains addressed by the model. First, the queries, could manipulate only the values (records) having a quality greater than a certain threshold. Second – a query could act over all the records but the result could provide also a measure for the quality of the respective result along given dimensions or as a total. Third - a quality measuring method could be devised for calculation of the current quality of a given table or of the whole database. Fourth – the introduction of quality tracking in the database will outreach the framework of the information system and will make the employees put greater emphasis on the quality of their work. As the users are in fact the ultimate judges of how high quality of the data they need, then they will best take care to consider and improve quality of the data on an on-going basis.

## References

9000:2005, ISO, Fundamentals and vocabulary. Quality Management systems (2005)

Kolev, B.: Intuitionistic Fuzzy PostgreSQL. Advanced Studies in Contemporary Mathematics 2(11), 163–177 (2005)

Batini, C., Scannapieco, M.: Data Quality - Concepts, Methodologies and Techniques. Springer, Heidelberg (2006)

Boyadzhieva, D., Kolev, B.: An Extension of the Relational Model to Intuitionistic Fuzzy Data Quality Attribute Model. In: Proceedings of 4th International IEEE Conference on Intelligent Systems, vol. 2, pp. 13-14 –13-19 (2008)

Tayi, G.K., Ballou, D.P.: Examining Data Quality. Communications of the ACM 41(2), 54–57 (1998)

Huang, K., Lee, Y.W.: Quality Information and Knowledge. Prentice-Hall, Englewood Cliffs (1999)

Jarke, M., Jeusfeld, M.A.: Architecture and Quality in Data Warehouses: An Extended Repository Approach. Information Systems 24(3), 229–253 (1999)

Juran, J.: The Quality Control Handbook, 3rd edn. McGraw-Hill, New York (1974)

Kolev, B., Chountas, P.: Representing Uncertainty and Ignorance in Probabilistic Data Using the Intuitionistic Fuzzy Relational Data Model. In: Issues in the Representation and Processing of Uncertain and Imprecise Information. Fuzzy Sets, Generalized Nets and Related Topics, pp. 198–208. Akademicka Oficyna Wydawnicza EXIT, Warszawa (2005)

Pipino, L.L., Lee, Y.W.: Data Quality Assessment. Communications of the ACM 45, 211–218 (2002)

Orr, K.: Data quality and systems theory. Communications of the ACM 41(2), 66–71 (1998)

Wang, R.Y., Strong, D.M.: Beyond Accuracy: What Data Quality Means to Data Consumers. Technical Report TDQM-94-10, Total Data Quality Management Research Program (1994)

Wang, R.Y., Gibbs, M.R.: Towards Quality Data: An attribute-based Approach. Decision Support Systems 13 (1995)

# Redundancy Detection and Removal Tool for Transparent Mamdani Systems

Andri Riid, Kalle Saastamoinen, and Ennu Rüstern

**Abstract.** In Mamdani systems, redundancy of fuzzy rule bases that derives from extensive sharing of a limited number of output membership functions among the rules, is often an overlooked property. In current study, means for detection and removal of such kind redundancy have been developed. Our experiments with case studies collected from literature and Mackey-Glass time series prediction models show error-free rule base reduction by 30-60% that partially cures the curse of dimensionality problem characteristic to fuzzy systems.

## 1 Motivation

One very acute problem that is marring the large scale applications of fuzzy logic is the combinatorial explosion of rules (curse of dimensionality). As the number of membership functions (MFs) and/or input variables increases, the upper bound on the count of fuzzy rules grows exponentially:

$$R_{max} = \prod_{i=1}^{N} S_i, \tag{1}$$

where $S_i$ is the number of MFs per $i$-th input variable ($i = 1, ..., N$).

Andri Riid
Laboratory of Proactive Technologies, Tallinn University of Technology,
Ehitajate tee 5, 19086, Tallinn, Estonia
e-mail: andri@dcc.ttu.ee

Kalle Saastamoinen
Department of Military Technology, National Defence University, P.O. Box 7 FI-00861, Helsinki, Finland
e-mail: kalle.saastamoinen@mil.fi

Ennu Rüstern
Department of Computer Control, Tallinn University of Technology, Ehitajate tee 5, 19086, Tallinn, Estonia
e-mail: ennu.rystern@dcc.ttu.ee

V. Sgurev et al. (Eds.): Intelligent Systems: From Theory to Practice, SCI 299, pp. 397–415.
springerlink.com 

In the ideal fuzzy system the number of fuzzy rules $R = R_{max}$, meaning that the rule base of the system is fully defined and contains all possible antecedent combinations. Situation $R > R_{max}$ indicates a failure in fuzzy system design - either redundant or contradictory rules are present, both of which are the signs of sloppy system design. In real life applications, however, the number of rules often remains well below $R_{max}$ for several reasons.

First of all, commonly there is not enough material (data) or immaterial (knowledge) evidence to cover the input space universally, not only because it would be too time consuming to collect exhaustive evidence in large scale applications but also because of potential inconsistency that certain antecedent combinations may present (an antecedent "IF sun is bright AND rain is heavy" could be one such example).

Moreover, it is common practice that for the sake of compactness, the rules with little relevance are excluded from the model (for all we know they may be based on few noisy samples). The exclusion decision of a given rule may be based on its contribution to approximation properties (using singular value decomposition, orthogonal transforms, etc. [1]) or on how often or to what degree a given rule is contributing to the output (this, for example, can be easily evaluated by computing cumulative rule activation degrees on available data).

On the whole, rule base reduction can be fitted under two categories - error-free reduction or degrading reduction. Error-free reduction searches for existing redundancies in the model. In other words, if error-free reduction is effective, it is actually an indicator that initial system design was not up to the standard.

With degrading simplification, the model is made less complex by removing non-redundant system parameters. Incidentally, this is achieved at the expense of system universality, accuracy etc.

Typically, reduction is carried out on initial complex model. However, with certain design methodologies unnecessary complexity is avoided by model design procedure. A typical example is the application of tree partitioning of the input space (instead of more common grid partitioning) but the most common constructive compactness-friendly approach these days (related primarily to 1st order Takagi-Sugeno systems [2]) is fuzzy clustering. With clustering, the rules are created in product space only in regions where data concentration is high. Interestingly enough, the side effect of that is the redundancy of cluster projections that are used as the prototypes for MFs of the model. The projections that become fuzzy sets may be highly similar to each other, similar to the universal set or reduced to singleton sets, which calls for adequate methods to deal with that [3]. Another feature of product space clustering is that $R$ is always a lot smaller than $R_{max}$ (in fact $R = S_i$ prior to simplification). For this reason and also from interpolational aspect, product space clustering is not very well suited for Mamdani modeling.

In Mamdani systems, a relatively small set of output MFs is typically shared among rules. This creates substantial redundancy potential, which can exploited for rule base reduction. For a special class of Mamdani systems (transparent Mamdani systems, more closely observed in Sect. 2) this reduction can actually be error-free, i.e. without any performance loss. In Sect. 3, practical redundancy detection and removal scenarios have been investigated. Sect. 4 considers the typical implementation issues when designing a computer program for reduction of Mamdani systems. The remainder of the paper presents application examples and performance analysis. Note that the current implementation of the reduction tool can be freely downloaded from http://www.dcc.ttu.ee/andri/rdart".

## 2 Transparent Mamdani Systems

Generally, fuzzy rules in Mamdani-type fuzzy systems are based on the disjunctive rule format

$$\begin{array}{c}\text{IF } x_1 \text{ is } A_{1r} \text{ AND } x_2 \text{ is } A_{2r} \text{ AND } \ldots \ldots \text{ AND } x_N \text{ is } A_{Nr} \text{ THEN } y \text{ is } B_r \\ \text{OR} \ldots\end{array} \tag{2}$$

where $A_{ir}$ denote the linguistic labels of the $i$-th input variable associated with the $r$-th rule ($i = 1, ..., N$), and $B_r$ is the linguistic label of the output variable, associated with the $r$-th rule.

Each $A_{ir}$ has its representation in the numerical domain - the membership function $\mu_{ir}$ (the same applies to $B_r$ that is represented by $\gamma_r$) and in general case the inference function that computes the fuzzy output $F(y)$ of the system (2) has the following form

$$F(y) = \bigcup_{r=1}^{R}\left(\left(\bigcap_{i=1}^{N} \mu_{ir}(x_i)\right) \cap \gamma_r\right), \tag{3}$$

where $\cup_r^R$ denotes the aggregation operator (corresponds to OR in (2), $\cap$ is the implication operator (THEN) and $\cap_i^N$ is the conjunction operator (AND). In order to obtain crisp output, (3) is generally defuzzified with center-of-gravity method

$$y = Y_{cog}(F(y)) = \frac{\int_Y yF(y)dy}{\int_Y F(y)dy}. \tag{4}$$

The results obtained in current paper are valid for a class of Mamdani systems that satisfy the following requirements:

- The inference operators used here are product and sum. With product-sum inference (4) reduces to

$$y = \frac{\sum_{r=1}^{R} \tau_r c_r s_r}{\sum_{r=1}^{R} \tau_r s_r}, \tag{5}$$

  where $\tau_r$ is the activation degree of $r$-th rule (computed with the conjunction operator (product)) and $c_r$ and $s_r$ are the center-of-gravity and area of $\gamma_r$, respectively (see [4]).
- The input MFs ($s = 1, ..., S_i$) are given by the following definition:

$$\mu_i^s(x_i) = \begin{cases} \frac{x_i - a_i^{s-1}}{a_i^s - a_i^{s-1}}, & a_i^{s-1} < x_i < a_i^s \\ \frac{a_i^{s+1} - x_i}{a_i^{s+1} - a_i^s}, & a_i^s < x_i < a_i^{s+1} \\ 0, & a_i^{s+1} \leq x_i \leq a_i^{s+1} \end{cases}, \tag{6}$$

Such definition of input MFs satisfies input transparency condition assumed for correct interpretation of Mamdani rules (see [5] for further details), however, in current paper we are more interested in its other property, namely

$$\sum_{s=1}^{S_i} \mu_i^s = 1. \tag{7}$$

- The number of output MFs is relatively small and they are shared among rules (this is the usual case in Mamdani systems).

## 3 Error-Free Rule Base Reduction Principles

Consider a pair of fuzzy rules that share the same output MF $B_\xi$

$$\begin{array}{c} \text{IF } x_1 \text{ is } A_1^{s_1} \text{ AND ... AND } x_i \text{ is } A_i^s \text{ ...... AND } x_N \text{ is } A_N^{s_N} \text{ THEN } y \text{ is } B_\xi \\ \text{IF } x_1 \text{ is } A_1^{s_1} \text{ AND ... AND } x_i \text{ is } A_i^{s+1} \text{ ...... AND } x_N \text{ is } A_N^{s_N} \text{ THEN } y \text{ is } B_\xi \end{array} \tag{8}$$

It is possible to replace these two rules by a single one:

$$\begin{array}{c} \text{IF } x_1 \text{ is } A_1^{s_1} \text{ AND ... AND } x_i \text{ is } (A_i^s \text{ OR } A_i^{s+1}) \text{ ...} \\ \text{... AND } x_N \text{ is } A_N^{s_N} \text{ THEN } y \text{ is } B_\xi \end{array} \tag{9}$$

This replacement can be validated very easily, as it derives from (5) that numerically, (8) is represented by (10).

$$\mu_i^s \prod_{j=1 j\neq i}^{N} \mu_j^{s_j} + \mu_i^{s+1} \prod_{j=1 j\neq i}^{N} \mu_j^{s_j}. \tag{10}$$

Obviously (10), is equivalent to (11)

$$(\mu_i^s + \mu_i^{s+1}) \prod_{j=1 j\neq i}^{N} \mu_j^{s_j}, \tag{11}$$

which is nothing else than a representation of (9), assuming that the OR operand is implemented through sum.

This line of logic, while hardly practical for the reduction of fuzzy systems (fuzzy logic software does not usually have any support for such constructions as (9) and numerically, (11) is not really an improvement over (10)), however, has three offsprings (or special cases) that can be really useful as evidenced below.

**Lemma 1.** Consider not a pair but a subset of fuzzy rules consisting of $S_i$ rules that share the same output MF $B_\xi$ so that

$$\begin{gathered}\text{IF } x_1 \text{ is } A_1^{s1} \text{ AND ... AND } x_i \text{ is } A_i^s \text{ ...... AND } x_N \text{ is } A_N^{sN} \text{ THEN } y \text{ is } B_\xi \\ s = 1,...,S_i\end{gathered} \tag{12}$$

Apparently, this would be equivalent to a rule

$$\begin{gathered}\text{IF } x_1 \text{ is } A_1^{s1} \text{ AND ... AND } x_i \text{ is } (A_i^1 \text{ OR } A_i^2 \text{ OR ... OR } A_i^{S_i}) \text{ ...} \\ \text{... AND } x_N \text{ is } A_N^{sN} \text{ THEN } y \text{ is } B_\xi\end{gathered} \tag{13}$$

We proceed by showing that (13) is equivalent to (14).

$$\begin{gathered}\text{IF } x_1 \text{ is } A_1^{s1} \text{ AND ... AND } x_{i-1} \text{ is } A_{i-1}^{s_{i-1}} \text{ AND } x_{i+1} \text{ is } A_{i+1}^{s_{i+1}} \text{ ...} \\ \text{... AND } x_N \text{ is } A_N^{sN} \text{ THEN } y \text{ is } B_\xi\end{gathered} \tag{14}$$

*Proof.* For proof we need to show that

$$\sum_{s=1}^{S_i} \mu_i^s \prod_{j=1 j\neq i}^{N} \mu_j^{s_j} = \prod_{j=1 j\neq i}^{N} \mu_j^{s_j}, \tag{15}$$

which is valid when

$$\sum_{s=1}^{S_i} \mu_i^s = 1, \tag{16}$$

which is ensured by (6) that concludes the proof.

**Example 1.** Consider three rules of a two-input tipping system:

$$\begin{gathered}\text{IF } \mathit{food} \text{ is } \mathit{bad} \text{ AND } \mathit{service} \text{ is } \mathit{bad} \text{ THEN } \mathit{tip} \text{ is } \mathit{zero} \\ \text{IF } \mathit{food} \text{ is } \mathit{OK} \text{ AND } \mathit{service} \text{ is } \mathit{bad} \text{ THEN } \mathit{tip} \text{ is } \mathit{zero} \\ \text{IF } \mathit{food} \text{ is } \mathit{good} \text{ AND } \mathit{service} \text{ is } \mathit{bad} \text{ THEN } \mathit{tip} \text{ is } \mathit{zero}\end{gathered} \tag{17}$$

If there are no more linguistic labels of food quality as Fig. 1 clearly implies, it is indeed the case that if service is good, output of the system (the amount of tip) is independent from food quality that can be expressed by the following single rule

$$\text{IF } \mathit{service} \text{ is } \mathit{bad} \text{ AND } \mathit{food} \text{ is } \mathit{whatever} \text{ THEN } \mathit{tip} \text{ is } \mathit{zero}, \tag{18}$$

where "whatever" (or "don't care") describes the situation that food quality may have any value in its domain without a slightest effect to the output and can thus be removed from the rule, resulting in a nice compressed formulation

$$\text{IF } \mathit{service} \text{ is } \mathit{bad} \text{ THEN } \mathit{tip} \text{ is } \mathit{zero} \tag{19}$$

**Lemma 2:** If a subset of fuzzy rules consisting of $S_i - 1$ rules share the same output MF

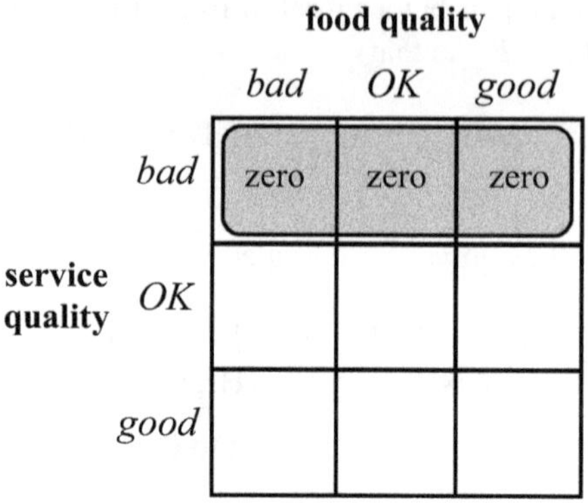


**Fig. 1** Redundancy of rules that makes rule compression possible

$$\text{IF } x_1 \text{ is } A_1^{s1} \text{ AND ... AND } x_i \text{ is } A_i^s \text{ ...... AND } x_N \text{ is } A_N^{sN} \text{ THEN } y \text{ is } B_\xi \qquad s = 1,...,S_i, s \neq t \tag{20}$$

then this group of rules can be replaced by a following single rule.

$$\text{IF } x_1 \text{ is } A_1^{s1} \text{ AND ... AND } x_i \text{ is NOT } A_i^t \text{ ...... AND } x_N \text{ is } A_N^{sN} \text{ THEN } y \text{ is } B_\xi \tag{21}$$

*Proof.* To prove that we need to show that

$$\sum_{s=1 s\neq t}^{S_i} \mu_i^s \prod_{j=1 j\neq i}^{N} \mu_j^{s_j} = (1-\mu_i^t) \prod_{j=1 j\neq i}^{N} \mu_j^{s_j}, \tag{22}$$

where $1-\mu_i^t$ represents the negation of $A_i^t$.

It is easy to see that $\sum_{s=1 s\neq t}^{S_i} \mu_i^s = 1-\mu_i^t$ if MFs of the $i$-th input variable add up to one (7), which completes the proof.

For the remainder of the paper we term replacement schemes (14) and (21) as rule compression scenarios A and B, respectively.

**Example 2:** Consider two rules of a hypothetical fuzzy system

$$\begin{array}{c}\text{IF } \textit{food} \text{ is } \textit{good} \text{ AND } \textit{service} \text{ is } \textit{OK} \text{ THEN } \textit{tip} \text{ is } \textit{large} \\ \text{IF } \textit{food} \text{ is } \textit{good} \text{ AND } \textit{service} \text{ is } \textit{good} \text{ THEN } \textit{tip} \text{ is } \textit{large}\end{array} \tag{23}$$

(23) implies that the amount of tip is independent from service quality if food quality is good and service quality is “anything else than bad” or simply “NOT bad”:

$$\text{IF } \textit{food} \text{ is } \textit{good} \text{ AND } \textit{service} \text{ is NOT } \textit{bad} \text{ THEN } \textit{tip} \text{ is } \textit{large} \tag{24}$$

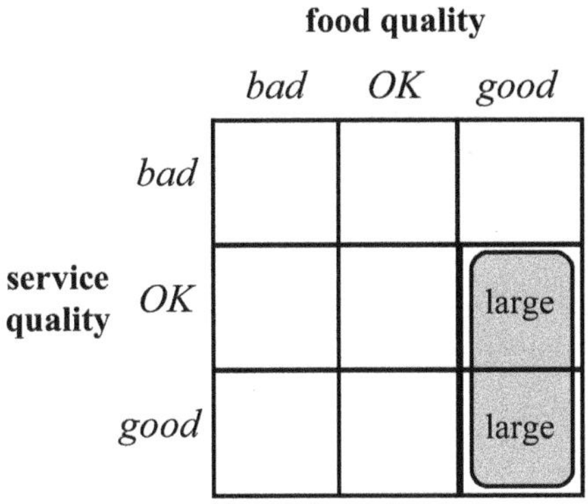


**Fig. 2** Rule base configuration allowing NOT construction

Note that it would be also possible to write (24) as

$$\text{IF } food \text{ is } good \text{ THEN } tip \text{ is } large \text{ UNLESS } food \text{ is } bad \tag{25}$$

**Lemma 3.** Consider a pair of rules (8) and assume that there are $R' = \prod_{j=1, j\neq i}^{N} S_j$ similar pairs that share the output MFs $B_\xi$ within the pair ($\xi \in [1, ..., T]$). If this is the case, the MFs $\mu_i^s$ and $\mu_i^{s+1}$ can be merged into $\mu_i^{s \cup s+1} = \mu_i^s + \mu_i^{s+1}$ by the means of summation, consequently each rule pair (8) will reduce to

$$\text{IF } x_1 \text{ is } A_1^{s_1} \text{ AND ... AND } x_i \text{ is } A_i^{s \cup s+1} \text{ ...... AND } x_N \text{ is } A_N^{s_N} \text{ THEN } y \text{ is } B_\xi \tag{26}$$

,

*Proof.* Any rule pair (8) is represented by (27) with ($\xi \in [1, ..., T]$).

$$(\mu_i^s + \mu_i^{s+1}) \prod_{j=1 j\neq i}^{N} \mu_j^{s_j} \sum_{\xi=1}^{R'} c_\xi s_\xi \tag{27}$$

Obviously, the common term $\mu_i^s + \mu_i^{s+1}$ can be permanently replaced by $\mu i^{s \cup s+1} = \mu_i^s + \mu_i^{s+1}$.

**Example 3.** The six rules depicted in Fig. 3 can be replaced by three rules

$$\begin{array}{c} \text{IF } food \text{ is } bad \text{ AND } service \text{ is } bad \text{ THEN } tip \text{ is } zero \\ \text{IF } food \text{ is } bad \text{ AND } service \text{ is } OK \text{ THEN } tip \text{ is } normal \\ \text{IF } food \text{ is } bad \text{ AND } service \text{ is } good \text{ THEN } tip \text{ is } large \end{array} \tag{28}$$

where "bad" is the name for the MF that combines former "very bad" and "bad".

Note that the merge of two triangles of (6) by sum would result in a trapezoid MF and the updated partition would still satisfy (7).

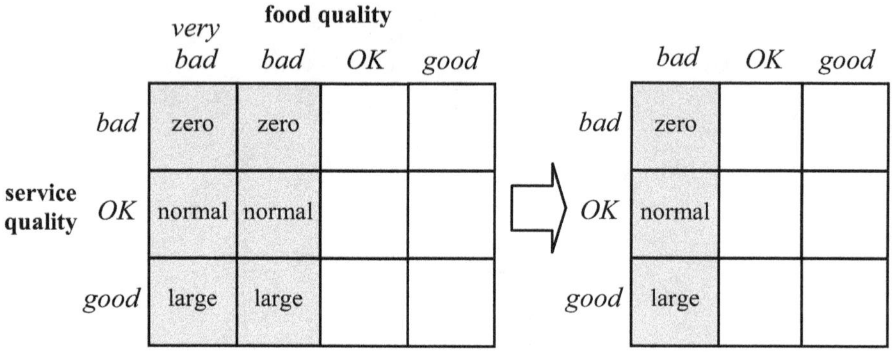


**Fig. 3** Redundancy of MFs revealed by rule base analysis

## 4 Implementation

For fuzzy logic software tools the rule base information is generally stored in a separate $R \times (N+1)$ dimensional matrix (MATLAB Fuzzy Logic Toolbox uses a variant of this) that accommodates the identifiers of MFs associated with fuzzy rules. Each row in the matrix represents an individual rule and column specifies the input variable (output variable in the last column, which is written in bold in the examples below) to which the identifier in current column is assigned to. Note that NOT-operator is conveniently represented by a minus sign and 0 represents a removed antecedent variable, e.g. $r$-th line 1 2 0 -1 **4** would be equivalent to a rule

$$\text{IF } x_1 \text{ is } A_{1r} \text{ AND } x_2 \text{ is } A_{2r} \text{ AND } x_4 \text{ is NOT } A_{1r} \text{ THEN } y \text{ is } B_4 \tag{29}$$

Implementation of rule base reduction schemes described in Sect. 3 is based on the analysis of the rule matrix and subsequent manipulation of it, which is described with the following algorithm (except the detection of redundant MFs that follows directly the logic under Lemma 3).

1. Fix $i$-th input variable (e.g. input No. 3 in Fig. 4)
2. Delete (temporarily) the indices corresponding to $i$-th variable from the rule matrix.
3. Split the rule matrix into submatrices/subsets of rules so that remaining input variables have fixed MFs throughout the subset.
4. Pick a submatrix.
   a. If the output MF associated with the rules is the same throughout the submatrix (like in (12)) apply rule compression scenario A by picking one of the rules, inserting zero into the blank spot and deleting the remaining rules.
   b. If there are two output MFs associated with the rules and one of them is used only once apply rule compression scenario B by picking the rule with the output MF used only once and restoring its deleted index. Then pick one rule

from the rest of the rules and insert negative value of the index just restored to the blank spot and delete the remaining rules.

c. If none of the above is true just restore the deleted indices within the submatrix.

5. Pick another submatrix (step 4) or, alternatively, if all submatrices have been treated, combine submatrices into one rule matrix.
6. Pick another variable (step 1) or, alternatively, if all input variables have been treated end.

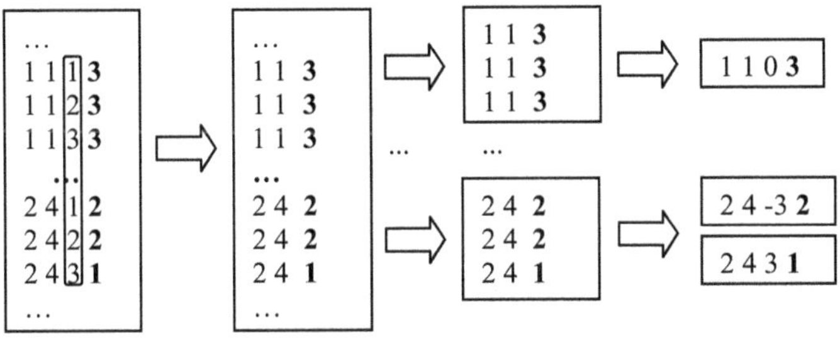


**Fig. 4** Principal steps of rule compression algorithm

## 4.1 Higher-Order Compression and Decompression

Though higher-order redundancies are less frequently encountered, the algorithm must be able to handle such situations, i.e. to be able to detect redundancies between already compressed rules. Consider a generous tipping example depicted in Fig. 5.

| service quality \ food quality | *bad* | *OK* | *good* |
|---|---|---|---|
| *bad* | zero | zero | zero |
| *OK* | normal | large | large |
| *good* | normal | large | large |

**Fig. 5** Rule base with higher order redundancies

In the first run we will come up with following compressed rules

$$\begin{array}{c} \text{IF } \textit{service} \text{ is } \textit{bad} \text{ THEN } \textit{tip} \text{ is } \textit{zero} \\ \text{IF } \textit{food} \text{ is } \textit{bad} \text{ AND } \textit{service} \text{ is NOT } \textit{bad} \text{ THEN } \textit{tip} \text{ is } \textit{normal} \\ \text{IF } \textit{food} \text{ is } \textit{OK} \text{ AND } \textit{service} \text{ is NOT } \textit{bad} \text{ THEN } \textit{tip} \text{ is } \textit{large} \\ \text{IF } \textit{food} \text{ is } \textit{good} \text{ AND } \textit{service} \text{ is NOT } \textit{bad} \text{ THEN } \textit{tip} \text{ is } \textit{large} \end{array} \tag{30}$$

For seeking for further compression we run the algorithm once again (or $N$ times in general case) on compressed rule set. Fig. 6 depicts the rule matrix before (left) and after second compression (right). The rules corresponding to the latter are given by (31).

$$\begin{bmatrix} 1 & 0 & \mathbf{1} \\ -1 & 1 & \mathbf{2} \\ 2 & -1 & \mathbf{3} \\ 3 & -1 & \mathbf{3} \end{bmatrix} \Rightarrow \begin{bmatrix} 1 & 0 & \mathbf{1} \\ -1 & 1 & \mathbf{2} \\ -1 & -1 & \mathbf{3} \end{bmatrix}$$

**Fig. 6** Rule base reduction with higher order redundancies

$$\begin{array}{c} \text{IF } \textit{service} \text{ is bad THEN } \textit{tip} \text{ is zero} \\ \text{IF } \textit{food} \text{ is bad AND } \textit{service} \text{ is NOT bad THEN } \textit{tip} \text{ is normal} \\ \text{IF } \textit{food} \text{ is NOT } \textit{bad} \text{ AND } \textit{service} \text{ is NOT } \textit{bad} \text{ THEN } \textit{tip} \text{ is } \textit{large} \end{array} \tag{31}$$

The inverse procedure to rule compression - decompression - is even easier to implement. The premise part of the rule base is scanned for zero and negative indices and if such is found, each rule containing a zero index in $i$-th position is replaced by $S_i$ rules so that all indices from 1 to $S_i$ are represented in the $i$-th position. If a negative index is found then $S_i - 1$ rules are generated, the index at $i$-th position running from 1 to $S_i$, except for the index that is the absolute value of found negative index. The scan is carried on until there are no more nonzero or negative indices in the rule base. For example, rule (29) has been decompressed in Fig. 7 (we assume that $S_3 = 3, S_4 = 4$). We can see that a deceptively innocent rule can have a large "family" of offsprings when decompressed.

## 4.2 *Preserving the Integrity of the Rule Base*

Naturally enough, one would expect that the decompression of a compressed rule set would return us to the initial rule base. To ensure that, rule subsets subject to compression must be non-overlapping, i.e. each original rule can only serve once as "raw material" for the compression procedure and cannot be "recycled". Consider the following, "conservative tipping" example. From the initial rule set in Fig. 8 we can extract a compressed rule "IF *service* is *bad* THEN *tip* is zero" or "IF *food* is *bad* THEN *tip* is *zero*" but not both simultaneously because this means that the rule

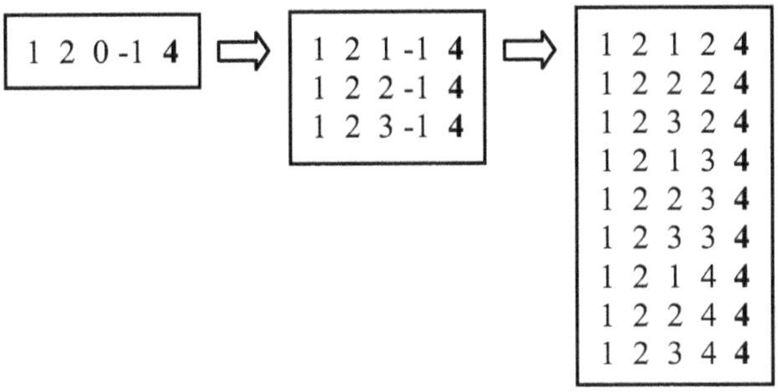


**Fig. 7** Inverse procedure to the compression of rules

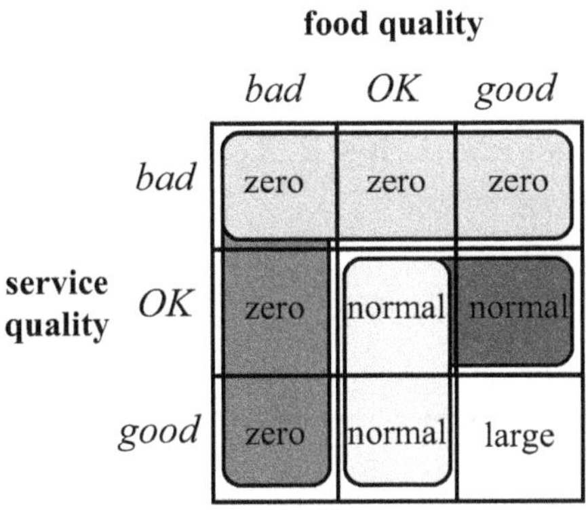


**Fig. 8** Conflicting simplification scenarios

"IF *service* is *bad* AND *food* is *bad* THEN *tip* is *zero*" has contributed twice for the improved rule base and would be also decompressed twice. [1]

Similarly, if we decide in favor of the first compressed rule then next we choose "IF *service* is *good* AND *food* is *good* THEN *tip* is *large*" and from the remainder we can extract "IF *service* is NOT *bad* and *food* is *OK* THEN *tip* is *normal*" but not simultaneously with "IF *service* is *OK* AND *food* is NOT *bad* THEN *tip* is *normal*". Due to the need to validate the compressions, the execution of rule compression becomes much more complicated than could be understood in first place and is controlled with the following mechanism described below.

When we fix an input variable, and extract a subset of rules, feasibility of compression is verified first against the internal conditions within the subset. When this test is passed, necessary compression is carried out and its feasibility is verified against the initial rule base (by simply looking for duplicates between the initial rule base and decompressed compressed subset). If the duplicates are found, the

[1] These two rules together are technically equivalent to "IF *service* is *bad* OR *food* is *bad* THEN *tip* is *zero*", Our reasoning above implies that disjunctive antecedents are prohibited.

subset is returned to the initial set and next subset is extracted. Only if there are no duplicates, the compression is actually executed, the compressed rule is added to the set of compressed rules (if available). The source rules, however, are returned to the working rule set. After that a new subset is handled. In the end of the cycle (all input variables have been picked one by one) the working rule set minus all these rules that were used for the compression, becomes a so-called reserve set (which will be temporarily added to the working rule set when duplicates are being sought) and the set of compressed rules becomes a new working rule set for higher order compression. In the very end (the rule compression algorithm has been applied $N$ times) the reserve set plus the compressed set of $N$-th cycle become the final rule base.

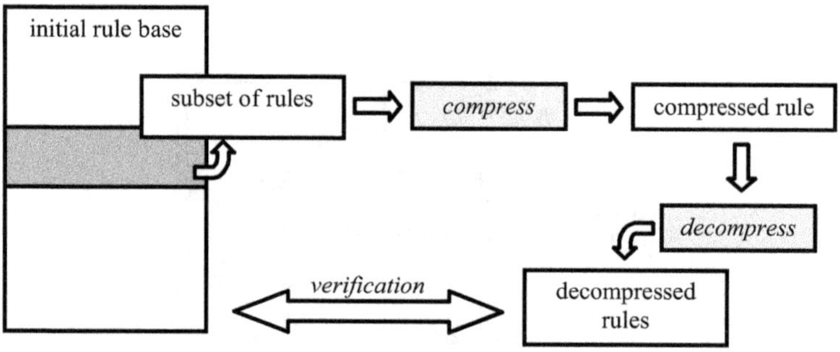


**Fig. 9** Compression validation

## 4.3 Incomplete Rule Bases

The reasoning throughout the section so far is based on the assumption that the rule base is complete. Yet in many applications we must deal with incomplete rule bases (the number of rules $R < \prod_{i=1}^{N} S_i$). Incompleteness of the rule base arises the question how to treat the blank spots in the rule base. Could we use them to our advantage so as to minimize the number of rules? Or should we maintain status quo and integrity of the rule base? To explain this dilemma in finer detail let us look at another tipping system in Fig. 10 that has two undefined rules. In first (optimistic) approach where we are about to make advantage of rule base incompleteness, we would end up with five rules including two compressed rules "IF *service* is *bad* THEN *tip* is *zero*" and "IF *service* is *OK* THEN *tip* is *normal*", ignoring the fact that it would actually mean writing *zero* and *normal*, respectively, into two blank spots and thus changing the original rule base. In conservative approach, though the number of rules would be the same after compression, two compressed rules would be "IF *service* is *bad* AND *food* is NOT *OK* THEN *tip* is *zero*" and "IF *service* is *OK* AND *food* is NOT *bad* THEN *tip* is *normal*", leaving the blank spots unmodified. Conservative approach seems somewhat closer to the spirit of error-free reduction, however, when we look at the problem at numerical level, the pros and cons are not so obvious.

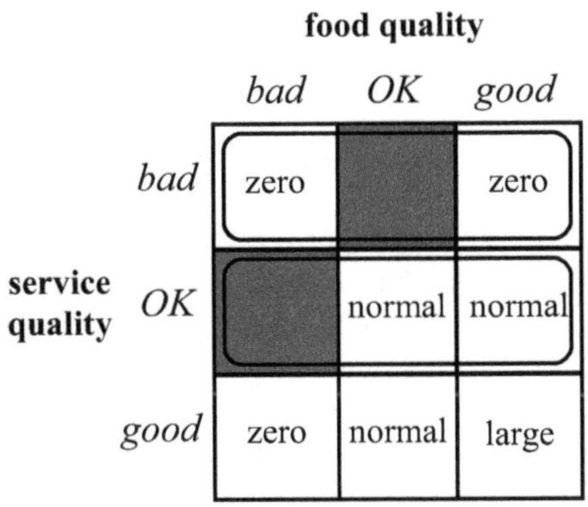


**Fig. 10** An incomplete rule base: another challenge

Each undefined rule means that there is some area in the input space for which we cannot compute matching output values. In practice some pre-specified value (usually average of the domain of the output variable) is used in this case to maintain continuity. If we use the blank spot to our advantage we typically use a neighboring rule as the prototype. Both may and may not be adequate guesses for the missing rule, neither is clearly better. Therefore, the simplification tool must have a built-in option to determine if we take optimistic or conservative approach when treating incomplete rule bases. In the following, we use the notation $I_c = 1$ for the conservative and $I_c = 0$ for the optimistic approach.

Taking above considerations into account, rule compression algorithm becomes more complex. Whereas with optimistic strategy nothing changes - we can apply rule compression scenario A when there is one output MF throughout the subset (see the algorithm in Sect. 4) and scenario B when there are two (and one of them is used only once), with conservative strategy we must also take into account how many rules are there in the subset. The selection map given in Table 1, where $N_o$ denotes the number of unique output MFs within the subset (once again, 2 MFs must satisfy the condition than one of them cannot be used more than once).

**Table 1** Selecting between different compression scenarios (conservative strategy)

| $N_o$ | $S_i$ rules | $S_i - 1$ rules |
|---|---|---|
| 1 | A | B |
| 2 | B | - |

# 5 Applications

The proposed approach is validated on three applications from literature that come from different areas of engineering - truck backer-upper control, skin permeability modeling and fuzzy decision support systems. Additionally, a thorough experiment on simplifying Mackey-Glass time series prediction models is carried out to provide analysis material.

## 5.1 Simplification of Systems from Literature

In first case study the simplification algorithm was applied to the fuzzy trajectory management unit (TMU) of truck backer-upper control system from [6] that originally uses 28 rules that specify the optimal truck angle $\Phi_r$ in respect to its coordinates $x$ and $y$. Application of the algorithm reveals that the original controller is heavily redundant as the number of its rules can be reduced to 11 without any loss in control quality that means almost 60% reduction in size (see Fig. 11). Incidentally, the biggest contribution to size reduction comes from detection and merging redundant MFs (13 rules), rule compression scenario A removes 2 and scenario B further 2 rules. As the original rule base is complete (as it is the case with two remaining case studies), applying the simplification tool with either (optimistic or conservative) option produces the same exact result.

In second case study, we undertook the task of reducing the rule base of the model developed in [10] that has octanol-water partition coefficient ($\log K_{ow}$), molecular weight ($M_w$) and temperature ($T$) as its inputs and skin permeability coefficient ($\log K_p$) as the output. The inputs are partitioned into 4, 3 and 3 fuzzy sets, respectively and output has three MFs. Because the MFs in [10] do not satisfy (7) (custom MFs functions were used in the application, moreover, for this system a special inference scheme has been developed because the model is expected to behave as a classifier) rule base reduction would not be error-free but nevertheless, the number of rules could potentially be reduced to 20 from 36 (56% reduction) - 8 by rule compression scenarios A and B each.

In third case study, the simplification tool was applied to the fuzzy decision support system [11] that has three inputs - detection, frequency and severity (all of which have been partitioned into five fuzzy sets), an output - fuzzy risk priority category - that has nine MFs and 125 rules. We are able to bring the rule count down to 75 (40% reduction) - out of which 25 disappear by merging two of the MFs of severity, 16 can be compressed by scenario A and further 9 by scenario B.

## 5.2 Mackey-Glass Time Series Prediction

The example includes prediction of time series that is generated by the Mackey-Glass [7] time-delay differential equation

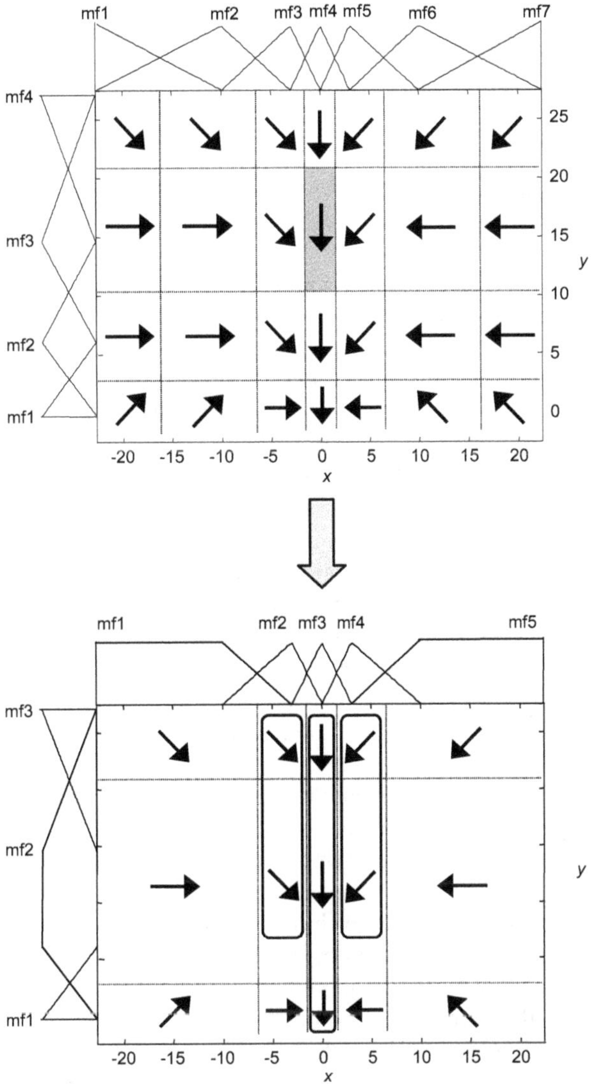


**Fig. 11** TMU of the truck backer-upper before (above) and after (below) the simplification

$$\dot{x} = \frac{0.2x(t-\tau)}{1+x^{10}(t-\tau)} - 0.1x(t), \tag{32}$$

and subsequent simplification of the prediction model.

To obtain the time series value at integer points, the numerical solution to the above MG equation is found using the fourth-order Runge-Kutta method. We assume $x(0) = 1.2, \tau = 17$, and $x(t) = 0$ for $t < 0$.

We use up to 4 known values of the time series in time, to predict the value in the future. For each t, the input training data is a four-dimensional vector of the following form.

$$x(t+6) = f(x(t-18), x(t-12), x(t-6), x(t)) \tag{33}$$

There are 1000 input/output data values. We use the first 500 data values for training, while the others are used as checking data for validating the identified fuzzy model.

To obtain the prediction model, we apply a simplistic modeling algorithm [8] that provides a crude predictor of the phenomenon (we are more interested in the performance of our simplification algorithm than in modeling accuracy). This method assumes predefined input-output partition - we are using an uniform one for the sake of simplicity - and finds out the best matching set of fuzzy rules for this partition on the basis of training data samples. For each potential rule we identify the sample $[x_1(k)x_2(k)x_3(k)x_4(k)y(k)]$ that yields maximum rule activation degree $\tau_r(k)$ and use $y(k)$ to determine matching output MF $\gamma_j, j \in 1, ..., T$ (the one that produces $\max(\gamma_j(y(k))))$.

To observe the effects of simplification we employ models of different sizes by varying the number of MFs (and even input variables - both 3- and 4-input models are being used). The results are given in Table 2, where first column specifies the $S_i$s of each input variable, the second the number of output MFs ($T$). Further columns contain modeling errors on training and checking data ($\varepsilon_{tr}$ and $\varepsilon_{ch}$, respectively), the number of rules before ($R_0$) and after simplification ($R_f$) and rule reduction rates ($\eta$).

The results reveal some general characteristics of the simplification algorithm. It can be seen that rule reduction rate is higher if the number of output MFs is small and number of input MFs is high, which is rather logical, because with a small number of output MFs these are shared more extensively and thus redundancy is more likely to exist. Large number of input MFs on the other hand means that the number of rules that can be replaced by a single rule is generally larger. However, with incomplete rule bases, the large number of input MFs increases the number of undefined rules thus limiting algorithm's capability when we take conservative approach regarding completeness of the rule base. It is, however, clearly evident that this option does not have any effect to the modeling error neither with training nor checking data.

It also turns out that redundancy of input MFs is a rather rare phenomenon as it was detected in none of the above models. Rule compression scenario A contributes more to $\eta$ if $I_c = 1$ and scenario B is mostly responsible for redundancy removal if $I_c = 0$.

As for comparison of the modeling accuracy - ANFIS [9] with two generalized bell membership functions on each of the four inputs and containing 16 Takagi-Sugeno rules shows $\varepsilon_{tr} = \varepsilon_{ch} = 0.0025$ after 10 training epochs.

**Table 2** results of MG time series prediction and model simplification

| $S_i$ | $T$ | $\varepsilon_{tr}$ | $\varepsilon_{ch}$ | $R_0$ | $R_f$ | $I_c$ | $\eta$ |
|---|---|---|---|---|---|---|---|
| 4×4×4 | 5 | 0.0691 | 0.0687 | 57 | 29 | 1 | 49.1% |
| 4×4×4 | 5 | 0.0691 | 0.0687 | 57 | 26 | 0 | 54.4% |
| 4×4×4 | 9 | 0.0623 | 0.0620 | 57 | 44 | 1 | 22.8% |
| 4×4×4 | 9 | 0.0623 | 0.0620 | 57 | 40 | 0 | 30.0% |
| 5×5×5 | 5 | 0.0599 | 0.0593 | 88 | 56 | 1 | 36.4% |
| 5×5×5 | 5 | 0.0599 | 0.0593 | 88 | 37 | 0 | 58.0% |
| 5×5×5 | 9 | 0.0426 | 0.0420 | 88 | 79 | 1 | 10.2% |
| 5×5×5 | 9 | 0.0426 | 0.0420 | 88 | 60 | 0 | 31.8% |
| 6×6×6 | 5 | 0.0483 | 0.0479 | 128 | 112 | 1 | 12.5% |
| 6×6×6 | 5 | 0.0483 | 0.0479 | 128 | 60 | 0 | 53.1% |
| 6×6×6 | 9 | 0.0347 | 0.0346 | 128 | 128 | 1 | 0% |
| 6×6×6 | 9 | 0.0347 | 0.0346 | 128 | 90 | 0 | 29.7% |
| 3×3×3×3 | 5 | 0.0639 | 0.0627 | 68 | 36 | 1 | 47.1% |
| 3×3×3×3 | 5 | 0.0639 | 0.0627 | 68 | 41 | 0 | 39.7% |
| 3×3×3×3 | 9 | 0.0619 | 0.0607 | 68 | 43 | 1 | 36.8% |
| 3×3×3×3 | 9 | 0.0619 | 0.0607 | 68 | 44 | 0 | 35.3% |
| 4×4×4×4 | 5 | 0.0384 | 0.0376 | 181 | 110 | 1 | 39.2% |
| 4×4×4×4 | 5 | 0.0384 | 0.0376 | 181 | 96 | 0 | 47.0% |
| 4×4×4×4 | 9 | 0.0404 | 0.0397 | 181 | 131 | 1 | 27.6% |
| 4×4×4×4 | 9 | 0.0404 | 0.0397 | 181 | 115 | 0 | 36.5% |
| 5×5×5×5 | 5 | 0.0451 | 0.0442 | 275 | 227 | 1 | 17.5% |
| 5×5×5×5 | 5 | 0.0451 | 0.0442 | 275 | 128 | 0 | 53.5% |
| 5×5×5×5 | 9 | 0.0354 | 0.0345 | 275 | 254 | 1 | 7.6% |
| 5×5×5×5 | 9 | 0.0354 | 0.0345 | 275 | 162 | 0 | 41.1% |

## 6 Conclusions

In this paper we presented the basis and working principles for the tool for redundancy detection and removal for a special class of Mamdani systems and also demonstrated that the implementation of these ideas indeed reduces complexity of fuzzy systems from different areas of engineering by 30-60% without any loss of accuracy. The major factors that influence the reduction rate by rule compression are the number of input MFs (positive correlation) and the number of output MFs (negative correlation) In certain cases (if the number of input variables is relatively small) MF redundancy may also be the case. Additionally, it was found out that with optimistic strategy, scenario A is mostly responsible for rule base reduction, whereas scenario B plays the key role with conservative strategy; the latter also tends to be ineffective if the number of input variables is large.

However, there may be cases such as the one depicted in Fig. 12, which is a primitive version of standard McVicar-Whelan rule base [12] where, even if the number of unique output MFs is well below the number of rules implying output MF sharing and redundancy potential, the homogeneous rule subsets are oriented so that it

| | | **change in error** | | |
|---|---|---|---|---|
| | | *N* | *Z* | *P* |
| | *N* | NB | NS | Z |
| **error** | *Z* | NS | Z | PS |
| | *P* | Z | PS | PB |

**Fig. 12** Compact version of McVicar-Whelan rule base where N stands for "negative", P for "positive", Z for "zero", PB for "positive big", PS for "positive small", NS for "negative small" and NB for "negative big".

is really impossible to apply any scheme of rule compression or MF merge. Consequently, orthogonal orientation of homogeneous rule subsets is another, somewhat hidden but nevertheless important requirement for redundancy removal. Investigation, if there are means for redundancy reduction for such fuzzy rule bases is a matter of future research.

**Acknowledgements.** This work has been partially supported by Estonian Science Foundation, grant no. 6837.

## References

1. Yen, J., Wang, L.: Simplifying Fuzzy Rule-Based Models Using Orthogonal Transformation Methods. IEEE Trans. Systems, Man, Cybern. Part B 29(1), 13–24 (1999)
2. Takagi, T., Sugeno, M.: Fuzzy identification of systems and its applications to modeling and control. IEEE Trans. Systems Man and Cybern. 15, 116–132 (1985)
3. Roubos, H., Setnes, M.: Compact and Transparent Fuzzy Models and Classifiers Through Iterative Complexity Reduction. IEEE Trans. Fuzzy Systems 9(4), 516–524 (2001)
4. Riid, A., Rüstern, E.: On the Interpretability and Representation of Linguistic Fuzzy Systems. In: Proc. IASTED International Conference on Artificial Intelligence and Applications, Benalmadena, Spain, pp. 88–93 (2003)
5. Riid, A., Rüstern, E.: Transparent Fuzzy Systems in Modeling and Control. In: Casillas, J., Cordon, O., Herrera, F., Magdalena, L. (eds.) Interpretability Issues in Fuzzy Modeling, pp. 452–476. Springer, New York (2003)
6. Riid, A., Rüstern, E.: Fuzzy logic in control: truck backer-upper problem revisited. In: Proc. IEEE Int. Conf. Fuzzy Systems, Melbourne, Australia, vol. 1, pp. 513–516 (2001)
7. Mackey, M.C., Glass, L.: Oscillation and chaos in physiological control systems. Science 197, 287–289 (1977)

8. Wang, L.X., Mendel, J.M.: Generating fuzzy rules by learning from examples. IEEE Trans. on Systems, Man and Cybern. 22(6), 1414–1427 (1992)
9. Jang, J.-S.R.: ANFIS: Adaptive-Network-Based Fuzzy Inference System. IEEE Trans. on Systems, Man and Cybern. 23(3), 665–685 (1993)
10. Keshwania, D.R., Jonesb, D.D., Meyerb, G.E., Brand, R.M.: Rule-based Mamdani-type fuzzy modeling of skin permeability. Applied Soft Computing 8(1), 285–294 (2008)
11. Puente, J., Pino, R., Priore, P., Fuente, D.D.L.: A decision support system for applying failure mode and effects analysis. Int. J. Quality and Reliability Mgt 19(2), 137–150 (2002)
12. MacVicar-Whelan, P.J.: Fuzzy Sets for Man-Machine Interaction. Int. J. Man-Machine Studies 8, 687–697 (1976)

# Optimization of Linear Objective Function under Fuzzy Equation Constraint in $BL-$ Algebras – Theory, Algorithm and Software

Ketty Peeva and Dobromir Petrov

**Abstract.** We study optimization problem with linear objective function subject to fuzzy linear system of equations as constraint, when the composition is $inf - \rightarrow$ in $BL-$algebra . The algorithm for solving fuzzy linear system of equations is provided by algebraic-logical properties of the solutions. We present algorithms for computing the extremal solutions of fuzzy linear system of equations and implement the results for solving the linear optimization problem.

**Keywords:** Linear optimization, fuzzy relational equations, $inf - \rightarrow$ composition.

## 1 Introduction

The main problem that we solve here is to optimize the linear objective function

$$Z = \sum_{j=1}^{n} c_j x_j, c_j \in \mathbb{R}, 0 \le x_j \le 1, 1 \le j \le n, \tag{1}$$

with traditional addition and multiplication, if $c = (c_1, ..., c_n)$ is the cost vector, subject to fuzzy linear system of equations as constraint

$$A \odot X = B \ , \tag{2}$$

where $A = (a_{ij})_{m \times n}$ stands for the matrix of coefficients, $X = (x_j)_{n \times 1}$ stands for the matrix of unknowns, $B = (b_i)_{m \times 1}$ is the right-hand side of the system and for each $i, 1 \le i \le m$, and for each $j, 1 \le j \le n$, we have $a_{ij}$, $b_i$, $x_j \in [0, 1]$. The composition written as $\odot$ is $inf - \rightarrow$. The aim is to minimize or maximize (1) subject to constraint (2). The results for solving this linear optimization problem are provided

Ketty Peeva · Dobromir Petrov
Faculty of Appl. Math. and Informat., Technical University of Sofia,
8, Kl. Ohridski St. Sofia 1000
e-mail: kgp@tu-sofia.bg, dspetrov84@abv.bg

V. Sgurev et al. (Eds.): Intelligent Systems: From Theory to Practice, SCI 299, pp. 417–431.
springerlink.com 

by the inverse problem resolution for fuzzy linear system of equations (FLSE) with $inf - \rightarrow$ composition as presented in [15] and next developed here for optimization. The main results for solving FLSE with $\max - \min$, $\min - \max$, $\max -$product, $inf - \rightarrow$, $sup - \star$ composition are published in [2] – [5], [13] – [17]. The results concern finding the extremal solutions, estimating time complexity of the problem and applications in optimization problems are given in [7], [9] – [11].

In Section 2 we introduce basic notions for $BL-$algebras and FLSE. Section 3 covers fuzzy linear systems of equations in Gödel algebra, Goguen algebra and Łukasiewicz algebra. We describe a method and algorithm for solving them, following [15]. Rather than work with the system (2), we use a matrix, whose elements capture all the properties of the equations in (2). A sequence of algebraic-logical simplification rules are performed on this matrix, resulting all maximal solutions and none redundant solutions. When the system (2) is consistent, its solution set is completely determined by the unique minimal solution and a finite number of maximal solutions. Since the solution set can be non-convex, traditional linear programming methods cannot be applied to optimization problem. In Section 4 we propose method, algorithm and software for solving linear optimization problem (1) subject to constraint (2). Section 5 describes the codes and software realization for solving linear optimization problem.

Terminology for computational complexity and algorithms is as in [1], for fuzzy sets and fuzzy relations is according to [3], [5], [8], [16], [18], for algebra - as in [6], [12].

## 2 Basic Notions

Partial order relation on a partially ordered set (poset) $P$ is denoted by the symbol $\leq$. By a *greatest element* of a poset $P$ we mean an element $b \in P$ such that $x \leq b$ for all $x \in P$. The *least element* of $P$ is defined dually.

The algebraic structure

$$BL = \langle L, \vee, \wedge, *, \rightarrow, 0, 1 \rangle$$

is called $BL-$algebra [17], where $\vee, \wedge, *, \rightarrow$ are binary operations, 0, 1 are constants and:

i) $L = \langle L, \vee, \wedge, 0, 1 \rangle$ is a lattice with universal bounds 0 and 1;
ii) $L = \langle L, *, 1 \rangle$ is a commutative semigroup;
iii) $*$ and $\rightarrow$ establish an adjoint couple:

$$z \leq (x \rightarrow y) \Leftrightarrow x * z \leq y, \forall x, y, z \in L.$$

iv) for all $x, y \in L$

$$x * (x \rightarrow y) = x \wedge y \quad \text{and} \quad (x \rightarrow y) \vee (y \rightarrow x) = 1.$$

The negation $\neg$ is defined as $\neg x = x \rightarrow 0$.
The following algebras are examples for $BL-$algebras.

1. Gödel algebra

$$BL_G = \langle [0,1], \vee, \wedge, \rightarrow_G, 0, 1 \rangle,$$

where $* = \wedge$ and

$$x \rightarrow_G y = \begin{cases} 1 \text{ if } x \leq y \\ y \text{ if } x > y \end{cases}, \neg x = \begin{cases} 1 \text{ if } x = 0 \\ 0 \text{ if } x > 0 \end{cases}.$$

2. Product (Goguen) algebra

$$BL_P = \langle [0,1], \vee, \wedge, \circ, \rightarrow_P, 0, 1 \rangle,$$

where $\circ$ is conventional real number multiplication and

$$x \rightarrow_P y = \begin{cases} 1 \text{ if } x \leq y \\ \frac{y}{x} \text{ if } x > y \end{cases}, \neg x = \begin{cases} 1 \text{ if } x = 0 \\ 0 \text{ if } x > 0 \end{cases}.$$

The laws of cancelation and contradiction are valid in Product algebra, i.e.

$$x * z = y * z \Rightarrow x = y \text{ if } z \neq 0, \;\; x \wedge \neg x = 0.$$

3. Łukasiewicz algebra

$$BL_L = \langle [0,1], \vee, \wedge, \otimes, \rightarrow_L, 0, 1 \rangle,$$

where

$$x \otimes y = 0 \vee (x + y - 1),$$
$$x \rightarrow_L y = 1 \wedge (1 - x + y), \; \neg x = 1 - x.$$

4. Boolean algebra is also $BL-$algebra.

A matrix $A = (a_{ij})_{m \times n}$, with $a_{ij} \in [0,1]$ for each $i, j, 1 \leq i \leq m, 1 \leq j \leq n$, is called a *membership matrix* [8]. In what follows we write "matrix" instead of "membership matrix".

**Definition 1.** Let the matrices $A = (a_{ij})_{m \times p}$ and $B = (b_{ij})_{p \times n}$ be given. The matrix $C = (c_{ij})_{m \times n} = A \odot B$ is called $inf-\rightarrow$ *product* of $A$ and $B$ if $c_{ij} = \inf_{k=1}^{p} (a_{ik} \rightarrow b_{kj})$, when $1 \leq i \leq m$, $1 \leq j \leq n$.

## 3 Fuzzy Linear Systems

We first give the solution set of fuzzy linear systems of equations (2) with $inf-\longrightarrow$ composition. The system (2) has the following form:

$$\left| \begin{array}{ccccc} (a_{11} \to x_1) & \wedge\cdots\wedge & (a_{1n} \to x_n) & = & b_1 \\ \cdots & \cdots & \cdots & & \cdots\cdots \\ (a_{m1} \to x_1) & \wedge\cdots\wedge & (a_{mn} \to x_n) & = & b_m \end{array} \right. , \tag{3}$$

written in the equivalent matrix form (2) :

$$A \odot X = B.$$

Here $A = (a_{ij})_{m\times n}$ stands for the matrix of coefficients, $X = (x_j)_{n\times 1}$ stands for the matrix of unknowns, $B = (b_i)_{m\times 1}$ is the right-hand side of the system. For each $i, 1 \le i \le m$ and for each $j, 1 \le j \le n$, we have $a_{ij}$, $b_i$, $x_j \in [0, 1]$ and the $inf-\longrightarrow$ composition is written as $\odot$.

For $X = (x_j)_{n\times 1}$ and $Y = (y_j)_{n\times 1}$ the inequality $X \le Y$ means $x_j \le y_j$ for each $j, 1 \le j \le n$.

**Definition 2.** Let the FLSE in $n$ unknowns be given.

i) $X^0 = (x_j^0)_{n\times 1}$ with $x_j^0 \in [0,1]$, when $1 \le j \le n$, is called a *solution* of (2) if $A \odot X^0 = B$ holds.

ii) The set of all solutions $\mathbb{X}^0$ of (2) is called *complete solution set*. If $\mathbb{X}^0 \neq \emptyset$ the FLSE is called *consistent,* otherwise it is *inconsistent.*

*iii)* A solution $X^0_{\text{low}} \in \mathbb{X}^0$ is called a *lower (minimal) solution* of FLSE if for any $X^0 \in \mathbb{X}^0$ the relation $X^0 \le X^0_{\text{low}}$ implies $X^0 = X^0_{\text{low}}$, where $\le$ denotes the partial order in $\mathbb{X}^0$. Dually, a solution $X^0_{\text{u}} \in \mathbb{X}^0$ is called an *upper (maximal) solution* of FLSE if for any $X^0 \in \mathbb{X}$ the relation $X^0_{\text{u}} \le X^0$ implics $X^0 = X^0_{\text{u}}$. When the lower solution is unique, it is called *least* or *minimal* solution.

We consider inhomogeneous systems ($b_i = 0$ for each $i = 1,...,m$ makes the problem uninteresting).

The solution set of (2) is determined by all maximal solutions (they are finite number) and the minimal one. Properties of the solution set are studied in [17].

**Theorem 1.** *[5] Any consistent FLSE has unique lower (minimal) solution* $\check{X} = (\check{x}_j)_{n\times 1}$ *with components:*

$$\check{x}_j = \vee_{i=1}^{m}(a_{ij} * b_i),\ j = 1,\ldots,n. \tag{4}$$

We denote by $\check{X}$ the minimal solution of (2) with components determined by (4).

We denote by $\hat{X}$ a maximal solution of (2).

**Theorem 2.** *[15] If the system (2) is consistent and* $\hat{X} = (\hat{x}_j)_{n\times 1}$ *is its maximal solution then* $\hat{x}_j = 1$ *or* $\hat{x}_j = \check{x}_j$ *when* $j = 1,...,n$.

In order to obtain the complete solution set of a consistent system (2) it would be sufficient to obtain minimal solution and all maximal solutions. For a general $BL-$algebra this problem is complicated and requires a ramified study of a relationship between coefficients of the equation and its right-hand side [17]. We restrict our investigation to the case when (2) is over Gödel algebra, Goguen algebra and Łukasiewicz algebra as described in Examples 1-3.

## 3.1 Main Steps in Solving (2)

We develop algorithm and codes for solving (2) according to next simplifications and also using the algebraic-logical properties of the solution set of (2) as described below.

### 3.1.1 Simplifying Steps

We describe steps for simplifying (2) in order to find the complete solution set easier. For more details see [15].

**Step 1.** Calculate $\check{X}$ according to (4).

**Step 2.** Establish consistency of the system (2). If $A \odot \check{X} \neq B$ the system is inconsistent – end.

**Step 3.** If $b_i = 1$ for each $i = 1, ..., n$ then the system has unique maximal solution $\hat{X} = (1, ..., 1)$ – end.

**Step 4.** Remove all equations with $b_i = 1$ – they do not influence the maximal solutions. Upgrade $m$ to be equal to the new number of equations, all of them with right-hand side $b_i \neq 1$.

**Step 5.** Create a matrix $P = (p_{ij})_{m\times n}$ with elements $p_{ij} = a_{ij} * b_i$.

**Step 6.** Create a matrix $C = (c_{ij})_{m\times n}$ with elements

$$c_{ij} = \begin{cases} 0, & \text{if } p_{ij} = \check{x}_j \\ 1, & \text{if } p_{ij} \neq \check{x}_j \end{cases}. \tag{5}$$

**Remark.** The matrix $C$ distinguishes coefficients in the consistent system that may contribute to find maximal solutions (marked by zero in $C$ ) from these coefficients that do not contribute to find maximal solutions (they are marked by 1 in $C$).

**Step 7.** Upgrade $C$ to remove redundant $c_{ij}$: for each $\check{x}_j$ if $c_{ij} = 0$ but $\neg a_{ij} > b_i$ put $c_{ij} = 1$.

**Step 8.** Use the upgraded matrix $C = (c_{ij})_{m\times n}$ to compute the maximal solutions of the system. For any $i$, $1 \leq i \leq m$, the element $c_{ij} = 0$ in $C = (c_{ij})_{m\times n}$ marks a way $\langle j_i \rangle$ to satisfy the $i-$th equation of the FLSE.

### 3.1.2 Finding Maximal Solutions – Algebraic-Logical Approach

In this subsection we propose algebraic-logical approach to find all different ways to satisfy simultaneously equations of the FLSE.

We symbolize logical sum of different ways $\langle j_i \rangle$ for fixed $i$ as $\sum_{1 \le j \le n} \langle j_i \rangle$. For $\langle j_i \rangle$ we often write $\langle j \rangle$ when there does not exist danger of misunderstanding.

We have to consider equations simultaneously, i.e., to compute the *concatenation* $W$ of all ways, symbolized by the sign $\prod$:

$$W = \prod_{1 \le i \le m} \left( \sum_{1 \le j \le n} \langle j_i \rangle \right). \tag{6}$$

In order to compute complete solution set, it is important to determine different ways to satisfy simultaneously equations of the system. To achieve this aim we list the properties of concatenation (6).

Concatenation is *distributive* with respect to addition, i.e.

$$\langle j_{i_1} \rangle \left[ \langle j_{i_2} \rangle + \langle j_{i_3} \rangle \right] = \langle j_{i_1} \rangle \langle j_{i_2} \rangle + \langle j_{i_1} \rangle \langle j_{i_3} \rangle . \tag{7}$$

We expand the parentheses in (6) and obtain the set of ways, from which we extract the maximal solutions:

$$W = \sum_{(j_1, \cdots, j_m)} \langle j_1 \rangle \langle j_2 \rangle \cdots \langle j_m \rangle. \tag{8}$$

Any term $\langle j_1 \rangle \langle j_2 \rangle \cdots \langle j_m \rangle$ defines a solution with components $\check{x}_{j_i}$; for the missing $j$ we put $x_j^0 = 1$. The expression (8) contains all maximal solutions of the system.

Concatenation is *commutative*:

$$\langle j_{i_1} \rangle \langle j_{i_2} \rangle = \langle j_{i_2} \rangle \langle j_{i_1} \rangle . \tag{9}$$

The sum of the ways (8) satisfies *absorptions* (10) and (11):

$$\langle j_{i_1} \rangle \langle j_{i_2} \rangle = \begin{cases} \langle j_{i_1} \rangle \text{ if } \langle j_{i_1} \rangle = \langle j_{i_2} \rangle , \\ \text{unchanged, if } \langle j_{i_1} \rangle \neq \langle j_{i_2} \rangle \end{cases} , \tag{10}$$

$$\langle j_{i_1} \rangle + \langle j_{i_1} \rangle \langle j_{i_2} \rangle = \langle j_{i_1} \rangle . \tag{11}$$

This leads to simplification rule:

$$\langle j_{i_1} \rangle \left( \sum_{i=2}^{m} \langle j_i \rangle \right) = \begin{cases} \text{unchanged, if } \langle j_{i_1} \rangle \neq \langle j_i \rangle , i = 2, \cdots, m \\ \langle j_{i_1} \rangle \text{ otherwise} \end{cases} . \tag{12}$$

**Step 9.** Compute the maximal solutions by simplifying (6) according to (7), (9) – (12).

### 3.1.3 Main Results

Let the system (2) be consistent. The time complexity function for establishing the consistency of the FLSE and for computing $\check{X}$ is $O(m^2n)$. The resulting terms after steps 1–9 correspond exactly to the maximal solutions – the final result does not contain redundant (or extra) terms. The maximal solutions are computable and finite number.

In [15] we propose code for solving (2) . It answers the following questions:

(i) Is the system consistent or not.
(ii) If it is inconsistent - the equations that can not be satisfied by the least solution are revealed.
(iii) If it is consistent: what is its complete solution set.

## 4 Linear Optimization Problem – The Algorithm

Our aim is to solve the optimization problem, when the linear objective function (1) is subject to the constraints (2). We first decompose the linear objective function $Z$ in two functions $Z'$ and $Z''$ by separating the nonnegative and negative coefficients (as it is proposed in [9] for instance). Using the extremal solutions for constraint and the above two functions, we solve the optimization problem, as described below.

The linear objective function

$$Z = \sum_{j=1}^{n} c_j x_j, \; c_j \in \mathbb{R}, \; 0 \le x_j \le 1, \; 1 \le j \le n, \tag{13}$$

determines a cost vector $Z = (c_1, c_2, ..., c_n)$. We decompose $Z$ into two vectors with suitable components

$$Z' = (c'_1, c'_2, ..., c'_n) \text{ and } Z'' = (c''_1, c''_2, ..., c''_n),$$

such that the objective value is

$$Z = Z' + Z''$$

and cost vector components are:

$$c_j = c'_j + c''_j, \text{ for each } j = 1, ..., n,$$

where

$$c'_j = \begin{cases} c_j, & \text{if } c_j \ge 0, \\ 0, & \text{if } c_j < 0 \end{cases}, \tag{14}$$

$$c''_j = \begin{cases} 0, & \text{if } c_j \ge 0, \\ c_j, & \text{if } c_j < 0 \end{cases}. \tag{15}$$

Hence the components of $Z'$ are non-negative, the components of $Z''$ are non-positive.

We study how to minimize (maximize, respectively) the linear objective function (13), subject to the constraint (2).

In this section we present the algorithm that covers following optimization problems:

- Maximize the linear objective function (13), subject to constraint (2).
- Minimize the linear objective function (13), subject to constraint (2).

### 4.1 *Maximize the Linear Objective Function, Subject to Constraint (2)*

The original problem: to maximize $Z$ subject to constraint (2) splits into two problems, namely to maximize both

$$Z' = \sum_{j=1}^{n} c'_j x_j \tag{16}$$

with constraint (2) and

$$Z'' = \sum_{j=1}^{n} c''_j x_j \tag{17}$$

with constraint (2), i.e. for the problem (13) $Z$ takes its maximum when both $Z'$ and $Z''$ take maximum.

Since the components $c'_j$, $1 \le j \le n$, in $Z'$ are non-negative, $Z'$ takes its maximum among the maximal solutions of (2). Hence for the problem (16) the optimal solution is among the maximal solutions of the system (2).

Since the components $c''_j$, $1 \le j \le n$, in $Z''$ are non-positive, $Z''$ takes its maximum for the least solution of (2). Hence for the problem (17) the optimal solution is $\check{X} = (\check{x}_1, ..., \check{x}_n)$.

The optimal solution of the problem (13) with constraint (2) is

$$X^* = (x_1^*, ..., x_n^*),$$

where

$$x_i^* = \begin{cases} \check{x}_i, & \text{if } c_i < 0 \\ \hat{x}_i, & \text{if } c_i \ge 0 \end{cases} \tag{18}$$

and the optimal value is

$$Z^* = \sum_{j=1}^{n} c_j x_j^* = \sum_{j=1}^{n} c'_j \hat{x}_j + c''_j \check{x}_j. \tag{19}$$

### 4.2 Minimize the Linear Objective Function, Subject to Constraint (2)

If the aim is to minimize the linear objective function (13), we again split it, but now for $Z''$ the optimal solution is among the maximal solutions of the system (2), for $Z'$ the optimal solution is $\check{X}$. In this case the optimal solution of the problem is

$$X^* = (x_1^*, ..., x_n^*),$$

where

$$x_j^* = \begin{cases} \hat{x}_j, & \text{if } c_j < 0 \\ \check{x}_j, & \text{if } c_j \geq 0 \end{cases}. \tag{20}$$

and the optimal value is

$$Z^* = \sum_{j=1}^{n} c_j x_j^* = \sum_{j=1}^{n} c'_j \check{x}_j + c''_j \hat{x}_j. \tag{21}$$

### 4.3 Algorithm for Finding Optimal Solutions

1. Enter the matrices $A_{m\times n}$, $B_{m\times 1}$ and the cost vector $C_{1\times n}$.
2. Establish consistency of the system (2). If the system is inconsistent go to step 8.
3. Compute $\check{X}$ and all maximal solutions of (2), using software from [15].
4. If finding $Z_{\min}$ go to Step 6.
5. For finding $Z_{\max}$ compute $x_j^*$, $j = 1, ..., n$ according to (18). Go to Step 7.
6. For finding $Z_{\min}$ compute $x_j^*$, $j = 1, ..., n$ according to (20).
7. Compute the optimal value according to (19) (for maximizing) or (21) (for minimizing).
8. End.

### 4.4 Examle

We solve the maximizing optimization problem with the following given matrices in Łukasiewicz algebra:

$$c = \begin{pmatrix} -2 & 1 & 1.5 \end{pmatrix},$$

$$A = \begin{pmatrix} 0.8 & 0.3 & 0.1 \\ 0.7 & 0.2 & 0.6 \\ 0 & 0 & 0.3 \\ 0.5 & 0.6 & 0.9 \end{pmatrix}, \quad B = \begin{pmatrix} 0.4 \\ 0.5 \\ 1 \\ 0.5 \end{pmatrix}.$$

First the constraint system is solved. We illustrate its solution following Subsection 3.1.We begin by finding $\check{X}_{3\times 1} = (\, 0.2 \; 0.1 \; 0.4 \,)^t$ according to (4):

$$\begin{aligned}
\check{x}_1 &= (0.8 \otimes 0.4) \vee (0.7 \otimes 0.5) \vee (0 \otimes 1) \vee (0.5 \otimes 0.5) = 0.2 \vee 0.2 \vee 0 \vee 0 = 0.2\\
\check{x}_2 &= (0.3 \otimes 0.4) \vee (0.2 \otimes 0.5) \vee (0 \otimes 1) \vee (0.6 \otimes 0.5) = 0 \vee 0 \vee 0 \vee 0.1 = 0.1\\
\check{x}_3 &= (0.1 \otimes 0.4) \vee (0.6 \otimes 0.5) \vee (0.3 \otimes 1) \vee (0.9 \otimes 0.5) = 0 \vee 0.1 \vee 0.3 \vee 0.4 = 0.4.
\end{aligned}$$

The calculated vector is used to check if the system is consistent:

$$\begin{pmatrix}
(0.8 \to_L 0.2) \wedge (0.3 \to_L 0.1) \wedge (0.1 \to_L 0.4) = 0.4 \wedge 0.8 \wedge 1 = 0.4\\
(0.7 \to_L 0.2) \wedge (0.2 \to_L 0.1) \wedge (0.6 \to_L 0.4) = 0.5 \wedge 0.9 \wedge 0.8 = 0.5\\
(0 \to_L 0.2) \wedge (0 \to_L 0.1) \wedge (0.3 \to_L 0.4) = 1 \wedge 1 \wedge 1 = 1\\
(0.5 \to_L 0.2) \wedge (0.6 \to_L 0.1) \wedge (0.9 \to_L 0.4) = 0.7 \wedge 0.5 \wedge 0.5 = 0.5.
\end{pmatrix}$$

Therefore the system is solvable and $\check{X}$ is its minimal solution. The third equation has right-hand side 1 so it is removed. Next the matrix $P_{3\times 3}$ is calculated:

$$P = \begin{pmatrix} 0.2 & 0 & 0\\ 0.2 & 0 & 0.1\\ 0 & 0.1 & 0.4 \end{pmatrix}$$

with the elements:

$$\begin{aligned}
p_{11} &= (0.8 \otimes 0.4) = 0.2, & p_{12} &= (0.3 \otimes 0.4) = 0, & p_{13} &= (0.1 \otimes 0.4) = 0,\\
p_{21} &= (0.7 \otimes 0.5) = 0.2, & p_{22} &= (0.2 \otimes 0.5) = 0, & p_{23} &= (0.6 \otimes 0.5) = 0.1,\\
p_{31} &= (0.5 \otimes 0.5) = 0, & p_{32} &= (0.6 \otimes 0.5) = 0.1, & p_{33} &= (0.9 \otimes 0.5) = 0.4.
\end{aligned}$$

Using $P$ we create the matrix $C_{3\times 3}$ as described in Step 6:

$$C = \begin{pmatrix} 0 & 1 & 1\\ 0 & 1 & 1\\ 1 & 0 & 0 \end{pmatrix}.$$

There are no elements satisfying the conditions from Step 7 and $C$ is not changed. Then we start computing $W$ according to the mentioned laws. In fact we are interested only in zeros: for the first and the second equations the only zeros stand in the first column and they form the first two terms $\langle 1\rangle$ in $W$; in the third equation, there are two zeros – in the second and in the third columns, and it forms the last term $(\langle 2\rangle + \langle 3\rangle)$ in

$$W = \langle 1\rangle \langle 1\rangle (\langle 2\rangle + \langle 3\rangle).$$

Using (10) and (7) $W$ is simplified:

$$W = \langle 1\rangle \langle 1\rangle (\langle 2\rangle + \langle 3\rangle) = \langle 1\rangle (\langle 2\rangle + \langle 3\rangle) = \langle 1\rangle \langle 2\rangle + \langle 1\rangle \langle 3\rangle .$$

The two maximal solutions are $\hat{X}^1_{3\times 1} = (\,0.2\ 0.1\ 1\,)^t$ and $\hat{X}^2_{3\times 1} = (\,0.2\ 1\ 0.4\,)^t$. For both of the maximal solutions of the constraint an optimal solution of the problem is calculated according (18):

$$\begin{aligned}
X_1^* &= (\,0.2\ 0.1\ 1\,)\\
X_2^* &= (\,0.2\ 1\ 0.4\,).
\end{aligned}$$

The optimal values for them are:

$$\begin{aligned} Z_1^* &= -2*0.2+1*0.1+1.5*1 = -0.4+0.1+1.5 = 1.2 \\ Z_2^* &= -2*0.2+1*1+1.5*0.4 = -0.4+1+0.6 = 1.2. \end{aligned}$$

Since the optimal values for both maximal solutions are equal, the two found solutions are optimal. Finally, the optimal value of the optimization problem is 1.2. We should note that the maximal solutions of the system and the optimal solutions of the optimizaion problem are the same. The reason for this is that the first coefficient of the cost vector is negative and this leads to using elements from the minimal solution for its place, while in the maximal solution we have the same elements at the first position.

## 5 Software

Software for solving linear optimization problem (1) subject to constraint (2) is developed using the .NET Framework and the language C#, based on the described algorithms in Sections 3 and 4. It can be obtained free by contact with any of the authors.

The organization of the source code has two parts. The first part involves the classes for the interface of the program. The second implements the working logic that solves the problems. Only the latter is of interest in this paper. The classes used to model the problem and realize the algorithms for its solution are five. The most important class is called *FuzzySystem* and represents the linear objective function (1) and the constraint FLSE (2). It contains all the data and methods used to solve the optimization problem. When an object of this class is created, characteristics of the problem, such as its dimensions and problem type, must be introduced (the problem type can be optimization problem, also direct or inverse problem in case the user wants just the solution of a FLSE). One of the fields in the class represents an object that specifies the algebra in which we are working and is described below. The input matrices also must be entered. We have only one public method – *SolveSystem*, which is called to solve the desired problem. For optimization it realizes all the steps given in Section 3 by calling private methods and then calls another method which finds the optimal solutions and calculates the optimal value.

The other four classes form a hierarchy that models the operations in the three considered algebras. There is an abstract class *BL_algebra* which implements the operations that are same for the three algebras – minimum and maximum. The other operations are put as abstract methods. They are implemented in the three subclasses – *GodelAlgebra*, *GoguenAlgebra* and *LukasiewiczAlgebra*. As a result the modeling of algebras' operations is done by using polymorphism. In the class *FuzzySystem* we have a (mentioned before) field of type *BL_algebra*, which has an object of type one of the three subclasses. The methods in *FuzzySystem* do not care in which algebra the calculations are done – they use this object's methods. A flexible feature of this code structure is the ability to add algebras for which these algorithms yield

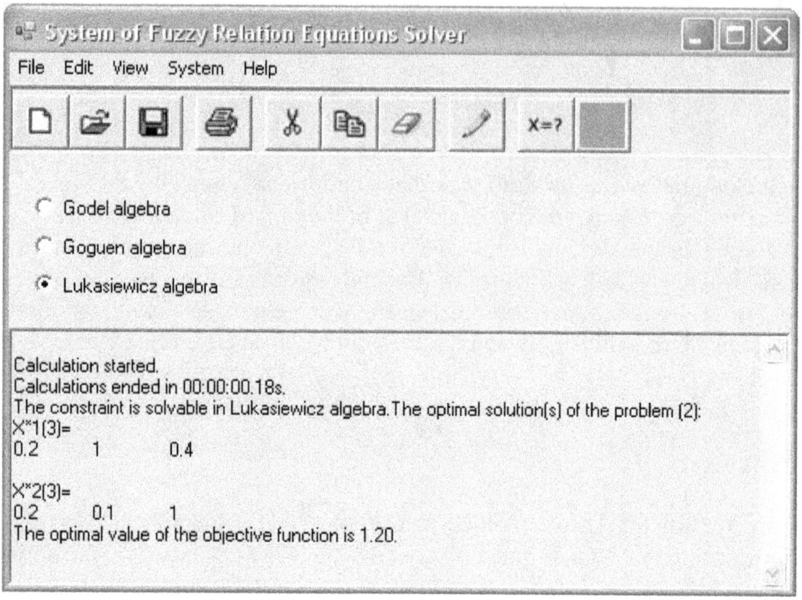


**Fig. 1** Main form

the solution. Only a subclass of *BL_algebra* should be added and no changes in *FuzzySystem* must be done for the algorithms to work.

The interface of the program consists of three forms – one main and two auxiliary for the input and editing of the data.

The main form (Fig. 1) has all the necessary components for controlling the process of solving the problem. A menu and a toolbar are placed in the upper part of the form. The user also chooses the desired algebra there. The area left below is used for the output box, where all the messages and the results are printed. There are two ways for data input for the system. The first is to enter it manually in a dialog box and the other is to load it from a file. When entering a new problem manually the other two forms are used. One of them is for choosing the type of the problem and the dimensions of the system. The other form is shown afterwards, where the user enters the elements of the input matrices. This form is also used for editing the data when it is entered or loaded from a file. The system can also be saved in a file.

An interesting feature of the program is the presence of two modes of the dialog for entering/editing the data – the *Normal View* and the *Simple View* modes. The *Normal View* (Fig. 2) is available for the optimization problem and the inverse problem and it cannot be used for solving the direct problem. This view mode shows the objective function and equations with all the symbols for the operations and all unknowns.

In the *Simple View* mode (Fig. 3) only boxes for the elements of the matrices are displayed. The normal view is orientated to small problems and its aim is to give the user a better representation of the input data. However, when the system is large, the

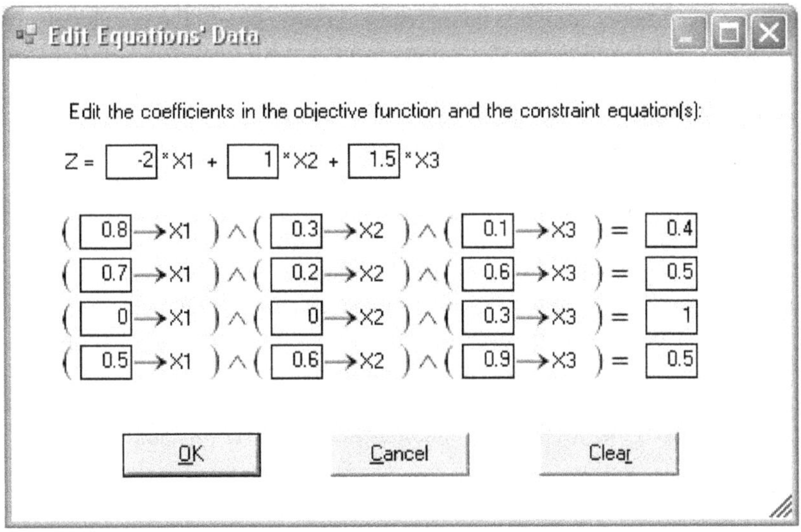


**Fig. 2** Normal View mode

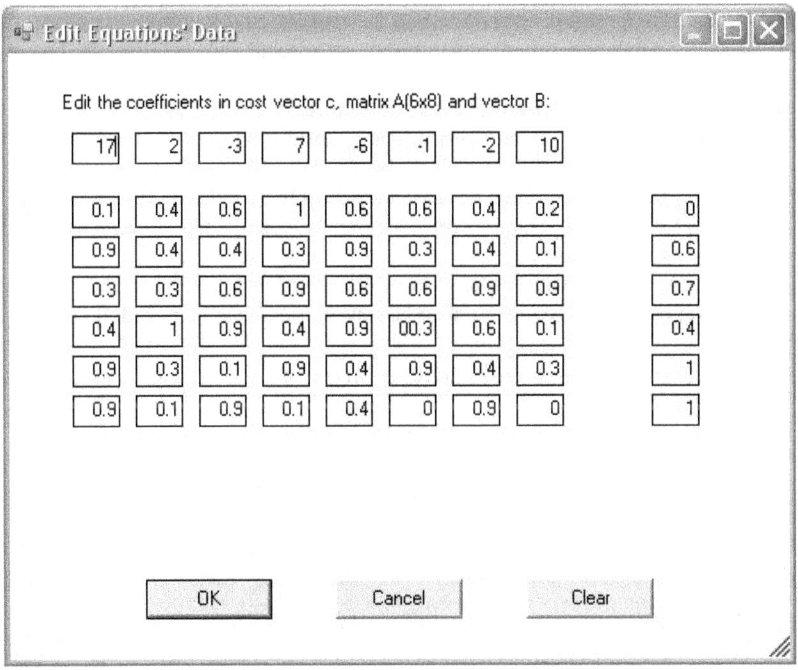


**Fig. 3** Simple View mode

dialog becomes "heavy" and it is inconvenient to enter or edit the coefficients in that way. Another point is that the PC resources needed for displaying all the symbols are increasing much. So the program automatically switches to *Simple View* mode when the size of the system is above a certain threshold.

When the user has entered the data, the software is ready to solve the problem. The calculations are started by clicking a button. The program is multithread and the process of solving the problem is in a thread different from the one controlling the interface. This gives the user the control over the solver during the calculations. A timer is displayed showing how much time has passed since the solving started and a button to stop the calculations becomes enabled. In that way all the data is preserved from being lost because without being able to stop the process of computing, the user would have to force the program to stop.

On successful end of the solution process the results are displayed in the output area. The messages displayed there can be printed or saved in a text file.

## References


1. Aho, A., Hopcroft, J., Ullman, J.: The Design and Analysis of Computer Algorithms. Addison-Wesley Publ. Co., London (1976)
2. Bourke, M.M., Fisher, D.G.: Solution algorithms for fuzzy relational equations with max-product composition. Fuzzy Sets and Systems 94, 61–69 (1998)
3. De Baets, B.: Analytical solution methods for fuzzy relational equations. In: Dubois, D., Prade, H. (eds.) Fundamentals of Fuzzy Sets. Handbooks of Fuzzy Sets Series, vol. 1, pp. 291–340. Kluwer Academic Publishers, Dordrecht (2000)
4. Di Nola, A., Lettieri, A.: Relation Equations in Residuated Lattices. Rendiconti del Circolo Matematico di Palermo, s. II, XXXVIII, pp. 246–256 (1989)
5. Di Nola, A., Pedrycz, W., Sessa, S., Sanchez, E.: Fuzzy Relation Equations and Their Application to Knowledge Engineering. Kluwer Academic Press, Dordrecht (1989)
6. Grätzer, G.: General Lattice Theory. Akademie-Verlag, Berlin (1978)
7. Guu, S.M., Wu, Y.-K.: Minimizing a linear objective function with fuzzy relation equation constraints. Fuzzy Optimization and Decision Making 4(1), 347–360 (2002)
8. Klir, G., Clair, U.H.S., Bo, Y.: Fuzzy Set Theory Foundations and Applications. Prentice Hall PRT, Englewood Cliffs (1977)
9. Loetamonphong, J., Fang, S.-C.: An efficient solution procedure for fuzzy relational equations with max-product composition. IEEE Transactions on Fuzzy Systems 7(4), 441–445 (1999)
10. Loetamonphong, J., Fang, S.-C.: Optimization of fuzzy relation equations with max-product composition. Fuzzy Sets and Systems 118(3), 509–517 (2001)
11. Loetamonphong, J., Fang, S.-C., Young, R.E.: Multi-objective optimization problems with fuzzy relation equation consrtaints. Fuzzy Sets and Systems 127(3), 141–164 (2002)
12. MacLane, S., Birkhoff, G.: Algebra. Macmillan, New York (1979)
13. Noskova, L., Perfilieva, I.: System of fuzzy relation equations with $\sup - * -$composition in semi-linear spaces: minimal solutions. In: Proc. FUZZ-IEEE Conf. on Fuzzy Systems, London, July 23-26, pp. 1520–1525 (2007)
14. Peeva, K.: Universal algorithm for solving fuzzy relational equations. Italian Journal of Pure and Applied Mathematics 19, 9–20 (2006)

15. Peeva, K., Petrov, D.: Algorithm and Software for Solving Fuzzy Relational Equations in some BL-algebras. In: 2008 IVth International IEEE Conference "Intelligent Systems", Varna, September 2008, vol. 1, pp. 2-63–2-68 (2008) ISBN 978-I-4244-1739
16. Peeva, K., Kyosev, Y.: Fuzzy Relational Calculus-Theory, Applications and Software (with CD-ROM). In: The series Advances in Fuzzy Systems - Applications and Theory, vol. 22. World Scientific Publishing Company, Singapore (2004)
17. Perfilieva, I., Noskova, L.: System of fuzzy relation equations with $\inf - \rightarrow$ composition: complete sets of solutions. Fuzzy Sets and Systems 150(17), 2256–2271
18. Sanchez, E.: Resolution of composite fuzzy relation equations. Information and Control 30, 38–48 (1976)

# Electric Generator Automation and Protection System Fuzzy Safety Analysis

Mariana Dumitrescu

**Abstract.** In a fault-tolerant power system the failures must be detected and isolated saving the operational state of the system. So we propose a model for performance evaluation of fail safe behavior concerning automation and protection systems. Fuzzy Safety measures are computed for the most significant protection and automations types. The paper explains fuzzy safety analysis for the electric generator (EG) protection and automation system. Power system electric generator (EG) is protected to various types of faults and abnormal workings. The protection and automation system (PAS) is composed of waiting subsystems, which must properly respond to each kind of dangerous events. An original fuzzy logic- system enables us to analyze the qualitative evaluation of the event –tree, modeling PAS behavior. Fuzzy - set logic is used to account for imprecision and uncertainty in data while employing event-tree analysis. The fuzzy event-tree logic allows the use of verbal statement for the probabilities and consequences, such as very high, moderate and low probability.

**Index Terms:** power system, safety, fuzzy logic, critical analysis.

## 1 Introduction

Reliability information can be best expressed using fuzzy sets, because seldom it can be crispy, and the use of natural language expressions about reliability offers a powerful approach handling the uncertainties more effectively [1], [2], [8]. Fuzzy - set logic is used to account for imprecision and uncertainty in data while employing a safety analysis. Fuzzy logic provides an intuitively appealing way of handling this uncertainty by treating the probability of failure as a fuzzy number. This allows the analyst to specify a range of values with an associated possibility distribution for the failure probabilities. If it is associated a triangular membership function with the interval, this implies that the analyst is "more confident" that the actual parameter lies near the center of the interval than at the edges [2]. Using fuzzy numbers for the verbal statements of an event probability means fuzzy probability event evaluation [4], [5], [2].

---

Mariana Dumitrescu
Member IEEE

V. Sgurev et al. (Eds.): Intelligent Systems: From Theory to Practice, SCI 299, pp. 433–444.
springerlink.com 

In a qualitative analysis event trees (ET) give the sequences of events and their probabilities of occurrence. They start with some initiate event (say a failure of some kind) and then develop the possible sequences of events into a tree. At the end of each path of events the result (safe shutdown, the damage of equipment) is obtained. The probability of each result is computed using the probabilities of the events in the sequence leading to it [2], [6], [7].

The fuzzy probability of an event can be put into subcategories based on a range of probability, high-if probability is greater than 0,6 but less than 1,0; very low-if probability is greater than 0 but less than 0,2; etc. The fuzzy event-tree (FET) allows the use of verbal statement for the probabilities and consequences, such as very high, moderate and low probability. The occurrence probability of a path in the event tree is than calculated as the product of the event probabilities in the path [6], [7].

A first direction in event-tree analysis [5] uses fuzzy-set logic to account for imprecision and uncertainly in data while employing this analysis. The fuzzy event-tree allows: 1) uncertainty in the probability of failure and 2) verbal statements for the probabilities and consequences such as low, moderate and high for the impact of certain sequences of events such as normal, alert and abnormal.

A second direction in the event-tree analysis [2] uses a linguistic variable to evaluate the fuzzy failure probability. The linguistic values that can be assigned to the variable, called its "term sets", are generally defined as fuzzy sets that act as restrictions on the base values that it represents. Each of these fuzzy sets constitutes a possibility distribution over the domain of the base variable.

The paper uses the linguistic variable in fuzzy failure probability evaluation, because this concept is coupled with the possibility theory, which becomes an especially powerful tool for working with uncertainty [6], [7].

The paper explains fuzzy safety analysis for EG-PAS. Section 2 introduces how to use the fuzzy logic system in the event tree modeling. Section 3 gives information about EG-PAS critical analysis. Section 4 presents the conclusions of the paper.

## 2 Fuzzy Safety Analysis Using Event-Trees

Event-trees examine sequences of events and their probability of occurrence. They develop the possible sequences of events into a tree. For example: is the failure detected?, does a safety relay activate? The result at the end of each chain of events is then determined and the probability of each result calculated from the probabilities of the events in sequence leading to it.

The procedure for analysing a fault tree is precise, but the probabilities on which the methodology is based often are not [1], [2], [6]. We illustrated the technique, with the use of fuzzy probabilities in the analysis of an event- tree, for the electric power protection system. Usually, the fuzzy event-tree analysis has the following steps:

- fuzzy "failure" probability and fuzzy "operation" probability evaluation, for all reliability block diagram elements of the electric power protection system;

- fuzzy "occurrence" probability for each path (sequence of events) of the tree;
- fuzzy " consequence" on power system evaluation, after the events sequences achievement;
- fuzzy "risk" on power system for each path of the tree, evaluation depending on the path "occurrence" and path " consequence";
- the tree-paths hierarchy establishing, depending on the path " risk".

In order to see which one of these outcomes has the highest possibility some applications rank the outcomes on the basis of the maximum probabilities associated with outcomes. Others make the same thing on the basis of the probabilities having maximum degree of membership in the fuzzy probabilities. But both of these approaches may lead to improper ranking of the outcomes. The proper approach of the different outcomes is to consider both the maximum probability, associated with various outcomes and the degree of membership of the rating [5]. A tree - paths ranking, from " risk " point of view, is calculated considering both the maximum probability associated with various outcomes and the degree of membership of the rating.

The tree - paths ranking evaluation is not enough for the electric power protection system reliability calculation. A methodology to calculate a quantitative index for this kind of system is necessary to be developed. On this goal an adequate fuzzy - logic system (FLS), having the following steps, is proposed [3] (see steps 3, 4, 5, 6, 7 from Fig. 1.):

- the elaboration of the linguistic variables for FLS parameters "Occurrence" and "Severity";
- the evaluation of the FLS inputs: event tree path "Occurrence" (OC/ET path) and event tree path "Severity" (SV/ET path);
- FLS rule base proposal;
- FLS rule base evaluation and Fuzzy Conclusion for each Event Tree – path (FC/ET path) evaluation ("Safety" (SF) evaluation).

The fuzzy event tree analysis needs the following steps (see Fig. 1.):

1. - the elaboration of the reliability block diagram;
2. - the construction of the event tree paths;
3.,4.,5.,6.,7. - the FLS construction and fuzzy inference process realisation;
8. - the evaluation of the general fuzzy conclusion (GFC) for all tree - paths ("General Safety" (GSF));
9. - GFC defuzzification and "General Safety Degree" (GSFD) crisp value calculation.

Generally the FLS input parameters could be the *Occurrence* of the path tree event, the *Detectability* of the path tree event and the *Severity* of the path tree event. The FLS output parameter is the *Safety* of the analyzed system according to the path tree event. The fuzzification process of FLS input parameters uses the linguistic variables. These are elaborated with the proposed FLS input membership functions. The deffuzification process uses a linguistic variable elaborated with the proposed FLS output membership function (Fig.2).

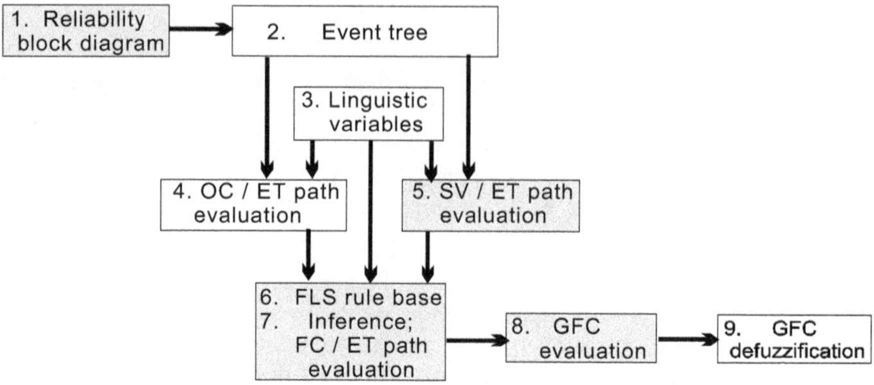


**Fig. 1** The steps of the fuzzy event tree analysis

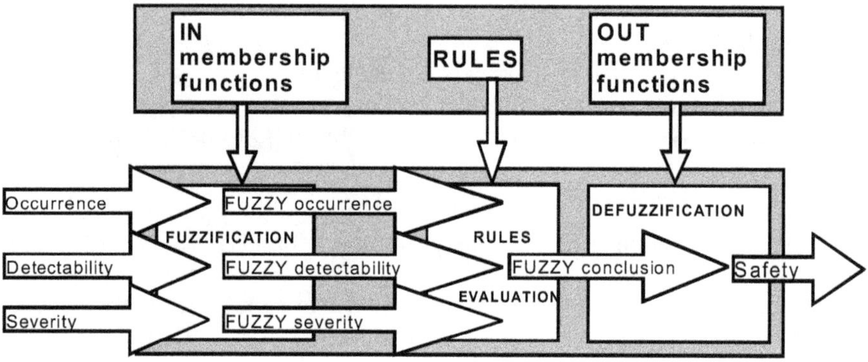


**Fig. 2** FLS associated to the fuzzy event tree analysis

It is very important to say that the calculation of the failure probability is not enough. Another important consideration is for example the severity of the effect of the failure. The risk associated with a failure increases as either the severity of the effect of the failure or the failure probability increases. The Severity of the effect of the failure, included in the FLS and ranked according to the seriousness of the effect of failure, allows modeling this judgment, by its very nature highly subjective, [4], [5].

The proposed FLS enables us to analyze the qualitative evaluation of the event –tree and to reach the independent safety analysis for AS. The technique allows to develop a fuzzy-event algorithm and to gain quantitative results, as the fuzzy set "General safety " (GSF) and the crisp value "General Safety Degree" (GSD) associates to all the paths in the tree.

For the fuzzy event-tree analysis of an electric- power protection-automation system we elaborated an efficient software tool "Fuzzy Event Tree Analysis" (FETA) to help us in the independent performance analysis of PAS and also in the fuzzy critical analysis. The computed safety measures are very important for power system safety analysis.

For example, to express the fuzzy set “Severity” and its equivalent fuzzy number, FETA develops a linguistic variable (see Fig. 3.). The “Severity” membership functions are “very low” (fs), “low” (s), “moderate” (m), “high” (î), “very high” (fî). The centroid value of the “Severity” fuzzy set is computed and also fuzzy set possibilities values î = 0.62045, fî = 1 are computed.

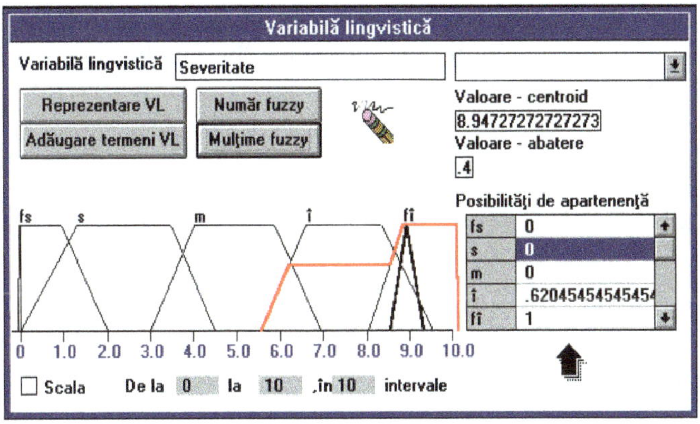


**Fig. 3** Linguistic variable “Severity” for proposed FLS

The program is able to do the independent analyze of PAS safety and also a complex safety analysis of the ensemble power system connected with its automation/protection system. In Fig 4 is shown the reliability block diagram for an automation system and the associate event tree with the 138th path selected.

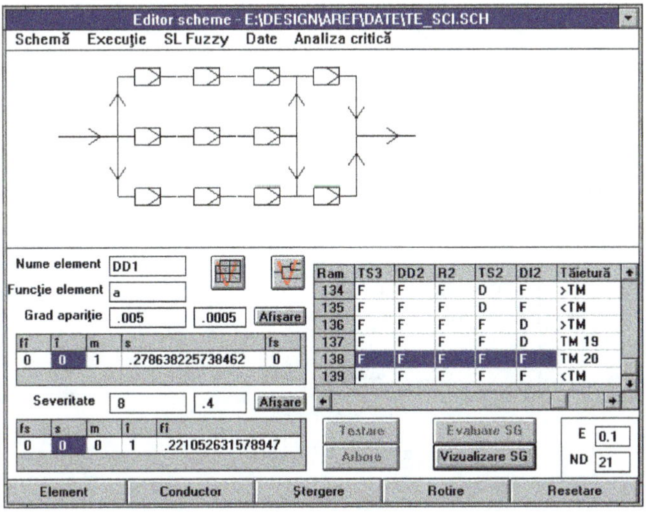


**Fig. 4** FETA- event tree and block diagram elaboration

## 3 Electric Generator Protection and Automation System

The electric generator (EG) protection system is sensible to the following failure types: low tension level (LT), negative power flow (NPF), external short-circuits (ESC), over-voltage (OV).

The power system presented in Fig. 5 has the electric generator protected by a circuit breaker (CB). CB operates by fault detectors (DD1 or DD2), a combined relay (R) and trip signal (TS) Fig. 5b. Supposing that a fault occurs on the electric generator, it is desirable to evaluate the probability of successful operation of the protection system. Generally power protection system involves the sequential operation of a set of components and devices. Fig. 6 shows the fuzzy event-tree for the network presented in Fig.5.

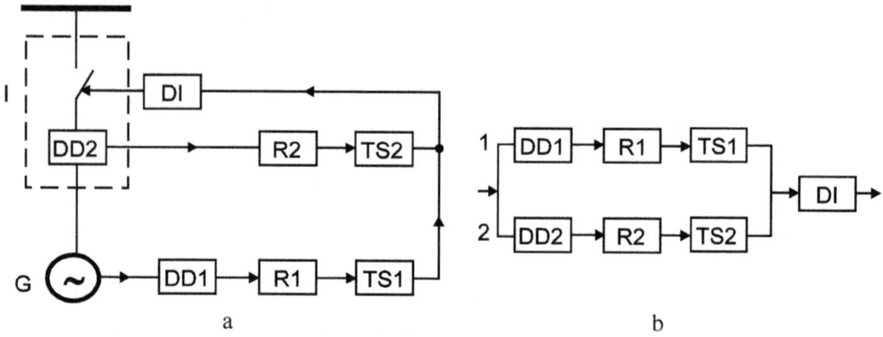


**Fig. 5** Electric generator protection system (a). Reliability block diagram (b).

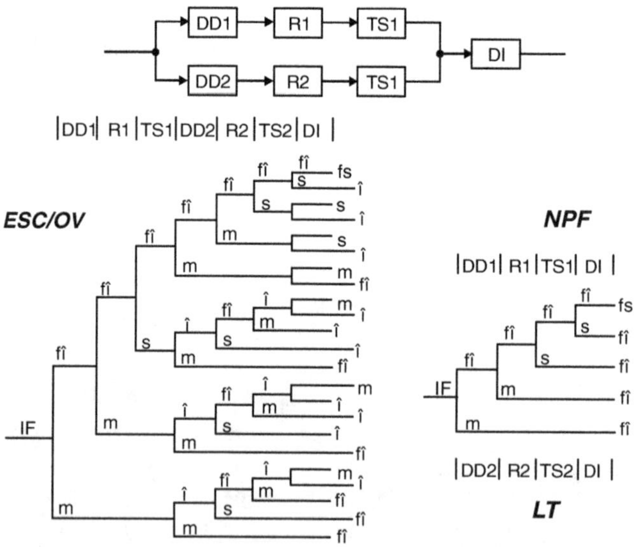


**Fig. 6** Fuzzy event- trees for low tension level (LT), negative power flow (NPF), external short-circuits (ESC), over-voltage (OV)

Starting with an initiating fault IF, a very low failure probability is considered for it, because the generator is a very safe equipment. The fault detecting system consists of current transformers, and high impedance relays (which from experience are reliable) excepting the case of faults with high currents. The relay/ trip signal device, consists of relay, trip coil, pneumatic systems, coming along with many moving parts, whose high probability is assumed to have a successful operation. Finally, since high technologies have been used in design and manufacturing of the circuit breakers (CB) their successful operation probability is considered to be very high.

For a non stop electric power supply of the consumers always there is a reserve electric power supply, able to be used when the normal electric power supply has a failure. Automatic closing reserve waits for the failure to come in the normal circuit of the electric power supply.

When failure appeared in the main circuit ACR system commands normal circuit breaker disconnection and reserve circuit breaker connection. The continuity of the electric power supply of the consumers is obtained.

ACR system has two important parts:

- a low tension protection acting when a failure appears on the main electric power supply of the consumer; it commands the disconnection of the main circuit breaker;
- automation elements to command the connection of the reserve circuit breaker.

Fig. 7.a presents the electric power reserve automatic supply for a diesel electric generator DG. When the main power supply fails the diesel electric generator takes its place and the consumers are supplied continuously. The block diagram of ACR for electric diesel generator case is presented in Fig. 7.b.

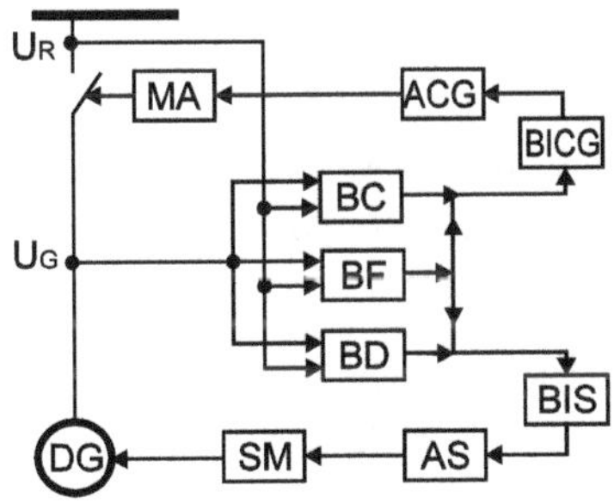


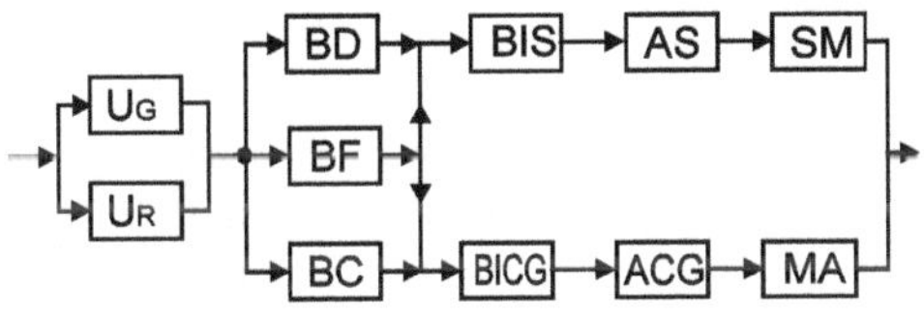


**Fig. 7** Electric power supply system (a) block diagram for ACR system in electric generator case (b)

Main electric circuit tension Ur and diesel electric generator tension Ug are input and output parameters for ACR feed back system in the EG case. For frequency control the block diagram has the following elements: BD element comparing generator frequency and main circuit frequency, pulse production element BF, BIS pulse element for synchronization, amplification element (AS), servomotor element (SM) for diesel frequency regulation, comparing tension element (BC), pulse production element for electric generator connection element (BICG).

## 4 Fuzzy Safety Analysis

The critical analysis uses the GSF fuzzy number and the GSFD crisp number associated to each analyzed protection system. This enables us to compute the Global Safety fuzzy set (GLSF) and the Global Safety Degree (GLSD) crisp value. The algorithm introduces the fuzzy parameter "Weight" associated to various types of faults and abnormal workings of the EG stopped by the associated PS.

To express the fuzzy number "Weight" we developed a linguistic variable. For example the electric generator (EG) protection system, sensible to the failure types low tension level (LT), negative power flow (NPF), external short-circuits (ESC), over-voltage (OV), gets a weight $W_i$ for each failure type, expressed as a fuzzy number and represented with the membership functions "Very low", "Low", "Moderate", "High", "Very high". The fuzzy number $W_i$ elaborated for each type of failure and the fuzzy number $GSF_i$ computed for each kind of PS are used for $GSF_j$ computation:

$$GSF_j = \sum_{i=1}^{D} W_{ji} \cdot GSF_i \quad , j = \overline{1,C} \tag{1}$$

D meaning the failure types number.

The "Global Safety" GLSF for PS is computed using the $GSF_j$ fuzzy numbers:

$$GLSF = \bigcup_{j=1}^{C} GSF_j \tag{2}$$

The fuzzy number GLSF is defuzzified and its centroid value "Global Safety Degree" GLSD is computed. We use GLSF and GLSD qualitative indexes as a result of fuzzy critical analysis (see Table 1 and Fig. 8).

We use full line to represent the General Safety Degree for each type of failure and interrupted line for the Global Safety Degree of the protection system. Because the Severity input parameter of the FLS is greater in the ESC failure type (ESC failure has the greatest Severity for the protected EG), it is obvious that GSFD qualitative index is lower in the ESC failure case. The most important safety degree is for LT failure type, but the global safety of the protection system is close to OV failure type, the most common of all and having the greatest occurrence level.

The detailed fuzzy safety measures, computed with FETA software, are presented in Table 2 for OV failure type. For all 23 paths of the event tree input fuzzy parameters OC and SV (possibilities, centered values and uncertainty) are presented. The active rules from rule base, the output fuzzy parameter SF (possibilities, centered value SDF and uncertainty) are also presented.

**Table 1** Fuzzy Critical Analysis Results for EG

| Failure type | Active block types | Path number | Possibilities $W_i$ | $GSF_i$ | $GSG_{sf}$ Uncertainty $GSF_i$ |
|---|---|---|---|---|---|
| NPF | DD1, R1, TS1 | 5 | fs 0.38636<br>s 1.00000 | im 0.1573<br>î 0.78712<br>fî 0.65839 | 9.3662<br>0.7203 |
| ESC | DD1(DD2), R1(R2), TS1(TS2) | 23 | s 1.00000<br>m 0.29655 | im 0.2766<br>î 0.86887<br>fî 0.63764 | 9.2538<br>0.9359 |
| OV | DD1(DD2), R1(R2), TS1(TS2) | 23 | s 0.63816<br>m 0.98710 | im 0.2612<br>î 0.85428<br>fî 0.65194 | 9.2840<br>0.9257 |
| LT | DD2, R2, TS2 | 5 | fs 0.82787<br>s 0.64336 | im 0.1479<br>î 0.77477<br>fî 0.6702 | 9.3867<br>0.3173 |
| "Global Safety" | (GLSF) fuzzy set possibilities: î: 0.61107, fî: 0.8411; | | | | |
| Global Safety | Degree (GLSD): 9.31206 | | | | |

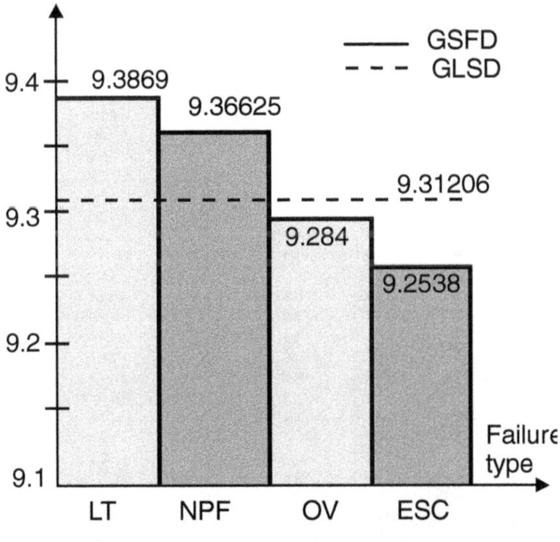


**Fig. 8** GSFD and GLSD indexes for the electric generator

**Table 2** Detailed Fuzzy Critical Analysis Results for OV Failure Type

| Path | Possibilities | | Centered/ Uncertainty | | Active rules | Possibilities | SFDi/ Uncertainty SGi |
|---|---|---|---|---|---|---|---|
| i | OCi | SVi | OCi | SVi | | SFi | |
| 1 | s 0.4425 | î 1.0000 | 2.5000E-5 | 8.1758 | 4, 4; 4, 5 | fs 0.3136 | 3.6972 |
| | fs 0.7633 | fî 0.3136 | 5.2500E-6 | 0.4000 | 5, 4; 5, 5 | s 0.4425 | 3.1618 |
| | | | | | | m 0.7633 | |
| 2 | s 0.9248 | î 0.8589 | 7.9600E-5 | 8.6046 | 4, 4; 4, 5 | fs 0.5392 | 3.0003 |
| | fs 0.3127 | fî 0.5392 | 3.5380E-5 | 0.4000 | 5, 4; 5, 5 | s 0.8589 | 3.2456 |
| | | | | | | m 0.3127 | |
| 3 | s 0.5686 | î 0.9328 | 3.4268E-5 | 8.4974 | 4, 4; 4, 5 | fs 0.4828 | 3.3892 |
| | fs 0.6126 | fî 0.4828 | 6.3449E-6 | 0.4000 | 5, 4; 5, 5 | s 0.5686 | 3.4201 |
| | | | | | | m 0.6126 | |
| 4 | s 0.2409 | î 0.6079 | 1.4583E-5 | 8.9686 | 4, 4; 4, 5 | fs 0.2409 | 3.6108 |
| | fs 1.0000 | fî 0.7308 | 4.2460E-6 | 0.4000 | 5, 4; 5, 5 | s 0.7308 | 3.3306 |
| | | | | | | m 0.6079 | |
| 5 | m 1.0000 | s 1.0000 | 4.8465E-3 | 2.3900 | 3, 2; 4, 2 | m 1.0000 | 5.6021 |
| | s 0.2911 | | 5.1983E-4 | 0.4000 | | im 0.2911 | 2.6111 |
| 6 | s 0.9006 | î 0.7850 | 7.4625E-5 | 8.7117 | 4, 4; 4, 5 | fs 0.5956 | 3.0106 |
| | fs 0.3475 | fî 0.5956 | 3.4880E-5 | 0.4000 | 5, 4; 5, 5 | s 0.7850 | 3.3099 |
| | | | | | | m 0.3475 | |
| 7 | s 1.0000 | î 0.7850 | 2.3761E-4 | 8.7117 | 4, 4; 4, 5 | fs 0.5956 | 2.3297 |
| | | fî 0.5956 | 1.7862E-4 | 0.4000 | | s 0.7850 | 2.6236 |
| 8 | s 1.0000 | î 0.7850 | 1.0229E-4 | 8.7117 | 4, 4; 4, 5 | fs 0.5956 | 2.7995 |
| | fs 0.2111 | fî 0.5956 | 4.4728E-5 | 0.4000 | 5, 4; 5, 5 | s 0.7850 | 3.0409 |
| | | | | | | m 0.2111 | |
| 9 | s 0.7024 | î 0.6079 | 4.3531E-5 | 8.9686 | 4, 4; 4, 5 | fs 0.7024 | 3.2101 |
| | fs 0.5808 | fî 0.7308 | 2.4631E-5 | 0.4000 | 5, 4; 5, 5 | s 0.6079 | 3.6176 |
| | | | | | | m 0.5808 | |
| 10 | m 1.0000 | s 1.0000 | 1.4467E-2 | 2.8900 | 3, 2; 3, 3 | s 0.1602 | 4.6026 |
| | s 0.0150 | m 0.1602 | 4.9593E-3 | 0.4000 | 4, 2; 4, 3 | m 1.0000 | 2.3510 |
| | | | | | | im 0.0150 | |
| 11 | s 0.5073 | î 0.8589 | 2.9402E-5 | 8.6046 | 4, 4; 4, 5 | fs 0.5073 | 3.4205 |
| | fs 0.6863 | fî 0.5392 | 5.8310E-6 | 0.4000 | 5, 4; 5, 5 | s 0.5392 | 3.4969 |
| | | | | | | m 0.6863 | |
| 12 | s 0.2417 | î 0.6079 | 1.4613E-5 | 8.9686 | 4, 4; 4, 5 | fs 0.2417 | 3.6097 |
| | fs 1.0000 | fî 0.7308 | 4.2545E-6 | 0.4000 | 5, 4; 5, 5 | s 0.7308 | 3.3317 |
| | | | | | | m 0.6079 | |
| 13 | m 1.0000 | s 1.0000 | 4.8564E-3 | 2.3900 | 3, 2; 4, 2 | m 1.0000 | 5.6011 |
| | s 0.2904 | | 5.2085E-4 | 0.4000 | | im 0.2904 | 2.6096 |
| 14 | s 0.9870 | î 0.8589 | 9.3617E-5 | 8.6046 | 4, 4; 4, 5 | fs 0.5392 | 2.8925 |
| | fs 0.2419 | fî 0.5392 | 4.0305E-5 | 0.4000 | 5, 4; 5, 5 | s 0.8589 | 3.1061 |
| | | | | | | m 0.2419 | |
| 15 | s 0.7256 | î 0.6079 | 4.6528E-5 | 8.9686 | 4, 4; 4, 5 | fs 0.7256 | 3.1726 |
| | fs 0.5492 | fî 0.7308 | 2.5187E-5 | 0.4000 | 5, 4; 5, 5 | s 0.6079 | 3.5762 |
| | | | | | | m 0.5492 | |
| 16 | m 1.0000 | s 1.0000 | 1.5463E-2 | 2.7900 | 3, 2; 3, 3 | s 0.1050 | 4.6746 |
| | | m 0.1050 | 4.9761E-3 | 0.4000 | | m 1.0000 | 2.2085 |
| 17 | s 0.6345 | î 0.8589 | 4.0302E-5 | 8.6046 | 4, 4; 4, 5 | fs 0.5392 | 3.2689 |
| | fs 0.5338 | fî 0.5392 | 7.0010E-6 | 0.4000 | 5, 4; 5, 5 | s 0.6345 | 3.4474 |
| | | | | | | m 0.5338 | |
| 18 | s 0.3628 | î 0.6079 | 2.0030E-5 | 8.9686 | 4, 4; 4, 5 | fs 0.3628 | 3.4690 |
| | fs 0.8694 | fî 0.7308 | 5.3007E-6 | 0.4000 | 5, 4; 5, 5 | s 0.7308 | 3.5018 |
| | | | | | | m 0.6079 | |
| 19 | m 1.0000 | s 1.0000 | 6.6567E-3 | 2.6900 | 3, 2; 3, 3 | s 0.0497 | 5.2541 |
| | s 0.1739 | m 0.0497 | 5.6259E-4 | 0.4000 | 4, 2; 4, 3 | m 1.0000 | 2.4592 |
| | | | | | | im 0.1739 | |
| 20 | s 0.3032 | î 0.6079 | 1.7151E-5 | 8.9686 | 4, 4; 4, 5 | fs 0.3032 | 3.5345 |
| | fs 0.9398 | fî 0.7308 | 4.7799E-6 | 0.4000 | 5, 4; 5, 5 | s 0.7308 | 3.4213 |
| | | | | | | m 0.6079 | |
| 21 | m 1.0000 | s 1.0000 | 5.7000E-3 | 2.7900 | 3, 2; 3, 3 | s 0.1050 | 5.2336 |
| | s 0.2315 | m 0.1050 | 5.5043E-4 | 0.4000 | 4, 2; 4, 3 | m 1.0000 | 2.6953 |
| | | | | | | im 0.2315 | |
| 22 | m 1.0000 | î 0.9310 | 2.8414E-3 | 8.5000 | 3, 4; 4, 4 | fs 0.9310 | 2.0734 |
| | s 0.4905 | fî 0.4842 | 5.1407E-4 | 0.4000 | 4, 5 | s 0.4905 | 2.3595 |
| 23 | fî 1.0000 | fs 1.0000 | 9.4430E-1 | 0.0250 | 1, 1; 2, 1 | î 0.0762 | 9.5117 |
| | î 0.0762 | | 1.2016E-2 | 0.0050 | | fî 1.0000 | 0.6097 |

GSF fuzzy set possibilities:
im:0.2366043; h:0.9017167; vh:0.6716917;
GSFD crisp value :9.28409

Fuzzy safety analysis for ACR system in electric generator case uses the FLS input Occurrence (OC), Severity (SV) parameters from Tables 3.

FETA software elaborates fuzzy event tree having 268 paths for block diagram of ACR system in electric generator case. OC and SV for each one of 268 paths are evaluated. The rule base is used and fuzzy safety conclusion SG for each path is computed. Finally fuzzy general conclusion GSFgen is elaborated and its central value GSDgen presented in Table 4 is concluded.

**Table 3** Input Parameters OC and SV for Block Diagram Parameters in ACR System for Electric Generator Case

| | $U_G$; $U_R$ | BD; BF; BC | BIS; BICG | AS; ACG | SM | MA |
|---|---|---|---|---|---|---|
| | m 1.0000 | m 1.0000 | m 0.9684 | m 06535 | m 1.000 | m 1.000 |
| OC | s 0.2788 | s 0.6205 | s 0.7250 | s 0.8646 | 0. 10290 | s 0.4700 |
| | î 1.000 | î 1.000 | î 1.000 | î 1.0000 | î 1.000 | î 1.000 |
| SV | fî 0.2211 | fî 0.2211 | fî 0.2211 | fî 0.2211 | fî 0.433 | fî 0.433 |

**Table 4** Fuzzy Safety Analysis Results for ACR System in Electric Generator Case

| | Block diagram elements | ET path number | Possibilities $GSF_{gen}$ | Central value $GSD_{gen}$ |
|---|---|---|---|---|
| ACR- EG | $U_G$, $U_R$, BD, BF, BC, BIS, BICG, AS, ACG, SM, MA | 268 | im 0.17301<br>î 0.842079<br>fî 0.634722 | 9.30924 |

## 5 Conclusions

In this paper an application of fuzzy logic for safety computing of an Electric Generator Protection-Automation System was presented. The author uses a fuzzy logic system adequate to fuzzy event-tree analysis. Event-trees are often used in power protection system quality computing, but the paper presents fuzzy sets and fuzzy logic also to do a proper model of the analyzed system. An efficient software toll FETA ("Fuzzy Event-Tree Analysis"), for independent analyzing of the power protection- automation system, elaborated by the author is used to achieve the proposed goal.

The FETA software uses four analyzing methodology steps and gives a global qualitative index for all the paths of the fuzzy event-tree. An adequate rules base allows computing the electric generator protection system "Safety" using the fuzzy logic. Fuzzy logic system elements, used for the fuzzy event- tree analysis of the electric generator protection are adequate to the proposed application. The FLS inputs "Occurrence" and "Severity" are associated to the tree- path and the "Safety" of the protection system, obtained as the FLS output, is used for fuzzy critical analysis of the PS presented in the paper.

The proposed FLS uses as an output element the power protection-automation system "Safety" instead of the usually "Risk" parameter, used in engineering applications (with, or without fuzzy logic elements). The "Safety" FLS output is necessary and may be used for the power protection system independent analysis. Also it could be used in other applications for the combined qualitative analysis of the protected power (electric generator for example) together with its protection system. This type of analysis implies the hybrid modeling of the combined system.

## References

1. Bastani, F.B., Chen, I.R.: Reliability of Systems with Fuzzy-Failure Criterion. In: Proc. Ann. Reliability and Maintainability Symposium, pp. 265–270 (1995)
2. Bowles, J.B., Pelaez, C.E.: Application of Fuzzy Logic to Reliability Engineering. Proceedings of the IEEE 3, 99–107 (1995)
3. Dumitrescu, M., et al.: Application of Fuzzy Logic in Safety Computing for a Power Protection System. In: Wang, L., Jiao, L., Shi, G., Li, X., Liu, J. (eds.) FSKD 2006. LNCS (LNAI), vol. 4223, pp. 980–989. Springer, Heidelberg (2006)
4. `http://www.springerlink.com/content/v04427137h03/?sortorder=asc&v=expanded&o=40`
5. Kenarangui, R.: Verbal rating of alternative sites using multiple critter weights. Transactions. ANS 33, 617–619 (1979)
6. Kenarangui, R.: Event - tree Analysis by Fuzzy Probability. IEEE Transactions on Reliability 1, 45–52 (1991)
7. Mendel, M.: Fuzzy Logic Systems for Engineering A Tutorial. Proceedings of the IEEE 3, 345–377 (1995)
8. Meyer, J.F.: On Evaluation the Performability of Degradable Computing Systems. IEEE Transactions on computers 29(8), 720–731 (1980)

# A Multi-purpose Time Series Data Standardization Method

Veselka Boeva and Elena Tsiporkova

**Abstract.** This work proposes a novel multi-purpose data standardization method inspired by gene-centric clustering approaches. The clustering is performed via template matching of expression profiles employing Dynamic Time Warping (DTW) alignment algorithm to measure the similarity between the profiles. In this way, for each gene profile a cluster consisting of a varying number of neighboring gene profiles (determined by the degree of similarity) is identified to be used in the subsequent standardization phase. The standardized profiles are extracted via a recursive aggregation algorithm, which reduces each cluster of neighboring expression profiles to a singe profile. The proposed data standardization method is validated on gene expression time series data coming from a study examining the global cell-cycle control of gene expression in fission yeast *Schizosaccharomyces pombe*.

## 1 Introduction

Data transformation techniques are commonly used in quantitative data analysis. The choice of a particular data transformation method is determined by the type of data analysis study to be performed. The microarray technologies allow to measure the expression of almost an entire genome simultaneously

Veselka Boeva
Computer Systems and Technologies Department
Technical University of Sofia - branch Plovdiv
Tsanko Dyustabanov 25, 4000 Plovdiv, Bulgaria
e-mail: vboeva@tu-plovdiv.bg

Elena Tsiporkova
Software Engineering and ICT group
Sirris, The Collective Center for the Belgian technological industry
Brussels, Belgium
e-mail: elena.tsiporkova@sirris.be

V. Sgurev et al. (Eds.): Intelligent Systems: From Theory to Practice, SCI 299, pp. 445–460.
springerlink.com 

nowadays. In order to make inter-array analysis and comparison possible, a whole arsenal of data transformation methodologies like background correction, normalization, summarization and standardization is typically applied to expression data. A review of most widely used data normalization and transformation techniques is presented in [13]. Some specific transformations, developed for analysis of data from particular platforms, are considered in [10, 11]. Speed [19] recommends to transform the expression data by a logarithmic transformation. The justification for this transformation is to make the distribution more symmetric and Gaussian-like. The log2-transformation involves normalization of the expression profile of each gene by the expression value at time point 0 and consequently taking log2 of the ratios. This transformation may be essential for performing between experiment or between species comparison of gene expression time series. The application of different statistical methods, as for instance regression analysis or permutation tests, is also usually preceded by log2-transformation. However, a high percentage of the time expression values at time zero may be affected by various stress response phenomena associated with the particular treatment (*e.g.* synchronization method) or experimental conditions. Therefore the choice of the first measurement as a reference expression value bears the danger of creating a distorted perception of the gene expression behavior during the whole time of sampling. The effect of logarithm transformation on the result of microarray data analysis is examined in [25]. It is shown that this transformation may affect results on selecting differentially expressed genes.

Another classical approach, almost as a rule applied before performing clustering, template matching or alignment, is to standardize the expression profiles via z-transformation. The expression profile of each gene is adjusted by subtracting the profile mean and dividing with the profile standard deviation. The z-transformation can be relevant when the general shape rather than the individual gene expression amplitudes at the different time points is important. In [3], z-transformation is used to compare different methods for predicting significantly expressed genes and it is shown to be a useful microarray analysis method in combination with z ratios or z tests. However, z-score transformation needs to be used with caution, baring in mind that the expression levels of low expressed genes will be amplified by it. Other transformations for the purpose of normalization are also possible [17, 18], such as square-root, Box-Cox [2], and arcsine transformations. A variance stabilization technique, which stabilizes the asymptotic variance of microarray data across the full range of data, is discussed in [5, 6]. Further Geller et al. [8] demonstrate how this stabilization can be applied on Affymetrix GeneChip data and provide a method for normalization of Affymetrix GeneChips, which produces a data set with constant variance and with symmetric errors.

The quality of microarray data is also affected by many other experimental artifacts, as for instance, occurrence of peak shifts due to lost of synchrony, poor signal to noise ratio for a set of sampling times resulting in partially fluctuating profiles, etc. Unfortunately, there is no an universal data transformation method that offers adequate corrections for all these. We propose here a novel data transformation method aiming at multi-purpose data standardization and inspired by gene-centric clustering approaches. The idea is to perform data standardization via template matching of each expression profile with the rest of the expression profiles employing Dynamic Time Warping (DTW) alignment algorithm to measure the similarity between the expression profiles. Template matching is usually employed in studies requiring gene-centric approaches since it allows mining gene expression time series for patterns that fit best a template expression profile. The DTW algorithm aims at aligning two sequences of feature vectors by warping the time axis iteratively until an optimal match (according to a suitable metrics) between the two sequences is found. It facilitates the identification of a cluster of genes whose expression profiles are related, possibly with a non-linear time shift, to the profile of the gene supplied as a template. For each gene profile a varying number (based on the degree of similarity) of neighboring gene profiles is identified to be used in the subsequent standardization phase. The latter uses a recursive aggregation algorithm in order to reduce the set of neighboring expression profiles to a singe profile representing the standardized version of the profile in question.

## 2 Data Standardization Procedure

Assume that a particular biological phenomenon is monitored in a high-throughput experiment. This experiment is supposed to measure the gene expression levels of $m$ genes in $n$ different time points, i.e. a matrix $m \times n$ will be produced. For each gene $i$ ($i = 1, \ldots, m$) of this expression matrix two distinctive steps will be performed: 1) selection of estimation genes; 2) calculation of standardized expression profile.

### *2.1 Selection of Estimation Genes*

A dedicated algorithm has been developed for the purpose of generating an estimation list for each gene profile. Such an estimation list consists of genes with expression profiles which exhibit certain (preliminary determined) similarity in terms of some distance measure to the expression profile of the gene in question. For each gene profile a varying number (based on the degree of similarity) of neighboring gene profiles is identified to be used in the subsequent standardization phase. The motivation behind this approach is that the expression values of each profile will be standardized by adjusting them relative to these expression profiles in the same microarray dataset, which appear to be closely related to the target profile.

The proposed standardization method is targeting adequate standardization of time series data. The Dynamic Time Warping (DTW) algorithm [15] (see Appendix) has been chosen to perform the template matching between the expression profiles. DTW aims at aligning two sequences of time vectors by warping the time axis iteratively until an optimal match (according to a suitable metric) between the two sequences is found. Thus, DTW is a much more robust distance measure for time series than classical distance metrics as Euclidean or a variation thereof since it allows similar shapes to match even if they are out of phase in the time axis.

Assume that a matrix $G$ of $m \times n$ $(m \gg n)$ contains the expression values of $m$ genes measured in $n$ time points,

$$G = \begin{bmatrix} g_1 \\ \vdots \\ g_m \end{bmatrix} = \begin{bmatrix} g_{11} & \cdots & g_{1n} \\ \vdots & & \vdots \\ g_{m1} & \cdots & g_{mn} \end{bmatrix},$$

where the row (vector) $g_i = [g_{i1}, \ldots, g_{in}]$ represents the time expression profile of the $i$-th gene. Formally, a gene estimation list $E_i$ needs to be constructed for each gene $i = 1, \ldots, m$. The values of the gene profiles in the constructed estimation list $E_i$ will subsequently be used to standardize the values of the expression profile $g_i$. The contribution of each gene in $E_i$ is weighted by the degree of similarity of its expression profile to the expression profile of gene $i$.

In order to standardize values in any location of gene $i$, a set of genes all at a certain (maximum) DTW distance from the profile $g_i$ needs to be identified. This maximum DTW distance is preliminary determined as a fraction (expressed as a global radius) of the mean DTW distance to the profile in question. In this process, all gene profiles are considered and a gene estimation list $E_i$, which is further used to standardize the values of gene $i$, is constructed.

Let us formally define the algorithm that builds this gene estimation list. Define a global radius $R \in (0, 1)$ (common for all genes) and consider an expression profile $g_i = [g_{i1}, \ldots, g_{in}]$. Construct an initial gene estimation list as $E_i = \{\text{all genes}\}$. Then for each gene $j = 1, \ldots, m$ calculate the DTW distance $\mathrm{dtw}_{ij}$ between gene $i$ and gene $j$. Remove from $E_i$ all genes $k$ for which

$$\frac{\mathrm{dtw}_{ik}}{\mathrm{mean}_j(\mathrm{dtw}_{ij})} \geq R.$$

The final estimation list contains only genes at a maximum $R \cdot \mathrm{mean}_j(\mathrm{dtw}_{ij})$-radius DTW distance from gene $i$. Let $m_i = \#E_i$. Consequently,

$$E_i = \{\text{genes } k_l | l = 1, \ldots, m_i \text{ and } \mathrm{dtw}_{ik_l} < R \cdot \mathrm{mean}_j(\mathrm{dtw}_{ij})\}. \quad (1)$$

Note that the use of double indexing $k_l$ for the elements of the estimation list $E_i$ is necessary since the gene order in the original data set is different from the one in the estimation list. Thus $k_l$ is the gene index in the original expression matrix $G$, i.e. $k_l$ refers to the gene expression profile, while $l$ merely refers to the gene position in $E_i$.

The values of the gene profiles in the estimation list $E_i$ are used in the following section to standardize the values of the expression profile $g_i$. The contribution of each gene $k_l \in E_i$ is weighted by the degree of similarity of its expression profile to the expression profile of gene $i$. Thus each gene $k_l \in E_i$ is assigned a weight $w_{k_l}$:

$$w_{k_l} = \left( 1 - \frac{\mathrm{dtw}_{ik_l}}{\sum_{l=1}^{m_i} \mathrm{dtw}_{ik_l}} \right) / (m_i - 1). \quad (2)$$

It can easily be checked that $\sum_{l=1}^{m_i} w_{k_l} = 1$. Moreover $\mathrm{dtw}_{ik_p} < \mathrm{dtw}_{ik_q}$ implies $w_{k_p} > w_{k_q}$, i.e. expression profiles closely matching the pattern of gene $i$ will always have a greater contribution to the standardized values than expression profiles which match the profile of gene $i$ to a lower extent.

The profile of gene $i$ will always be assigned the highest possible weight $w_i = 1/(m_i - 1)$ due to $dtw_{ii} = 0$. In case the estimation list of $i$ contains only one other profile besides the profile of $i$, i.e. $m_i = 2$, then $w_i$ will be 1. The latter implies that the second profile will not be taken into account during the standardization procedure and the profile of $i$ will remain unchanged. Only one other matching profile is therefore not sufficient to enforce data transformation since the profile of $i$ is then considered rather unique. Minimum 2 other profiles need to match closely the profile of $i$ in order to subject it to standardization and $w_i = 1/2$ will still be relatively high. The degree to which the closely matching profiles of $i$ will contribute to its standardization is thus determined by the size of the estimation list.

## 2.2 Calculation of Standardized Expression Profile

We discuss herein a recursive aggregation algorithm aiming at reducing a given data matrix (or a set of data vectors) into a single vector. This algorithm will be applied to obtain the standardized expression profile of an arbitrary gene by aggregating the expression profiles of the genes in its estimation list (see Section 2.1).

Consider an expression profile $g_i = [g_{i1}, \ldots, g_{in}]$ with an estimation list $E_i$ as defined in (1), consisting of the expression values of $m_i$ genes measured in $n$ time points. Let us associate a vector

$$t_j = \begin{bmatrix} g_{k_1 j} \\ \vdots \\ g_{k_{m_i} j} \end{bmatrix}$$

of $m_i$ expression values (one per gene) with each time point $j$ ($j = 1, \ldots, n$). Consequently, a matrix $G_i$ can be defined as follows $G_i = [t_1, \ldots, t_n]$. Additionally, each gene (row vector of expression values) $k_l$, $l = 1, \ldots, m_i$ is associated with a weight $w_{k_l}$ as defined in (2), expressing the relative degree of importance (contribution) assigned to profile $k_l$ in the aggregation process. Thus a vector $w = [w_{k_1}, \ldots, w_{k_{m_i}}]$, where $\sum_{l=1}^{m_i} w_{k_l} = 1$ and $w_{k_l} \in [0, 1]$ is given.

The ultimate goal of the aggregation algorithm is to transform the above matrix $G_i$ into a single vector $\mathbf{g}_i = [\mathbf{g}_{i1}, \ldots, \mathbf{g}_{in}]$, consisting of one (overall) value per time point and representing the standardized version of $g_i$. Each $\mathbf{g}_{ij}$, ($j = 1, \ldots, n$) can be interpreted as the trade-off value, agreed between the different genes, for the expression value of the time point in question. Naturally, the aggregated values are expected to take into account, in a suitable fashion, all the individual input values of vectors $t_j$ ($j = 1, \ldots, n$). The choice of aggregation operator is therefore crucial. Some aggregation operators can lead to a significant loss of information since their values can be greatly influenced by extreme scores (arithmetic mean), while others are penalizing too much for low-scoring outliers (geometric and harmonic means). A possible and quite straightforward solution to the described problem is to use different aggregation operators in order to find some trade-off between their conflicting behavior. In this way, different aspects of the input values will be taken into account during the aggregation process.

We suggest to apply a hybrid aggregation process, developed in [20] (see also [21, 22]), by employing a set of $k$ aggregation operators $A_1, \ldots, A_k$. The values of matrix $G_i$ can initially be combined in parallel with the weighted versions of these $k$ different aggregation operators. Consequently, a new matrix $G_i^{(0)}$ of $n$ column vectors, i.e. $G_i^{(0)} = [t_1^{(0)}, \ldots, t_n^{(0)}]$, is generated as follows:

$$t_j^{(0)} = \begin{bmatrix} A_1(w, t_j) \\ \vdots \\ A_k(w, t_j) \end{bmatrix}.$$

Thus a new vector of $k$ values (one per aggregation operator) is produced for each time point $j = 1, \ldots, n$ by aggregating the expression values of vector $t_j$ (see Fig. 1). The new matrix can be aggregated again, generating again a matrix $G_i^{(1)} = [t_1^{(1)}, \ldots, t_n^{(1)}]$, where

$$t_j^{(1)} = \begin{bmatrix} A_1(t_j^{(0)}) \\ \vdots \\ A_k(t_j^{(0)}) \end{bmatrix}.$$

In this fashion, each step is modeled via $k$ parallel aggregations applied over the results of the previous step, i.e. at step $q$ ($q = 1, 2, \ldots$) a matrix $G_i^{(q)} = [t_1^{(q)}, \ldots, t_n^{(q)}]$ is obtained and

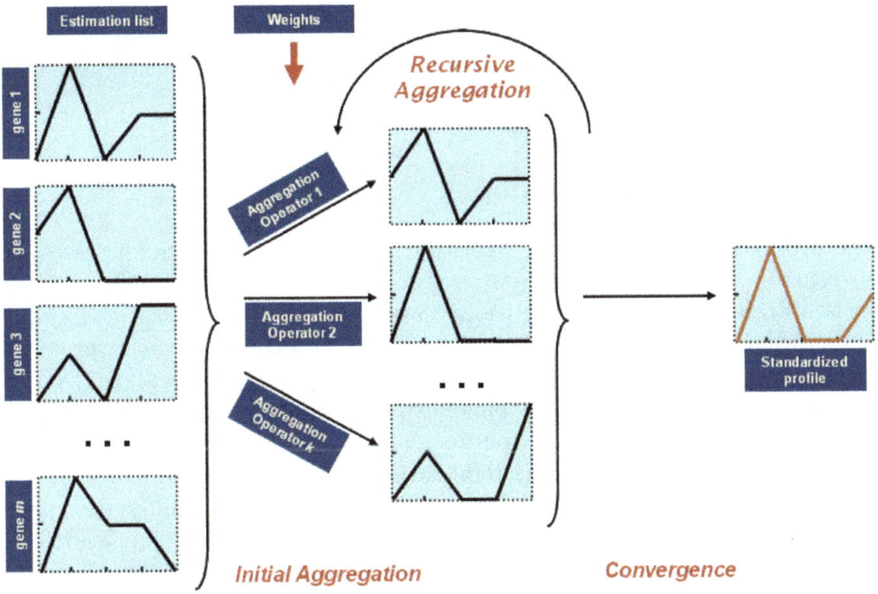


**Fig. 1** Recursive aggregation algorithm

$$t_j^{(q)} = \begin{bmatrix} A_1(t_j^{(q-1)}) \\ \vdots \\ A_k(t_j^{(q-1)}) \end{bmatrix},$$

for $j = 1, \ldots, n$.

Thus the standardized expression profile $\mathbf{g}_i = [\mathbf{g}_{i1}, \ldots, \mathbf{g}_{in}]$ of gene $i$ will be obtained by applying the foregoing recursive aggregation algorithm to the gene expression values (matrix $G_i$) of its estimation list $E_i$. The expression profiles included in matrix $G_i$ are initially combined in parallel with $k$ different weighted aggregation operators using a set of weights as defined in (2). In this way $k$ new expression profiles (one per aggregation operator) are produced and these new profiles are aggregated again this time with the non-parametric versions of the given aggregation operators. The latter process is repeated again and again until for each time point the difference between the aggregated values is small enough to stop further aggregation.

In [20, 21], we have shown that any recursive aggregation process, defined via a set of continuous and *strict-compensatory aggregation operators*, following the algorithm described herein is *convergent*. For instance, any weighted mean operator with non-zero weights is continuous and strict compensatory [7]. Thus, if $w_1, w_2, \ldots, w_n$ are positive real numbers such that $\sum_{i=1}^{n} w_i = 1$ then the *weighted arithmetic* $\mathbf{M_w} = \sum_{i=1}^{n} w_i x_i$, the *weighted geometric* $\mathbf{G_w} = \prod_{i=1}^{n} {x_i}^{w_i}$ and the *weighted harmonic* $\mathbf{H_w} = 1/(\sum_{i=1}^{n} w_i/x_i)$

means are continuous and strict compensatory. We have shown in [20] that a recursive aggregation process, defined via a combination of the above means is, in fact, an aggregation mean operator that compensates between the conflicting properties of the different mean operators.

## 3 Results and Discussion

The proposed standardization algorithm is evaluated and demonstrated on microarray datasets coming from a study examining the global cell-cycle control of gene expression in fission yeast *Schizosaccharomyces pombe* [14]. The study includes 8 independent time-course experiments synchronized respectively by 1) elutriation (three independent biological repeats), 2) cdc25 block-release (two independent biological repeats, where the second in two technical replicates, and one experiment in a sep1 mutant background), 3) a combination of both methods (elutriation and cdc25 block-release as well as elutriation and cdc10 block-release). Thus the following 9 different expression test sets are available: 1) elutriation1, 2) elutriation2, 3) elutriation3, 4) cdc25-1, 5) cdc25-2.1, 6) cdc25-2.2, 7) cdc25-sep1, 8) elutriation-cdc25-br, 9) elutriation-cdc10-br.

The normalized data for 9 experiments has been downloaded from the website of the Sanger Institute (http://www.sanger.ac.uk/PostGenomics/S_pombe/). Subsequently, the rows with more than 25% missing entries have been filtered out from each expression matrix and any other missing expression entries have been imputed with the DTWimpute algorithm [23]. In this way nine complete expression matrices have been obtained. Subsequently, the standardization method described in Section 2 has been applied to each complete matrix. For each gene profile occurring in such a matrix a gene estimation list has been created identifying a varying number of neighboring gene profiles (at maximum 20% of the mean DTW distance from the gene profile, i.e. $R = 0.2$) to be used in the calculation of the standardized expression profile. For each matrix the mean value of the DTW distances used for construction of the gene estimation lists has been recorded and the results for all nine experiments are summarized in Fig. 2c. In addition, the number of standardized genes and the mean number of selected estimation genes have been calculated for each matrix (see Fig. 2a and Fig. 2b, respectively). Fig. 2a unravels that the number of standardized genes is very different for the different experiments, *e.g.* it is only 14 for elutriation2 and more than 2500 for elutriation-cdc10-br. Similar phenomenon is observed in the number of the used estimation genes depicted in Fig. 2b. This is probably due to the more distant (unique) expression profiles in elutriation2 (elu2) experiment than those in elutriation-cdc10-br. This hypothesis is also supported by the mean DTW distances depicted in Fig. 2c. The number of standardized genes can eventually be considered as a measure for the data quality.

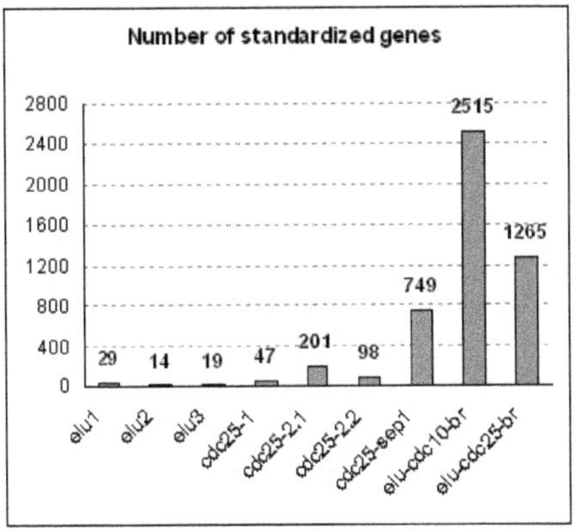


(a) Standardized genes

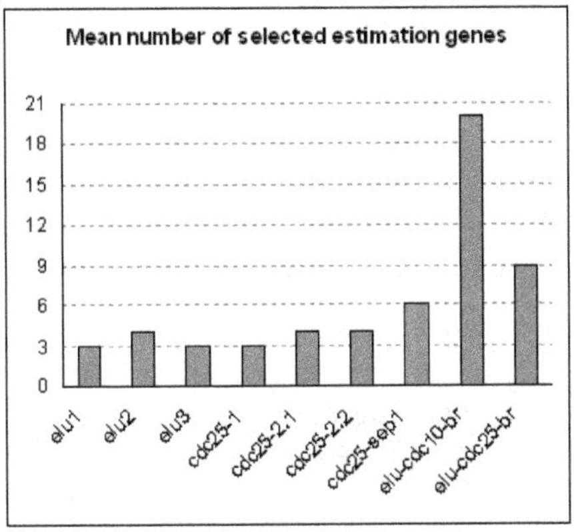


(b) Estimation genes

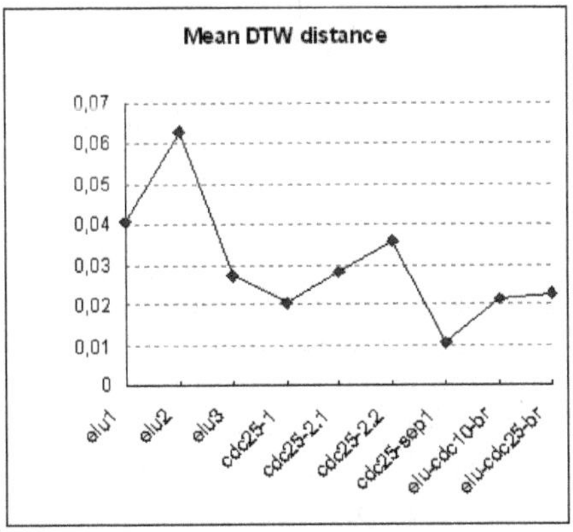


(c) Mean DTW distance

**Fig. 2** Number of standardized genes, number of selected estimation genes and mean DTW distances for the nine experiments

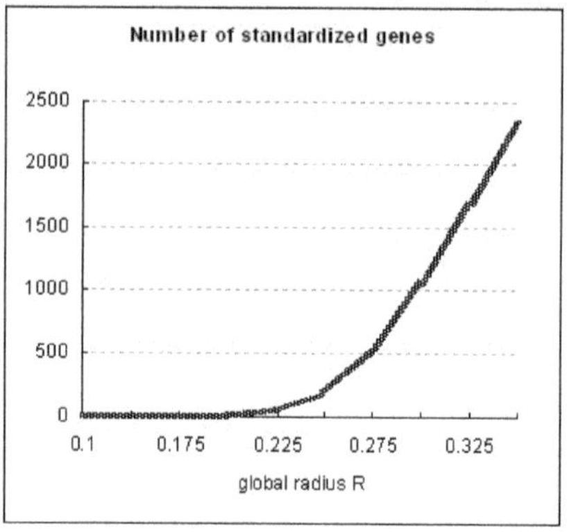


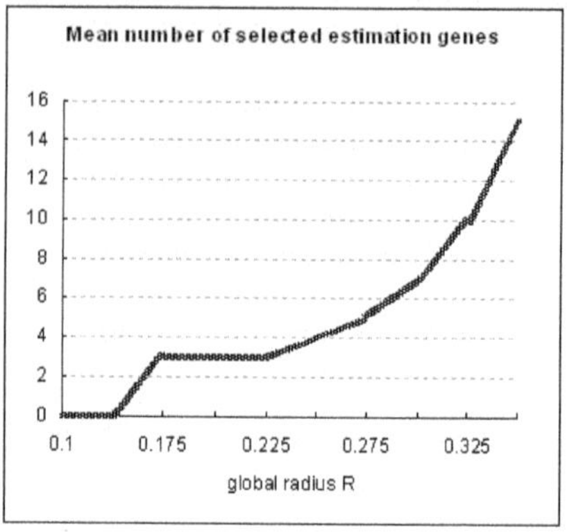


**Fig. 3** The number of the standardized genes and the mean number of the selected estimation genes as functions of global radius $R$

Fig. 3 depicts the number of standardized genes and the mean number of the selected estimation genes as functions of the global radius $R$. The presented results are obtained by applying the standardization method to elutriation3 experiment for a few different values of $R$. Notice that both functions are monotonic with respect to $R$, *i.e.* the number of standardized genes will increase for higher values of $R$ and analogously, the number of estimation genes used in the standardization process will increase too.

The recursive aggregation procedure, as defined in Section 2.2, has been applied to the gene expression values of the estimation list to calculate the standardized expression profiles. For the purpose of the hybrid aggregation procedure, three different aggregation operators have been selected: arithmetic, geometric and harmonic means. Their definitions can be found in Section 2.2. Each one of these aggregation operators exhibits certain shortcomings when used individually. For instance, the arithmetic mean values are strongly influenced by the presence of extremely low or extremely high values. This may lead in some cases to an averaged overall (standardized) value at some estimated time point, which does not adequately reflect the individual expression values at the corresponding time point of the estimation genes. In case of the geometric mean, the occurrence of a very low expression value (*e.g.* 0 or close to 0) in some position for a single estimation gene is sufficient to produce a low overall value for this position, no matter what the corresponding expression values for the rest of the estimation genes are. The harmonic mean behaves even more extremely in situations when single entries with very low values are present.

Fig. 4 depicts for 4 different genes the standardized and original expression profiles on the background of the estimation profiles used for the standardization of each original profile. In addition, each gene is presented with two

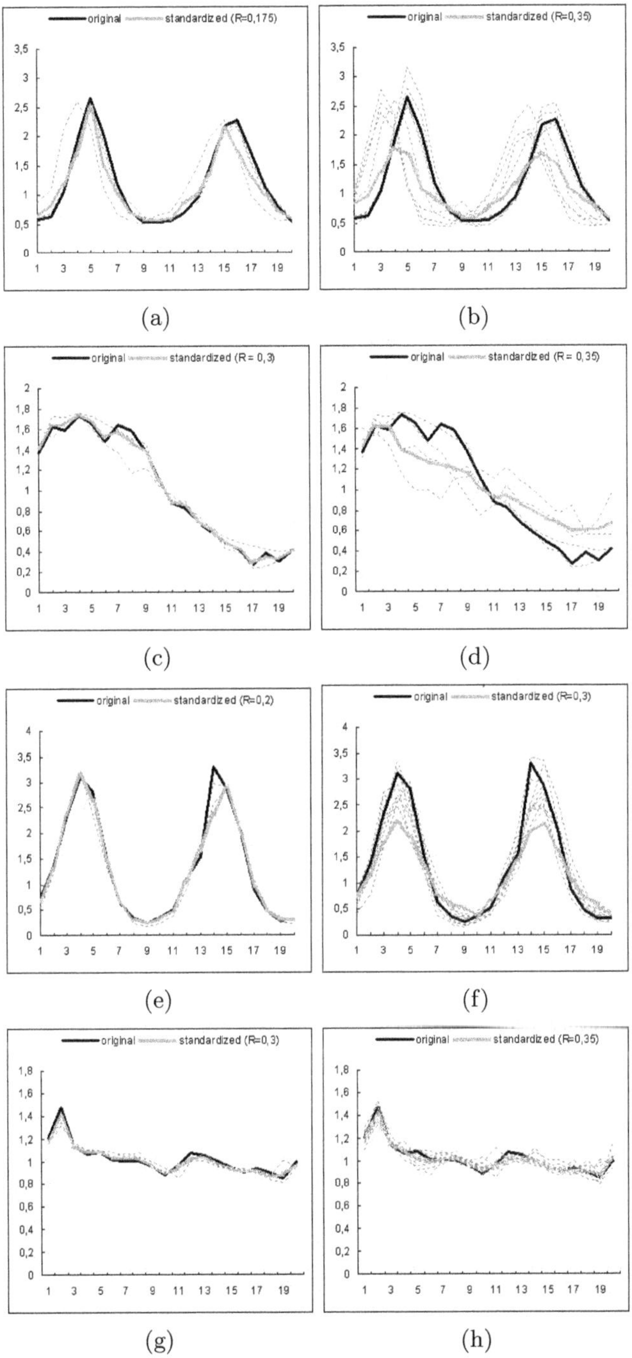


**Fig. 4** Original (thick black line) versus standardized (thick grey line) expression profiles. The profiles used for the standardization are in the background (dotted thin line).

plots, each one obtained for a different value of the global radius $R$. The radiuses used for the generation of the standardized profiles in the left column of the figure are the lowest ones for which the genes occur in the standardized profile list. The right column consists of plots generated for radiuses which do not identify more than 20 estimation genes. The standardized profiles depicted in Fig. 4a and Fig. 4e, exhibit correction for their second peak shifts of the original profiles with respect to their neighboring profiles. The profiles in plots Fig. 4c and Fig. 4d are clearly smoothed by the standardization process. In fact they are examples of a clear fluctuation reduction as a result of the standardization procedure. The latter can easily be noticed in the upper and down parts of the standardized profiles. In plots Fig. 4g and Fig. 4h, the depicted standardized profiles almost repeat the original ones, which is obviously due to the closer match between the original profile and the profiles used for the standardization. Finally, the profiles in Fig. 4b and Fig. 4f have been somewhat reduced in amplitude during their two peaks. In general, the presented results in Fig. 4 demonstrate that the standardization procedure operates as a sort of data correction for *e.g.*, peak shifts, amplitude range, fluctuations, etc.

In order to investigate whether the proposed standardization technique affects significantly the selection of differentially expressed genes, we have designed an experiment, which extracts significant genes from the original and standardized matrices (see Fig. 5). Two different computational methods for identification of cell cycle regulated genes have been applied: statistical tests for regulation and statical tests for periodicity, both described in [12]. Our benchmark set is composed of the list of cell cycle regulated genes (p-value lower than 0.05) identified in [24]. The performance of each computational method on the identification of significant genes from the benchmark set is measured as follows:

$$C = N^2/M \cdot M_b,$$

where $N$ is the number of overlapping genes across the two sets (identified and benchmark), $M$ and $M_b$ are the number of genes in the newly obtained and benchmark sets, respectively. $C$ is referred to as coverage. It can easily be checked that the coverage will be zero ($C = 0$) in case of empty set identified and 1 when the two sets are identical. In addition, the above formula will reduce to $N/M_b$ when all identified genes are from the benchmark set, but their number is below the cardinality of the set and to $N/M$ when the obtained set contains all benchmark genes, but its cardinality is greater than that of the benchmark one. Moreover, $N_1 > N_2$ implies $C_1 > C_2$ in all situations except for the case when $N_1^2/N_2^2 < M_1/M_2$, i.e. the cardinality $M_1$ of the extracted set is much surpassing the number of identified overlapping genes $N_1$. Consequently, higher values of the fraction (coverage) will imply better performance of the underlying computational method on the corresponding test matrix.

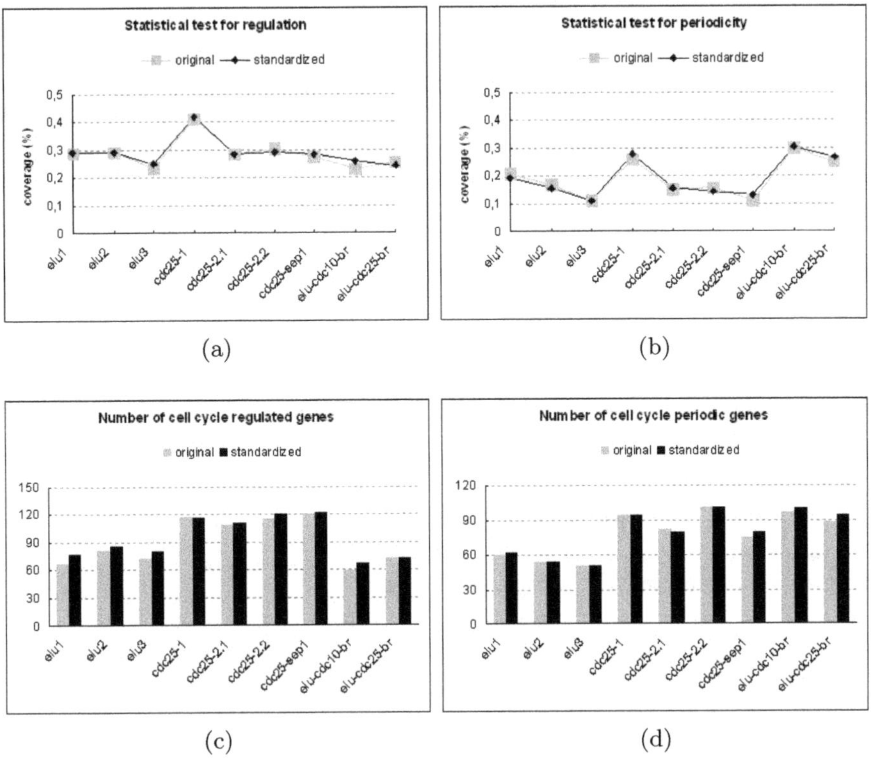


**Fig. 5** The significant gene results on the original and standardized matrices

Fig. 5a and Fig. 5b depict the calculated coverage values for the identified significant genes on the original and standardized matrices for the two computational methods, respectively. The presented results have been obtained for $R = 0.25$ and the obtained coverage values do not seem to be significantly influenced by the standardization procedure. In addition, Fig. 5c and Fig. 5d present the exact number of overlapping genes across the identified and benchmark sets as obtained for the original and standardized matrices. It can be observed that in majority of the cases, the standardized data produces higher overlap figures. The latter is probably due to the fact that the achieved noise reduction in the standardized dataset has a positive effect on the identification of false positives.

## 4 Conclusion

We have proposed here a novel data transformation method aiming at multi-purpose data standardization and inspired by gene-centric clustering approaches. The method performs data standardization via template matching of each expression profile with the rest of the expression profiles employing DTW alignment algorithm to measure the similarity between the expression profiles. For each gene profile a varying number (based on the degree of similarity) of neighboring gene profiles is identified to be used in the standardization phase. Subsequently, a recursive aggregation algorithm is applied in order to transform the identified neighboring profiles into a single standardized profile. The proposed transformation method has been evaluated on gene expression time series data coming from a study examining the global cell-cycle control of gene expression in fission yeast *Schizosaccharomyces pombe*. It has been demonstrated to be an adequate data standardization procedure aiming at fluctuation reduction, peak and amplitude correction and profile smoothing in general. In addition, the positive effect of the standardization procedure on the identification of differentially expressed genes has also been demonstrated experimentally.

## 5 Supplementary Materials

The test datasets and the software are available at `http://cst.tu-plovdiv.bg/bi/DataStandardization/`.

## References


1. Aach, J., Church, G.M.: Aligning gene expression time series with time warping algorithms. Bioinformatics 17, 495–508 (2001)
2. Box, G.E.P., Cox, D.R.: An analysis of transformation. Journal of R. Stat. Society B. 26, 211–243 (1964)
3. Cheadle, C., Vawter, M.P., Freed, W.J., Becker, K.G.: Analysis of microarray data using Z score transformation. Journal of Molecular Diagnostics 5(2), 73–81 (2003)
4. Criel, J., Tsiporkova, E.: Gene Time Expression Warper: A tool for alignment, template matching and visualization of gene expression time series. Bioinformatics 22, 251–252 (2006)
5. Durbin, B.P., Hardin, J.S., Hawkins, D.M., Rocke, D.M.: A variance-stabilizing transformation for gene-expression microarray data. Bioinformatics 18(suppl. 1), S105–S110 (2002)
6. Durbin, B.P., Rocke, D.M.: Estimation of transformation parameters for microarray data. Bioinformatics 19, 1360–1367 (2003)
7. Fodor, J.C., Roubens, M.: Fuzzy Preference Modelling and Multicriteria Decision Support. Kluwer Academic Publishers, Dordrecht (1994)
8. Geller, S.C., Gregg, J.P., Hagerman, P., Rocke, D.M.: Transformation and normalization of oligonucleotide microarray data. Bioinformatics 19(14), 1817–1823 (2003)
9. Hermans, F., Tsiporkova, E.: Merging microarray cell synchronization experiments through curve alignment. Bioinformatics 23, e64–e70 (2007)

10. Ideker, T., Thorsson, V., Siegel, A.F., Hood, L.E.: Testing for differentially-expressed genes by maximum-likelihood analysis of microarray data. Journal of Computational Biology 7, 805–817 (2001)
11. Li, C., Wong, W.: Model-based analysis of oligonucleotide arrays: expression index computation and outlier detection. Proc. National Academy Science USA 98, 31–36 (2001)
12. de Lichtenberg, U., Jensen, L.J., Fausbøll, A., Jensen, T.S., Bork, P., Brunak, S.: Comparison of computational methods for the identification of cell cycle-regulated genes. Bioinformatics 21(7), 1164–1171 (2004)
13. Quackembush, J.: Microarray data normalization and transformation. Nature Genetics Supplement 32, 496–501 (2002)
14. Rustici, G., Mata, J., Kivinen, K., Lio, P., Penkett, C.J., Burns, G., Hayles, J., Brazma, A., Nurse, P., Bähler, J.: Periodic gene expression program of the fission yeast cell cycle. Natural Genetics 36, 809–817 (2004)
15. Sakoe, H., Chiba, S.: Dynamic programming algorithm optimization for spoken word recognition. IEEE Trans. on Acoust., Speech, and Signal Proc. ASSP-26, 43–49 (1978)
16. Sankoff, D., Kruskal, J.: Time Warps, String Edits, and Macromolecules: The Theory and Practice of Sequence Comparison. AddisonWesley, Reading Mass (1983)
17. Smyth, G.K., Speed, T.P.: Normalization of cDNA microarray data. Methods 31, 265–273 (2003)
18. Sokal, R.R., Rohlf, F.J.: Biometry, 3rd edn. W.H. Freeman and Co., New York (1995)
19. Speed, T.: Always log spot intensities and ratio. Speed Group Microarray Page, `http://www.stat.berkeley.edu/users/terry/zarray/Html/log.html`
20. Tsiporkova, E., Boeva, V.: Nonparametric recursive aggregation process. Kybernetika. Journal of the Czech Society for Cybernetics and Information Sciencies 40(1), 51–70 (2004)
21. Tsiporkova, E., Boeva, V.: Multi-step ranking of alternatives in a multi-criteria and multi-expert decision making environment. Information Sciencies 76(18), 2673–2697 (2006)
22. Tsiporkova, E., Boeva, V.: Modelling and simulation of the genetic phenomena of additivity and dominance via gene networks of parallel aggregation processes. In: Hochreiter, S., Wagner, R. (eds.) BIRD 2007. LNCS (LNBI), vol. 4414, pp. 199–211. Springer, Heidelberg (2007)
23. Tsiporkova, E., Boeva, V.: Two-pass imputation algorithm for missing value estimation in gene expression time series. Journal of Bioinformatics and Computational Biology 5(5), 1005–1022 (2007)
24. Tsiporkova, E., Boeva, V.: Fusing Time Series Expression Data through Hybrid Aggregation and Hierarchical Merge. Bioinformatics 24(16), i63–i69 (2008)
25. Wentian, L., Suh, Y.J., Zhang, J.: Does Logarithm Transformation of Microarray Data Affect Ranking Order of Differentially Expressed Genes? In: Proc. Engineering in Medicine and Biology Society, EMBS 2006. 28th Annual International Conference of the IEEE, Suppl., pp. 6593–6596 (2006)

# Appendix

## A.1 Dynamic Time Warping Algorithm

The Dynamic Time Warping (DTW) alignment algorithm aims at aligning two sequences of feature vectors by warping the time axis iteratively until an optimal match (according to a suitable metrics) between the two sequences is found. It was developed originally for speech recognition applications [15]. Due to its flexibility, DTW has been widely used in many scientific disciplines including several computational biology studies [1, 4, 9]. A detail explanation of DTW algorithm can be found in [15, 9, 16]. Therefore the description following below is restricted to the important steps of the algorithm.

Two sequences of feature vectors:

$$A = [a_1, a_2, \ldots, a_n] \text{ and } B = [b_1, b_2, \ldots, b_m]$$

can be aligned against each other by arranging them on the sides of a grid, *e.g.* one on the top and the other on the left hand side. Then a distance measure, comparing the corresponding elements of the two sequences, can be placed inside each cell. To find the best match or alignment between these two sequences one needs to find a path through the grid $P = (1,1), \ldots, (i_s, j_s), \ldots, (n, m)$, $(1 \leq i_s \leq n$ and $1 \leq j_s \leq m)$, which minimizes the total distance between $A$ and $B$. Thus the procedure for finding the best alignment between $A$ and $B$ involves finding all possible routes through the grid and for each one compute the overall distance, which is defined as the sum of the distances between the individual elements on the warping path. Consequently, the final DTW distance between $A$ and $B$ is the minimum overall distance over all possible warping paths:

$$\mathrm{dtw}(A, B) = \frac{1}{n+m} \min_P \left( \sum_{s=1}^{k} \mathrm{dist}(i_s, j_s) \right).$$

It is apparent that for any pair of considerably long sequences the number of possible paths through the grid will be very large. However, the power of the DTW algorithm resides in the fact that instead of finding all possible routes through the grid, the DTW algorithm makes use of dynamic programming and works by keeping track of the cost of the best path at each point in the grid.

# Classification of Coronary Damage in Chronic Chagasic Patients

Sergio Escalera, Oriol Pujol, Eric Laciar, Jordi Vitrià,
Esther Pueyo, and Petia Radeva

**Abstract.** American Trypanosomiasis, or Chagas' disease is an infectious illness caused by the parasite Tripanosoma Cruzi. This disease is endemic in all Latin America, affecting millions of people in the continent. In order to diagnose and treat the chagas' disease, it is important to detect and measure the coronary damage of the patient. In this paper, we analyze and categorize patients into different groups based on the coronary damage produced by the disease. Based on the features of the heart cycle extracted using high resolution ECG, a multi-class scheme of Error-Correcting Output Codes (ECOC) is formulated and successfully applied. The results show that the proposed scheme obtains significant performance improvements compared to previous works and state-of-the-art ECOC designs.

## 1 Introduction

American Trypanosomiasis, or Chagas' disease is an infectious illness caused by the parasite Tripanosoma Cruzi, which is commonly transmitted to humans through the feces of blood-sucking bugs of the subfamily Triatominae [1] and much less frequently by blood transfusion, organ transplantation, congenital transmission, breast milk, contaminated food or accidental laboratory exposure [2]. More than 120

Sergio Escalera · Oriol Pujol · Jordi Vitrià · Petia Radeva
Dept. Matemàtica Aplicada i Anàlisi, Universitat de Barcelona,
Gran Via de les Corts Catalanes 585, 08007, Barcelona, Spain
Computer Vision Center, Campus UAB, Edifici O, 08193, Bellaterra, Spain
e-mail: sergio@maia.ub.es, oriol@maia.ub.es,
jordi@maia.ub.es, petia.ivanova@ub.edu

Eric Laciar
Gabinete de Tecnología Médica, Facultad de Ingeniería, Universidad Nacional de San Juan,
Av. San Martín 1109 (O), 5400, San Juan, Argentina
e-mail: laciar@gateme.unsj.edu.ar

Esther Pueyo
Instituto de Investigacón en Ingeniería de Aragón and CIBER-BBN,
Universidad de Zaragoza, Spain

V. Sgurev et al. (Eds.): Intelligent Systems: From Theory to Practice, SCI 299, pp. 461–477.
springerlink.com 

species of Triatominae bugs live in the most diverse habitats and some are well adapted to the human houses [3], constituting a serious problem of public health in Latin American countries (from Mexico to Argentina). The World Health Organization estimates that 16 to 18 million people in Latin American countries are already infected by the parasite and other 100 million people are at risk of being infected [4].

In general terms, two different stages of Chagas' disease can be distinguished: the acute phase which appears shortly after the initial infection, and the chronic phase which appears after a silent and asymptomatic period that may last several years [1].

The acute stage lasts for 1 or 2 months of parasitical infection. It usually occurs unnoticed because it is symptoms free or exhibits only mild and unspecified symptoms like fever, fatigue, headache, rash, loss of appetite, diarrhea and vomiting. Occasionally, this phase also produces mild enlargement of the liver or spleen, swollen glands and swelling of the eyelids. Even if these symptoms appear, they usually resolve spontaneously within 3-8 weeks in 90% of individuals. Although the symptoms resolve, the infection, if untreated, persists. Rarely, patients die during this stage by complications produced by a severe inflammation of the heart (myocarditis) or brain (meningoencephalitis).

Several years or even decades after initial infection, an estimated 30% of infected people will develop the chronic stage over the course of their lives. The lesions of the chronic phase affect the heart, the esophagus, the colon and the peripheral nervous system. Particularly, cardiac involvement is characterized by a progressive inflammation of cardiac muscle (Chagas' myocarditis) that produces a destruction of cardiac fibers, a fibrosis in multiple areas of the myocardium and a malfunctioning in the propagation of the electrical impulse [5]. This myocarditis, if untreated, may cause during the following years a blundle branch block, congestive heart failure, hypertrophy, thromboembolism, atrioventricular block, ventricular tachycardia and sudden death. In areas where the illness is endemic, Chagas' cardiomyopathy represents the first cause of cardiovascular death [24].

In order to optimize treatment for chronic chagasic patients, it is essential to make use of an effective diagnosis tool able to determine the existence of cardiac damage and, if positive, its magnitude. Clinical diagnosis is usually based on different non-invasive tests such as chest x-rays, echocardiogram, or ElectroCardioGram (ECG), which can be either Holter ECG or conventional rest ECG. The use of High-Resolution ElectroCardioGraphy (HRECG) has been reported in the literature as a useful tool for clinical assessment of Chagas' disease [6, 11]. This electrocardiographic technique is oriented specifically to the detection of cardiac micropotentials, such as Ventricular Late Potentials (VLP). They are very low-level high-frequency cardiac signals found within the terminal part of the QRS complex and the beginning of the ST segment. The standard method for VLP detection is based on the evaluation of different temporal indexes computed on QRS complex from a temporally averaged beat [10]. Using this standard method, the presence of VLP has been detected in signal-averaged HRECG recordings of chronic chagasic patients [14, 22]. A different approach has been proposed in another study [22], in which the temporal beat-to-beat variability of the QRS complex duration on HRECG recording has been measured, and it has been shown that such a variability is more accentuated in chronic chagasic patients, particularly when their degree of myocardial damage is severe. Since Chagas' myocarditis frequently leads to alterations in the heart's electrical conduction, the measurement of upward and downward slopes of QRS complex has been also proposed in order to determine the myocardial damage associated with the disease [30].

Based on the temporal indices and slopes of QRS complex as extracted features, an automatic system that categorized patients into different groups is presented. To perform a multi-classification system able to learn the level of damage produced by the disease, we focus on Error-Correcting Output Codes. ECOC were born as a general framework to combine binary problems to address the multi-class problem [13]. Based on the error correcting principles and because of its ability to correct the bias and variance errors of the base classifiers [21], ECOC has been successfully applied to a wide range of Computer Vision applications, such as face recognition [35], face verification [20], text recognition [18] or manuscript digit classification [37].

The ECOC technique can be broken down into two distinct stages: encoding and decoding. Given a set of classes, the coding stage designs a codeword[1] for each class based on different binary problems. The decoding stage makes a classification decision for a given test sample based on the value of the output code. Many coding designs have been proposed to codify an ECOC coding matrix, obtaining successful results [15][32]. However, the use of a proper decoding strategy is still an open issue. In this paper, we propose the Loss-Weighted decoding strategy, witch exploits the information provided at the coding stage to perform a successful classification of the level of coronary damage of chronic chagasic patients. The results show that the present ECOC scheme outperforms the state-of-the-art on decoding designs, at same time that obtains significant performance improvements characterizing the level of damage of patients with the Chagas' disease.

The paper is organized as follows: Section 2 explains the feature extraction from QRS complex of chronic chagasic patients. Section 3 presents the Loss-Weighted decoding strategy to decode any ECOC design. Section 4 shows the experimental results of the multi-class categorization system. Finally, section 5 concludes the paper.

## 2 QRS Features

In order to obtain the features to evaluate the degree of myocardial damage associated with the disease, temporal indices and slopes of QRS are analyzed for all the HRECG recordings of 107 individuals from the Chagas database recorded at Simón Bolívar University (Venezuela).

Standard temporal QRS indices defined to detect the presence of VLP in HRECG recordings [10] are evaluated in this work. Previous studies in the literature have shown the ability of those indices to determine the severity of Chagas' myocarditis [14, 22]. They are computed from the QRS complex of the vector magnitude $vm(n)$ of the filtered averaged signals of X, Y and Z leads. Figure 1 illustrates the

[1] The codeword is a sequence of bits of a code representing each class, where each bit identifies the membership of the class for a given binary classifier.

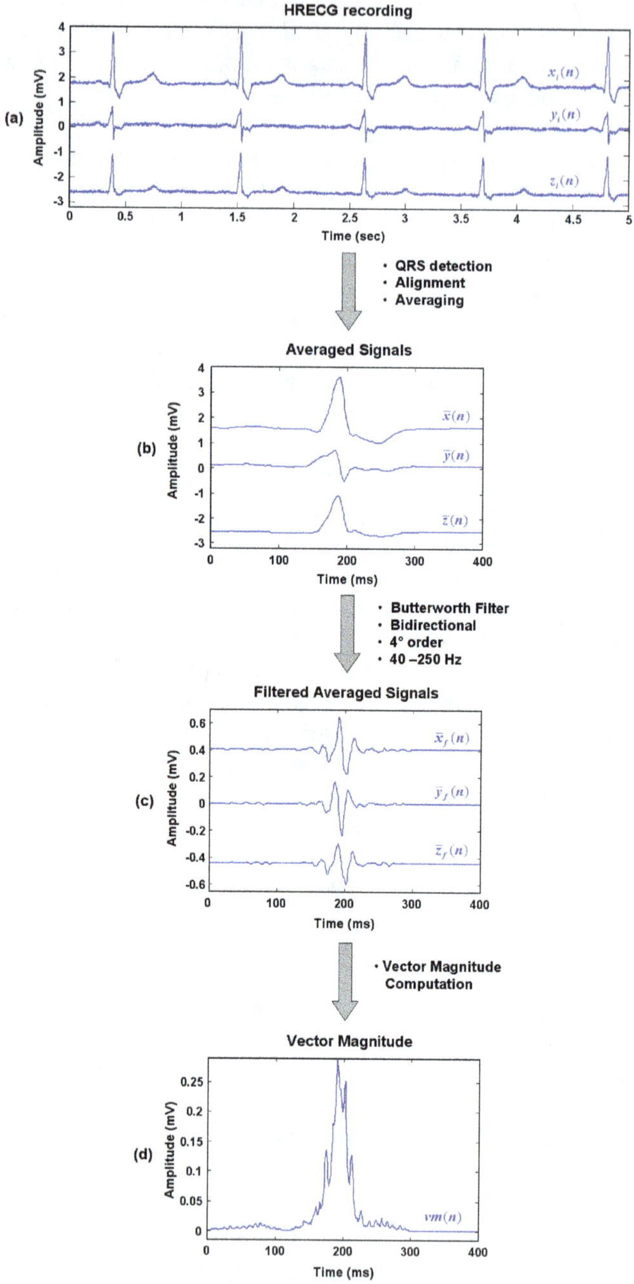


**Fig. 1** Computation of the vector magnitude: (a) Temporal segment of a HRECG recording, (b) Averaged signals, (c) Filtered averaged signals, and (d) Vector magnitude

process of computation of the signal $vm(n)$. Its upper panel Fig. 1(a) shows a temporal segment with X, Y, and Z leads of a HRECG recording acquired in a chronic chagasic patient with severe myocardial damage.

For the HRECG recording, let us denote $x_i(n)$, the $i$-th beat of lead X, where $i = 1,..,I$ and $n = 0,..,N$, where $I$ is the number of normal beats to be averaged and $N$ is the length of averaging window. Analogously, let us denote $y_i(n)$ and $z_i(n)$ the $i$-th beat of leads Y and Z, respectively. After applying to this record different algorithms of QRS detection, alignment and averaging [7, 8] and following the standard recommendation described in [10], averaged signals $\bar{x}(n)$, $\bar{y}(n)$, and $\bar{z}(n)$ are obtained as the temporally averaging of all normal beats $i = 1,..,I$ of the recording. Ectopic and grossly noisy beats were excluded of the averaging process [8]. As it is suggested in the standard document [10], the averaged signals $\bar{x}(n)$, $\bar{y}(n)$, and $\bar{z}(n)$ (Fig. 1(b)) are then filtered using a bi-directional 4th-order Butterworth filter with a passband between 40 and 250 Hz. The resultant filtered averaged signals $\bar{x}_f(n)$, $\bar{y}_f(n)$, and $\bar{z}_f(n)$ (Fig. 1(c)) are finally combined into a vector magnitude $vm(n)$ (Fig. 1(d)), defined as follows:

$$vm(n) = \sqrt{\overline{x_f}^2(n) + \overline{y_f}^2(n) + \overline{z_f}^2(n)} \tag{1}$$

On the signal $vm(n)$ three temporal QRS indices defined to detect VLP are computed based on previous identification of time instants $n_b$ and $n_e$ corresponding to the beginning and the end of the QRS complex [10]. These indices are: (a) the total QRS duration (QRSD), (b) the root mean square voltage of the last 40 ms of the QRS complex (RMS40), and (c) the duration of the terminal low amplitude of $vm(n)$ signal below 40 $\mu V$ (LAS40). They are defined as follows (see Figure 2):

$$QRSD = n_e - n_b \tag{2}$$

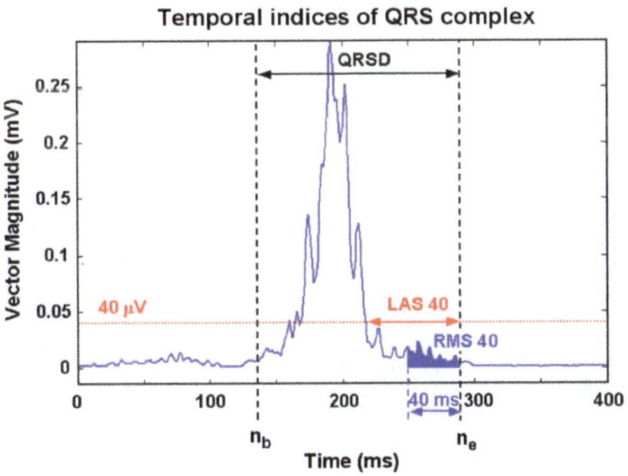


**Fig. 2** Temporal indices of QRS complex computed from vector magnitude

$$RMS40 = \sqrt{\frac{1}{n_2 - n_1} \sum_{n=n_1}^{n_2} vm^2(n)}, \quad n_1 = n_e - 40ms, \quad n_2 = n_e \tag{3}$$

$$LAS40 = n_e - argmax\{n|vm(n) \geq 40\mu V\} \tag{4}$$

Another temporal index $\Delta QRSD$ is measured to take into account the temporal beat-to-beat variability of QRS duration in HRECG recording. This index proposed in other study [22] has shown that it is more accentuated in chronic chagasic patients with severe myocardial degree. This index is computed on the set of vector magnitude functions $vm_i(n)$ of the filtered (non-averaged) signals $(x_{i,f}(n), y_{i,f}(n), z_{i,f}(n))$, defined as follows:

$$vm_i(n) = \sqrt{x^2_{i,f}(n) + y^2_{i,f}(n) + z^2_{i,f}(n)} \tag{5}$$

On each signal $vm_i(n)$, $i = 0,..,I$, the duration of its complex QRS is estimated and denoted by $QRSD_i$. The index $\Delta QRSD$ is defined as the standard deviation of the beat-to-beat $QRSD_i$ series [23] that is:

$$\Delta QRSD = \sqrt{\frac{\Sigma_{i=1}^{I}(QRSD_i - \overline{QRSD}^2)}{I-1}}, \text{ where } \overline{QRSD} = \frac{\Sigma_{i=1}^{I} QRSD_i}{I} \tag{6}$$

In addition to temporal QRS indices described above, the slopes of QRS complex are also measured in order to determine the myocardial damage associated with the disease [30]. Consequently, we use QRS slopes in conjunction with the QRS indices. A three-step process is applied to compute the upward QRS slope ($\alpha_{US}$) and the downward QRS slope ($\alpha_{DS}$) on each averaged signal $\overline{x}(n)$, $\overline{y}(n)$, and $\overline{z}(n)$. The computation of both slopes is illustrated in Figure 3 and explained next for $\overline{x}(n)$

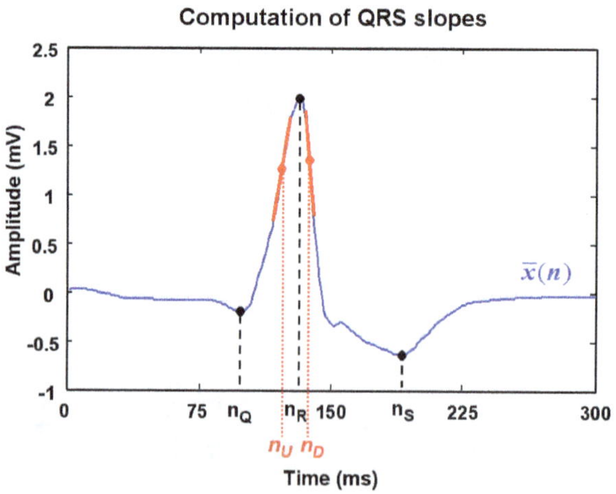


**Fig. 3** Computation of QRS slopes on averaged signal $\overline{x}(n)$

signal, a similar procedure is made for $\overline{y}(n)$ and $\overline{z}(n)$. In the first step, a delineation is performed using a wavelet-based technique [25] that determines the temporal locations Q, R, and S wave peaks, which are denoted by $n_Q$, $n_R$, and $n_S$, respectively [31]. The second step identifies the time instant $n_U$ associated with maximum slope of the ECG signal (i.e., global maximum of its derivative) between $n_Q$ and $n_R$. Analogously, the time instant $n_D$ corresponding to minimum slope of the ECG signal between $n_R$ and $n_S$ is identified. As a final step, a line is fitted in the least squares sense to the ECG signal in a window of 15ms around $n_U$, and the slope of that line is defined as $\alpha_{US}$. In the same manner, $\alpha_{DS}$ is defined as the slope of a line fitted in a 15ms window around $n_D$.

Based on the previous features, we present a design of Error-Correcting Output Codes [13] that automatically diagnoses the level of damage of patients with the Chaga's disease.

## 3 Error-Correcting Output Codes

Given a set of $N_c$ classes (in our case, $N_c$ levels of Chaga's disease) to be learned, at the coding step of the ECOC framework, $n$ different bi-partitions (groups of classes) are formed, and $n$ binary problems (dichotomies) are trained. As a result, a codeword of length $n$ is obtained for each class, where each bin of the code corresponds to a response of a given dichotomy. Arranging the codewords as rows of a matrix, we define a "coding matrix" $M$, where $M \in \{-1,0,1\}^{N_c \times n}$ in the ternary case. Joining classes in sets, each dichotomy, that defined a partition of classes, codes by $\{+1,-1\}$ according to their class set membership, or 0 if the class is not considered by the dichotomy. In fig.4 we show an example of a ternary matrix $M$. The matrix is coded using 7 dichotomies $\{h_1,...,h_7\}$ for a four class problem ($c_1$, $c_2$, $c_3$, and $c_4$). The white regions are coded by 1 (considered as positive for its respective dichotomy, $h_i$), the dark regions by -1 (considered as negative), and the grey regions correspond to the zero symbol (not considered classes by the current dichotomy). For example, the first classifier ($h_1$) is trained to discriminate $c_3$ versus $c_1$ and $c_2$ ignoring $c_1$, the second one classifies $c_2$ versus $c_1$, $c_3$ and $c_4$, and so on.

During the decoding process, applying the $n$ trained binary classifiers, a code $x$ is obtained for each data point in the test set. This code is compared to the base codewords of each class $\{y_1,...,y_4\}$ defined in the matrix $M$, and the data point is assigned to the class with the "closest" codeword [9][36].

### 3.1 Decoding Designs

The decoding step decides the final category of an input test by comparing the codewords. In this way, a robust decoding strategy is required to obtain accurate results. Several techniques for the binary decoding step have been proposed in the literature [36][19][29][12], the most common ones are the Hamming ($HD$) and the Euclidean ($ED$) approaches [36]. In fig.4, a new test input $x$ is evaluated by all the classifiers and the method assigns label $c_1$ with the closest decoding distances. Note that in the

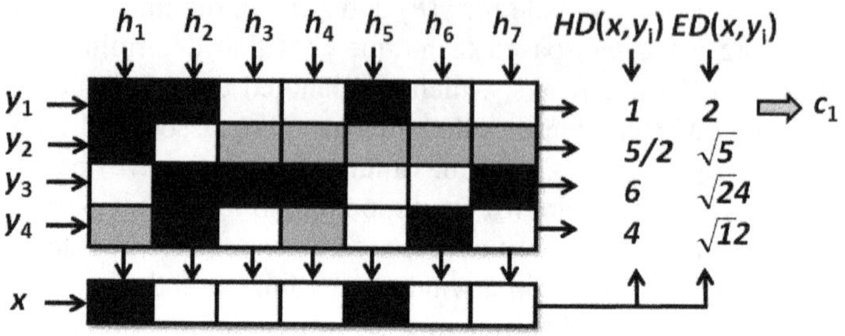


**Fig. 4** Example of ternary matrix $M$ for a 4-class problem. A new test codeword is classified by class $c_1$ when using the traditional Hamming and Euclidean decoding strategies.

particular example of fig. 4 both distances agree. In the work of [32], authors showed that the Euclidean distance was usually more suitable than the traditional Hamming distance in both the binary and the ternary cases. Nevertheless, little attention has been paid to the ternary decoding approaches.

In [9], the authors propose a Loss-based technique when a confidence on the classifier output is available. For each row of $M$ and each data sample $\wp$, the authors compute the similarity between $f^j(\wp)$ and $M(i,j)$, where $f^j$ is the $j^{th}$ dichotomy of the set of hypothesis $F$, considering a loss estimation on their scalar product, as follows:

$$D(\wp, y_i) = \sum_{j=1}^{n} L(M(i,j) \cdot f^j(\wp)) \tag{7}$$

where $L$ is a loss function that depends on the nature of the binary classifier. The most common loss functions are the linear and the exponential one. The final decision is achieved by assigning a label to example $\wp$ according to the class $c_i$ with the minimal distance.

Recently, the authors of [29] proposed a probabilistic decoding strategy based on the margin of the output of the classifier to deal with the ternary decoding. The decoding measure is given by:

$$D(y_i, F) = -log\left(\prod_{j\in[1,n]:M(i,j)\neq 0} P(x^j = M(i,j)|f^j) + \alpha\right) \tag{8}$$

where $\alpha$ is a constant factor that collects the probability mass dispersed on the invalid codes, and the probability $P(x^j = M(i,j)|f^j)$ is estimated by means of:

$$P(x^j = y_i^j|f^j) = \frac{1}{1 + exp(y_i^j(A^j f^j + B^j))} \tag{9}$$

Vectors $A$ and $B$ are obtained by solving an optimization problem [29].

## 4 Loss-Weighted Decoding (*LW*)

In this section, we present the multi-class scheme of Error-Correcting Output Codes proposed to learn the QRS complex features described in section 2.

The ternary symbol-base ECOC allows to increase the number of bi-partitions of classes (thus, the number of possible binary classifiers) to be considered, resulting in a higher number of binary problems to be learned. However, the effect of the ternary symbol is still an open issue. Since a zero symbol means that the corresponding classifier is not trained on a certain class, to consider the "decision" of this classifier on those zero coded position does not make sense. Moreover, the response of the classifier on a test sample will always be different to 0, so it will register an error. Let us return to fig. 4, where an example about the effect of the 0 symbol is shown. The classification result using the Hamming distance as well as the Euclidean distance is class $c_1$. On the other hand, class $c_2$ has only coded first both positions, thus it is the only information provided about class $c_2$. The first two coded locations of the test codeword $x$ correspond exactly to these positions. Note that each position of the codeword coded by 0 means that both -1 and +1 values are possible. Hence the correct classification should be class $c_2$ instead of $c_1$. The use of standard decoding techniques that do not consider the effect of the third symbol (zero) frequently fails. In the figure, the $HD$ and $ED$ strategies accumulate an error value proportional to the number of zero symbols by row, and finally miss-classify the sample $x$.

To solve the commented problems, we propose a Loss-Weighted decoding. The main objective is to find a weighting matrix $M_W$ that weights a loss function to adjust the decisions of the classifiers, either in the binary and in the ternary ECOC frameworks. To obtain the weighting matrix $M_W$, we assign to each position $(i, j)$ of the matrix of hypothesis $H$ a continuous value that corresponds to the accuracy of the dichotomy $h_j$ classifying the samples of class $i$ (10). We make $H$ to have zero probability at those positions corresponding to unconsidered classes (11), since these positions do not have representative information. The next step is to normalize each row of the matrix $H$ so that $M_W$ can be considered as a discrete probability density function (12). This step is very important since we assume that the probability of considering each class for the final classification is the same (independently of number of zero symbols) in the case of not having *a priori* information ($P(c_1) = ... = P(c_{N_c})$). In fig. 5 a weighting matrix $M_W$ for a 3-class problem with four hypothesis is estimated. Figure 5(a) shows the coding matrix $M$. The matrix $H$ of fig. 5(b) represents the accuracy of the hypothesis classifying the instances of the training set. The normalization of $H$ results in the weighting matrix $M_W$ of fig. 5(c)[2].

The Loss-weighted algorithm is shown in table 1. As commented before, the loss functions applied in equation (12) can be the linear or the exponential ones. The linear function is defined by $L(\theta) = \theta$, and the exponential loss function by $L(\theta) = e^{-\theta}$, where in our case $\theta$ corresponds to $M(i, j) \cdot f^j(\wp)$. Function $f^j(\wp)$ may return either the binary label or the confidence value of applying the $j^{th}$ ECOC classifier to the sample $\wp$.

[2] Note that the presented Weighting Matrix $M_W$ can also be applied over any decoding strategy.

$$M = \begin{bmatrix} 1 & 1 & -1 & 0 \\ 1 & -1 & 0 & 0 \\ 1 & 1 & 1 & -1 \end{bmatrix} \quad H = \begin{bmatrix} 0.955 & 0.955 & 1.000 & 0.000 \\ 0.900 & 0.800 & 0.000 & 0.000 \\ 1.000 & 0.905 & 0.805 & 0.805 \end{bmatrix} \quad M_W = \begin{bmatrix} 0.328 & 0.328 & 0.344 & 0.000 \\ 0.529 & 0.471 & 0.000 & 0.000 \\ 0.285 & 0.257 & 0.229 & 0.229 \end{bmatrix}$$

(a) (b) (c)

**Fig. 5** (a) Coding matrix $M$ of four hypotheses for a 3-class problem. (b) Matrix $H$ of hypothesis accuracy. (c) Weighting matrix $M_W$.

**Table 1** Loss-Weighted algorithm

Given a coding matrix $M$,

1) Calculate the matrix of hypothesis $H$:

$$H(i,j) = \frac{1}{m_i} \sum_{k=1}^{m_i} \gamma(h_j(\wp_k^i), i, j) \tag{10}$$

$$\text{based on} \quad \gamma(x_j, i, j) = \begin{cases} 1, & \text{if} \quad x_j = M(i,j) \\ 0, & \text{otherwise.} \end{cases} \tag{11}$$

2) Normalize $H$ so that $\sum_{j=1}^{n} M_W(i,j) = 1, \forall i = 1, ..., N_c$:

$$M_W(i,j) = \frac{H(i,j)}{\sum_{j=1}^{n} H(i,j)},$$
$$\forall i \in [1,...,N_c], \quad \forall j \in [1,...,n]$$

Given a test input $\wp$, decode based on:

$$d(\wp, i) = \sum_{j=1}^{n} M_W(i,j) L(M(i,j) \cdot f(\wp, j)) \tag{12}$$

## 5 Results

Before the experimental results are presented, we comment the data, methods, and evaluation measurements.

• *Data*: In this work, we analyzed a population composed of 107 individuals from the Chagas database recorded at Simón Bolívar University (Venezuela). For each individual, a continuous 10-minute HRECG was recorded using orthogonal XYZ lead configuration. All the recordings were digitalized with a sampling frequency of 1 kHz and amplitude resolution of 16 bits.

Out of the total 107 individuals of the study population, 96 are chagasic patients with positive serology for Trypanosoma Crucy, clinically classified into three different groups according on their degree of cardiac damage (Groups I, II, and III). This grouping is based on the clinical history, Machado-Guerreiro test, conventional ECG of twelve derivations, Holter ECG of 24 hours, and myocardiograph study for each patient. The other 11 individuals are healthy subjects with negative serology taken as a control group (Group 0). All individuals of the database are described with a features vector of 16 features based on the previous analysis of section 2. The four analyzed groups are described in detail next:

· Group 0: 11 healthy subjects in the age 33.6±10.9 years, 9 men and 2 women.

· Group I: 41 total patients with the Chagas' disease in the age of 41.4±8.1 years, 21 men and 20 women, but without evidences of cardiac damage in cardiographic study.

· Group II: 39 total patients with the Chagas' disease in the age of 45.8±8.8 years, 19 men and 20 women, with normal cardiographic study and some evidences of weak or moderate cardiac damage registered in the conventional ECG or in the Holter ECG of 24 hours.

· Group III: 16 total patients with the Chagas' disease in the age of 53.6±9.3 years, 9 men and 7 women, with significant evidences of cardiac damage detected in the conventional ECG, premature ventricular contractions and/or cases of ventricular tachycardiac registered in the Holter ECG and reduced fraction of ejection estimated in the cardiographic study.

• *Methods*: We compare our results with the performances reported in [30] for the same data. Moreover, we compare different ECOC designs: the one-versus-one ECOC coding strategy [33] applied with the Hamming [13], Euclidean [15], Probabilistic [29], and the presented Loss-Weighted decoding strategies. We selected the one-versus-one ECOC coding strategy because the individual classifiers are usually smaller in size than they would be in the rest of ECOC approaches, and the problems to be learned are usually easier, since the classes have less overlap. Each ECOC configuration is evaluated for three different base classifiers: Fisher Linear Discriminant Analysis (*FLDA*) with a previous 99.9% of Principal Components [16], Discrete Adaboost with 50 runs of Decision Stumps [17], and Linear Support Vector Machines with the regularization parameter $C$ set to 1 [34][28].

• *Evaluation measurements*: To evaluate the methodology we apply leave-one-patient-out classification on the Chagas data set. We also apply the Nemenyi test to look for statistical differences among the method performances [38].

## *5.1 Chagas Data Set Categorization*

We divide the Chagas categorization problem into two experiments. First, we classify the features obtained from the 107 patients considering the four groups in a leave-one-patient-out experiment for the different ECOC configurations and base classifiers. Since each patient is described with a vector of 16 features, 107 tests are performed. And second, the same experiment is evaluated over the 96 patients with

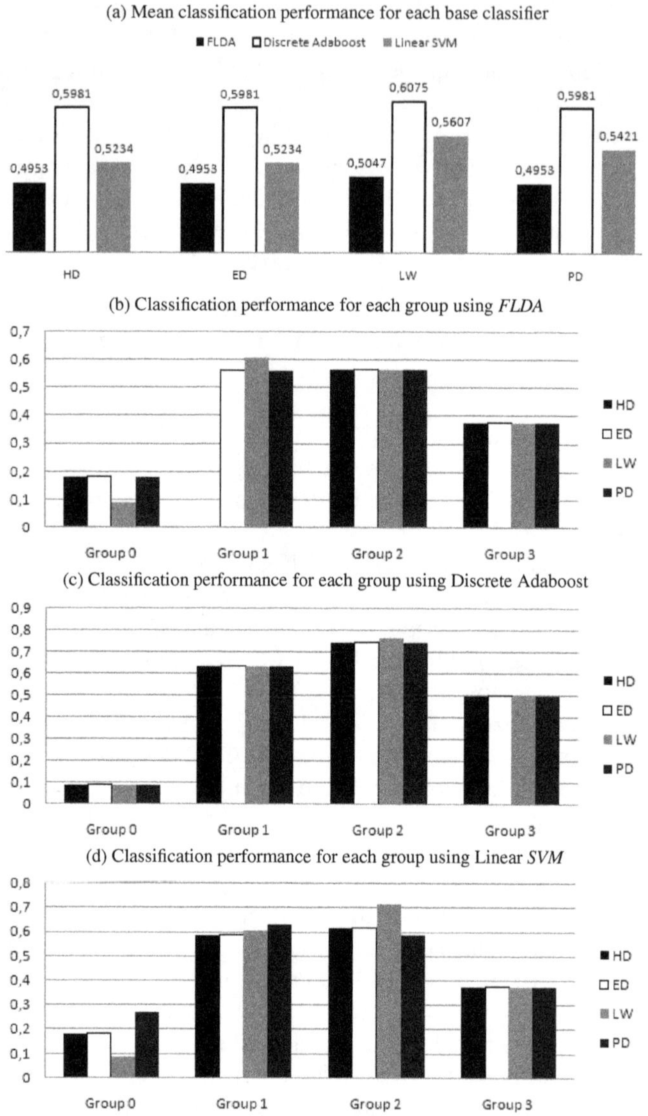


**Fig. 6** Leave-one-patient-out classification using one-versus-one ECOC design (HD: Hamming decoding, ED: Euclidean decoding, LW: Loss-Weighted decoding, PD: Probabilistic decoding) for the four groups with and without Chagas' disease.

the Chagas' disease from groups I, II, and III. This second experiment is more useful in practice since the splitting of healthy people from the patients with the Chagas' disease is solved with an accuracy upon 99.8% using the Machado-Guerreiro test.

### 5.1.1 4-Class Characterization

The results of categorization for the four groups of patients reported by [30] are shown in fig. 7. Considering the number of patients from each group, the mean classification accuracy of [30] is of 57%. The results using the different ECOC configurations for the same four groups are shown in fig. 6. In fig. 6(a), the mean accuracy for each base classifier and decoding strategy is shown. The individual performances of each group of patients for each base classifier are shown in fig. 6(b), fig. 6(c), and fig. 6(d), respectively. Observing the mean results of fig. 6(a), one can see that any ECOC configuration outperforms the results reported by [30]. Moreover, even if we use *FLDA*, Discrete Adaboost, or Linear *SVM* in the one-versus-one ECOC design, the best performance is always obtained with the proposed Loss-Weighted decoding strategy. In particular, the one-versus-one ECOC coding with Discrete Adaboost as the base classifier and Loss-Weighted decoding attains the best performance, with a classification accuracy upon 60% considering the four groups of patients.

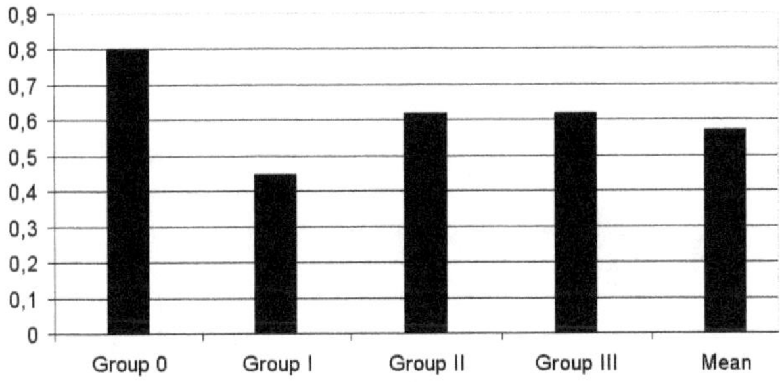


**Fig. 7** Classification performance reported by [30] for the four groups of patients

### 5.1.2 3-Class Characterization

Now, we evaluate the same strategies on the three groups of patients with the Chagas' disease, without considering the healthy people. The new results are shown in fig. 8. In fig. 8(a), the mean accuracy for each base classifier and decoding strategy is shown. The individual performances of each group of patients for each base classifier are shown in fig. 8(b), fig. 8(c), and fig. 8(d), respectively. In the mean results of fig. 8(a), one can see that independently of the base classifier applied, the Loss-Weighted decoding strategy attains the best performances. In this example, the one-versus-one ECOC coding with Discrete Adaboost as the base classifier and Loss-Weighted decoding also attains the best results, with a classification accuracy about 72% distinguishing among three levels of patients with the Chagas' disease.

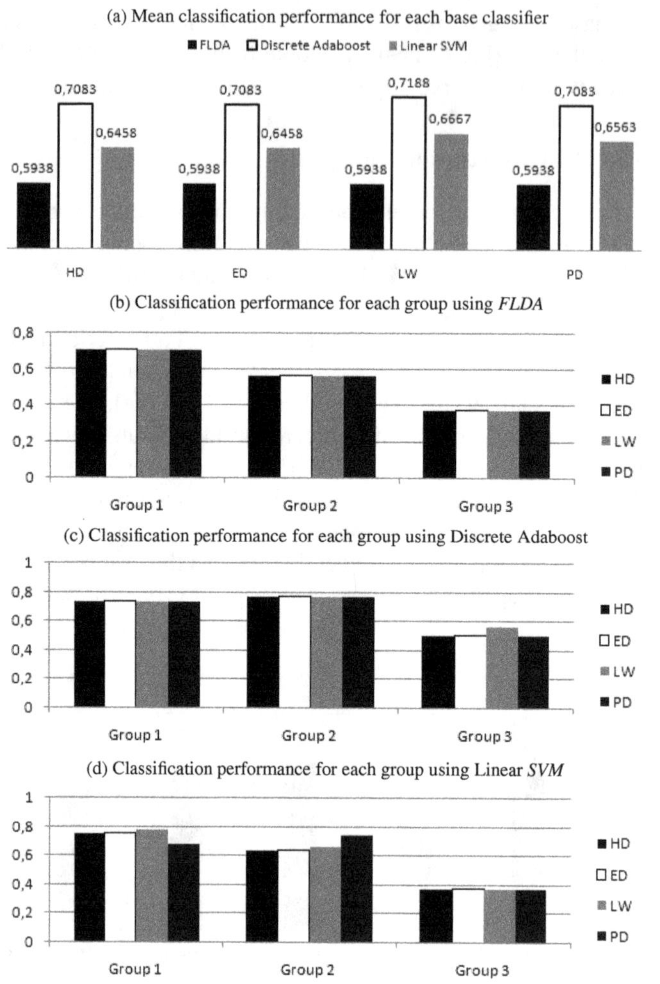


**Fig. 8** Leave-one-patient-out classification using one-versus-one ECOC design (HD: Hamming decoding, ED: Euclidean decoding, LW: Loss-Weighted decoding, PD: Probabilistic decoding) for the three groups with Chagas' disease.

In order to determine if there exists statistically significance differences among the method performances, table 2 shows the mean rank of each ECOC decoding strategy considering the six different experiments: three classifications for four classes and three classifications for the three classes given the three different base classifiers. The rankings are obtained estimating each particular ranking $r_i^j$ for each problem $i$ and each decoding $j$, and computing the mean ranking $R$ for each decoding as $R_j = \frac{1}{N}\sum_i r_i^j$, where $N$ is the total number of problems (3 base classifiers $\times$ 2 databases). One can see that the Loss-Weighted ECOC strategy attains the best position for all experiments. To analyze if the difference between methods ranks are

**Table 2** Mean rank for each ECOC decoding strategy over all the experiments

| ECOC decoding design | HD | ED | LW | PD |
|---|---|---|---|---|
| Mean rank | 3.50 | 3.50 | 1.00 | 3.33 |

statistically significant, we apply the Nemenyi test - two techniques are significantly different if the corresponding average ranks differ by at least the critical difference value (CD):

$$CD = q_\alpha \sqrt{\frac{k(k+1)}{6N}} \tag{13}$$

where $q_\alpha$ is based on the Studentized range statistic divided by $\sqrt{2}$. In our case, when comparing four methods with a confidence value $\alpha = 0.10$, $q_{0.10} = 1.44$. Substituting in eq.13, we obtain a critical difference value of 1.07. Since the difference of any technique rank with the Loss-Weighted rank is higher than the *CD*, we can infer that the Loss-Weighted approach is significantly better than the rest with a confidence of 90% in the present experiments.

## 6 Conclusions

In this paper, we characterized patients with the Chagas' disease based on the coronary damage produced by the disease. We used the features extracted using the ECG of high resolution from the heart cycle of 107 patients, and presented a decoding strategy of Error-Correcting Output Codes lo learn a multi-class system. The results show that the proposed scheme outperforms previous works characterizing patients with different coronary damage produced by the Chagas' disease (upon 10% performance improvements), at the same time that it achieves better results compared with the state-of-the-art ECOC designs for different base classifiers.

**Acknowledgements.** This work has been partially supported by the projects TIN2006-15694-C02, CONSOLIDER-INGENIO 2010 (CSD2007-00018), and the Consejo Nacional de Investigaciones Científicas y Técnicas of Argentina.

## References

1. Morris, S., Tanowitz, H., Wittner, M., Bilezikian, J.: Pathophysiological insights into the Cardiomyopathy of Chagas disease. Circulation 82 (1990)
2. da Silva Valente, S.A., de Costa Valente, V., Neto, H.F.: Considerations on the epidemiology and transmission of Chagas disease in the Brazilian Amazon. Mem. Inst. Oswaldo Cruz 94, 395–402 (1999)
3. Schofield, C.: Triatominae, Biology and Control. Eurocommunica Publications, West Sussex (1994)
4. World Health Organization (WHO). Report of the Scientific Working Group on Chagas Disease (2005), `http://www.who.int/tdr/diseases/chagas/swg_chagas.htm`

5. Rassi Jr., A., Rassi, A., Little, W.: Chagas heart disease. Clinical Cardiology 23, 883–892 (2000)
6. Madoery, C., Guindo, J., Esparza, E., Viñolas, X., Zareba, W., Martinez-Rubio, A., Mautner, B., Madoery, R., Breithardt, G., Bayes de Luna, A.: Signal-averaged ECG in Chagas disease. Incidence of late potentials and relationship to cardiac involvement. J. Am. Coll. Cardiol. 19, 324A (1992)
7. Laciar, E., Jané, R., Brooks, D.H.: Improved alignment method for noisy high-resolution ECG and Holter records using multiscale cross-correlation. IEEE Trans. Biomed. Eng. 50, 344–353 (2003)
8. Laciar, E., Jané, R.: An improved weighted signal averaging method for high-resolution ECG signals. Comput. Cardiol. 28, 69–72 (2001)
9. Allwein, E., Schapire, R., Singer, Y.: Reducing multiclass to binary: A unifying approach for margin classifiers. JMLR 1, 113–141 (2002)
10. Breithardt, G., Cain, M.E., El-Sherif, N., Flowers, N.C., Hombach, V., Janse, M., Simson, M.B., Steinbeck, G.: Standards for analysis of ventricular late potentials using high-resolution or signal-averaged electrocardiography. Circulation 83, 1481–1488 (1991)
11. Carrasco, H., Jugo, D., Medina, R., Castillo, C., Miranda, P.: Electrocardiograma de alta resolució y variabilidad de la frecuencia cardiaca en pacientes chagásicos crónicos. Arch. Inst. Cardiol. Mex. 67, 277–285 (1997)
12. Dekel, O., Singer, Y.: Multiclass learning by probabilistic embeddings. In: NIPS, vol. 15 (2002)
13. Dietterich, T., Bakiri, G.: Solving multiclass learning problems via error-correcting output codes. Journal of Artificial Intelligence Research 2, 263–282 (1995)
14. Dopico, L., Nadal, J., Infantosi, A.: Analysis of late potentials in the high-resolution electrocardiogram of patients with chagas disease using weighted coherent average. Revista Brasileira de Engenharia Biomédica 16, 49–59 (2000)
15. Escalera, S., Pujol, O., Radeva, P.: Boosted landmarks of contextual descriptors and forest-ecoc: A novel framework to detect and classify objects in clutter scenes. Pattern Recognition Letters 28, 1759–1768 (2007)
16. T. N. Faculty of Applied Physics, Delft University of Technology, `http://www.prtools.org/`
17. Friedman, J., Hastie, T., Tibshirani, R.: Additive logistic regression: a statistical view of boosting, Technical Report (1998)
18. Ghani, R.: Combining labeled and unlabeled data for text classification with a large number of categories. Data Mining, 597–598 (2001)
19. Ishii, N., Tsuchiya, E., Bao, Y., Yamaguchi, N.: Combining classification improvements by ensemble processing. In: ACIS, pp. 240–246 (2005)
20. Kittler, J., Ghaderi, R., Windeatt, T., Matas, J.: Face verification using error correcting output codes. In: CVPR, vol. 1, pp. 755–760 (2001)
21. Kong, E.B., Dietterich, T.G.: Error-correcting output coding corrects bias and variance. In: ICML, pp. 313–321 (1995)
22. Laciar, E., Jane, R., Brooks, D.H.: Evaluation of myocardial damage in chagasic patients from the signal-averaged and beat-to-beat analysis of the high resolution electrocardiogram. Computers in Cardiology 33, 25–28 (2006)
23. Laciar, E., Jane, R., Brooks, D.H., Torres, A.: Análisis de senal promediada y latido a latido del ecg de alta resolución en pacientes con mal de chagas. In: XXIV Congreso Anual de la Sociedad Espanola de Ingenieria Biomédica, pp. 169–172 (2006)
24. Maguire, J.H., Hoff, R., Sherlock, I., Guimaraes, A.C., Sleigh, A.C., Ramos, N.B., Mott, K.E., Seller, T.H.: Cardiac morbidity and mortality due to chagas disease. Prospective electrocardiographic study of a brazilian community circulation 75, 1140–1145 (1987)

25. Martinez, J.P., Almeida, R., Olmos, S., Rocha, A.P., Laguna, P.: A wavelet-based ecg delineator: evaluation on standard databases. IEEE Trans. Biomed. Eng. 51, 570–581 (2004)
26. Mora, F., Gomis, P., Passariello, G.: Senales electrocardiográficas de alta resolución en chagas. El proyecto SEARCH Acta Cientifica Venezolana 50, 187–194 (1999)
27. W.D. of Control of Tropical Diseases. Chagas disease elimination. burden and trends. WHO web site, `http://www.who.int/ctd/html/chagburtre.html`
28. OSU-SVM-TOOLBOX, `http://svm.sourceforge.net`
29. Passerini, A., Pontil, M., Frasconi, P.: New results on error correcting output codes of kernel machines. IEEE Transactions on Neural Networks 15, 45–54 (2004)
30. Pueyo, E., Anzuola, E., Laciar, E., Laguna, P., Jane, R.: Evaluation of QRS slopes for determination of myocardial damage in chronic chagasic patients. Computers in Cardiology 34, 725–728 (2007)
31. Pueyo, E., Sornmo, L., Laguna, P.: QRS slopes for detection and characterization of myocardial ischemia. IEEE Trans. Biomed. Eng. 55, 468–477 (2008)
32. Pujol, O., Radeva, P., Vitrià, J.: Discriminant ecoc: A heuristic method for application dependent design of error correcting output codes. PAMI 28, 1001–1007 (2006)
33. Hastie, T., Tibshirani, R.: Classification by pairwise grouping. In: NIPS, vol. 26, pp. 451–471 (1998)
34. Vapnik, V.: The nature of statistical learning theory. Springer, Heidelberg (1995)
35. Windeatt, T., Ardeshir, G.: Boosted ecoc ensembles for face recognition. In: International Conference on Visual Information Engineering, pp. 165–168 (2003)
36. Windeatt, T., Ghaderi, R.: Coding and decoding for multiclass learning problems. Information Fusion 1, 11–21 (2003)
37. Zhou, J., Suen, C.: Unconstrained numeral pair recognition using enhanced error correcting output coding: a holistic approach. In: Proc. in Conf. on Doc. Anal. and Rec., vol. 1, pp. 484–488 (2005)
38. Demsar, J.: Statistical Comparisons of Classifiers over Multiple Data Sets. JMLR 7, 1–30 (2006)

# Action-Planning and Execution from Multimodal Cues: An Integrated Cognitive Model for Artificial Autonomous Systems

Zenon Mathews, Sergi Bermúdez i Badia, and Paul F.M.J. Verschure

**Abstract.** Using multimodal sensors to perceive the environment and subsequently performing intelligent sensor/motor allocation is of crucial interest for building autonomous systems. Such a capability should allow autonomous entities to (re)allocate their resources for solving their most critical tasks depending on their current state, sensory input and knowledge about the world. Architectures of artificial real-world systems with internal representation of the world and such dynamic motor allocation capabilities are invaluable for systems with limited resources. Based upon recent advances in attention research and psychophysiology we propose a general purpose selective attention mechanism that supports the construction of a world model and subsequent intelligent motor control. We implement and test this architecture including its selective attention mechanism, to build a probabilistic world model. The constructed world-model is used to select actions by means of a Bayesian inference method. Our method is tested in a multi-robot task, both in simulation and in the real world, including a coordination mission involving aerial and ground vehicles.

## 1 Introduction

The rapid task-dependent processing of sensory information is among the many phenomenal capabilities of biological nervous systems. Biomimetic robotics aims

Zenon Mathews
SPECS, Institut Universitari de l'Audiovisual, Universitat Pompeu Fabra, Barcelona, Spain
e-mail: `zenon.mathews@upf.edu`

Sergi Bermúdez i Badia
SPECS, Institut Universitari de l'Audiovisual, Universitat Pompeu Fabra, Barcelona, Spain
e-mail: `sergi.bermudez@upf.edu`

Paul F.M.J. Verschure
SPECS, Institut Universitari de l'Audiovisual, Universitat Pompeu Fabra, Barcelona, Spain
Institució Catalana de Recerca i Estudis Avançats (ICREA) Barcelona, Spain
e-mail: `paul.verschure@upf.edu`

V. Sgurev et al. (Eds.): Intelligent Systems: From Theory to Practice, SCI 299, pp. 479–497.
springerlink.com 

at capturing these kinds of capabilities of biological systems to construct more advanced artificial systems. Indeed, in recent years, the control design of artificial autonomous systems have seen a shift from mere symbolic artificial intelligence (sense-plan-act) to newer concepts like embodiment, situatedness and contextual intelligence [27]. In this context, visual attention and selective attention for task-related performance have been relocated as a functional basis for behavior-based control [27, 6]. Selective attention systems usually rely on the notion of an attentional window or "spotlight of attention", which is defined as a subset of the sensory data perceived by the information processing system [21]. This "spotlight of attention" is forwarding a selected subset of sensory data to higher-order processes that plan and trigger the responses of the system. The constraints imposed by limited resources make such solutions to information bottlenecks of great interest for artificial autonomous systems. Common autonomous system tasks in robotics such as collission-avoidance, navigation and object manipulation all give a prime role of machine attention to find points of interest. At the same time, streams of multimodal sensory data provide new challenges for systems of selective attention. E.g. currently available humanoid robots such as the iCub robot have more than five sensor modalities and more than fifty degrees of movement freedom [3]. At the same time the design of novel autonomous information processing systems has seen an increasing interest in mimicking the mechanisms of selective attention observed in biological systems [20, 31, 17]. Such systems are sometimes designed to learn by acting in the real-world, under utilization of attentional strategies [20, 27]. However, autonomous system selective attention systems are still in their early stages [14, 27]. Moreover, the interplay among the different components of such complex systems are yet to be formalized.

In this context, we propose a general-purpose architecture for autonomous systems with multimodal sensors, employing biologically inspired mechanisms enabling it to alternate between volitional, top-down and reflex level actions to maintain coherence of action. Our model was inspired by the Distributed Adaptive Control (DAC) architecture proposed earlier for controlling behavioral systems both in simulations and in the real-world [29, 30]. In our experiments we explore how, for a given complex task, attention guided world model is used to perform actions that are either computed using Bayesian inference or a stochastic path planning method. A push-pull mechanism for selective attention, similar to the one hypothesized recently for the extrastriate cortex based on psychophysiological studies [22], is integrated in our model to allow for optimal data-flow between the different subsystems supporting load balancing. We propose attentional modulation of sensory data inside the DAC framework. In the next sections we lay the mathematical foundations for data association, world model building and decision making and introduce our neural network implementation for selective attention. Finally we discuss a real-world multi-robot task and a robot swarm simulation task to validate the systems performance.

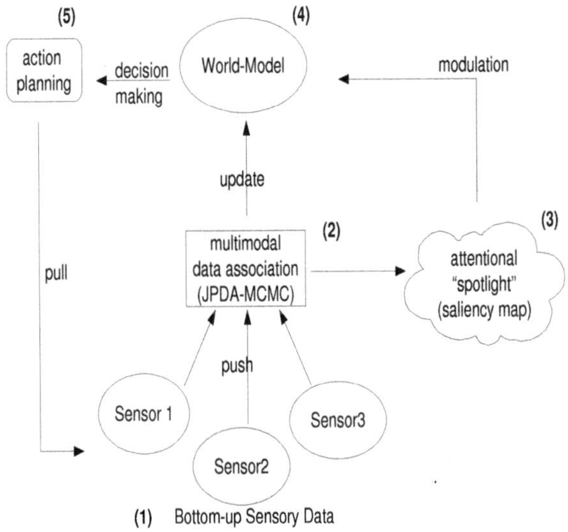


**Fig. 1** Model Architecture and the Push-Pull Data Flow: 1) Multimodal sensory data are forwarded/*pushed* by the sensors in real-time. 2) Multimodal stimuli are associated to already existing targets or new targets are created. 3) Attentional mechanism mechanism modulates the relevance of target representations in the world model depending on the current task. 4) A probabilistic representation of the relevance of the targets is maintained in the world-model. Action decision making is based on this world model and generates motor actions using a concrete action generation mechanism.

## 2 Methods

### 2.1 *Model Architecture*

Our model is capable of filtering the currently relevant information for a given task from the multimodal sensory input and then select an optimal action in the Bayesian fashion, thereby updating its existing world model. The bottom-up multimodal sensory data is continuously *pushed* by the individual sensors to the data association mechanism, which associates the multimodal stimuli to already existing targets or creates new targets. The result of this is forwarded to the world-model but also to the saliency computation module. In parallel the goal-oriented attentional spotlight generation modulates the relevance of target representations in the world model so that depending on the current task the representation of relevant targets is enhanced. In the world model the relevance of the individual targets is represented probabilistically by means of Gaussian distributions. The decision making module operates on this world model and selects motor actions which are then sent to a planning process.

## 2.2 Managing Bottom-Up Multimodal Sensory Data

The question we address in this section is how an autonomous entity manages the amount of multimodal sensory information it receives continuously. In the following, *stimulus* refers to a single modal observation (or data unit) and *target* means a well-defined physical object that also exists in the same space as the autonomous entity. Targets are perceived by the autonomous entity through the multimodal stimuli they evoke. In this section we discuss the so called data association (or data alignment) problem.

### 2.2.1 Joint Probabilistic Data Association

JPDA has been successfully used for solving data association problems in a various fields such as computer vision, surveillance, mobile robots, etc. JPDA is a single-scan approximation to the optimal Bayesian filter, which associates observations to known targets sequentially. JPDA thereby enumerates all possible associations between observations and targets at each time step and computes the association probabilities $\beta_{jk}$, which is the probability that the $j$-th observation was caused by the $k$-th target. Once such association probabilities can be computed, the target state can be estimated by Kalman filtering [9]. Such a conditional expectation of the state is weighed by the association probability. In the following, let $x_t^k$ indicate the state of target $k$ at time step $t$, $\omega_{jk}$ the association event where the observation $j$ is associated to target $k$ and $Y_{1:t}$ stays for all the observations from time step 1 to time step $t$. Then the state of the target can be estimated as

$$E(x_t^k|Y_{1:t}) = \sum_{\omega} E(x_t^k|\omega, Y_{1:t})P(\omega|Y_{1:t}) \tag{1}$$

$$= \sum_{j} E(x_t^k|\omega_{jk}, Y_{1:t})P(\omega_{jk}|Y_{1:t}) \tag{2}$$

where $\omega_{jk}$ denotes the association event where observation $j$ is associated to target $k$ and $\omega_{0k}$ denotes the event that no observation is associated to target $k$. Therefore the event association probability is

$$\beta_{jk} = P(\omega_{jk}|Y_{1:t}) \tag{3}$$

JPDA uses the notion of a validation gate and only observations inside the validation gate for each target are considered. A validation gate is computed for each target using the kalman innovation of new observations. For further mathematical details of JPDA see [9].

$\beta_{jk}$ can be computed by summing over the posterior probabilities and the exact calculation is NP-hard, which is the major drawback of JPDA [15]. This is due to the fact that the number of association events rise exponentially in relation to the number of observations. We therefore implemented a Markov Chain Monte Carlo method to compute $\beta_{jk}$ in polynomial time [23] similar to the proposal by Oh and Sastry in [26].

The Markov Chain Monte Carlo (MCMC) method is used in our system to estimate the association event probabilities $\beta_{jk}$ in polynomial time and with good stability. We only consider feasible mappings from data to target, i.e. the ones that respect the validation gate criteria for the JPDA. The algorithm starts with one such feasible mapping and a Markov chain is generated. MCMC is used for computing $\beta_{jk}$ in real-time as its time complexity is polynomial with respect to the number of targets. For details of the MCMC approximation of $\beta_{jk}$, its convergence and stability see Oh and Sastry [26]. The stimuli have to be associated to existing targets, or if the stimuli is spatiotemporally distant from existing targets, i.e. outside all validation gates, new targets have to be created.

## *2.3 Goal Oriented Selective Attention*

Biological nervous systems are still intriguingly superior to what can technically be implemented today for sensory data processing. For instance, recent research has shown that the retina transmits between one and ten million bits per second, which is about the same rate as an Ethernet connection could support, to the brain [4]. Here we explore how attentional selection can add functional advantages to behavioral systems that deal with large amounts of sensory data. In particular, we consider here goal-oriented top-down selective attention as an information bottleneck that filters the most relevant sensory data, depending on the current task of the system [5, 22]. Such an information bottleneck, that changes dynamically with the system's task, is critical for the survival of biological organisms as the incoming sensory data clearly overwhelms the available limited computational resources. Psychophysiological research suggests that the selective attention is load-dependend, i.e. how many unattended stimuli are processed depends on the degree to which attentional resources are engaged by an attended stimulus [**?**]. This delivers evidence for a load-dependent push-pull protocol of selective attention operating at intermediate processing stages of the sensory-data. Such a push-pull protocol has behavioral effects for an autonomous system: when the attentional load is low the system can allocate motor and computational resources for unattended targets. Our architecture makes use of this load-dependend push-pull mechanism and this allows the acting system to switch between volitional, reflexive and explorative behaviors.

For the implementation of the selective attention mechanism we use the IQR system for distributed large-scale real-time real-world neuronal simulations [10]. IQR allows implementing large neural networks for real-time applications and interfacing them to real world devices [8, 28]. As suggested by Itti and Koch [21], we implemented a set of neuronal feature filters with excitatory, inhibitory and time-delayed connections between them for the computation of salient points. The feature filters is modulated by the current state of the system. E.g. if the system is running out of power, the feature filters for the charger have stronger excitatory influence on the salience computation. This computation delivers goal-dependend salient target locations which are a subset of the total number of targets the data association mechanism has computed before (see figure 2). In the experiment section examples of such feature filters are discussed in more detail.

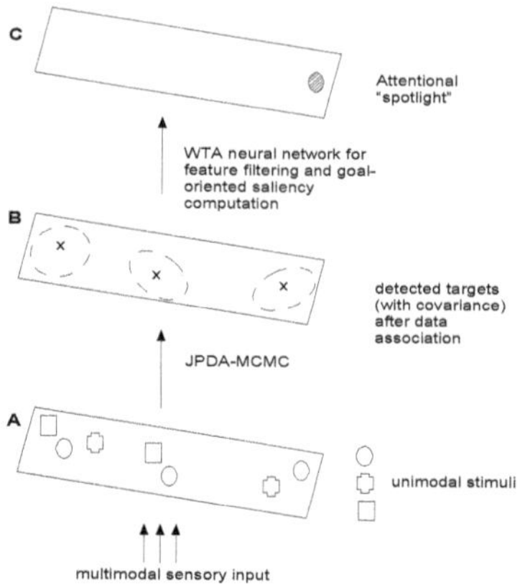


**Fig. 2** From multimodal sensory input to attentional saliency: The multimodal sensory input **(A)** to the JPDA-MCMC algorithm is associated to targets in the world-model(**B**). The goal-dependend saliency computation filters output the most salient targets depending on the current state of the system and the task at hand **(C)** .

## 2.4 *World Model and Decision Making*

The world model of an attention-guided behaving system should ideally consist of the targets it attends to, but also the unattended targets. Such a world model, or dynamic memory, allows the system to plan its actions depending on the top-down attentional load and the bottom-up sensory input. In this section we discuss the building, maintenance and the use of such a world model for decision making.

The data association and the attentional mechanisms deliver a constant input to the world model. Our world model contains the spatial and temporal information of a total set of targets with the attended ones being represented more relevantly than the unattended ones. We define $\Theta_s^t$ as the relevance of a certain target $s$ at time $t$, and we are interested in the following conditional probability:

$$P(\Theta_s^t | F_s^t(\Theta_s^{t-1}) A_t(s)) \tag{4}$$

where $F_s^t(\Theta_s^{t-1})$ and $A_t(s)$ are two time-dependend functions which weigh the target $s$. For example $F_s^t(\Theta_s^{t-1})$ evaluates the spatial proximity of the target if there is at least one onset stimulus associated to this target and decays the current weight of the target otherwise. Whereas $A_t(s)$ evaluates the goal-dependend attentional saliency of this target. By computing the joint distribution of these relevance probabilities for all targets $s$ the system can perform the motor action appropriate for the most relevant targets. The following subsection elaborates the update of these relevance probabilities.

Let us assume that we can compute relevance probabilities of individual targets as shown above in eq. 4. Given these individual target relevances we are interested in the fused relevance distribution:

$$P(\Theta^t | F^t(\Theta^{t-1}) A_t) \tag{5}$$

We express this probability as the normalized sum of probabilities of individual relevances:

$$P(\Theta^t | F^t(\Theta^{t-1}) A_t) = \sum_S P(s) P(\Theta_S^t | F_S^t(\Theta_S^{t-1}) A_t(s) S) \tag{6}$$

where random variable $S \in 1...n$, $n$ being the number of targets and $P(s)$ indicates the probability of this target. As $P(s)$ is uniformly distributed over all targets this allows for normalization.

Given the relevance distribution a decision that is optimal in the Bayesian sense is computed. We are therefore interested in the following probability distribution of action:

$$P(Action | F_S^t(\Theta_S^{t-1}) A^t(s)) \tag{7}$$

where $F$ implements a time-dependend decay function for utility. This probability distribution can be computed using the Bayesian rule, given apriori information about the environment the autonomous system is acting in.

## 2.5 *Action Planning and Execution*

### 2.5.1 Bayesian Optimal Action

The world model of an attention-guided behaving system should ideally consist of the *items* it perceives at the moment, but also possibly *items* perceived in the past [18, 13, 12]. We formulate an optimal Bayesian decision making method for generating actions based on a transient/dynamic memory that allows the system to plan its actions depending on current and past stimuli.

Multimodal stimuli from different sensors of the autonomous system is associated using the JPDA method discussed above. This method creates *items* in a memory of which the utility probability is computed. In the following we derive the equations from general equations 4, 5 and 6.

Let us assume that the motor action consists simply of choosing a direction of motion $\gamma \in 0..360$ and a travel distance $\psi \in 1..10$. The best action is then chosen in the direction $\gamma$ of the most relevant *item* at distance $\psi$ in the world-model. As in the general equation 7, here are interested in computing the most relevant direction of motion $\gamma$ and distance $\psi$. Therefore we are interested in the probability:

$$P(\gamma\psi | F_S^t(\Theta_S^{t-1}) A^t(s)) \tag{8}$$

As $F$ is a function of distance $d$ and time $t$ we can express $F$ as the known distance $d$ to the *item*, the time $t$ since the time of a previous stimulus associated to this *item*, the relative orientation $\gamma_i$ and the attentional weights $a_i$ for each *item*. For $n$ *items* we formalize the above probability as:

$$P(\gamma\psi|d_1,...d_nt_1,...t_n\gamma_1,...\gamma_na_1,..,a_n) \tag{9}$$

We consider the conditional probability 9 as if there were only one *item i* and without attentional inputs $a_i$. Supposing conditional independence for angle and distance domains we do the following decomposition:

$$P(\gamma\psi|d_it_i\gamma_i) = P(\gamma|d_it_i\gamma_i)P(\psi|d_it_i\gamma_i)P(t_id_i\gamma_i)$$

We formulate the probability distributions $P(\gamma|d_it_i\gamma_i)$ and $P(\psi|d_it_i\gamma_i)$ as Gaussian distributions:

$$P(\gamma|d_it_i\gamma_i) = \mathcal{N}(\gamma_i, \frac{d_it_i}{c_1}) \tag{10}$$

where the Gaussian is centered on the angle $\gamma_i$ at which the *item i* is located. The standard deviation is a function of time $t_i$ at which this *item* was last perceived and the distance $d_i$ at which this *item* is. This allows gradualforgetting (time decay) what has been perceived in the past, as past information is always prone to changes in a dynamic world.

Similarly for the distance domain, the Gaussian is centered on the distance $d_i$ of the *item* and the standard deviation is again a function of time $t_i$, allowing a time decay.

$$P(\psi|d_it_i\gamma_i) = \mathcal{N}(c_2d_i, c_3t_id_i) \tag{11}$$

And we assume the uniform distribution for the joint probability $P(t_id_i\gamma_i)$ as we do not have any prior information about possible correlations between those random variables.

$$P(t_id_i\gamma_i) = \mathcal{U} \tag{12}$$

where $c_1$, $c_2$ and $c_3$ are constants.

We now take the utilities of all the *items* into account for the computation of the total utility as shown in equation 6. We include the attentional components $a_i$ and consider the following conditional probability distribution:

$$P(\gamma\psi|d_1,...d_nt_1,...t_n\gamma_1,...\gamma_na_1,..,a_n) = \tag{13}$$

$$\sum_i \frac{a_i}{a_{tot}} P(\gamma|d_it_i\gamma_i)P(\psi|d_it_i\gamma_i)P(t_id_i\gamma_i) \tag{14}$$

where $a_{tot}$ is the sum of all attentional components $a_i$, which are the attentional saliencies for individual *items* depending on their detected remaining charge.

According to this formulation the attentional components $a_i$ weigh the shares of the individual *items* to the joint conditional probability distribution, i.e. attention modulates the world model, which is expressed as a probability distribution that changes in each step with the sensory input.

### 2.5.2 Stochastic Action Execution

In order to navigate through an arbitrary environment, an autonomous robot should make use of the knowledge about the environment. We evaluate our model in a testbed which contains dangerous objects such as mines and obstacles, where the autonomous robot should navigate from a given point A to point B avoiding the obstacles. In our approach, we decided to employ a metric or grid-based approach [19, 25]. In this framework, each cell in a grid represents a part of the field. Cells in an occupancy grid contain information about the presence of an obstacle. The cells are then updated relying on the information received by sensors. The value associated to a cell represents the degree of belief of the presence of an obstacle. However, in our case we had to extend this environment representation and adapt it to our needs. In fact, as previously explained, we have to deal with some issues. In particular, our environment is characterized by:

1. A high degree of uncertainty regarding the positions of both the robot and of the obstacles due to the position information resolution
2. Obstacles can be added on the fly provided updated information
3. We need to decide the cruise speed of the robots within each cell according to the probability of finding obstacles

Hence, instead of associating to each cell the likelihood of a cell being an obstacle, we associate to it the probability of colliding with an obstacle/mine in the part of the field represented by the cell. We subsequently employ such a probability to control the speed of the autonomous agent.

We are interested in reaching a known goal position while minimizing the probability of incurring in obstacles and maximizing exploration to improve the knowledge about our environment. Nevertheless, we have to consider two important factors inherent to our problem. The former is that that the available time to complete a task is limited, and the latter is that it may not exist a path not containing dangerous zones. Therefore, the two objectives of shortening the path to the solution, and minimizing the probability of incurring in mines could be in contrast. Hence, the output of our planning algorithm should be a sufficiently short path that reduces as much as possible the probability of entering into dangerous zones.

In order to implement the path-finder we employed a variation of a stochastic hill-climbing (or stochastic gradient decent) [11] algorithm boosted via a taboo search technique [16]. These algorithms are guided by a heuristic that combines the danger probability with the distance to the goal. In particular, the value of the heuristic function at each point of the grid is a weighted sum of the probability of encountering mines with the distance to the goal. This allows combining the fact that we aim at avoiding dangerous zones while not increasing too much the length of the path.

The Stochastic Hill-Climbing (SHC) algorithm works as follows. Departing from an initial state (a cell in our grid) a set of possible successors states are generated. Each of such successors has associated a value of the heuristic function estimating how good such a state is (how close to the goal and how dangerous it is).

However, the HC algorithm suffers from the limitation that local minima are very likely to produce an infinite loop. In order to overcome such a limitation SHC has been introduced. In SHC the successor that is chosen is not always the one minimizing a given heuristic function, but rather there is a given probability that the second-best, third-best, or one of the others is selected. This algorithm continues until the goal is reached. In this way, most of the local minima can be escaped and a random exploratory behavior is introduced in the path planning. Nevertheless, broader local minima are still a big problem. This may happen when we move on states that are cyclically connected. The mechanism underlying taboo search keeps track of some of the states that have already been visited and marks them as taboo, that is forbidden.

## *2.6 Test Scenarios*

### 2.6.1 A Combined Micro-Aerial (MAV), Unmanned Ground Vehicle (UGV) and Human Rescue Mission

In this testbed we test

multirobot coordination, construction of a world-model from local perceptions and its online update using a complex human-robot interactive system for a real-world task.

In this testbed we use a heterogeneous group of UGVs and MAVs in which multiple robots are used to create a shared knowledge base about the world they are interacting in. The mission plan will consist of a sampling phase in which robots equipped with camera systems (UGV and MAV) will be driven to sample the environment and gather information. The cognitive system will generate the plan instructions to be executed autonomously by the UGVs that can also be monitored and manually updated using the world model interface and 3D representation. The system allows for an eventual manual control of all robots.The MAV is a commercial quadcopter made by Ascending technologies, Germany (Fig 1). It consists of a cross like structure with four independent motor controllers that run 4 propellers that provide the necessary lift force during approx. 30 minutes continuous operation. The total weight of the MAV is about 600 grams with the battery pack. The quadcopter has been equipped with an additional wireless camera system for remote pilotage and inspection. With the additional 150 grams of the camera system, the autonomy is reduced to about 10-15 minutes. The range of the wireless video link is approx. 800 meters. However, a UGV has been designed as a mobile repeater station for all video signals providing additional 800 meters. On the base, a pilot, using a Head Mounted Display system, controls the robot from the image provided from its camera. The MAV can be remotely turned off during flight operation, turning it ballistic on demand.

Two custom made tracked robots (50 x 30 x 20) and a standard RC wheeled vehicle (50 x 40 x 30), equipped with wireless camera systems constitute the Unmanned Ground Vehicle (UGV) support team. The tracked robots are equipped with standard GPS, compass, Metal-thin oxide chemo-sensors (supplied by Alpha MOS SA, France) and ultrasonic sensors to provide way-point based navigation, and to generate a world model and planning (cognitive layer) (fig.2).

The communication among all robots goes through the automatically generated world map. This map lives in a base station, which is used to communicate via a radio link and instruct the different robots. Additionally, the world model has a user interface that allows the operators to contribute to by adding supplementary information such as about obstacles and mines that will be taken into account by the cognitive system while generating the path planning.

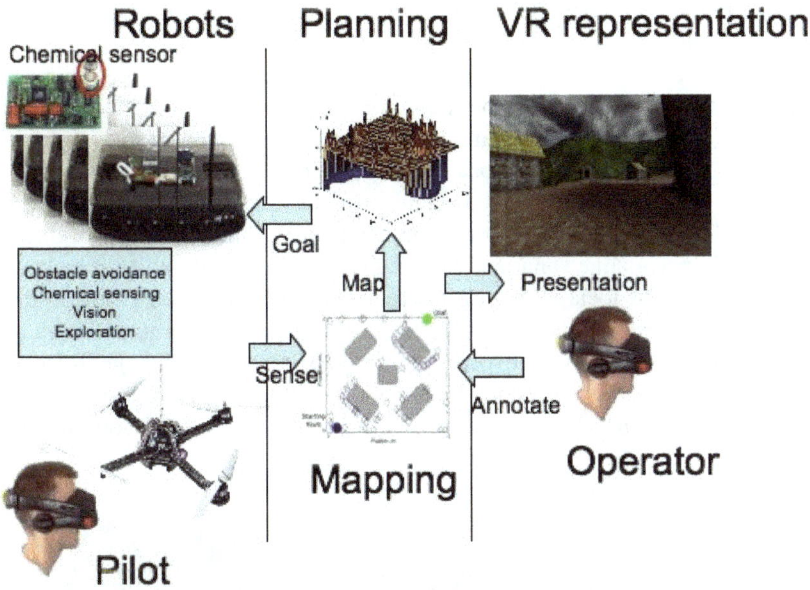


**Fig. 3** Grand scheme of the integrated autonomous dynamic mapping and planning approach. In this task, robots are used as autonomous dynamic sensing platforms that contribute information to a central mapping and planning stage (cognitive system). The planning stage defines goal positions the robots should attempt to reach using their local proximal sensory-motor capabilities, e.g. collision avoidance, mine detection, etc. The aerial vehicle is guided by a human pilot, and the information gathered by this method is added to the world model. The state of the world model is transformed into a 3D representation of the task area. To this representation objects and terrain features are added in real-time. The human operator inspects the 3D model and makes decisions on future actions while making his own annotation in the language of the virtual world. See text for further explanation.

The whole mission and world model status is online represented on a 3D model of the mission that allows the operators to freely navigate through the whole field and have a remote visualization of the scenario from the base station.

### 2.6.2 Simulation of Robotic Swarm for Rescue Missions

The objective of this testbed is to test

world-model construction and the exploitation of the world-model for robot coordination and action generation that is optimal in the Bayesian sense.

In testbed 1 we tested the capability of our model for multirobot coordination, creating and maintaining a world model, whereas here we specifically test how from local perception a world-model can be constructed, which is subsequently used to generate actions, which are optimal when solving a multiple goal task under limited resources constraints. For this purpose, we consider the following robot swarm scenario. A swarm of robots are on a common mission in a given environment. A rescue robot is equipped with our model, and is involved in the specific task of aiding expired (i.e. broken-down or out of charge) agents. This means that the rescue robot first has to localize the expired agents using its sensors and approach them for repair or recharge. The rescue robot is equipped with a limited number of distance-measurement sensors like sonar and laser range scanners, with which it has to scan the environment and localize the agents to accomplish the given task. From time to time, also the rescue robot has to go back to the base station to recharge itself. Solving this multiple goal task involves multimodal data association, goal-driven selective attention generation for attending to the most vital subtask at the moment and maintaining a dynamic world model, which is used to compute the optimal action

**Fig. 4** The MAV and UGV platforms. **(A)** Quadcopter (Ascending Technologies) [1] and **(B)** custom build tracked ground vehicle. A wireless link connects the robotic platforms to the computing unit and/or human operators. We incorporated a camera, a GPS, ultrasonic sensors, a LI-PO battery, a compass and chemo-sensors. See text for further explanation. **(C)** A group of mini UGVs used for indoor testing of the multi-robot mission. We use the EPuck robot which features multimodal sensors and wireless communication [2].

in the Bayesian sense. We implemented this testbed in a simulation with $N$ robots and $M$ sensors for the rescue robot.

*items* in this testbed are robots involved in the common mission. *Attentional saliency* here is proportional to the detected remaining power of this *item*, giving a high utility for reapproaching nearly expired *items*. The rescue robot computes the overall utility distribution of going in a certain angle and distance from the individual utilities of *items*. We simulate the data from different range sensors and use this as multimodal stimuli to the rescue robot. This creates *items* of which the utility probability has to be computed.

# 3 Results

## 3.1 Static World, Multirobot Coordination

In the first testbed we look at a static real-world environment which has to be explored using multiple robots. As discussed earlier we have outdoor ground and aerial vehicles for an exploratory mission involving avoiding dangerous mines. Multirobot collaborative exploration is achieved using our model. First we test the multirobot mission using the e-Puck robots in an indoor environment which allows scaling and testing of individual components of our model. We setup a 2 x 2 meter arena where a single robot was said to reach a feeder located at the center position from randomly starting points. Thus we measured the mean positioning error resulting from the PID controller after 10 runs, being it about 4 cm. Consequently, some objects where placed in the arena to obstruct the direct path from the starting point to the goal position, figure 5. In this case, both the PID and the obstacle avoidance (the reactive layer of the Distributed Adaptive Control) were necessary to accomplish this task. We employed multiple robots to perform this task. The results show that the multirobot autonomous control permitted the robot to reach its goal in all cases. Nevertheless, the resulting paths were not the optimal ones as evidenced by the run durations (median 170 seconds). Additionally, we observe that the gain of using a number of robots to explore the environment does not report very good results if there is no strategy on how to use the acquired information and how can this be share among robots.

In order to improve the performance of both exploration and goal oriented behavior, we implemented the previously described autonomous control architecture. Preliminary tests were done by letting a single robot explore the environment. While the robot was performing the task, the world model was created with its sensory information (proximity sensors) and was also improved online with new information. The then generated world model contains in this case the detected contours of elements within the test environment, figure 6. This information is therefore very valuable in order to plan strategies at collective of individual level, which are then executed by the group of robots controlled by the multirobot autonomous control system.

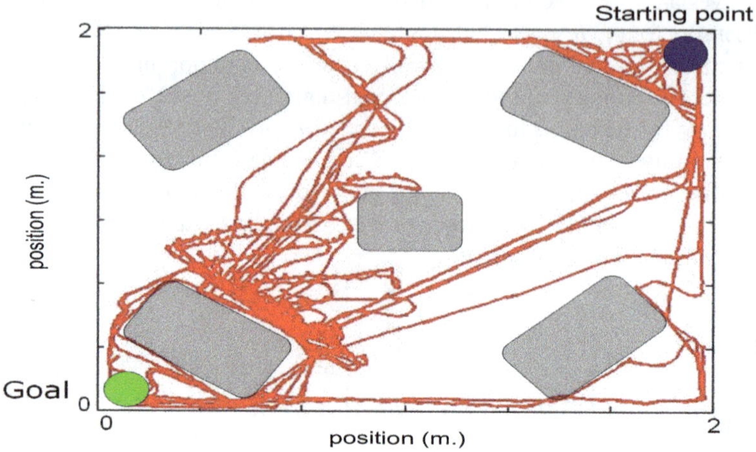


**Fig. 5 Multirobot runs:** Traces of multiple robots and runs under the control of the robot autonomous layer. The robots are released at the starting point and their autonomous (PID + reactive) layer allow them to reach their goals in the presence of obstacles.

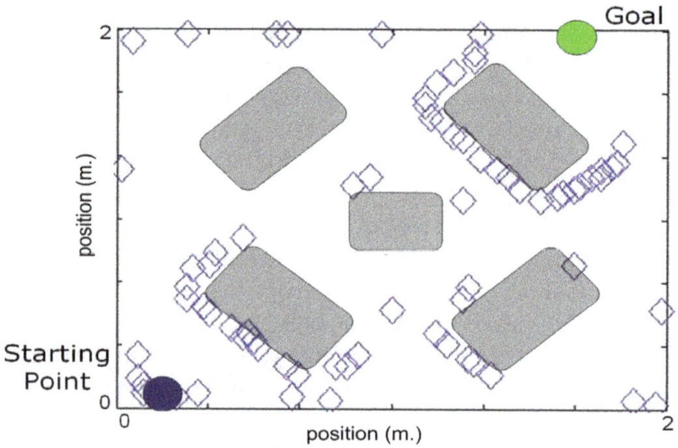


**Fig. 6 World model generation from multirobot coordination:** Generated world model resulting of the goal oriented and exploratory behavior of a robot after a number of test runs. Goal, robot position and objects are represented.

We tested the reliability of the world model with the path planning system in order to generate routes through the test arena. The experiment consisted in this case of a goal oriented exploration of the environment, in which a robot had to reach a goal. Every run the robot was performing the task, the multirobot autonomous model was improving its world model at the same time the planning could generate more reliable paths through the arena, figure 7A. The results show how over runs, the length of the robot trajectory gets shorter and closer to an optimum, figure 7B. In the case of multiple robots, the generation of the world model and therefore the planning strategy would improve even faster since all the robots would collaborate by contributing with their local sensory information to the multirobot autonomous control architecture.

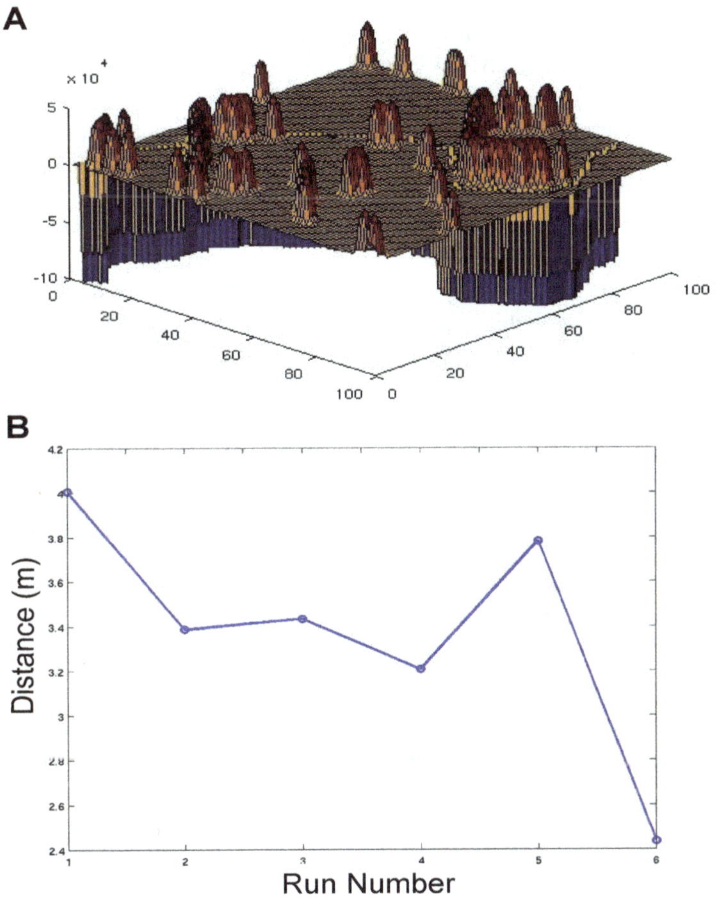


**Fig. 7 Exploration and Path Planning. A) Online generated robot trajectory:** The positive values indicate the cost related to go through a specific position of the arena, the computed trajectory is represented by negative values. **B) Task performance vs number of runs:** evolution of the performance of the planning with the number of test runs. The decrease in the traveled distances shows that the world model is more complete and accurate, and therefore it results in a more optimal robot path.

### 3.2 Dynamic World, Partial Perception

Subsequently we investigated how the system behaves in a dynamic environment using a single agent. The partial knowledge acquired by the agent is then used to construct the world-model and robot coordination and world-model exploitation in a robot swarm simulation is evaluated. We used 10 simulated agents who move around the experimental field of a 600 by 600 m. window, who start with a maximum speed and maximum energy but slow down as the energy drops. The energy-drop is proportional to the covered distance. Their direction of motion is arbitrary but always inside the arena. The rescue robot, controlled by our model, always starts from the base station and alternates between exploration and exploitational time slots. During exploration it moves about randomly in the field to detect the agents. Thereby the multimodal sensor fusion and attentional saliency computation delivers input to update its world model. During the exploitation time slot, the rescue robot performs the intelligent motor actions as described earlier. We compare the performance of the rescue robot using the world model in the exploitation phase with a system when not using it. When the world model is not used to compute an intelligent action, the rescue robot is in constant exploration. For each category 5 trials each with 5000 time-steps were carried out. A probabilistic world model computed as is shown in figure 8.

To assert the performance of the system, we evaluate the number of recharged agents during each trial and also at the total expiry time of all agents together in each run and observe a significant improvement when using the probabilistic world model and motor action selection. WM indicates the use of world model and non-

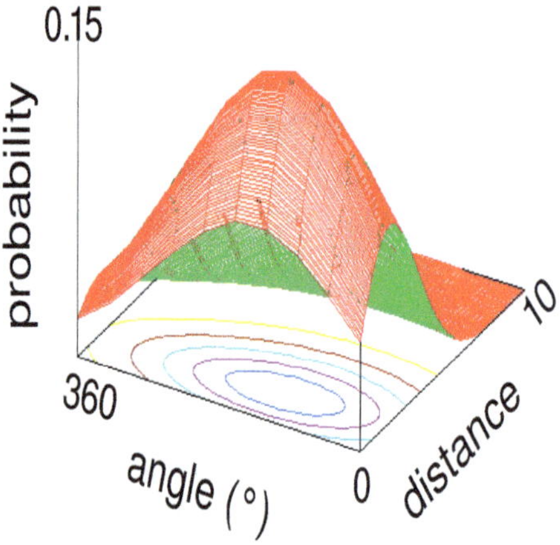


**Fig. 8 World model probability manifold example:** Angles range from 0..360 and distances range from 1 to 10. Higher salient experiences are represented with higher probabilities. This world model suggests the most probable action as the one that leads to the expired agents, which were perceived to be running slow in the past. This probability distribution is computed at each time step before an intelligent motor action decision is made.

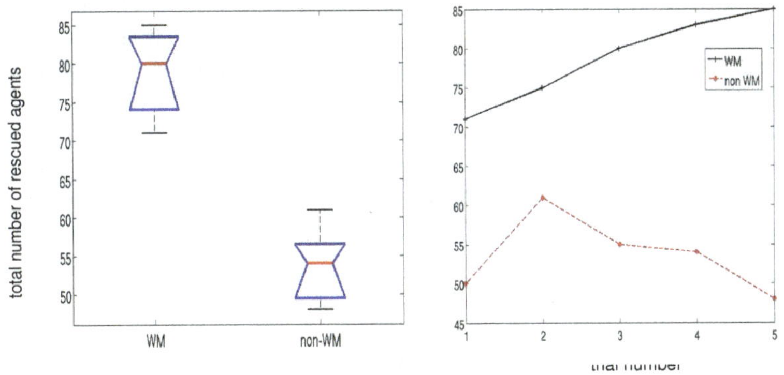


**Fig. 9 Number of recharged agents:** The use of our autonomous system control with the world-model achieves higher number of recharged agents in all trials when compared to a system that does not possess a world model.

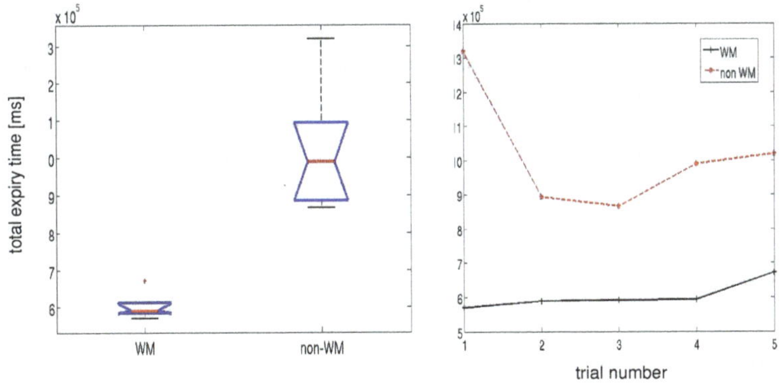


**Fig. 10 Total expiry time of agents:** The use of our autonomous system control with the world-model achieves much less expiry time of agents in all trials when compared to a system that does not possess a world model.

WM indicates the use of a reactive system that explores the robot arena without a world-model or attentional mechanisms; see figures 9 and 10.

## 4 Conclusions

We have proposed an integrated model of multimodal data association, attentional saliency computation, world-model construction and maintenance and action selection for artificial autonomous systems. Our model is based on biological principles of perception, information processing and action selection, and is incorporated in the Distributed Adaptive Control framework for automated control. Our model suggests how the different subsystems of an artificial autonomous system can interplay seamlessly in an integrated framework. We demonstrated the use of our model in a multirobot coordination task, where a common world-model for the multiple robots is created, maintained and used to compute optimal actions for the individual robots. We have shown how to generate trajectories for individual robots using a multirobot exploration of the environment and how the performance of the system augments with increasing exploration. The first testbed was for a static environments for multirobot collaboration. In the second testbed we evaluated the possibility of computing a global world-model from local perceptions in a robot swarm experiment where we used a dynamic, partially visible environment. Selective attention mechanisms are employed to focus the information processing capacities on the currently most relevant task. We have shown that our model performs significantly better than a system without a world-model in the given rescue mission. The modularity of our architecture allows for customizing the individual components of the model for the given task. In further work we will evaluate the capability of the model for the control of various autonomous systems such as our insect inspired robotic navigational model [24], and for humanoid robot control using the iCub robot [3].

**Acknowledgements.** The authors wish to thank Fabio Manzi, Ramón Loureiro, Andrea Giovanucci, Ivan Herreros, Armin Duff, Joan Reixach Sadurní, Riccardo Zucca, and Wendy Mansilla for the joint work on the multirobot coordination project. This work is supported by the European PRESENCCIA (IST-2006-27731) and EU SYNTHETIC FORAGER (FP7-217148) projects.

## References

1. `http://www.asctec.de`
2. `http://www.e-puck.org/`
3. `http://www.robotcub.org`
4. How much the eye tells the brain. Current Biololgy 16, 1428–1434 (2006)
5. Search goal tunes visual features optimally. Neuron 53, 605–617 (2007)
6. Arkin, R.: Behavior-based robotics (1998)
7. Bermúdez i Badia, S., Manzi, F., Mathews, Z., Mansilla, W., Duff, A., Giovannucci, A., Herreros, I., Loureiro, R., Reixac, J., Zucca, R., Verschure, P.F.M.J.: Collective machine cognition: Autonomous dynamic mapping and planning using a hybrid team of aerial and ground based robots. In: 1st US-Asian Demonstartion and Assessment of Micro-Aerial and Unmanned Ground Vehicle Technology (2008)
8. Bermúdez i Badia, S., Pyk, P., Verschure, P.F.M.J.: A biologically based flight control system for a blimp-based UAV. In: Proceedings of the 2005 IEEE International Conference on Robotics and Automation, ICRA 2005, pp. 3053–3059 (2005)
9. Bar-Shalom, Y., Fortmann, T.E.: Tracking and data association. Academic Press, Boston (1988)
10. Bernardet, U., Blanchard, M., Verschure, P.F.M.J.: Iqr: A distributed system for real-time real-world neuronal simulation. Neurocomputing, 1043–1048 (2002)
11. Blum, C., Roli, A.: Metaheuristics in combinatorial optimization: Overview and conceptual comparison. ACM Comput. Surv. 35(3), 268–308 (2003)

12. Botvinick, M.M., Plaut, D.C.: Short-term memory for serial order: a recurrent neural network model. Psychological review 113(2), 201–233 (2006)
13. Byrne, P., Becker, S., Burgess, N.: Remembering the past and imagining the future: a neural model of spatial memory and imagery. Psychological review 114(2), 340–375 (2007)
14. Coelho, J., Piater, J., Grupen, R.: Developing haptic and visual perceptual categories for reaching and grasping with a humanoid robot. Robotics and Autonomous Systems 37, 195–218 (2001)
15. Collins, J.B., Uhlmann, J.K.: Efficient gating in data association with multivariate gaussian distributed states. IEEE Transactions on Aerospace and Electronic Systems 28(3), 909–916 (1992)
16. Cvijovicacute, D., Klinowski, J.: Taboo search: An approach to the multiple minima problem. Science 267(5198), 664–666 (1995)
17. Dickinson, S.J., Christensen, H.I., Tsotsos, J.K., Olofsson, G.: Active object recognition integrating attention and viewpoint control. Computer Vision and Image Understanding 67, 239–260 (1997)
18. Dominey, P.F., Arbib, M.A.: A cortico-subcortical model for generation of spatially accurate sequential saccades. Cerebral cortex (New York, N.Y.: 1991) 2(2), 153–175 (1992)
19. Elfes, A.: Using occupancy grids for mobile robot perception and navigation. Computer 22(6), 46–57 (1989)
20. Billock, G., Koch, C., Psaltis, D.: Selective attention as an optimal computational strategy. Neurobiology of Attention, 18–23 (2005)
21. Itti, L., Koch, C.: Feature combination strategies for saliency-based visual attention systems. Journal of Electronic Imaging 10(1), 161 (2001)
22. Pinsk, M.A., Doniger, G., Kastner, S.: Push-pull mechanism of selective attention in human extrastriate cortex. Journal of Neurophysiology 92 (2004)
23. Mathews, Z., Bermúdez i Badia, S., Verschure, P.F.M.J.: Intelligent motor decision: From selective attention to a bayesian world model. In: 4th International IEEE Conference on Intelligent Systems, vol. 1 (2008)
24. Mathews, Z., Lechón, M., Calvo, J.M.B., Dhir, A., Duff, A., Bermúdez i Badia, S., Verschure, P.F.M.J.: Insect-like mapless navigation using contextual learning and chemo-visual sensors. In: IEEE/RSJ International Conference on Intelligent Robots and Systems, IROS 2009 (2009)
25. Moravec, H.: Sensor fusion in certainty grids for mobile robots. AI Mag. 9(2), 61–74 (1988)
26. Oh, S., Sastry, S.: A polynomial-time approximation algorithm for joint probabilistic data association, vol. 2, pp. 1283–1288 (2005)
27. Paletta, L., Rome, E., Buxton, H.: Attenion architectures for machnine vision and mobile robots. Neurobiology of Attention, 642–648 (2005)
28. Pyk, P., Bermúdez i Badia, S., Bernardet, U., Knüsel, P., Carlsson, M., Gu, J., Chanie, E., Hansson, B.S., Pearce, T.C., Verschure, P.F.M.J.: An artificial moth: Chemical source localization using a robot based neuronal model of moth optomotor anemotactic search. In: Autonomous Robots (2006)
29. Verschure, P.F.M.J., Voegtlin, T., Douglas, R.J.: Environmentally mediated synergy between perception and behaviour in mobile robots. Nature 425(6958), 620–624 (2003)
30. Verschure, P.F.M.J., Althaus, P.: A real-world rational agent: unifying old and new ai. Cognitive Science A Multidisciplinary Journal 27(4), 561–590 (2003)
31. Jiang, Y., Xiao, N., Zhang, L.: Towards an efficient contextual perception for humanoid robot: A selective attention-based approach. In: 6th World Congress on Intelligent Control and Automation (2006)

# Design of a Fuzzy Adaptive Controller for Uncertain Nonlinear Systems with Dead-Zone and Unknown Control Direction

A. Boulkroune*, M. M'Saad, M. Tadjine, and M. Farza

**Abstract.** In this paper, a fuzzy adaptive control system is investigated for a class of uncertain multi-input multi-output (MIMO) nonlinear systems with both unknown dead-zone and unknown sign of the control gain matrix (i.e. unknown control direction). To deal with the unknown sign of the control gain matrix, the Nussbaum-type function is used. In the designing of the fuzzy adaptive control scheme, we will fully exploit a decomposition property of the control gain matrix. To compensate for the effects of the dead-zone, we require neither the knowledge of dead-zone parameters nor the construction of its inverse. Simulation results demonstrate the effectiveness of the proposed control approach.

## 1 Introduction

Hard nonlinearities as dead-zones are ubiquitous in various components of control systems including sensors, amplifiers and actuators, especially in valve-controlled pneumatic actuators, in hydraulic components and in electric servo-motors. The dead-zone is a static "*memoryless*" nonlinearity which describes the components

---

A. Boulkroune
Faculty of Engineering Sciences, University of Jijel, BP. 98, Ouled-Aissa, 18000, Jijel, Algeria
e-mail: boulkroune2002@yahoo.fr
* Corresponding author.

M. M'Saad · M. Farza
GREYC, UMR 6072 CNRS, Université de Caen, ENSICAEN, 6 Bd Maréchal Juin, 14050 Caen Cedex, France
e-mail: [msaad,mfarza]@greyc.ensicean.fr

M. Tadjine
LCP, Electrical Engineering Department, ENP, EL-Harrach, Algiers, Algeria
e-mail: tadjine@yahoo.fr

V. Sgurev et al. (Eds.): Intelligent Systems: From Theory to Practice, SCI 299, pp. 499–517.
springerlink.com 

insensitivity to small signals. The presence of this nonlinearity severely limits the system performance. Proportional-derivative controllers have been observed to result in limit cycles if the actuators have dead-zone. The most straightforward way to cope with dead-zone nonlinearities is to cancel them by employing their inverses. However, this can be done only when the dead-zone nonlinearities are exactly known. The study of constructing adaptive dead-zone inverse was initiated by Tao and Kokotovic [1,2]. Continuous-time and discrete-time adaptive dead-zone inverses for linear systems with immeasurable dead-zone outputs were built respectively in [1] and [2]. Simulation results show that tracking performance is significantly improved by using dead-zone inverse. This work was extended in [3,4] and a perfect asymptotical adaptive cancellation of an unknown dead-zone was achieved with the condition that the dead-zone output is available for measurement. However, this condition can be very restrictive. In [5-7] fuzzy precompensators were proposed to deal with dead-zone in nonlinear industrial motion systems. In [8], the authors employed neural networks to construct a dead-zone inverse precompensator. Given a matching condition to reference model, an adaptive control with adaptive dead-zone inverse has been investigated in [9]. For a dead-zone with equal slopes, a robust adaptive control was developed, in [10], for a class of nonlinear systems without constructing the inverse of the dead-zone. In [11], a decentralized variable structure control was proposed for a class of uncertain large-scale systems with state time-delay and dead-zone input. However, some dead-zone parameters and gain signs need to be known. In [12], an adaptive output feedback control using backstepping and smooth inverse function of the dead-zone was proposed for a class of nonlinear systems with unknown dead-zone. However, in this adaptive scheme, the over-parameterization problem still exists.

In other respects, most systems involved in control engineering are multivariable in nature and exhibit uncertain nonlinear behavior, leading thereby to complex control problems. This explains the fact that only few potential solutions are available in the general case. Some adaptive fuzzy control schemes [13-17] have been developed for a class of MIMO nonlinear uncertain systems thanks to the universal approximation theorem [18]. The stability of the underlying closed-loop control system has been analyzed in Lyapunov sense. A key assumption in these fuzzy adaptive control schemes is that the sign of the control gain matrix is known a priori. When there is no a priori knowledge about the signs of the control gains, the design of adaptive controllers for MIMO nonlinear systems becomes more challenging. For a special class of MIMO nonlinear systems with unknown gain signs, adaptive neural and fuzzy control schemes have been respectively proposed in [19] and [20]. In these control schemes, the Nussbaum-type function [21] has been used to deal with the unknown control directions. Moreover, two restrictive modeling assumptions have been made to facilitate the stability analysis and the control design, namely a lower triangular control structure for the system under control and the boundedness of the so-called high-frequency control gains.

In this paper, we consider a class of uncertain MIMO nonlinear systems with both unknown dead-zone and unknown sign of the control gain matrix. To the best of our knowledge, there is only two works in the literature dealing with uncertain

MIMO nonlinear systems with unknown sign of high-frequency gains [19, 20]. The main contributions of this paper with respect to [19, 20] are the following:

**1.** The considered class of systems is more larger as the modeling assumptions made in [19,20] are relatively restrictive, namely *lower triangular control structure with bounded high-frequency control gains.* Such modeling requirements are mainly motivated by stability analysis and control design purposes.

**2.** A unique Nussbaum-type function [21] is used in order to estimate the true sign of the control gain matrix unlike in [19, 20] where many Nussbaum-type functions are used.

**3.** Motived by a matrix decomposition used in [22, 23], we decompose the control gain matrix into the product of a symmetric positive-definite (SPD) matrix, and a diagonal matrix with +1 or -1 on the diagonal (which are ratios of the signs of the leading minors of the control input gain matrix), and a unity upper triangular matrix.

**4.** The stability analysis is relatively simple and different from that pursued in [19, 20].

## 2 Problem Formulation and Definition

Consider the following class of unknown nonlinear MIMO systems with unknown dead-zone nonlinearity:

$$\begin{aligned} y_1^{(r_1)} &= f_1(x) + \sum_{j=1}^{p} g_{1j}(x) N_j(v_j), \\ &\vdots \\ y_p^{(r_p)} &= f_p(x) + \sum_{j=1}^{p} g_{pj}(x) N_j(v_j). \end{aligned} \tag{1}$$

where $x = \left[y_1, \dot{y}_1, \ldots, y_1^{(r_1-1)}, \ldots, y_p, \dot{y}_p, \ldots, y_p^{(r_p-1)}\right]^T \in R^r$ is the overall state vector which is assumed to be available for measurement and $r_1 + r_2 + \ldots + r_p = r$, $v = \left[v_1, \ldots, v_p\right]^T \in R^p$ is the control input vector, $N_i(v_i) = u_i : R \to R$ is the actuator nonlinearity which is assumed here an unknown dead-zone, $y = \left[y_1, \ldots, y_p\right]^T \in R^p$ is the output vector, and $f_i(x), i = 1, \ldots, p$ are continuous unknown nonlinear functions, and $g_{ij}(x), i, j = 1, \ldots, p$ are continuous unknown nonlinear $C^1$ functions.

Let us denote

$$y^{(r)} = \left[y_1^{(r_1)} \ldots y_p^{(r_p)}\right]^T, \quad F(x) = \left[f_1(x) \ldots f_p(x)\right]^T,$$

$$G(x)=\begin{bmatrix} g_{11}(x) & \cdots & g_{1p}(x) \\ \vdots & \ddots & \vdots \\ g_{p1}(x) & \cdots & g_{pp}(x) \end{bmatrix}, N(v)=[N_1(v_1),...,N_p(v_p)]^T .$$

Then, the system (1) can be rewritten in the following compact form:

$$y^{(r)} = F(x)+G(x)N(v) \tag{2}$$

where $F(.)\in R^p$ and $G(.)\in R^{p\times p}$.

The objective of this paper is to design a control law $v$ which ensures the boundedness of all variables in the closed-loop system and guarantees the output tracking of a specified desired trajectory $y_d = \left[y_{d1},..., y_{dp}\right]^T \in R^p$.

Note that the desired trajectory vector, $x_d =[y_{d1}, \dot{y}_{d1},..., y_{d1}^{(r_1-1)}, y_{d1}^{(r_1)},..., y_{dp}, \dot{y}_{dp},..., y_{dp}^{(r_p-1)}, y_{dp}^{(r_p)}]^T$, is supposed continuous, bounded and available for measurement. Then, $x_d \in \Omega_{x_d} \subset R^{r+p}$, with $\Omega_{x_d}$ is a known bounded compact set.

Let us define the tracking error as

$$\begin{aligned} e_1 &= y_{d1} - y_1 \\ &\vdots \\ e_p &= y_{dp} - y_p \end{aligned} \tag{3}$$

and the filtered tracking error as

$$S = \left[S_1,..., S_p\right]^T \tag{4}$$

with

$$S_i = \left[\frac{d}{dt}+\lambda_i\right]^{r_i-1} e_i, \text{ for } \lambda_i > 0,\ \forall i = 1,..., p . \tag{5}$$

Then, we can write (5) as follows

$$S_i = \lambda_i^{r_i-1} e_i+(r_i-1)\lambda_i^{r_i-2}\dot{e}_i+\cdots+(r_i-1)\lambda_i e_i^{(r_i-2)} + e_i^{(r_i-1)}, \tag{6}$$

with $i = 1,..., p$.

Notice that if we choose $\lambda_i > 0$, with $i = 1,..., p$, then the roots of polynomial $H_i(s) = \lambda_i^{r_i-1} + (r_i-1)\lambda_i^{r_i-2}s+ \cdots+(r_i-1)\lambda_i s^{r_i-2} + s^{r_i-1}$ related to the characteristic equation of $S_i = 0$ are all in the open left-half plane.

The relation (6) can be rewritten in the following compact form

$$S_i = C_i^T E_i \tag{7}$$

with

$$E_i = [e_i \ \dot{e}_i \ ...e_i^{(r_i-2)} \ e_i^{(r_i-1)}]^T \tag{8}$$

$$C_i^T = [\lambda_i^{r_i-1} \ (r_i-1)\lambda_i^{r_i-2} \ ... \ (r_i-1)\lambda_i \ 1] \tag{9}$$

Consequently, the vector $S$ takes the form:

$$S = C^T E \tag{10}$$

where

$$C^T = diag\left[C_1^T C_2^T \ \ldots \ C_p^T\right]_{(p\times r)} \tag{11}$$

$$E = \left[E_1^T E_2^T \ \ldots \ E_p^T\right]^T_{(r\times 1)} \tag{12}$$

The dynamic of $S_i$ is given by:

$$\dot{S}_i = C_{ri}^T E_i + e_i^{(r_i)}, \text{ and } i = 1,..., p \tag{13}$$

where $C_{ri}$ is given by

$$C_{ri}^T = [0 \ \lambda_i^{r_i-1} \ (r_i-1)\lambda_i^{r_i-2} \ ... \ 0.5(r_i-1)(r_i-2)\lambda_i^2 \ (r_i-1)\lambda_i] \tag{14}$$

and therefore the dynamic of $S$ can be written into the following compact form

$$\dot{S} = C_r^T E + e^{(r)} \tag{15}$$

where

$$C_r^T = diag\left[C_{r1}^T C_{r2}^T \ \ldots \ C_{rp}^T\right]_{(p\times r)} \tag{16}$$

$$e^{(r)} = \left[e_1^{(r_1)} \ e_2^{(r_2)} \ \cdots \ e_p^{(r_p)}\right]^T \tag{17}$$

$e^{(r)}$ is calculated by:

$$e^{(r)} = y_d^{(r)} - y^{(r)} \tag{18}$$

where $y^{(r)} = \left[y_1^{(r_1)} \ y_2^{(r_2)} \ \cdots \ y_p^{(r_p)}\right]^T$ is previously defined, and

$$y_d^{(r)} = \left[ y_{d1}^{(r_1)} \; y_{d2}^{(r_2)} \; \cdots \; y_{dp}^{(r_p)} \right]^T \tag{19}$$

From (18), we can write (15) as follows

$$\dot{S} = C_r^T E + y_d^{(r)} - y^{(r)} \tag{20}$$

Thereafter, (20) will be used in the development of the fuzzy controller and the stability analysis.

## 2.1 Dead-Zone Model

The dead-zone model with input $v_i$ and output $u_i$ in Fig.1 can be described as follows.

$$u_i = N_i(v_i) = \begin{cases} m_{ri}(v_i - b_{ri}), & \text{for } v_i \geq b_{ri} \\ 0, & \text{for } b_{li} < v_i < b_{ri} \\ m_{li}(v_i - b_{li}), & \text{for } v_i \leq b_{li} \end{cases} \tag{21}$$

where $b_{ri} > 0$, $b_{li} < 0$ and $m_{ri} > 0$, $m_{li} > 0$ are parameters and slopes of the dead-zone, respectively. In order to study the characteristics of the dead-zone in the control problems, the following assumptions are made:

***Assumption 1***

***a)*** *The dead-zone output $u_i$ (i.e. $N_i(v_i)$) is not available for measurement.*

***b)*** *The dead-zone slopes are same in left and right, i.e. $m_{ri} = m_{li} = m_i$.*

***c)*** *The dead-zone parameters $b_{ri}$, $b_{li}$, $m_i$ are unknown bounded constants, but their signs are known i.e. $b_{ri} > 0$, $b_{li} < 0$ and $m_i > 0$.*

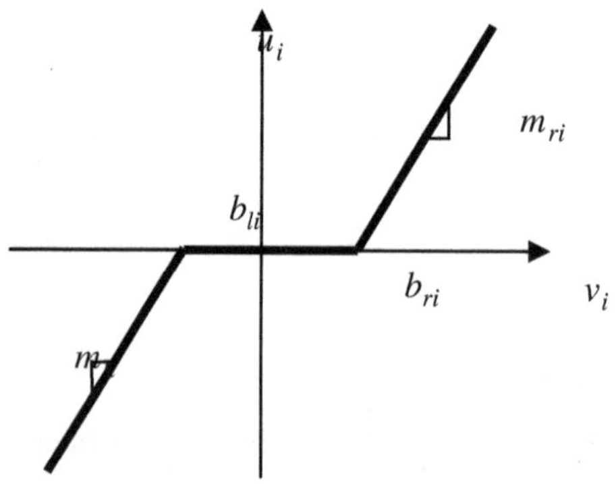


**Fig. 1** Dead-zone model

Based on the above features, we can redefine dead-zone model as follows

$$u_i = N_i(v_i) = m_i v_i + d_i(v_i) \tag{22}$$

where $d_i(v_i)$ is a bounded function defined as

$$d_i(v_i) = \begin{cases} -m_i b_{ri}, & \text{for } v_i \geq b_{ri} \\ -m_i v_i, & \text{for } b_{li} < v_i < b_{ri} \\ -m_i b_{li}, & \text{for } v_i \leq b_{li} \end{cases} \tag{23}$$

and $|d_i(v_i)| \leq d_i^*$, $d_i^*$ is an unknown positive constant.
Now, let us denote

$$d(v) = [d_1(v_1), d_2(v_2), ..., d_p(v_p)]^T,$$

$$d^* = [d_1^*, d_2^*, ..., d_p^*]^T,$$

$$M = diag[m_1, m_2, ..., m_p].$$

Then, the output vector of the dead-zone can be rewritten as follows:

$$u = Mv + d(v) \tag{24}$$

where $d(v)$ is an unknown bounded vector which can be treated as bounded disturbances. $u = [u_1, ..., u_p]^T$ = $[N_1(v_1), ..., N_p(v_p)]^T$ is the dead-zone output vector and recall that $v = [v_1, v_2, ..., v_p]^T$ is the input vector.

## 2.2 Decomposition of the Matrix G(.)

Motived by [22,23], in the control design, we need the following useful lemma.

***Lemma 1.*** [23] Any real matrix (symmetric or no-symmetric) $G(.) \in R^{p \times p}$ with non-zero leading principal minors can be decomposed as follows:

$$G(x) = G_s(x) D T(x) \tag{25}$$

where $G_s(x) \in R^{p \times p}$ is a SPD matrix, $D \in R^{p \times p}$ is a diagonal matrix with +1 or -1 on the diagonal, and $T(x) \in R^{p \times p}$ is a unity upper triangular.
***Proof of lemma 1.*** See [23] and [24].

It is worth noting that the decomposition of matrix $G(.)$ in (25) is very useful. In fact, the symmetric positive-definite $G_s(x)$ will be exploited in the Lyapunov-based stability, $D$ contains information on the sign of the original matrix $G(.)$, while the unity upper triangular matrix $T(x)$ allows for algebraic loop free sequential synthesis of control signals $v_i$, $\forall i = 1,2,\ldots, p$.

## 2.3 Nussbaum Function

In order to deal with the unknown sign of the control gain matrix, the Nussbaum gain technique will be used. A function $N(\zeta)$ is called a Nussbaum-function, if it has the following useful properties [21,25]:

$$1)\quad \lim_{s\to+\infty} \sup \frac{1}{s}\int_0^s N(\zeta)d\zeta = +\infty$$

$$2)\quad \lim_{s\to+\infty} \inf \frac{1}{s}\int_0^s N(\zeta)d\zeta = -\infty$$

***Example:*** The following functions are *Nussbaum functions* [25]:

$$N_1(\zeta) = \zeta^2 \cos(\zeta), \qquad N_2(\zeta) = \zeta \cos\left(\sqrt{|\zeta|}\right), \qquad N_3(\zeta) = \cos\left(\frac{\pi}{2}\zeta\right)e^{\zeta^2}.$$

Of course, the *cosine* in the above examples can be replaced by the *sine*. It is very easy to show that $N_1(\zeta)$, $N_2(\zeta)$ and $N_3(\zeta)$ are Nussbaum functions. For clarity, the even Nussbaum $N(\zeta) = \cos(0.5\pi\zeta)e^{\zeta^2}$ is used throughout this paper.

In the stability analysis, thereafter, we will need this lemma.
***Lemma 2.*** [20,25] Let $V(.)$ and $\zeta(.)$ be smooth functions defined on $[0,t_f)$, with $V(t)\geq 0$, $\forall t\in [0,t_f)$, and $N(.)$ be an even Nussbaum function. If the following inequality holds:

$$V(t) \leq c_0 + \int_0^t (gN(\zeta)+1)\dot{\zeta}\, d\tau, \ \forall t\in [0,t_f), \tag{26}$$

where $g$ is non-zero constant and $c_0$ represents some suitable constant, then $V(t)$, $\zeta(t)$ and $\int_0^t (gN(\zeta)+1)\dot{\zeta}d\tau$ must be bounded on $[0,t_f)$.
***Proof of Lemma 2.*** See the proof in [25].

## 2.4 Description of the Fuzzy Logic System

The basic configuration of a fuzzy logic system consists of a fuzzifier, some fuzzy IF-THEN rules, a fuzzy inference engine and a defuzzifier, as shown in Fig.2. The fuzzy inference engine uses the fuzzy IF-THEN rules to perform a mapping from an input vector $\bar{x}^T = [x_1, x_2,\ldots, x_n]\in R^n$ to an output $\hat{f}\in R$.

The $i$th fuzzy rule is written as

$$R^{(i)} : \text{if } x_1 \text{ is } A_1^i \text{ and} \ldots \text{and } x_n \text{ is } A_n^i \text{ then } \hat{f} \text{ is } f^i \quad (27)$$

where $A_1^i, A_2^i, \ldots,$ and $A_n^i$ are fuzzy sets and $f^i$ is the fuzzy singleton for the output in the $i$th rule. By using the singleton fuzzifier, product inference, and center-average defuzzifier, the output of the fuzzy system can be expressed as follows :

$$\hat{f}(\bar{x}) = \frac{\sum_{i=1}^{m} f^i \left( \prod_{j=1}^{n} \mu_{A_j^i}(x_j) \right)}{\sum_{i=1}^{m} \left( \prod_{j=1}^{n} \mu_{A_j^i}(x_j) \right)}$$

$$= \theta^T \psi(\bar{x}) \quad (28)$$

where $\mu_{A_j^i}(x_j)$ is the degree of membership of $x_j$ to $A_j^i$, $m$ is the number of fuzzy rules, $\theta^T = [f^1, f^2, \ldots, f^m]$ is the adjustable parameter vector (composed of consequent parameters), and $\psi^T = [\psi^1 \ \psi^2 \ \ldots \psi^m]$, where

$$\psi^i(\bar{x}) = \frac{\left( \prod_{j=1}^{n} \mu_{A_j^i}(x_j) \right)}{\sum_{i=1}^{m} \left( \prod_{j=1}^{n} \mu_{A_j^i}(x_j) \right)} \quad (29)$$

is the *fuzzy basis function (FBF)*.

It is worth noting that the fuzzy system (28) is the most frequently used in control applications. Following the universal approximation results [18], the fuzzy system (28) is able to approximate any nonlinear smooth function $f(\bar{x})$ on a compact operating space to any degree of accuracy. In this paper, like the majority of the available results, it is assumed that the structure of the fuzzy system (i.e. pertinent inputs, number of membership functions for each input, membership function type, and number of rules) and the membership function parameters are properly specified by the designer. As for the consequent parameters, i.e. $\theta$, they must be calculated by learning algorithms.

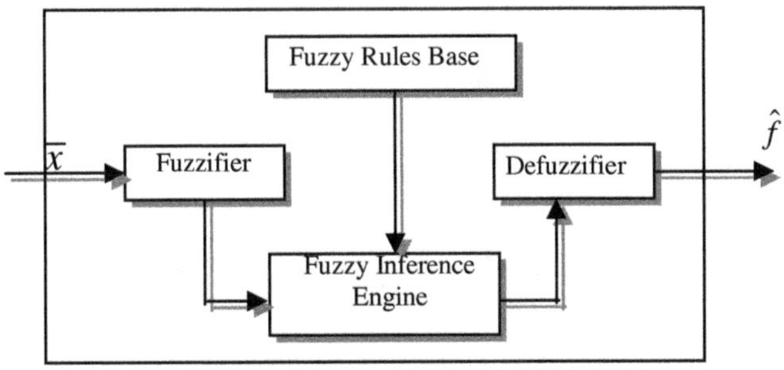


**Fig. 2** The basic configuration of a fuzzy logic system

## 3 Design of Fuzzy Adaptive Controller

Using the matrix decomposition (25), the system (2) can be rewritten as follows:

$$y^{(r)} = F(x) + G_s(x)DT(x)N(v) \tag{30}$$

To facilitate the control design and the stability analysis, the following realistic assumptions are considered [26].

***Assumption 2***

***a)*** *The sign of* $G(x)$ *is unknown. But, it must be positive-definite or negative-definite.*

***b)*** $G_s(x)$ *and* $\frac{d}{dt}G_s^{-1}(x)$ *are continuous functions.*

***c)*** $\partial g_{ij}(x) / \partial y_j^{(r_j-1)} = 0$, $\forall i = 1,2,\dots,p$, and $j = 1,2,\dots,p$.

***Remark 1***

***a)*** It is worth noting that all physical systems (MIMO or SISO) satisfy the assumption 2a.
***b)*** The assumption 2c means that the control gain matrix $G(x)$ depends only on the following vector $x_g = \left[y_1, \dot{y}_1, \dots, y_1^{(r_1-2)}, \dots, y_p, \dot{y}_p, \dots, y_p^{(r_p-2)}\right]^T \in R^{r-p}$. Consequently, matrices $G_s(x)$ and $T(x)$ are only functions of $x_g$. Assumption 2c is not restrictive as there are several physical (MIMO or SISO) systems of which the control gain matrix $G(x)$ satisfies this assumption, namely the manipulator robots, the electrical machines, the inverted pendulum, the chaotic systems. Note that Assumption 2c allows us to have $dG_s^{-1}(x)/dt$ which depends only on the state vector $x = \left[y_1, \dot{y}_1, \dots, y_1^{(r_1-1)}, \dots, y_p, \dot{y}_p, \dots, y_p^{(r_p-1)}\right]^T \in R^r$ [26].

From the equations (30) and (20), and since $G_s(x)$ is SPD, the dynamic of $S$ can be rewritten as follows

$$G_s^{-1}(x)\dot{S} = G_s^{-1}(x)[C_r^T E + y_d^{(r)} - F(x)] - DT(x)N(v) \tag{31}$$

Noting

$$G_1(x) = G_s^{-1}(x),$$
$$F_1(x,v) = G_s^{-1}(x)[C_r^T E + y_d^{(r)} - F(x)] - [DT(x) - D]N(v).$$

Equation (31) becomes

$$G_1(x)\dot{S} = F_1(x,v) - DN(v) \tag{32}$$

Using (24), (32) can be rearranged as follows

$$G_2(x)\dot{S} = -\frac{1}{2}\dot{G}_2 S + \alpha(\bar{z}) - Dv - M^{-1}Dd(v) \tag{33}$$

where

$$G_2(x) = M^{-1}G_1(x),$$

$$\alpha(\bar{z}) = \left[\alpha_1(\bar{z}_1), \alpha_2(\bar{z}_2), \ldots, \alpha_p(\bar{z}_P)\right]^T = M^{-1}F_1(x,v) + \frac{1}{2}\dot{G}_2(x)S,$$

with $\bar{z} = \left[\bar{z}_1^T, \bar{z}_2^T, \ldots, \bar{z}_p^T\right]^T$. The vector $\bar{z}_i$ will be defined later.

It is clear that since $M^{-1}$ is a diagonal positive-definite (PD) matrix and $G_1(x)$ is a SPD matrix, then the resultant matrix $G_2(x) = M^{-1}G_1(x)$ is also a PD matrix but not necessary symmetric. In order to preserve this useful propriety (i.e. the symmetry) which will be exploited later in the stability analysis, the following assumption is made on the matrix $M$ :

***Assumption 3.*** *All diagonal elements of M are equal, i.e.* $m_1 = m_2 = \ldots = m_p$.

By examining the expression of $F_1(x,v)$ and $\alpha(\bar{z})$, the vectors $\bar{z}_i$ can be determined as follows:

$$\begin{aligned}
\bar{z}_1 &= \left[x^T, S^T, v_2, \ldots, v_p\right]^T \\
\bar{z}_2 &= \left[x^T, S^T, v_3, \ldots, v_p\right]^T \\
&\vdots \\
\bar{z}_{p-1} &= \left[x^T, S^T, v_p\right]^T \\
\bar{z}_p &= \left[x^T, S^T\right]^T
\end{aligned} \tag{34}$$

It is very clear from the propriety of the matrix of $DT(x) - D$, that $\bar{z}_1$ depends on control inputs $v_2, \ldots,$ and $v_p$, $\bar{z}_2$ depends on $v_3, \ldots,$ and $v_p$, and so on. In fact, the structure of the nonlinearities $\alpha(\bar{z})$ is known under the name "*upper triangular control structure*". Recall that this useful structure allows for algebraic loop free sequential synthesis of control signals $v_i$, $\forall i = 1,2,\ldots,p$.

Define the compact sets as follows

$$\begin{aligned}
\Omega_{z_i} &= \left\{[x^T, S^T, v_{i+1}, \ldots, v_p]^T \middle| x \in \Omega_x \subset R^r, x_d \in \Omega_{x_d}\right\}, \quad i = 1,2,\ldots,p-1, \\
\Omega_{z_p} &= \left\{[x^T, S^T] \middle| x \in \Omega_x \subset R^r, x_d \in \Omega_{x_d}\right\}.
\end{aligned} \tag{35}$$

The unknown nonlinear function $\alpha_i(\bar{z}_i)$ can be approximated, on the compact set $\Omega_{z_i}$, by the fuzzy system (28), as follows:

$$\hat{\alpha}_i(\bar{z}_i,\theta_i)=\theta_i^T\psi_i(\bar{z}_i),\ i=1,...,p, \tag{36}$$

where $\psi_i(\bar{z}_i)$ is the FBF vector, which is fixed a priori by the designer, and $\theta_i$ is the adjustable parameter vector of the fuzzy system.

Let us define

$$\theta_i^*=\arg\min_{\theta_i}\left[\sup_{\bar{z}_i\in\Omega_{z_i}}\left|\alpha_i(\bar{z}_i)-\hat{\alpha}_i(\bar{z}_i,\theta_i)\right|\right] \tag{37}$$

as the optimal (or ideal) parameters of $\theta_i$. Note that the optimal parameters $\theta_i^*$ are *artificial constant quantities introduced* only for analysis purposes, and their values are not needed when implementing the controller.

Define

$$\tilde{\theta}_i=\theta_i-\theta_i^*,\ \text{with}\ i=1,...,p$$

as the parameter estimation error, and

$$\varepsilon_i(\bar{z}_i)=\alpha_i(\bar{z}_i)-\hat{\alpha}_i(\bar{z}_i,\theta_i^*) \tag{38}$$

is the fuzzy approximation error, where $\hat{\alpha}_i(\bar{z}_i,\theta_i^*)=\theta_i^{*T}\psi_i(\bar{z}_i)$.

As in literature [13-18, 26-28], we assume that the used fuzzy systems do not violate the universal approximator property on the compact set $\Omega_{z_i}$, which is assumed large enough so that input vector of the fuzzy system remains within $\Omega_{z_i}$ under closed-loop control system. So it is reasonable to assume that the fuzzy approximation error is bounded for all $\bar{z}_i\in\Omega_{z_i}$, i.e.

$$\left|\varepsilon_i(\bar{z}_i)\right|\le\bar{\varepsilon}_i,\ \forall\bar{z}_i\in\Omega_{z_i},$$

where $\bar{\varepsilon}_i$ is a given constant.

Now, let us denote

$$\hat{\alpha}(\bar{z},\theta)=\theta^T\psi(\bar{z})=[\hat{\alpha}_1(\bar{z}_1,\theta_1)...\hat{\alpha}_p(\bar{z}_p,\theta_p)]^T,$$
$$\varepsilon(\bar{z})=[\varepsilon_1(\bar{z}_1)...\varepsilon_p(\bar{z}_p)]^T,$$
$$\bar{\varepsilon}=[\bar{\varepsilon}_1...\bar{\varepsilon}_p]^T.$$

From the above analysis, we have

$$\begin{aligned}\hat{\alpha}(\bar{z},\theta)-\alpha(\bar{z})&=\hat{\alpha}(\bar{z},\theta)-\hat{\alpha}(\bar{z},\theta^*)+\hat{\alpha}(\bar{z},\theta^*)-\alpha(\bar{z}),\\&=\hat{\alpha}(\bar{z},\theta)-\hat{\alpha}(\bar{z},\theta^*)-\varepsilon(\bar{z}),\end{aligned}$$

$$= \tilde{\theta}^T \psi(\bar{z}) - \varepsilon(\bar{z}) . \tag{39}$$

where $\tilde{\theta}^T \psi(\bar{z}) = [\ \tilde{\theta}_1^T \psi_1(\bar{z}_1), \tilde{\theta}_2^T \psi_2(\bar{z}_2), \ldots, \tilde{\theta}_p^T \psi_p(\bar{z}_p)\ ]^T$, and $\tilde{\theta}_i = \theta_i - \theta_i^*$, $i = 1, \ldots, p$.

Consider the following control law which incorporates the Nussbaum function

$$\begin{aligned} v &= N(\zeta)[-\hat{\alpha}(\bar{z},\theta) - K_0 Sign(S) - K_1 S] \\ &= N(\zeta)[-\theta^T \psi(\bar{z}) - K_0 Sign(S) - K_1 S], \end{aligned} \tag{40}$$

and

$$\begin{aligned} \dot{\zeta} &= \sum_{i=1}^{p} [\theta_i^T \psi_i(\bar{z}_i) + k_{0i} Sign(S_i) + k_{1i} S_i ] S_i \\ &= S^T [\theta^T \psi(\bar{z}) + K_0 Sign(S) + K_1 S] \end{aligned} \tag{41}$$

where

$N(\zeta) = \cos(0.5\pi\zeta) e^{\zeta^2}$, $K_0 = Diag[k_{01}, k_{02}, \ldots, k_{0p}]$ and $K_1 = Diag[k_{11}, k_{12}, \ldots, k_{1p}]$.

Note that $k_{0i}$ is the online estimate of the uncertain term $0.5\sigma \left\| \theta_i^* \right\|^2 + m_i^{-1} d_i^* + \bar{\varepsilon}_i$ which will be later explained in details.

The adaptive laws are designed as follows:

$$\dot{\theta}_i = -\sigma_i \gamma_{1i} \left| S_i \right| \theta_i + \gamma_{1i} S_i \psi_i(\bar{z}_i) \tag{42}$$

$$\dot{k}_{0i} = \gamma_{2i} \left| S_i \right| \tag{43}$$

where $\gamma_{1i}, \gamma_{2i}, \sigma_i > 0$ are design constants. The term $\sigma_i \gamma_{1i} \left| S_i \right| \theta_i$, which is called *$e-$modification term*, is introduced in order to ensure both the parameters boundedness and the convergence of the tracking error to zero.

Note that the control law (40) is principally composed of the three control terms: a fuzzy adaptive term $\theta^T \psi(\bar{z})$ which is used to cancel the nonlinearities $\alpha(\bar{z})$, and a robust control term $K_0 Sign(S)$ which is introduced to compensate for the fuzzy approximation errors $\varepsilon_i(\bar{z}_i)$, and eliminate the effects of the dead-zone $m_i^{-1} d_i^*$ and that of the term $0.5\sigma_i \left\| \theta_i^* \right\|^2$ due to the use of the *$e-$modification* in the adaptation law (42). As for $K_1 S$, it is used for the stability purposes. Recall that the Nussbaum gain function $N(\zeta)$ is used to estimate the true control direction.

After substituting the control law (40) into tracking error dynamics (33) and using (39), we can get the following dynamics of the closed-loop system:

$$G_2(x)\dot{S} = -0.5\dot{G}_2 S - K_1 S - K_0 Sign(S) - \tilde{\theta}^T \psi(\bar{z}) + \varepsilon(\bar{z}) + [\theta^T \psi(\bar{z}) + K_0 Sign(S) + K_1 S] - Dv - M^{-1} Dd(v),$$

$$= -0.5\dot{G}_2 S - K_1 S - K_0 Sign(S) - \tilde{\theta}^T \psi(\bar{z}) + [\theta^T \psi(\bar{z}) + K_0 Sign(S) + K_1 S][1 + gN(\zeta)] + \varepsilon(\bar{z}) - M^{-1} Dd(v) \qquad (44)$$

where $g = D_{11} = ... = D_{pp}$, where $D_{ii}$ are diagonal terms of $D$.

Multiplying (44) by $S^T$, we have

$$S^T G_2(x)\dot{S} = -0.5 S^T \dot{G}_2 S - S^T K_1 S - \sum_{i=1}^{p} k_{0i} |S_i| - S^T \tilde{\theta}^T \psi(\bar{z}) + \sum_{i=1}^{p} (\varepsilon_i(\bar{z}_i) - m_i^{-1} g d_i(v_i)) S_i + \dot{\zeta} + gN(\zeta)\dot{\zeta} \qquad (45)$$

***Theorem.*** Consider the system (1) with Assumptions 1-3. Then, the control law defined by (40)-(41) with the adaptation law given by (42-43) guarantees the following properties:

- All signals in the closed loop system are bounded.
- The tracking errors and their derivatives decrease asymptotically to zero, i.e. $e_i^{(j)}(t) \to 0$ as $t \to \infty$ for $i = 1,..., p$ and $j = 0,1,..., r_i - 1$.

***Proof of Theorem***

Let us consider the following Lyapunov function candidate:

$$V = \frac{1}{2} S^T G_2(x) S + \frac{1}{2} \sum_{i=1}^{p} \frac{1}{\gamma_{1i}} \tilde{\theta}_i^T \tilde{\theta}_i + \frac{1}{2} \sum_{i=1}^{p} \frac{1}{\gamma_{2i}} \left(\tilde{k}_{0i}\right)^2 \qquad (46)$$

where $\tilde{k}_{0i} = k_{0i} - k_{0i}^*$, with $k_{0i}^* = 0.5\sigma_i \left\|\theta_i^*\right\|^2 + m_i^{-1} d_i^* + \bar{\varepsilon}_i$.

Its time derivative is given by

$$\dot{V} = S^T G_2(x)\dot{S} + \frac{1}{2} S^T \dot{G}_2(x) S + \sum_{i=1}^{p} \frac{1}{\gamma_{1i}} \tilde{\theta}_i^T \dot{\theta}_i + \sum_{i=1}^{p} \frac{1}{\gamma_{2i}} \tilde{k}_{0i} \dot{k}_{0i} \qquad (47)$$

Using (45) and (42-43), $\dot{V}$ can be bounded as follows

$$\dot{V} \le -\sum_{i=1}^{p} k_{1i} S_i^2 - \sum_{i=1}^{p} k_{0i} |S_i| + \sum_{i=1}^{p} \bar{\varepsilon}_i |S_i| + \sum_{i=1}^{p} m_i^{-1} d_i^* |S_i| + 0.5 \sum_{i=1}^{p} \sigma_i \left\|\theta_i^*\right\|^2 |S_i| + \dot{\zeta} + gN(\zeta)\dot{\zeta} + \sum_{i=1}^{p} \tilde{k}_{0i} |S_i|$$

$$\leq -\sum_{i=1}^{p} k_{1i} S_i^2 + \dot{\zeta} + gN(\zeta)\dot{\zeta} \tag{48}$$

Recall that $k_{0i}^* = 0.5\sigma_i \left\|\theta_i^*\right\|^2 + m_i^{-1} d_i^* + \bar{\varepsilon}_i$, $k_{0i}$ is the estimate of unknown parameter $k_{0i}^*$ and $\tilde{k}_{0i} = k_{0i} - k_{0i}^*$, and $|g| = |D_{ii}| = 1$.

Integrating (48) over $[0,t]$, we have

$$V(t) \leq V(t) + \int_0^t \sum_{i=1}^{p} k_{1i} S_i^2 dt \leq V(0) + \int_0^t (\dot{\zeta} + gN(\zeta)\dot{\zeta}) d\tau \tag{49}$$

According to Lemma 2, [20,25], we have $V(t)$, $\int_0^t (1 + gN(\zeta))\dot{\zeta} d\tau$, $\zeta$ are bounded in $[0, t_f)$. Similar to discussion in [20,25], we know that the above discussion is also true for $t_f = +\infty$. Therefore $S_i$, $\tilde{\theta}_i$, $\tilde{k}_{0i} \in L_\infty$. Then, from the boundedness of $S_i$, $\tilde{\theta}_i$, $\tilde{k}_{0i}$ and $\zeta$, we can easily conclude about the boundedness of $\theta_i$, $k_{0i}$ and $v$. From (48) and since $\int_0^\infty (1+gN(\zeta))\dot{\zeta} d\tau$ is bounded, it is very easy to show that $\int_0^\infty \sum_{i=1}^{p} S_i^2 dt$ exists, i.e. $S_i \in L_2$.

In order to show the boundedness of $\dot{S}_i$, we must rearrange Equation (44) as follows

$$\dot{S} = G_2^{-1}(x)[-0.5\dot{G}_2 S - K_1 S - K_0 Sign(S) - \tilde{\theta}^T \psi(\bar{z}) + \varepsilon(\bar{z}) + [\theta^T \psi(\bar{z}) + K_0 Sign(S) + K_1 S] - Dv - M^{-1} D d(v)] \tag{50}$$

From (50) and since $S_i$, $\tilde{\theta}_i$ $\theta_i$, $K_0$, $v$, $x$, $\varepsilon(\bar{z})$, $d(v) \in L_\infty$, $G_2^{-1}(x)$ is positive-definite matrix (i.e. $\exists \sigma_0 > 0$, such as $\left\|G_2^{-1}(x)\right\| \geq \sigma_0$) and $G_2^{-1}(x)$ and $\dot{G}_1(x)$ are continuous functions, we can easily show that $\dot{S}_i \in L_\infty$.

Finally, since $S_i \in L_2 \cap L_\infty$ and $\dot{S}_i \in L_\infty$, by using Barbalat's lemma, we can conclude that $S_i(t) \to 0$ as $t \to \infty$. Therefore, the tracking errors and their derivatives converge asymptotically to zero, i.e. $e_i^{(j)}(t) \to 0$ as $t \to \infty$ for $i = 1...p$ and $j = 0,1,...,r_i - 1$. □

***Remark 2.*** The choice of the vectors $\bar{z}_i$ (input arguments of the unknown functions $\alpha_i$) is not unique. In fact, since we known that $S$ and $v$ are functions of state $x$ and $x_d$, then it can be seen quite simply that all $\bar{z}_i$ are implicitly functions of $x$ and $x_d$ (e.g. we can chose $\bar{z}_i = [x^T, x_d^T]^T$ ,or $\bar{z}_i = [x^T, E^T]^T$, with $i = 1,..., p$ ).

## 4 Simulation Results

In this section, we present simulation results showing the tracking performances of the proposed control design approach applied to a two-link rigid robot manipulator which moves in a horizontal plane. The dynamic equations of this MIMO system are given by:

$$\begin{pmatrix} \ddot{q}_1 \\ \ddot{q}_2 \end{pmatrix} = \begin{pmatrix} M_{11} & M_{12} \\ M_{21} & M_{22} \end{pmatrix}^{-1} \left\{ \begin{pmatrix} u_1 \\ u_2 \end{pmatrix} - \begin{pmatrix} -h\dot{q}_2 & -h(\dot{q}_1 + \dot{q}_2) \\ h\dot{q}_1 & 0 \end{pmatrix} \begin{pmatrix} \dot{q}_1 \\ \dot{q}_2 \end{pmatrix} \right\},$$

$$\begin{pmatrix} u_1 \\ u_2 \end{pmatrix} = \begin{pmatrix} N_1(v_1) \\ N_2(v_2) \end{pmatrix} \tag{51}$$

where

$$M_{11} = a_1 + 2a_3\cos(q_2) + 2a_4\sin(q_2),\ M_{22} = a_2,$$
$$M_{21} = M_{12} = a_2 + a_3\cos(q_2) + a_4\sin(q_2),\ h = a_3\sin(q_2) - a_4\cos(q_2)$$

with $a_1 = I_1 + m_1 l_{c1}^2 + I_e + m_e l_{ce}^2 + m_e l_1^2$, $a_2 = I_e + m_e l_{ce}^2$, $a_3 = m_e l_1 l_{ce}\cos(\delta_e)$, $a_4 = m_e l_1 l_{ce}\sin(\delta_e)$.

In the simulation, the following parameter values are used $m_1 = 1, m_e = 2, l_1 = 1, l_{c1} = 0.5, l_{ce} = 0.6, I_1 = 0.12, I_e = 0.25,\ \delta_e = 30°$.

The control objective is to force the system outputs $q_1$ and $q_2$ to track the sinusoidal desired trajectories $y_{d1} = \sin(t)$ and $y_{d2} = \sin(t)$.

The fuzzy system $\theta_2^T\psi_2(\bar{z}_2)$ has $q_1, \dot{q}_1, q_2, \dot{q}_2$ as inputs, while $\theta_1^T\psi_1(\bar{z}_1)$ has $q_1, \dot{q}_1, q_2, \dot{q}_2, v_2$ as inputs. For each variable of the entries of the fuzzy systems, one defines three (triangular and trapezoidal [27]) membership functions uniformly distributed on the intervals $[-2,2]$ for $q_1, \dot{q}_1, q_2, \dot{q}_2$, and $[-25,25]$ for $v_2$.

The design parameters used in all simulations are chosen as follows: $\gamma_{11} = \gamma_{12} = 100$ $\sigma_1 = \sigma_2 = 0.1$, $\gamma_{21} = \gamma_{22} = 35$, $\lambda_1 = \lambda_2 = 2$, $k_{11} = k_{12} = 0.2$, $b_{r1} = b_{r2} = 3$, $b_{l1} = b_{l2} = -2.25$, $m_1 = m_2 = 2$.

The initial conditions are selected as $\underline{x}(0) = [0.5\ \ 0\ \ 0.5\ \ 0]^T$, $\theta_{1i}(0) = 0$, $\theta_{2i}(0) = 0$, $k_{01}(0) = k_{02}(0) = 0$. Note that, in this simulation, the discontinuous function $sign(S_i)$ has been replaced by a smooth function $\tanh(k_{si}S_i)$, with $k_{si} = 20$, $i = 1,2$.

The simulation results in Fig.3 show a good tracking performance. Fig.3(a) and Fig.3(b) show that the tracking errors are bounded and converge to zero. Fig.3(c) presents the dead-zone outputs (i.e. $u_i$). Fig.3(d) illustrates that the control signals ($v_i$) are bounded.

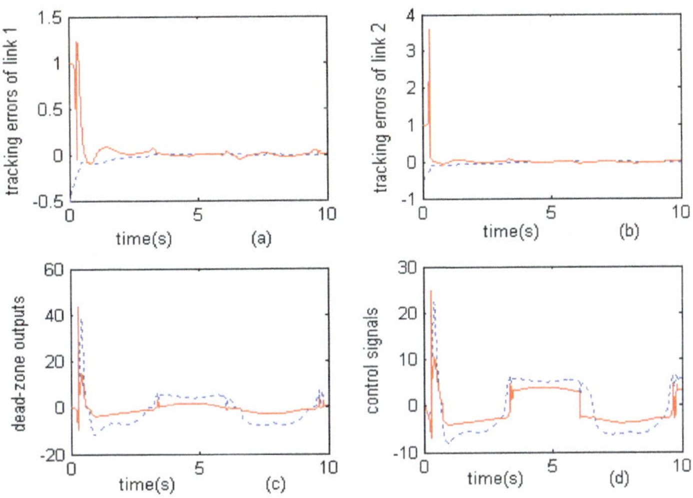


**Fig. 3** Simulation results. (a) Tracking errors of link 1: $e_1$ (dotted line) and $\dot{e}_1$ (solid line). (b) Tracking errors of of link 2 : $e_2$ (dotted line) and $\dot{e}_2$ (solid line). (c) Dead-zone outputs: $u_1$ (dotted line) and $u_2$ (solid line) (d). Control signals: $v_1$ (dotted line) and $v_2$ (solid line).

## 5 Conclusion

In this paper, a fuzzy adaptive controller, for a class of MIMO unknown nonlinear systems with both unknown dead-zone and unknown sign of the control gain matrix, has been presented. The Nussbaum-type function has been particularly used to deal with the unknown sign of the control gain matrix. The decomposition property of the control gain matrix has been fully exploited in the control design. A fundamental result has been obtained. It concerns the closed-loop control system stability as well as the convergence of the tracking error to zero. Simulation results have been reported to emphasize the performances of the proposed controller.

## References


1. Tao, G., Kokotovic, P.V.: Adaptive sliding control of plants with unknown dead-zone. IEEE Trans. Automat. Contr. 39, 59–68 (1994)
2. Tao, G., Kokotovic, P.V.: Discrete-time adaptive control of systems with unknown dead-zone. Int. J. Contr. 61, 1–17 (1995)
3. Cho, H.Y., Bai, E.W.: Convergence results for an adaptive dead zone inverse. Int. J. Adaptive Contr. Signal Process. 12, 451–466 (1998)
4. Bai, E.W.: Adaptive dead-zone inverse for possibly nonlinear control systems. In: Tao, G., Lewis, F.L. (eds.) Adaptive control of nonsmooth dynamic systems. Springer, New work (2001)
5. Kim, J.H., Park, J.H., Lee, S.W., et al.: A two-layered fuzzy logic controller for systems with dead-zones. IEEE Trans. Ind. Electr. 41, 155–161 (1994)
6. Lewis, F.L., Tim, W.K., Wang, L.Z., et al.: Dead-zone compensation in motion control systems using adaptive fuzzy logic control. IEEE Trans. Contr. Syst. Tech. 7, 731–741 (1999)
7. Jang, J.O.: A dead-zone compensator of a DC motor system using fuzzy logic control. IEEE Trans. Sys. Man Cybern. C. 31, 42–47 (2001)
8. Selmic, R.R., Lewis, F.L.: Dead-zone compensation in motion control systems using neural networks. IEEE Trans. Automat. Contr. 45, 602–613 (2000)
9. Wang, X.S., Hong, H., Su, C.Y.: Model reference adaptive control of continuous-time systems with an unknown dead-zone. IEE Proc. Control Theory Appl. 150, 261–266 (2003)
10. Wang, X.S., Su, C.Y., Hong, H.: Robust adaptive control of a class of linear systems with unknown dead-zone. Automatica 40, 407–413 (2004)
11. Shyu, K.K., Liu, W.J., Hsu, K.C.: Design of large-scale time-delayed systems with dead-zone input via variable structure control. Automatica 41, 1239–1246 (2005)
12. Zhou, J., Wen, C., Zhang, Y.: Adaptive output control of nonlinear systems with uncertain dead-zone nonlinearity. IEEE Trans. Automat. Contr. 51, 504–511 (2006)
13. Chang, Y.C.: Robust tracking control for nonlinear MIMO systems via fuzzy approaches. Automatica 36, 1535–1545 (2000)
14. Li, H.X., Tong, S.C.: A hybrid adaptive fuzzy control for a class of nonlinear MIMO systems. IEEE Trans Fuzzy Syst 11, 24–34 (2003)
15. Ordonez, R., Passino, K.M.: Stable multi-input multi-output adaptive fuzzy/neural control. IEEE Trans Fuzzy Syst. 7, 345–353 (1999)
16. Tong, S.C., Li, H.X.: Fuzzy adaptive sliding model control for MIMO nonlinear systems. IEEE Trans. Fuzzy Syst. 11, 354–360 (2003)
17. Tong, S.C., Bin, C., Wang, Y.: Fuzzy adaptive output feedback control for MIMO nonlinear systems. Fuzzy Sets Syst. 156, 285–299 (2005)
18. Wang, L.X.: Adaptive Fuzzy Systems and Control: Design and Stability Analysis. Prentice-Hall, Englewood Cliffs (1994)
19. Zhang, T.P., Ge, S.S.: Adaptive neural control of MIMO nonlinear state time-varying delay systems with unknown dead-zones and gain signs. Automatica 43, 1021–1033 (2007)
20. Zhang, T.P., Yi, Y.: Adaptive Fuzzy Control for a Class of MIMO Nonlinear Systems with Unknown Dead-zones. Acta Automatica Sinica 33, 96–99 (2007)
21. Nussbaum, R.D.: Some remarks on the conjecture in parameter adaptive control. Syst. Control Lett. 1, 243–246 (1983)

22. Chen, J., Behal, A., Dawson, D.M.: Adaptive Output Feedback Control for a Class of MIMO Nonlinear Systems. In: Proc. of the American Control Conf., June 2006, pp. 5300–5305 (2006)
23. Costa, R.R., Hsu, L., Imai, A.K., et al.: Lyapunov-based adaptive control of MIMO systems. Automatica 39, 1251–1257 (2003)
24. Strang, G.: Linear Algebra and its applications, 2nd edn. Academic press, New Work (1980)
25. Ge, S.S., Wang, J.: Robust adaptive neural control for a class of perturbed strict feedback nonlinear systems. IEEE Trans. Neural Netw. 13, 1409–1419 (2002)
26. Boulkroune, A., M'Saad, M., Tadjine, M., Farza, M.: Adaptive fuzzy control for MIMO nonlinear systems with unknown dead-zone. In: Proc. 4th Int. IEEE Conf. on Intelligent Systems, Varna, Bulgaria, September 2008, pp. 450–455 (2008)
27. Boulkroune, A., Tadjine, M., M'Saad, M., Farza, M.: How to design a fuzzy adaptive control based on observers for uncertain affine nonlinear systems. Fuzzy Sets Syst. 159, 926–948 (2008)
28. Boulkroune, A., Tadjine, M., M'Saad, M., Farza, M.: General adaptive observer-based fuzzy control of uncertain nonaffine systems. Archives of Control Sciences 16, 363–390 (2006)

# An Approach for the Development of a Context-Aware and Adaptive eLearning Middleware*

Stanimir Stoyanov, Ivan Ganchev, Ivan Popchev, and Máirtín O'Droma

**Abstract.** This chapter describes a generic, service-oriented and agent-based approach for the development of eLearning intelligent system architectures providing wireless access to electronic services (eServices) and electronic content (eContent) for users equipped with mobile devices, via a set of InfoStations deployed in key points around a University Campus. The approach adopts the ideas suggested by the Model Driven Architecture (MDA) specification of the Object Management Group (OMG). The architectural levels and iterations of the approach are discussed in detail and the resultant context-aware, adaptive middleware architecture is presented. The classification and models of the supporting agents are presented as well.

## 1 Introduction

One of the main characteristics of the eLearning systems today is the 'anytime-anywhere-anyhow' delivery of electronic content (eContent), personalized and customized for each individual user [1], [2]. To satisfy this requirement new types of context-aware and adaptive software architectures are needed, which are enabled to sense aspects of the environment and use this information to adapt their

* The authors wish to acknowledge the support of the Bulgarian Science Fund (Research Project Ref. No. ДО02-149/2008) and the Telecommunications Research Centre, University of Limerick, Ireland.

Stanimir Stoyanov
Department of Computer Systems, Plovdiv University "Paisij Hilendarski", Plovdiv, Bulgaria

Ivan Ganchev · Máirtín O'Droma
Telecommunications Research Centre, University of Limerick, Ireland

Ivan Popchev
Bulgarian Academy of Sciences

V. Sgurev et al. (Eds.): Intelligent Systems: From Theory to Practice, SCI 299, pp. 519–535.
springerlink.com 

behavior in response to changing situation. In conformity with [3], a context is any information that can be used to characterize the situation of an entity. An entity may be a person, a place, or an object that is considered relevant to the interaction between a user and an application, including the user and the application themselves.

One of the main goals of the Distributed eLearning Centre (DeLC) project [4], [5] is the development of such an architecture and corresponding software that could be used efficiently for on-line eLearning distance education. The approach adopted for the design and development of the system architecture is of essential importance for the success of this project. Our approach is focused on the development of a service-oriented and agent-based intelligent system architecture providing wireless access to electronic services (eServices) and eContent for users equipped with mobile devices, via a set of InfoStations deployed in key points around a University Campus. The approach is based on the ideas suggested by the Model Driven Architecture (MDA) of the Object Management Group (OMG) [6].

This chapter provides a general description of our approach including its architectural levels and iterations. A context-aware and adaptive middleware architecture developed as a result of this approach is presented. Furthermore the classification and models of the supporting agents are presented as well.

## 2 InfoStation-Based Network Architecture

The utilized InfoStation-based network architecture provides wireless access to eServices and eContent for users equipped with mobile devices, via a set of InfoStations deployed in key points around a University Campus [7], [8]. The InfoStation paradigm is an extension of the wireless Internet as outlined in [9], where mobile clients interact directly with Web service providers (i.e. InfoStations). The InfoStation-based network architecture consists of the following basic building entities as depicted in Figure 1: user mobile devices (mobile phones, PDAs, laptops/notebooks), InfoStations, and an InfoStation Center (ISC).

The users request services (through their mobile devices) from the nearest InfoStation via available Bluetooth, WiFi/WLAN, or WiMAX/WMAN connections. The InfoStation-based system employs the principles of the distributed control, where the InfoStations act as intelligent wireless access points providing services to users at first instance. Only if an InfoStation cannot fully satisfy the user request, the request is forwarded to the InfoStation Center, which decides on the most appropriate, quickest and cheapest way of delivering the service to the user according to his/her current individual location and mobile device's capabilities (specified in the user profile). The InfoStation Center maintains an up-to-date repository of all profiles and eContent. The InfoStations themselves maintain cached copies of all recently accessed (and changed) user profiles and service profiles, and act also as local repositories of cached eContent.

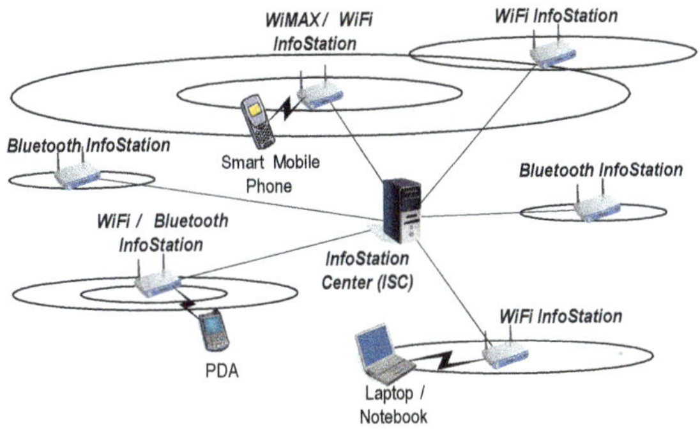


**Fig. 1** The InfoStation-based network architecture

# 3 DeLC Approach

For the development of our eLearning system we use a software development approach based on some fundamental OMG-MDA [6] ideas by taking into account the specifics of the InfoStation infrastructure.

## *3.1 Model Driven Architecture (MDA)*

The development of architectures and information systems satisfying the requirements of the modern eLearning distance education is a complex and sophisticated process. Decompositional approaches could facilitate this process better and lead to faster overall problem solution by allowing a complicated problem to be decomposed into simpler sub-problems. Each sub-problem solution could then be designed and implemented as a separate system component. However, integration of components and common control/management are required in order to realize the total functionality of the entire system. Another factor, which influences the complexity of the modern software design, is the re-usability of the existing components. MDA offers a suitable approach for coping with this situation. The approach is based on the OMG modelling standards that allow systematically to encompass and understand the entire problem before solving it. The MDA approach for the implementation of complex applications is presented in this subsection.

The core of our architecture is based on the following modelling standards proposed by OMG:

- *Unified Modeling Language (UML)* [10]: UML allows models to be created, considered, developed, and processed in a standard way starting from the initial analytical phase and going through all phases up to the final design and development. The UML models allow an application to be developed, assessed, and evaluated before starting to write the actual code. This way all necessary changes in the application could be made much easier and the cost of the overall design process could be reduced significantly;

- *Meta-Object Facility (MOF)*: MOF is a standardized way for managing the models using repositories;
- *Common Warehouse Meta-model (CWM)*: CWM standardizes the models representation as databases, schemes of transformational models, On-Line Analytical Processing (OLAP) and data mining models, etc.

The core models can be specified in a form of UML profiles. The core models are independent of the middleware platform used. Their number is relatively small because each of them presents features that are common for a particular category of problems. In this sense, the core models are meta-models for different categories. Three types of core models are used:

- *Models of business applications* with component architecture and transactional behaviour;
- *Models of real-time systems* with special requirements for resource control and management;
- *Models of other specialized systems.*

MDA allows applying a common standardized approach for the development of applications independently of the objective middleware platform used. The approach is based on the following steps:

- *Creation of a Platform-Independent Model (PIM)* in UML;
- *Creation of a Platform-Specific Model (PSM)* by mapping of PIM to the actual platform. PSM truly reflects the semantics of the functionality as well as the semantics of the application run-time. This is still a UML model but presented in one of the UML dialects (profiles), which reflects accurately the technical run-time elements of the target platform;
- *Writing the code of the application* that realizes the specific business logic and operational environment. Different types of code and corresponding configuration files are created for different component-oriented environments, e.g. interface files, definition and configuration files, files with primary code, configuration files for integration, etc.

Two main mappings are considered in the MDA approach for the realization of the alliance (coupling) of different levels (models):

- *Mapping of PIM to PSM:* Specialists (knowing in depth and in detail the requirements of the target platform) map the common model of the application (PIM) to a model, which is pursuant to and reflecting the specifics of the platform. Due to a variety of reasons this process could hardly be automated. Despite that, however, there are some automated tools (e.g. CCM, EJB, and MTS) that may facilitate mainly the mappings to standard platforms. Improvements in the automation of this type of mapping are currently hot research topics because they allow reducing significantly the amount of manually performed work. This is especially true for specialized platforms;

- *Mapping of PSM to a working application:* An automatically generated code is complemented with a manually written code, which is specific for the application.

## 3.2 Architectural Levels

In our case, we want to be able to model functionality of eLearning services independently of the utilized InfoStation network (as PIMs). On the other hand, the services should be deployable for provision (PIM mapping) in an InfoStation environment (PSM). In addition we must take into account yet another circumstance, namely the possible changes in the environment during the operation of the system. These changes have to be detected and identified by the system architecture and their effect on the service provision to be taken into consideration. To achieve this, here we propose a more sophisticated structure of PSM, which encompasses the InfoStation environment and the middleware needed for ensuring the required architecture's awareness. The middleware is developed independently as much as possible of the technical details and specifics of the InfoStation network.

Our approach envisages the existence of three architectural levels presented in the next subsections and depicted in Figure 2.

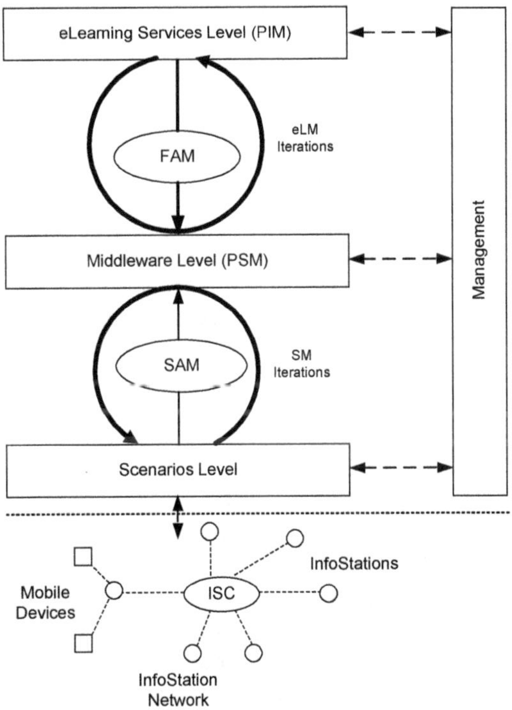


**Fig. 2** DeLC approach: architectural levels and iterations

### 3.2.1 eLearning Services Level

This level represents and models the functionality of the eLearning services provided by the system as specified in the eLearning Framework (ELF) [12]. ELF is based on a service-oriented factoring of a set of distributed core services required to support eLearning applications, portals and other user agents. Each service defined by the framework is envisaged as being provided as a networked service within an organization.

The service functionality is modelled in UML by taking into account the fact that the service realization is not directly unfolded by the system software but rather is processed by the middleware. The middleware acts as a kind of a virtual machine for the eLearning services. That's why it is very important to present the service as a composition of smaller parameterized activities, which could be navigated in different way. The actual navigation and parameterization depend on the environmental changes identified by the middleware during the provision of the corresponding service.

### 3.2.2 Middleware Level

This is an agent-based multi-layered level playing a mediator role between the services level and the scenarios level. It offers shared functionality: on one hand, it contains agents needed for the execution of different scenarios; on the other hand, it specifies a set of agents assuring the proper provision of eLearning services. In the light of the MDA philosophy, this level could be considered as PSM, which delivers a virtual (software) environment for service provision.

The main goal of the middleware level is to allow the architecture to execute/satisfy the user requests for eLearning services in a context-aware fashion. The two main tasks related to this are:

1) Detection and identification of all important environmental changes, i.e. the delivery of the relevant context for the provision of the requested services;
2) Adaptation of the architecture (in correspondence to the delivered context) as to support the provision of the requested services in the most efficient and convenient way.

### 3.2.3 Scenarios Level

This level presents the features and specifics of the InfoStation infrastructure in the form of different scenarios executed for the provision of eLearning services. The main task of the scenarios is to make transparent to the middleware level all the hardware characteristics of network nodes and details of communication in the InfoStation network. Scenarios reflect the main situations that are possible to happen in the InfoStation environment and related to the main task of the middleware, i.e. ensuring context-aware execution of user requests for eLearning services. Due to device mobility (i.e. moving between geographically intermittent InfoStation cells) and user mobility (i.e. shifting to another mobile device) the following four basic scenarios are possible [11]:

1) 'No change' scenario: If the local InfoStation can fulfil the user service request, the result is returned to the user. However, if the InfoStation is unable to meet the demands of the user, the request is forwarded onto the InfoStation Center, which retrieves the required eContent from a repository and sends it back to the InfoStation. The InfoStation may reformat/adapt the eContent in accordance with the user profile and then sends the adapted eContent to the user mobile device. The InfoStation also stores a copy the new eContent in its cache, in case if another user requests the same eContent.
2) 'Change of device' scenario: Due to the user mobility, it is entirely possible that during a service provision, the user may shift to another mobile device. For instance, by switching to a device with greater capabilities (for example from a PDA to a laptop), the user may experience a much richer service environment and utilize a wider range of resources. In this scenario, the mobile device sends a notification of device change to the InfoStation, detailing the make and model parameters of the new device. Then the InfoStation reformats the service eContent into a new format, which best suits the capabilities of the new user device.
3) 'Change of InfoStation' scenario: Within the InfoStation paradigm, the connection between the InfoStations and user mobile devices is by definition geographically intermittent. With a number of InfoStations positioned around the University Campus, the users may pass through a number of InfoStation cells during the service session. This transition between InfoStation cells must be completely transparent to the user, ensuring the user has apparent un-interrupted access to the service. As the user moves away from the footnote (service area) of an InfoStation, the user mobile device requests user de-registration from the current InfoStation. The device also requests one last user service profile update before leaving the coverage area of the current InfoStation. The InfoStation de-registers the user, updates the cached profile, and forwards the profile update to the InfoStation Center to make necessary changes in the Master Profile Repository. Meanwhile the execution of the user's request continues (for example reading through the downloaded eContent or completing the tests at the end of lecture's sections). When the user arrives within the coverage area of another InfoStation, the service execution continues from the last (synch) point reached by the user.
4) 'Change of device & InfoStation' scenario: We have outlined the separate instances where the user may switch his/her access device or pass between InfoStation cells during a service session. However, a situation may arise where the user may change the device simultaneously with the change of an InfoStation. In this scenario, both procedures for device change and InfoStation change may be considered as autonomic procedures, independent of each other. Hence each of these may be executed and completed at any point inside the other procedure without a hindrance to it.

## *3.3 Iterations*

A flexible approach is needed for the development of a context-aware and adaptive architecture in order to examine different development aspects and be able to extend the architecture step by step. The main idea behind our approach is to consider the system development as a process of iterations. The term *iteration* – borrowed from the Unified Software Development Process [13] – means a workflow or cooperation between the developers at different levels so as to be able to use and share particular products and artefacts.

There are two distinguished types of iterations in our approach (Figure 2):

- *SM iterations* - between the scenarios level and the middleware level. During each SM iteration, new scenarios that present/reflect particular aspects of the possible states and changes in the environment are developed (or the existing scenarios are modified and/or re-developed in more details). This way using the (formalized) presentation of scenarios, all corresponding middleware components needed for the support of these scenarios are developed step by step.
- *eLM iterations* - mappings of the eLearning services onto the middleware level, where the navigation model and parameterization of services are specified.

For the middleware development we plan the realization of six main SM iterations as described in the following subsections.

### 3.3.1 Basic Architecture

The first iteration aims at the development of the basic scenarios (presented in the previous subsection), which reflect the main changes that may happen in the InfoStation-based environment due to device mobility and user mobility. Based on these scenarios, a basic eLearning architecture has been developed. This architecture is presented in more detail in the next section.

### 3.3.2 Time-Based Management

Some important changes in the context during the user service request's execution – e.g. the device mobility – can be detected and identified by the system only if the temporal aspects of this process are taken into consideration. Thus the goal of this iteration is to develop concepts and formal models allowing a temporal adaptation of the processes supported by the middleware.

### 3.3.3 Adaptation

This iteration is concerned with problems related to strengthening the architecture, e.g. to support adaptability. In our opinion, personalized eLearning could be fully realized only by means of adaptive architectures, whereby the eLearning content is clearly distinguished from the three models influencing the learning process – the user model, the domain model, and the pedagogical model. The user model presents all information related to the learner's individuality, which is essential for the realization of a personalized learning process. The domain model presents the structure of the topic/subject for study/learning. In addition, in our architecture we want to support a goal-driven learning process, whereby in case of a learner's request sent to the system, a concrete pedagogical goal is generated based on the pedagogical model. The entire management of the user session afterwards obeys this pedagogical goal. These three models are supported explicitly in our architecture. They represent a strong foundation for seeking opportunities for adaptation to environmental/context changes so as to ensure more efficient personalized learning (in this sense we aim at realization of a user/domain/pedagogical model-driven optimization).

### 3.3.4 Resource Deficit

In some cases the user requests for particular services cannot be satisfied fully by the local InfoStation due to resource deficit (e.g. when information needed to satisfy the service request is unavailable in the database of the local InfoStation). In these cases the service provision must be globalized in a manner involving other InfoStations (through the InfoStation Center). The software needed to support this type of InfoStations interaction is developed as part of this iteration. The resource deficit in the serving InfoStations is caused not by dynamic factors but rather by the static deployment of resources on network nodes.

### 3.3.5 Collaboration

In many cases the execution of particular service requests requires interaction between the middleware agents. Usually information that is needed for making the decision is gathered locally, whereas the decision must be made by means of communication, cooperation, and collaboration between agents (centralized management of electronic resources and services is not envisaged in DeLC). During this iteration, the development of a common concept, models and supporting means for both local (within the service area of an InfoStation) and global (within the entire InfoStation network) agents' collaboration is envisaged.

### 3.3.6 Optimization

This iteration investigates the possibilities for optimal functioning of the middleware and proposes relevant corrections and extensions to the architecture. Different possibilities in the proposed InfoStation-based infrastructure exist for seeking the optimal solutions, e.g., the development of intelligent agents with new abilities (cloning, copying, mobility), which could be used to balance the workload out on the network nodes.

## 4 Basic System Architecture

This section presents the basic architecture, which was developed during the first SM iteration described in the previous section.

## 4.1 Tiers and Layers

In keeping with the principles of the InfoStation network, which can support a context-aware service provision, we develop a software for the three tiers of the architecture, namely for the mobile devices, InfoStations, and InfoStation Center [14]. In the standard InfoStation architecture, mobile devices use InfoStations only as mediators for accessing the services offered by the InfoStation Center. In our concept we foresee the spreading role of the InfoStations, which (besides the mediation role) act as hosts for the local eLearning services (LeS) and for preparation, adaptation, and conclusive operations of global eLearning services (GeS). This way the service provision is distributed across the whole architecture in an efficient way. The layered system architecture is depicted in Figure 3.

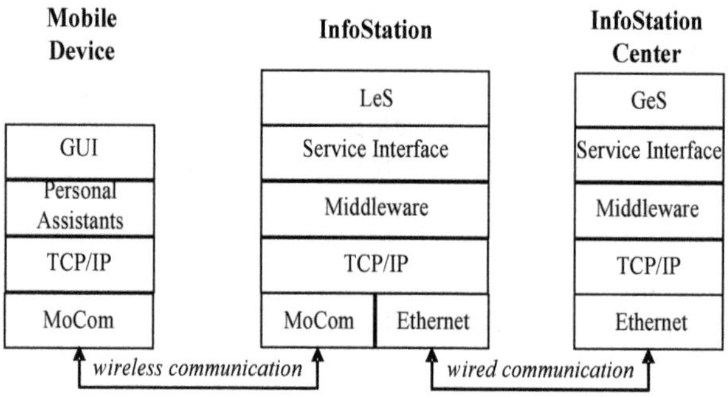


**Fig. 3** The layered system architecture

Different phases of a particular service provision may be carried out on different tiers of the architecture according to the scenario, which is currently executed. Mobile devices are provided with wireless access to services, offered by the InfoStations and/or InfoStation Center. Conceptually, the architecture required for maintaining the InfoStation configurations, is decomposed into the following logical layers: communications layers (Ethernet, mobile communications – MoCom, TCP/IP), middleware layer, service interface layer, and service layer (Figure 3). The middleware layer is responsible for detecting and identifying all the changes in the environment that may affect the provision of services requested by users and for the relevant adaptation of the architecture in response to these changes (i.e. this layer supports the contex-awareness aspect of the architecture). The service layer selects and activates the requested service. Details of the middleware layer are presented in the next subsections.

## 4.2 Middleware Agents

In order to facilitate the context-aware service provision, the middleware consists of different intelligent agents (deployed at different tiers of the InfoStation network), which interact to each other so as to satisfy in the 'best' possible way any user request they might encounter. Here we present the different types of middleware agents operating on different nodes of the InforStation network (i.e. on user mobile devices, InfoStations, InfoStation Center). The classification of the agent types is presented as an Agent-based Unified Modeling Language (AUML) diagram [15], (Figure 4). AUML is an extension of UML and is used for modelling agent-based architectures.

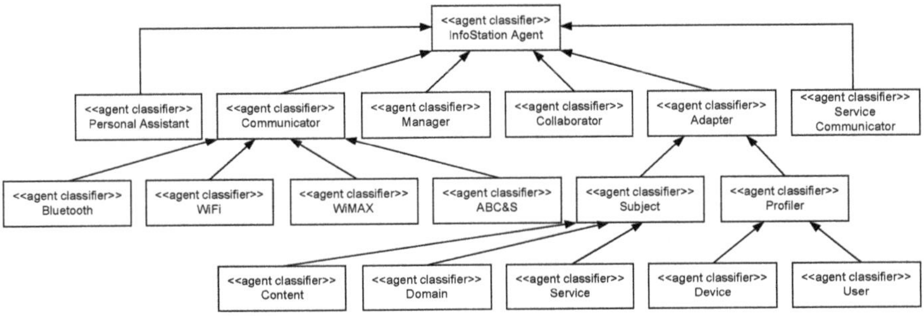


**Fig. 4** The middleware agent classification.

The functionality of each class of middleware agents is described in the next subsections.

### 4.2.1 Personal Assistant Class

This class encompasses the personal assistants installed on the user mobile devices (smart phones, PDAs, laptops, etc). The task of these agents is to help users request and use different services when working with the system.

### 4.2.2 Communicator Class

The task of this class of agents is to provide communication between different tiers of the InfoStation architecture. The main types of wireless communication used within the InfoStation environment are Bluetooth, WiFi, and WiMAX. Separate agents are developed for each of these. In addition, in accordance with the Always Best Connected and best Served (ABC&S) communication paradigm [20, 21], ABC&S agents help to choose (transparently to users) always the best connection available for each particular service requested by the user in each particular moment depending on the current context (e.g. the noise in communication channel, error rate, network congestion level, etc.).

The model of the Bluetooth communication agent is presented here as an example. This agent helps discovering the services, searching for other Bluetooth devices, establishing a connection with them, and detecting any lost of connection (e.g. out-of-radio-range problem). The main class here is the *BluetoothBehaviour* class (Figure 5).

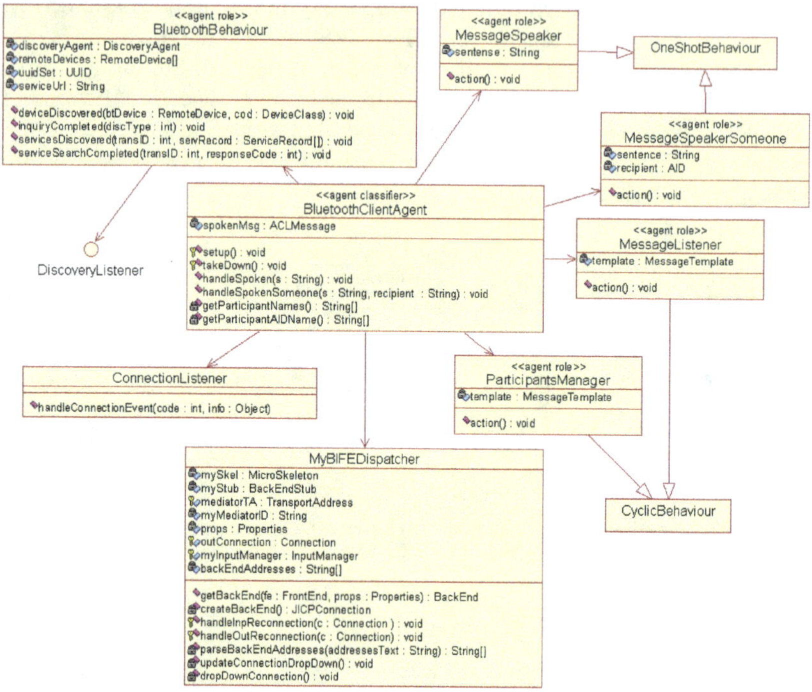


**Fig. 5** The Bluetooth communication agent

Additional classes are:

- *MessageSpeaker, MessageSpeakerSomeone* – used to send messages to one or many agents simultaneously;
- *ParticipantsManager* – used to receive up-to-date information about other agents currently available in the InfoStation environment;
- *MessageListener* – with an ability to capture messages from a particular type of agents bound to the same InfoStation;
- *CyclicBehaviour, OneShotBehaviour* – used for the realization of the cyclic behaviour and the one-shot behaviour, respectively;
- *ConnectionListener* – an interface helping to track the status of the connection with the JADE platform [18] (more precisely the connection between Front Ends and Back Ends);
- *MyBIFEDispatcher* – supports the IP communication between different InfoStations;

- *BluetoothClientAgent* – offers different methods needed for the maintenance of personal assistants, e.g. initial setup, end of work (takedown), different behaviour activations (handleSpoken, handleSpokenSomeone), processing of list of agents registered on a particular InfoStation, etc.

### 4.2.3 Manager Class

The agents in this class ensure proper detection and identification of all important events in the environment (e.g. events related to device mobility and user mobility) and deliver the actual context for the execution of the requested service. In doing this, the agents take into account the time characteristics of the events.

### 4.2.4 Adapter Class

These agents ensure the necessary adaptability of the architecture in response to the provided context from the manager agents. The adaptation model distinguishes two main groups of artifacts:

- *Adaptation objects*: These are defined data structures that must be changed in certain way (depending on adaptation subjects) before being offered to the users. The three main types of adaptation objects are: *content*, *domain*, and *service*;
- *Adaptation subjects*: These are the system users and their mobile devices. These are sources for different limitations/restrictions towards the adaptation objects. The restrictive conditions of the subjects are generalized and presented as profiles. Two main profiles are supported – user profiles and device profiles. Using the information stored in these profiles, the eContent can be adapted and customized according to the user preferences and capabilities of the user device currently in use. For instance, the user mobile device (a cellular phone) may be limited in its capabilities to play video content in which case video components are sent in a format that best suits the device, or they may be simply omitted. The user may choose to access the full capabilities of the eContent later, when using a device with greater capabilities (e.g. a laptop).

The *Adapter* class consists of two subclasses - *Subject* class and *Profiler* class. The *Subject* class provides three specialized agent types respectively for adaptation of: content (*Content* class), courses/modules (*Domain* class), and eLearning services (*Service* class). The *Profiler* class utilizes the "Composite Capabilities/Preference Profile" (CC/PP) standard [16]. The Master Profile repository in the InfoStation Center contains descriptions of all registered user mobile devices, i.e. their capabilities and technical characteristics. During the initialization, the user's personal assistant sends as parameters the make and model of the user device. An agent working on the InfoStation (or the InfoStation Center) reads the corresponding device's description from the repository and according to this, selects the 'best' format of the eContent, which is then forwarded to the user. For the support and processing of profiles we use two separate agent classes – *Device* class and *User* class.

#### 4.2.5 Collaborator Class

Collaboration (like adaptation) must be designed and built into the system from the start [17]; it cannot be patched on. This special agent class is required for the support and control of the run-time collaboration model. Besides the specification of agents needed for the support of the possible scenarios' execution, a specification of the possible relationship and interactions between agents is also needed. The agent collaboration has the potential to enhance the effectiveness of teamwork within the DeLC infrastructure. The roles played by the participants in a collaborative eLearning activity are important factors in achieving successfully the learning outcomes.

#### 4.2.6 Service Communicator Class

The main designation of these agents is to provide an interface to the services that represent the main system functionality. This class of agents realizes the service interface layer in Figure 3.

### 4.3 *Middleware Architecture*

As mentioned before, during the first SM iteration a basic version of the InfoStation's middleware architecture was developed and implemented as a set of cooperating agents (Figure 6). The agents perform different actions such as: searching for and finding mobile devices within the range of an InfoStation, creating a list of services required by mobile devices, initiation of a wireless connection with mobile devices, data transfer to- and from mobile devices, etc.

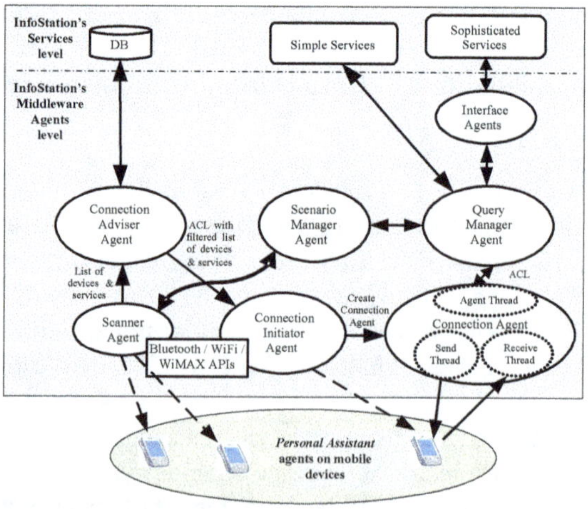


**Fig. 6** The InfoStation's middleware architecture

A short description of different agent types is provided in the subsections below.

### 4.3.1 Scanner Agent

This agent searches and finds mobile devices (within the range of an InfoStation) with enabled/activated wireless interface corresponding to the type of the InfoStation (Bluetooth/WiFi/WiMAX or mixed). In addition, this agent retrieves a list of services required by users (services are registered on mobile devices upon installation of the client part of the application and started automatically by the InfoStation agents).

### 4.3.2 Connection Adviser Agent

The main task of this agent is to filter the list (received from the *Scanner* agent) of mobile devices and services. The filtration is carried out with respect to a given (usually heuristic) criterion. Information needed for the filtration is stored in a special database (DB). The *Connection Adviser* agent sends the filtered list to the *Connection Initiator* agent.

### 4.3.3 Connection Initiator Agent

This agent initiates a communication required for obtaining the service(s) requested by the user. This agent generates the so-called Connection Object, through which a communication with the mobile device is established via Bluetooth or WiFi or WiMAX connection. In addition, for each active mobile device it generates a corresponding *Connection* agent, to which it handovers the control of the established wireless connection with this device.

### 4.3.4 Connection agent

The internal architecture of this agent contains three threads: an *Agent Thread* used for communication with the *Query Manager* agent, and a *Send Thread* and a *Receive Thread*, which support a bi-directional wireless communication with the mobile device.

### 4.3.5 Query Manager Agent

This agent is one of the most complicated components of the InfoStation's architecture. On one hand, the *Query Manager* prepares and determines where information received from the mobile device is to be directed, e.g. to simple services, or to sophisticated services via *Interface* agents. For this purpose, this agent transforms the messages coming from the *Connection* agent into messages of the corresponding protocols, e.g. UDDI or SOAP for simple services. For direct activation of simple services (e.g. Web services) there is no need for *Interface* agents. The latter are designed to maintain communication with more complicated services by using more sophisticated, semantic-oriented protocols (e.g. OWL-S [19]). In this case, the *Query Manager* acts as a mediator. In the opposite direction, this agent transforms the service execution's results into messages understandable by the agents. This operation is needed because results must be returned to the relevant *Connection* agent, which has requested the provision of the service on behalf of the user.

#### 4.3.6 Scenario Manager Agent

This agent performs the time-based scenario management based on a suitable way for formalizing the scenario presentation. For instance, the '*Change of InfoStation*' scenario could be specified mainly through the user (device) movement from one InfoStation to another. To detect this event, however, two local events must be detected during the run-time: (i) the user/device leaving the service area of the current InfoStation; (ii) the same user/device entering the service area of another InfoStation. The scenario identification cannot be centralized due to the necessity to detect these two local events. This is achieved through a message exchange between the two *Scenario Manager* agents running on the two InfoStations. The messages include the start time of events, mobile device's identification, and other parameters.

## 5 Conclusion

This chapter has presented an approach for the development of service-oriented and agent-based eLearning intelligent system architectures supporting wireless access to electronic services (eServices) and electronic content (eContent) for users equipped with mobile devices, via a set of InfoStations deployed in key points around a University Campus. The approach adopts the ideas suggested by the Model Driven Architecture (MDA) specification of the Object Management Group (OMG). The architectural levels and iterations of the approach have been discussed in detail. The first version of the resultant context-aware and adaptive middleware architecture developed accordingly to this approach by means of the agent-oriented platform JADE has been presented.

## References

[1] Barker, P.: Designing Teaching Webs: Advantages, Problems and Pitfalls. In: Educational Multimedia, Hypermedia & Telecommunication Association for the Advancement of Computing in Education, Charlottesville, VA, pp. 54–59 (2000)

[2] Maurer, H., Sapper, M.: E-Learning Has to be Seen as Part of General Knowledge Management. In: Proc. of ED-MEDIA 2001 World Conference on Educational Multimedia, Hypermedia & Telecommunications, Tampere, AACE, Charlottesville, VA, pp. 1249–1253 (2001)

[3] Dey, A.K., Abowd, G.D.: Towards a better understanding of context and context-awareness". In: Proceedings of the Workshop on the What, Who, Where, When and How of Context-Awareness. ACM Press, New York (2000)

[4] Stoyanov, S., Ganchev, I., Popchev, I., O'Droma, M.: From CBT to e-Learning. Journal Information Technologies and Control, No. 4/2005, Year III, pp. 2–10, ISSN 1312-2622

[5] Ganchev, I., Stojanov, S., O'Droma, M.: Mobile Distributed e-Learning Center. In: Proc. of the 5th IEEE International Conference on Advanced Learning Technologies (IEEE ICALT 2005), Kaohsiung, Taiwan, July 5-8, pp. 593–594 (2005), doi:10.1109/ICALT.2005.199
[6] http://www.omg.org/mda/ (to date)
[7] Ganchev, I., Stojanov, S., O'Droma, M., Meere, D.: An InfoStation-Based University Campus System Supporting Intelligent Mobile Services. Journal of Computers 2(3), 21–33 (2007)
[8] Ganchev, I., Stojanov, S., O'Droma, M., Meere, D.: Adaptable InfoStation-based mLecture Service Provision within a University Campus. In: Proc. of the 7th IEEE International Conference on Advanced Learning Technologies (IEEE ICALT 2007), Niigata, Japan, July 18-20, pp. 165–169 (2007) ISBN 0-7695-2916-X
[9] Adaçal, M., Bener, A.: Mobile Web Services: A New Agent-Based Framework. IEEE Internet Computing 10(3), 58–65 (2006)
[10] http://www.uml.org (to date)
[11] Ganchev, I., Stojanov, S., O'Droma, M., Meere, D.: InfoStation-Based Adaptable Provision of M-Learning Services: Main Scenarios. International Journal Information Technologies and Knowledge 2, 475–482 (2008)
[12] http://www.elframework.org/ (to date)
[13] Jacobson, I., Booch, G., Rumbaugh, J.: The Unified Software Development Process. Addison-Wesley, Reading (1999)
[14] Stoyanov, S., Ganchev, I., O'Droma, M., Mitev, D., Minov, I.: Multi-Agent Architecture for Context-Aware mLearning Provision via InfoStations. In: Proc. of the International Workshop on Context-Aware Mobile Learning (CAML 2008), Cergy-Pontoise, Paris, France, October 28-31, pp. 549–552. ACM, New York (2008)
[15] http://www.auml.org (to date)
[16] Kiss, C.: Composite Capability/Preference Profiles (CC/PP): Structure and Vocabularies 2.0. C. Kiss, W3C (2006)
[17] Grosz, B.J.: AI Magazine, pp. 67–85 (Summer 1996)
[18] JADE - Java Agent DEvelopment framework (to date), http://jade.cselt.it
[19] OWL-S: Semantic Markup for Web Services (to date), http://www.w3.org/Submission/OWL-S/
[20] O'Droma, M., Ganchev, I.: Toward a Ubiquitous Consumer Wireless World. IEEE Wireless Communications 14(1), 52–63 (2007)
[21] O'Droma, M., Ganchev, I., Chaouchi, H., Aghvami, H., Friderikos, V.: Always Best Connected and Served' Vision for a Future Wireless World. Journal of Information Technologies and Control, Year IV 3(4), 25–37+42 (2006)

# New Strategies Based on Multithreading Methodology in Implementing Ant Colony Optimization Schemes for Improving Resource Management in Large Scale Wireless Communication Systems

P.M. Papazoglou, D.A. Karras, and R.C. Papademetriou

**Abstract.** A great challenge in large scale wireless networks is the resource management adaptability to dynamic network traffic conditions, which can be formulated as a discrete optimization problem. Several approaches such as genetic algorithms and multi agent techniques have been applied so far in the literature for solving the channel allocation problem focusing mainly on network base-stations representation. A very promising computational intelligence technique known as ant colony optimization which constitutes a special form of swarm intelligence has been used for solving routing problems. This approach has been introduced by the authors for improving channel allocation in large scale wireless networks, focusing on network procedures as the basic model unit and not on network nodes as so far reported in the literature. In this paper, a novel channel allocation algorithm based on ant colony optimization and multi agents is thoroughly analyzed as well as important implementation issues based on the multi agent and multi threading concepts are developed. Finally, the experimental simulation results show clearly the impact of the proposed system for improving the channel allocation performance in generic large scale wireless communication systems.

**Index Terms:** wireless network, channel allocation, multi-agents, ant colony optimization, multi-threading.

---

P.M. Papazoglou
Lamia Institute of Technology Greece, University of Portsmouth, UK,
ECE Dept., Anglesea Road, Portsmouth, United Kingdom, PO1 3DJ
e-mail: papaz@teilam.gr

D.A. Karras
Chalkis Institute of Technology, Greece, Automation Dept., Psachna, Evoia,
Hellas (Greece) P.C. 34400
e-mail: dakarras@ieee.org, dakarras@teihal.gr

R.C. Papademetriou
University of Portsmouth, UK, ECE Department, Anglesea Road, Portsmouth,
United Kingdom, PO1 3DJ

V. Sgurev et al. (Eds.): Intelligent Systems: From Theory to Practice, SCI 299, pp. 537–578.
springerlink.com 

# I Introduction

*A. The Channel Allocation Problem in Wireless Communication Systems*

The capacity of a cellular system can be described in terms of the number of available channels, or the number of Mobile Users (MUs) that the system is able to support at the same time. The total number of channels made available to a system depends on the allocated spectrum and the bandwidth of each channel. The available frequency spectrum is limited and the number of MUs is increasing day by day, hence the channels must be reused as much as possible to increase the system capacity. The allocation of channels to cells or mobiles is one of the most fundamental resource management issues in a mobile communication system. The role of a channel allocation scheme is to allocate channels to cells or mobiles in such a way that it minimizes:

a) the probability of the incoming calls being dropped,
b) the probability of the ongoing calls being dropped, and
c) the probability of the carrier-to-interference ratio of any call falling below a pre-specified value.

In the literature, many channel allocation schemes have been widely investigated with the goal to maximize the frequency reuse. The channel allocation schemes are classified into three categories: Fixed Channel Allocation (FCA) [1-5], Dynamic Channel Allocation (DCA) [1,6-9], and the Hybrid Channel Allocation (HCA) [1, 10]. In FCA, a set of channels are permanently allocated to each cell based on a pre-estimated traffic intensity. The FCA scheme is simple but does not adapt to changing traffic conditions and MU distribution. Moreover, the frequency planning becomes more difficult in a microcellular environment as it is based on the accurate knowledge of traffic and interference conditions. These deficiencies can be overcome by DCA, however FCA outperforms most known DCA schemes under heavy load conditions. In DCA, there is no permanent allocation of channels to cells. Rather, the entire set of available channels is accessible to all the cells, and the channels are assigned on a call-by-call basis in a dynamic manner. One of the objectives in DCA is to develop a channel allocation strategy, which minimizes the total number of blocked calls. DCA schemes can be implemented as centralized or distributed. In the centralized approach all requests for channel allocation are forwarded to a channel controller that has access to system wide channel usage information. The central controller then assigns the channel by maintaining the required signal quality. In distributed DCA, the decision regarding the channel acquisition and release is taken by the concerned BS on the basis of the information coming from the surrounding cells. As the decision is not based on the global status of the network, it can achieve only suboptimal allocation compared to the centralized DCA and may cause forced termination of ongoing calls.

On the contrary, distributed channel allocation is done using local and neighboring cell information.

*B. Intelligent Techniques in Wireless Communication Systems*

*Multi-Agent Systems (MASs)*

The MAS technology has been used for the solution of the resource allocation problem in several studies. In the developed models of the above studies, various network components such as cells, Base Stations (BSs), etc have been modeled as agents. In [11], an overview of agent technology in communication systems is presented. This overview is concentrated on software agents that are used in communications management. More specifically, agents can be used to cope with some important issues such as network complexity, MU mobility and network management. A MAS for resource management in wireless mobile multimedia networks is presented in [12]. Based on the proposed MAS [12], the call dropping probability is low while the wireless network offers high average bandwidth utilization. According to [12], the final decision for call admission is based on the participation of neighboring cells. Thus, an agent runs in each BS or cell. A cooperative negotiation in a MAS for supporting real-time load balancing of a mobile cellular network is described in [13].

*Genetic Algorithms (GAs)*

GAs are widespread solutions to optimization problems [14-16]. The concept of a GA, is that superfit offsprings with higher fitness to the new environment can be produced as compared to their parents. This fitness is achieved by combining only good selected characteristics from different ancestors. The initial population and the solution evaluation constitute the two steps of a GA. The whole GA is an iterative procedure that terminates when the energy function constraints are met. The population contains possible solutions. These solutions are evaluated through fitness functions (in each iteration). The most suitable solutions (strings) are selected for the next generation in order to create the new population. This population is produced through two selected strings that are recombined using crossover and mutation. Crossover represents re-combinations between the two selected strings and mutation represents local modification in each string. Figure 1 shows the required steps that constitute an iterative GA.

```
Create initial population (candidate solutions)
LOOP
Evaluate population (use of fitness function)
Sufficient solution?
  YES
    Show the result, END
  NO
    Maximum iterations reached?
          YES
   STOP
          NO
   Select a pair (from population)
   Mutation
   Crossover
END LOOP (next iteration)
```

**Fig. 1** General structure of a GA

GAs have been widely used in wireless communication systems for addressing the channel allocation problem. In [17] two improved GAs which manipulates differently the allocated channels during the calls are proposed. These new algorithms have better performance than the general GA approach. Similar studies for solving the channel allocation problem based on GAs can be found in [18,19,17, 20, 21].

## 2 The Proposed Computational Intelligent Model Adapted to Large Scale Wireless Network

*A. Swarm Intelligence and Ant Colony Optimization*

The core idea of the swarm intelligence comes from the behavior of swarms of birds, fish, etc. The concept that stems from this idea is that the group complexity can be based on individual agents without the need of any central control. A definition of swarm intelligence according to [22] is as follows:

"*Swarm intelligence (SI) is the property of a system whereby the collective behaviors of (unsophisticated) agents interacting locally with their environment cause coherent functional global patterns to emerge*". Swarm intelligence has been used in a wide range of optimization problems [23-25]. The ant colony optimization (ACO) is a specific form of swarm intelligence based on ant behavior. Despite the simplicity of ant's behavior, an ant colony is highly structured [24]. ACO has been used for routing problems [26] in networks, due to the fact that some ants search for food while others follow the shortest path to the food. Figure 2 shows two different paths from nest N to food F. Between points N and F, there is an obstacle and so the length of each path is different.

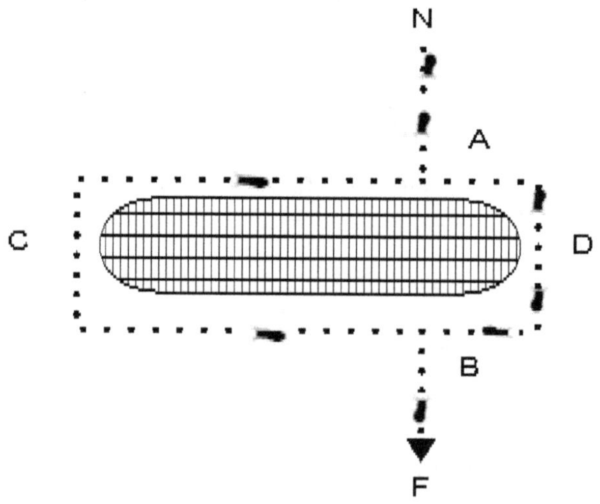


**Fig. 2** Path selection between nest and food source

When an ant reaches point A or B for the first time, the probabilities for a left or right turn are equal. Ants return faster to the nest through the path BDA, so more pheromone is left to the path. The intensity of pheromone leads ants to select more frequently the shortest path. After time passes, the shortest path will be followed by all ants. The complex collective behavior can be modelled and analyzed according to the individual behavior level (e.g. ant).

In [26] a routing algorithm based on ant algorithms applied to a simulated network is presented. In the above study an ant-inspired routing algorithm for package routing over package switching point-to-point network is examined. The used ants are artificial ants and not real ants [26] because artificial ants have different capabilities such as memory (e.g. for passed paths) and decision making algorithms based on distributions known as stochastic state transitions (e.g. expressing randomness). An ACO algorithm handles three main procedures which are:

(a) Ants generation
(b) Ants activity
(c) Pheromone trail update

The corresponding algorithm runs for a number of iterations until the final solution is reached. The most critical phase of the whole algorithm is the step (c) due to the fact that the update mechanism is the key for reaching the desired solution. The current status is qualified at each iteration and the update procedure will slightly change the model behavior towards the final solution.

***An ACO algorithm for solving the channel allocation problem has not been proposed so far in the literature.***

*B. The supported network services and channel allocation schemes by the experimental simulation model*

The evaluation of the whole network simulation model is based on the performance that is derived from the supported network services. These services are:

- New call arrival (new call admission) (NC)
- Call reallocation (handoff) (RC)
- User movement (MC)
- Call termination (FC)

Additionally, when a new channel is needed, a number of criteria must be fulfilled for a successful allocation:

- Channel availability
- Carrier strength (between MU and BS)
- CNR (signal purity)
- Signal to Noise plus interference ratio CNIR (interference from other connected MUs)

More precisely, after a new call arrival (within a new cell/cluster), several actions take place in turn:

a) Check if the maximum MU capacity in the cell/neighbor has been reached
b) Calculate a random MU position in the mesh (specific location within the cell)
c) Place the new MU according to cell's BS position and mesh spot (MU coordinates)
d) Calculate the signal strength between BS and new MU in the call initiated cell.

Firstly, the shadow attenuation [27,28] is obtained as follows

$$sh = 10^{\frac{\sigma \cdot n}{10}} \tag{1}$$

where $\sigma$ is the standard deviation of shadowing and $n$ is a number from the normal distribution. Using the shadow attenuation and distance between MU and BS, the distance attenuation *dw* can be derived. The CNR is calculated between MU and BS [27,28]

$$cn = 10^{\frac{cnedge}{10}} \cdot dw \tag{2}$$

where cnedge is the CNR on cell edge (dB).

e) Calculate interference among the new MU and other co-channel MUs that use the same channel
f) Check if C/(N+I) ratio is acceptable according to predefined threshold
g) If C/(N+I) is acceptable, establish the new call and update User Registry (UR), otherwise use any alternative selected Dynamic Channel Allocation (DCA) variation.

Formulas (1) and (2) are used for calculating the corresponding CNIR. The final CNIR, is derived from the formula [27,28]

$$R_{cni} = \frac{AP_0 d_0^{-\alpha} 10^{\frac{\xi_0}{10}}}{N + \sum_{i-1}^{n} AP_i d_i^{-\alpha} 10^{\frac{\xi_i}{10}}} \tag{3}$$

The formula (3) is based on the empirical formula:

$$\mathrm{P_r} = P_0 \left( \frac{d}{d_0} \right)^{-n} \tag{4}$$

The empirical formula (4) constitutes the most popular and general method among engineers for calculating the corresponding path loss [27,29,30]. Moreover, the shadow attenuation is modelled as a lognormal random variable [30].
Moreover, the supported channel allocation schemes are [31]:

- Unbalanced (UDCA)
- Balanced (BDCA)
- Best CNR (CDCA)
- Round Blocking (RDCA)
- CNR and Balanced hybrid (CBDCA)
- Hybrid DCA (HDCA)

In Unbalanced version, the network makes one try for user connection within the initiated cell (where the new call occurred). Round blocking scheme is an extension of unbalanced variation which searches also for an accepted channel in the neighbour cells. The algorithm stops when a successful channel assignment is made. To maintain balanced network conditions, the Balanced variation is developed. According to this algorithm, the final attempt for a MU connection is made within the cell (initiated or neighbour) with the minimum congestion. In Best CNR variation, the system calculates the CNR between MU and BS of initiated or neighbour cell. The final attempt for connection is made within the cell with maximum CNR between BS and MU. Thus, we achieve better shield of MU from interference. The goal of CNR and Balanced hybrid variation is to shield more the channel from interference and at the same time to maintain balanced traffic conditions in the network. The hybrid DCA algorithm, exhausts all the possibilities for a successful channel assignment in the neighbour and maintains balanced traffic conditions in the network.

*C. User request generation*

The number or MUs is large, the calls by each MU are limited and so the call arrivals can be assumed as random and independent. In the simulation program, the new calls result from a random or a Poisson distribution according to a predefined daily model. In the case of multimedia services, multiple channels are allocated.

The NC service allocates channels for every newly arrived MU. Figure 3 Shows how the NC procedure works.

```
//NC procedure
  Get Poisson probability (Poisson number Pn, lambda) Px
  If Px>random number then calculate candidate MUs based on the Pn and
maximum allowed MUs
  Loop (∀ candidate MU i)
    Calculate new MU position in the cell mesh
    Generate service type
    //voice
      Try for new voice channel connection based on the selected DCA scheme
    //data
      Generate the Required file size to be transferred
      Try for new data channel connection based on the selected DCA scheme
    //video
      Try for new video service connection based on the selected DCA scheme
```

**Fig. 3** User request generation

*D. Conventional call reallocation*

The computations are based on the signal strength and the way it is affected by other connected MUs in neighbor cells. If a MU signal does not fulfill the CNIR threshold the procedure tries to find another appropriate channel. At first, the algorithm calculates the signal strength between MU and BS and later on it calculates any interference coming from other connected MUs. If an accepted channel is found, it is allocated to the new MU, otherwise the call is dropped. In the case of multimedia services, partial channel reallocation is also supported.

Especially for video service handoffs, the reallocation can be performed only in some of the allocated channels due to unaccepted CNIR. The logic structure of the RC procedure is analyzed in figure 4.

```
//RC procedure
∀ MU i ∈ UR
Check connection status
If MU i is connected
  //voice
    Calculate Current Carrier to Noise plus Interference Ratio (CCNIR)
    If CCNIR<CNIR threshold then
      (disconnect MU i, try for new voice channel connection, based on selected DCA scheme)
  //data
    Calculate current CNIR (CCNIR)
    If CCNIR<CNIR threshold then
      (disconnect MU i, try for new data channel connection, based on selected DCA scheme)
  //video
    Get connection details (channels, initial requested capacity, etc) from UR
    Store UR record of MU i to temporary UR
    // check allocated channels one-by-one (work in temporary UR)
      Loop (∀ch ∈ n allocated channels)
        Calculate CCNIR for current channel ch
       If CCNIR<CNIR threshold then disconnect ch
      End Loop
    Calculated New Offered Capacity (NOC) (only from connected channels)
    If NOC<Initial Requested Capacity (IRC) then
      Calculate Additional Required Capacity (ARC=IRC-∑ connected channels capacity)
    Capacity=0
    Loop (while capacity<ARC || maximum number of tries is reached)
      Try for new video channel connection (based on selected DCA scheme)
      If channel is allocated successfully then Capacity+=capacity of the new channel
    End Loop
    If Capacity ≥ ARC then the video reallocation is successful
      Else
        Check if the new Total Offered Capacity (TOC) is accepted based on
      the initial requested QoS and the QoS tolerance threshold
     If TOC is acceptable then update UR from temporary UR
```

**Fig. 4** Call reallocation procedure

*E. The new approach for Channel allocation and reallocation based on the ACO algorithm*

In this section an ant inspired algorithm will be proposed for solving the channel allocation problem in a large scale environment. The basic objective of this novel approach, is to show through the proposed simulation models the adaptation of an ACO algorithm to a large scale wireless network focused on the problem of channel allocation. This heuristic channel allocation approach has a high grade of adaptation in dynamically changing wireless environment conditions such as traffic and signal quality. Figure 5 illustrates the basic concept of the proposed ACO algorithm.

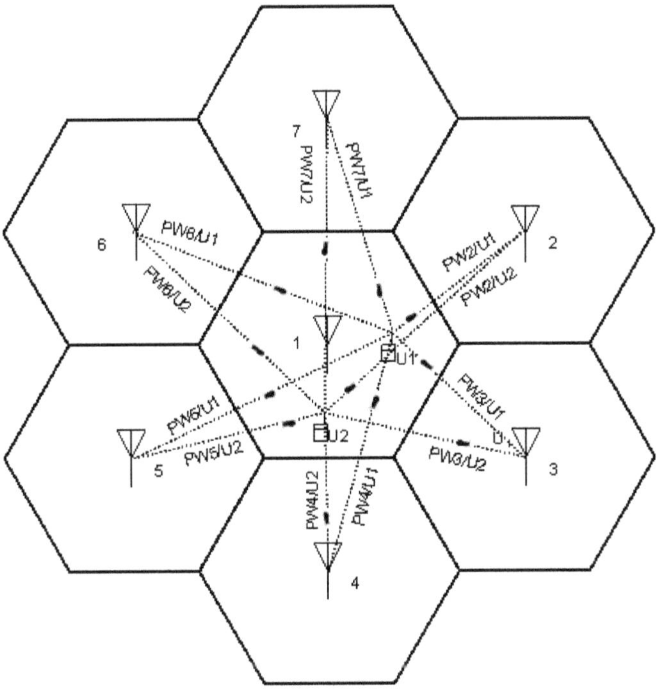


**Fig. 5** Possible ant paths between MUs and existing BSs in the neighbor

Assume that two MUs exist (U1,U2) in the center cell of the network. There are seven virtual paths between each MU (U1 or U2) and the neighbor BSs (including current cell). Thus, fourteen paths exist in the example cellular network of figure 4. Each path has a specific weight based on the signal and channel conditions of the BS that exists in its end. An ant starts from the MU position and decides which path to follow based on seven different probabilities that derive from the corresponding weight of each path. After the channel allocation for a set of MUs, the signal and channel conditions are changed in the current cell and neighbour, therefore the path weights are changed also. Thus, the ant decisions are based on different probability distributions at every simulation step and every channel allocation procedure. As user activities are increasing the network conditions also changing.

The resulting environment is a dynamic system where the network is self optimized. The basic characteristics and parameters of the above approach are:

### Characteristics

- $\forall\ MU\ \exists\ n$ paths and $n$ ants, where $n$ is the number of BSs in the neighbour. A path connects in a logically way each candidate or connected MU with the corresponding BS in the neighbour including the initiated cell.
- $\forall\ path\ \exists\ w = f\left(channels, CNIR\right)$, where $w$ is the path weight. The path weight is a probability for selecting the most suitable cell and BS for channel allocation (for a particular MU).
- $\forall\ BS\ \exists\ Cn$ channels. These channels are available from the channel set.

### Parameters

- *Total number of paths U x n*, where *U* is the total number of MUs in current cell (e.g. 7 ants for the 7 Bs and 3 MUs = 7 x 3 = 21 paths).
- *Transform:* Environment conditions to path weight (pheromone). Environment conditions correspond to current wireless conditions (propagation and signal quality) and channel availability (e.g. due to cell congestion). The path distances do not play any role in the operation of the ACO algorithm.
- *Set of Base Stations* BS={$BS_1$, … , $BS_n$}. Number of BSs in the cluster or in an area of clusters.
- *Set of Mobile Users* U={$MU_1$, … , $MU_x$}. Number of candidate or connected MUs.
- *Set of Requests* R={$MU_1R$, … , $MU_xR$}. Set of request types for a candidate or connected MU.
- *Set of pairs* (Channels in Base Stations) BSCh={$BS_1Ch_1$, $BS_1Ch_z$, … , $BS_nCh_z$}. The active channels for each individual BS.
- Subsets of Mobile Users $USet_1 \subseteq U, \ldots, USet_y \subseteq U$, $U = USet_1 \cup \ldots \cup USet_y$. Any MU set as a subset of the maximum allowed users in the network.
- *Set of virtual paths* P={$P_1$, … , $P_w$} based on the number of MUs (candidate and connected) and BS in the neighbor.

Figure 6 shows the basic structure of the proposed ACO algorithm.

```
Loop ( ∀ MU SubSet ∈ U )
    Loop
  (
  ∀ MU (Request ∈ R) ⊆ MU Subset ∈ U
  )
  //Check and Store Channel Availability, 0=free, 1=occupied
  AvMat[BSi][Chj]=Ch_Availability
  ∀ BS_i ∈ BS and ∀ Ch_j ∈ Ch
  //Check and Store only the number of available channels in each BS
  BSAvMat[BSi]= Σ_{j=1}^{z} AvMat[i][j]
  //Check and Store Average CNIR in each Base Station
  AvCNIR[BSi]= (1/z) Σ_{j=1}^{z} CNIR Ch_j ∈ BSCh_i
  //Calculate and Store Path Weights
  PathWMat[BSi]=WeightFunction(BSAvMat[BSi], AvCNIR[BSi])
  //Assign probability for each path
  AntPathProb[Pathi]=ProbFunction(PathWMat[BSi])
  Try to allocate channels
  ∀ MU (Request ∈ R) ⊆ MU Subset ∈ U
    End Loop
  End Loop
```

**Fig. 6** Abstract structure of the ACO algorithm for channel allocation

Assume that in the centre cell of a cluster two MUs exist. These MUs request a channel from the cellular network. Based on the ACO algorithm, the network has to assign a channel to those MUs from the available BSs in the cluster using the dynamically changing path weights between MUs and BSs. There are seven paths for each MU. The algorithm for channel allocation is as follows (fig. 7):

According to algorithm of figure 7 the free channels in the BSs of the cluster plays equal role (normalized to 50%) with the corresponding CNIR conditions. Initially, the percentage of free channels and the channels above the CNIR threshold for each path in the cluster is combined (Init $P_i$ Probability) in order to find out the total conditions in

```
factorA=0.5    //normalization factor
factorB=0.5    //normalization factor
number of BSs=7
//Set path weights
Loop (∀ path i∈ P)
    //W_i=f(P_i)=f(free channels at P_i end, CNIR conditions at P_i end)
    Free channels (%) in Path_i=(free channels at BS_i/total channels at BS_i)
    Free channel factor in Path_i=factorA * Free channels (%) in Path_i
    CNIR above threshold (%) in Path_i=
            (Free channels above threshold in BS_i/ Free channels in BS_i)
    CNIR factor in Path_i=factorB * CNIR above threshold (%) in Path_i
    Init P_i Probability= Free channel factor in Path_i+ CNIR factor in Path_i
    Final P_i Probability=(1/number of BSs) * Init P_i Probability
End Loop

//for MU_1 & MU_2
MU_1:
choose randomly from three paths with the highest probability from all paths
try for connection

MU_2:
choose randomly from three paths with the highest probability from all paths
try for connection
```

**Fig. 7** Channel allocation in ACO algorithm

each BS. Finally, all the path weights are normalized equally to 100% in order to produce a percentage matrix for the weights. When a new MU requests a channel, the network assigns randomly one available channel from the three paths with the highest weight between all the paths. This procedure is repeated for each defined set of MUs. Thus, at each iteration, the network conditions are different and so the area where the next channel assignments will be performed.

*D. The Multi-Agent model adapted to large scale wireless network*

Each of the above offered services (NC, RC, etc) are modeled as autonomous agents which interact with the wireless environment. Thus the corresponding agents are NCA for NC agent, etc. Moreover, the basic capabilities of the modeled agents are:

- ***Reactivity.*** NCA for example perceives network performance and a) gives priority in new calls, b) negotiates with RCA for the best agreement in order to satisfy its design objective which is the minimization of blocking probability (network performance optimization).

- ***Proactiveness.*** Network Agents, send messages to other agents (NCA to RCA and vice versa) in order to get performance benefits for the network.
- ***Social ability.*** Interaction with other agents for achieving the design objectives. NCA interacts with RCA for achieving the design objectives.

The proposed Multi-Agent model architecture can be analyzed in a cluster level basis and consists of four basic network agents. These agents are organized in layers according to the corresponding goals. Figure 8 shows the layered architecture.

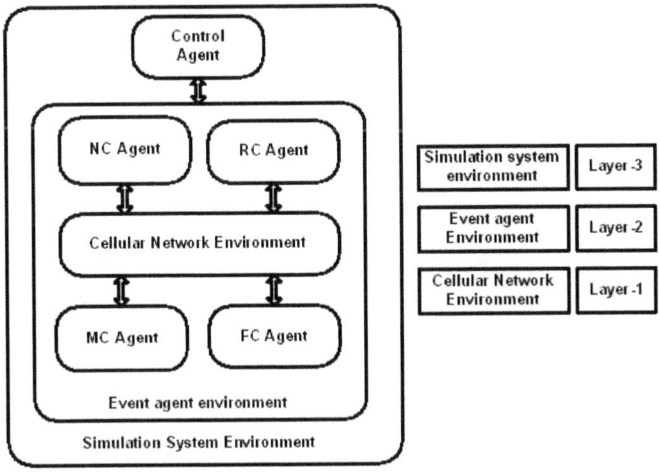


**Fig. 8** The proposed multi-layered / multi-agent architectural model

The layers illustrated in the above figure can be summarized as follows:

- ***Layer 1.*** (core). Cellular network structure.
- ***Layer 2.*** (Agents). Network behaviour.
- ***Layer 3.*** (Control Agent). Agent synchronization.

*Agent structure*
The agent structural components constitute a subsystem that works for a defined goal. Thus, each agent consists of some basic components (fig. 9) which are:

- ***Problem Solver.*** This is the core of each agent. This component supports each network service such as RC in a hand-off situation.
- ***Execution.*** Agent code activation according to clock signals.
- ***Communication.*** Exchange information with control agent (in case of centralized negotiation), or other agents.
- ***Control.*** Controls each active network component such as cell, MU, etc

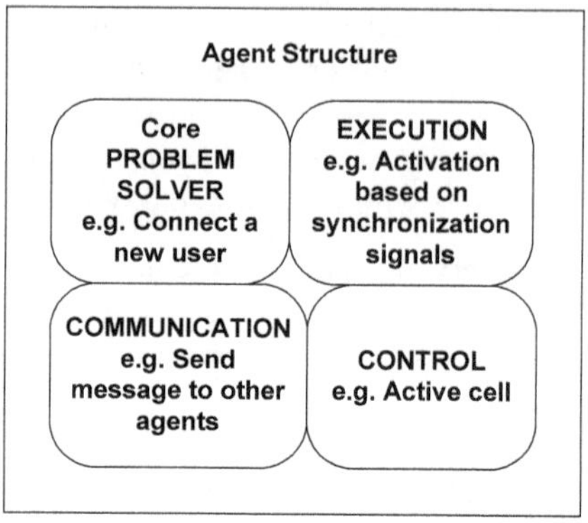


**Fig. 9** Agent Structure

*Multi-Agent control and synchronization*

For the correct completion of each simulation step, supplementary and main tasks must be activated in the right order. Thus, a control and synchronization mechanism exists.

The control agent synchronizes the agent actions. Figure 10 shows the synchronization operation. The internal clock operation can be summarized as follows:

- ***Clock=1.*** The needed supplementary actions are activated (initial procedures) while other agents and procedures are disabled. These supplementary actions include the lambda (of Poisson distribution) calculation for each new simulation step, etc.
- ***Clock=2.*** Network agents are activated while other actions are disabled.
- ***Clock=3.*** The final procedures for each simulation step are activated while other actions are disabled. This procedure calculates statistical data for the current completed simulation step, etc.

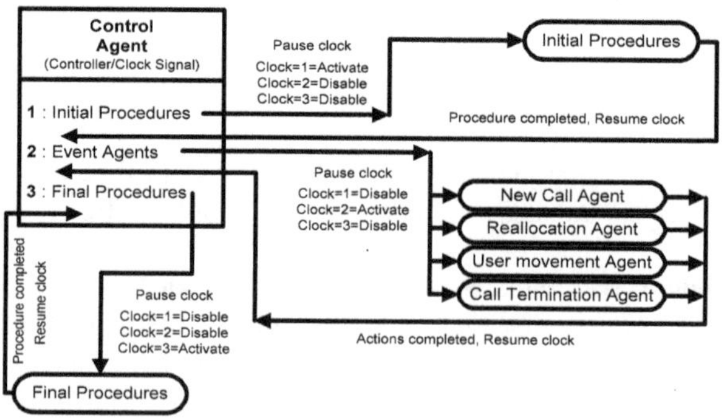


**Fig. 10** Agent and supplementary Procedure synchronization

Having analyzed the simulation model in a conceptual level, the next step is the integration to the large scale wireless network model which will provide a clear view of its scaling capabilities. As mentioned before in the MA basic model, four agents represent the offered services by the network. The main agents are the NCA and RCA due to the fact that affect directly the overall network performance in terms of blocking and dropping probability. When the network under investigation is a large scale network, the proposed network agents have to be distributed in the whole network. Assuming that the cellular network has ***N*** cells distributed in cell clusters where each cluster contains ***i*** cells, the total number of clusters is ***N/i***. Each set of the four agents is duplicated in every cluster. Thus, the total required agents are ***4*(N/i)***. In order to achieve acceptable adaptation of the MAS to the current traffic conditions and to improve the network performance the agent negotiation takes place between NCA and RCA of a set of clusters that belong in the same influence area.

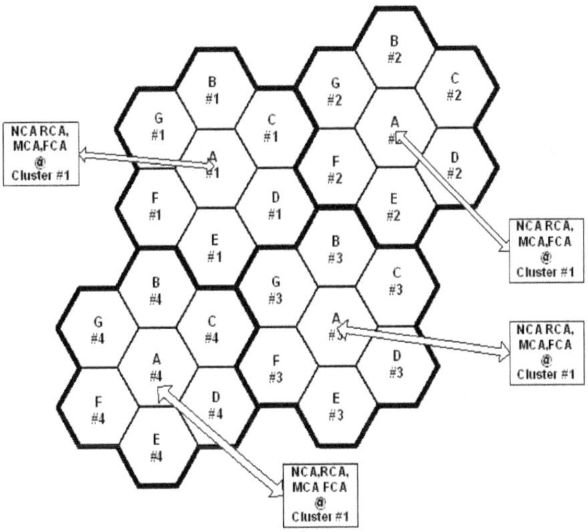


**Fig. 11** Scaling up (multiple agent sets)

For a wireless network with 28 cells (fig. 11), the total number of 7-cell clusters is 28/7=4. The total number of needed agents is 4*4=16 and especially for NCA and RCA the needed agents are 2*4=8.

*E. Network agent interaction*

It is known that an agent perceives the environment and acts on it. These two distinct activities are represented by two functions respectively (fig. 12).

Similarly, NCA and RCA, perceive environment and act on it. These two distinct activities for NCA and RCA are represented by two functions respectively (fig. 13).

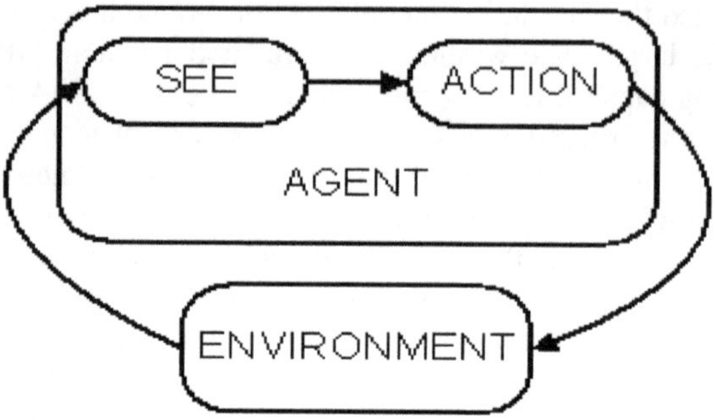


**Fig. 12** Agent interaction

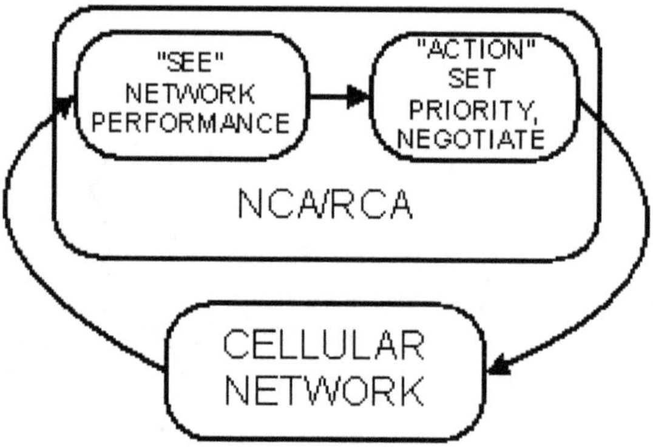


**Fig. 13** Network Agent interaction

The "see" function maps environment states to perception and "action" maps sequences of percepts to actions. As a concrete example of the above approach let x represent the statement "metric $M_1$ is acceptable" and let y represent the statement "metric $M_2$ is acceptable". Thus, the set E contains four combinations of x and y. This set can be expressed as follows:

$$E = \{\{\overline{x}, \overline{y}\}, \{\overline{x}, y\}, \{x, \overline{y}\}, \{x, y\}\} \tag{5}$$

$$e_1 = \{\overline{x}, \overline{y}\}, e_2 = \{\overline{x}, y\}, e_3 = \{x, \overline{y}\}, e_4 = \{x, y\} \tag{6}$$

Network behavior is evaluated via two basic statistical metrics which are the blocking and dropping probability. Thus, (5) and (6) will be expressed in terms of the above metrics as follows:

$$E = \{\{\overline{B}, \overline{D}\}, \{\overline{B}, D\}, \{B, \overline{D}\}, \{B, D\}\} \tag{7}$$

$$e_1 = \{\overline{B}, \overline{D}\}, e_2 = \{\overline{B}, D\}, e_3 = \{B, \overline{D}\}, e_4 = \{B, D\} \tag{8}$$

where B represents the statement "Blocking probability is acceptable" and D represents the statement "Dropping probability is acceptable". Now, the set E contains four combinations of B and D. According to (5, 6), the "see" function of the agent, will have two perceives in its range, $P_1$ and $P_2$ that indicate if the metric $M_1$ is acceptable or not. The behaviour of the "see" function can be described as follows:

$$see(e) = \begin{cases} p1 & if \quad e = e1 \quad or \quad e = e2 \\ p2 & if \quad e = e3 \quad or \quad e = e4 \end{cases} \tag{9}$$

According to (7, 8), the "see" function of the NCA, will have two percepts in its range, $P_1$ and $P_2$ that indicate if the blocking probability is acceptable or not. The behaviour of the "see" function can be described as follows:

$$see(e) = \begin{cases} P_1 & if \quad e = e_1 \quad or \quad e = e_2 \quad bad \\ P_2 & if \quad e = e_3 \quad or \quad e = e_4 \quad good \end{cases} \tag{10}$$

Similarly for RCA, the see(e) is formulated as follows:

$$see(e) = \begin{cases} P_1 & if \quad e = e_1 \quad or \quad e = e_3 \quad bad \\ P_2 & if \quad e = e_2 \quad or \quad e = e_4 \quad good \end{cases} \tag{11}$$

With two given environment states $e \in E$ and $e' \in E$, then e~e' can be written only if see(e)=see(e'). An agent has perfect perception if the different environment states are equal to distinct perceives. In this case

$$|\sim| = |E| \tag{12}$$

when the perception of an agent does not exist,

$$|\sim| = 1 \tag{13}$$

*Network Agent Utility functions*

A different utility function can be assigned to each agent (NCA, RCA) in order to "measure" how good the corresponding outcome is (blocking/dropping). This function, assigns a real number to each outcome, indicating how good is the outcome for the selected agent. In other words, this utility function defines the preference ordering over all the outcomes.

Let $\omega$ and $\omega'$, two possible outcomes in $\Omega$ set (where $\Omega=\{\omega_1,\omega_2,\ldots\}$), with utility functions that give

$$u_i(\omega) \geq u_i(\omega') \tag{14}$$

Thus, agent *i*, prefers better the $\omega$ outcome.

*Multi-Agent encounters*

In an environment with more than one agent, apart from the environment itself, the agent encounters have to be modelled as well. Assume that there are two possible actions by two agents such as "*C*" which means "*Cooperate*" and "*D*" which means "*Defect*" and let the action set $A_c=\{C,D\}$. Based on the above, the environment behavior can be modelled with the function

$$\tau : Ac_i \times Ac_j \rightarrow \Omega \tag{15}$$

where, $Ac_i$ and $Ac_j$ represent the actions of the two agents *i* and *j* respectively.

Cooperation (CR) and Competition (CT), are the two possible actions for NCA and RCA agents and so the action set becomes Ac={CR,CT}. Now the environment functions are:

$$\begin{aligned} &\tau(CT,CT)=\omega_1, \tau(CT,CR)=\omega_2, \\ &\tau(CR,CT)=\omega_3, \tau(CR,CR)=\omega_4 \end{aligned} \tag{16}$$

According to (16), each action is mapped to a different outcome.

Using utility functions for the ω in (16), the corresponding utility functions (NCA, RCA) are:

$$\begin{aligned} &u_{NC}(\omega_1)=1 \quad u_{NC}(\omega_2)=1 \quad u_{NC}(\omega_3)=2 \quad u_{NC}(\omega_4)=2 \\ &u_{RC}(\omega_1)=1 \quad u_{RC}(\omega_2)=2 \quad u_{RC}(\omega_3)=1 \quad u_{RC}(\omega_4)=2 \end{aligned} \tag{17}$$

The above numbers are indicative only and illustrate that the corresponding outcomes for an agent can be good or not (for NCA, a good result is the minimization of blocking probability). In other words, only two numbers are needed to coding the corresponding results. In practice, any two different numbers can be assigned just to indicate an acceptable or not result and to distinguish between possible outcomes.

Combining now (16) and (17) for NCA and RCA, the outcomes can be expressed as:

$$\begin{aligned} &u_{NC}(\tau(CT,CT))=1 \quad u_{NC}(\tau(CT,CR))=1 \quad u_{NC}(\tau(CR,CT))=2 \quad u_{NC}(\tau(CR,CR))=2 \\ &u_{RC}(\tau(CT,CT))=1 \quad u_{RC}(\tau(CT,CR))=2 \quad u_{RC}(\tau(CR,CT))=1 \quad u_{RC}(\tau(CR,CR))=2 \end{aligned} \tag{18}$$

The possible NCA, RCA actions based on (18) are

$$CR,CR \geq_{NC} CR,CT >_{NC} CT,CR \geq_{NC} CT,CT \tag{19}$$

It is obvious from (18) and (19) which action will be selected from each agent and thus each agent knows exactly what to do.

### *F. Network Agent Negotiation*

As mentioned before, in a MA environment each agent interacts with the environment, adapts its behavior according to current conditions, collaborates with other agents and finally makes decisions autonomously. The negotiation modules are supplementary components for the dialog between NCA and RCA. Figure 14 illustrates the extended multi-layered MA architecture.

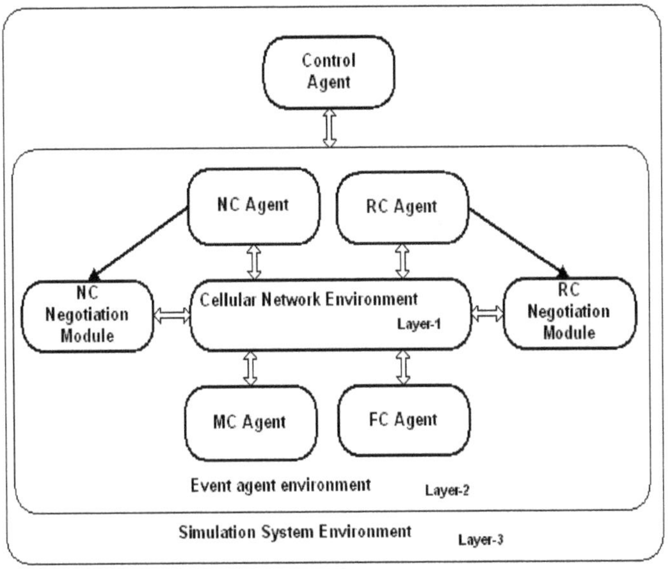


**Fig. 14** Extended multi layered / multi agent Architecture

*Extended agent control and synchronization for supporting MA negotiation*

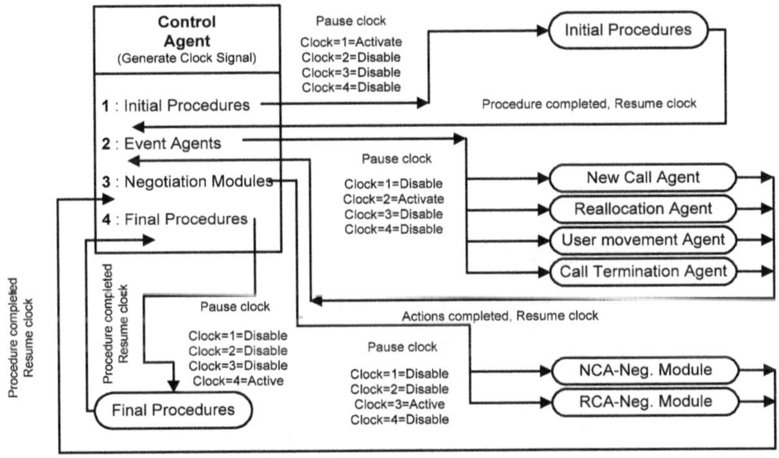


**Fig. 15** Extended synchronization mechanism

As shown in figure 15, two more components have been added for supporting MA negotiation. The negotiation modules are activated just after deactivation of the main agents. Thus the negotiation result creates new network conditions (due to different priorities of the network services) before the next simulation step.

Figures 16 and 17 show the algorithms that describe the NCA and RCA behavior respectively. In the following examples it is assumed that the current network performance is *HL* (High, Low) regarding the successfulness of call admission and handoffs (low blocking probability, high dropping probability). The following algorithms development is based on the multi agent formal analysis which defines the corresponding steps for network behavior. According to figure 16, the NCA decreases its priority to help the RCA to improve its performance regarding the handoff successfulness in order to minimize the dropping probability and hence to increase the corresponding network performance. On the other hand, figure 17 shows the algorithm for the RCA. In this figure, the RCA tries to improve its performance by increasing self priority and by sending a request to the NCA.

Each NM (NA-NM or RC-NM) checks for incoming message from the other agent and makes decisions according to this incoming message and its current status. The behavior of the RCA is analogous to NCA.

As mentioned before, if the network performance gets worse for a specific metric (e.g. blocking, dropping probability, reallocations needed) related with NCA for example (fig. 18), the NCA sends a message Request_RC_PRIC_DEC asking for priority decrement from the side of RCA. The RC-NM checks for incoming message. If a request for priority decrement exists from the NCA and RCA status is stable or good and current priority is not minimum, the request is accepted otherwise is rejected.

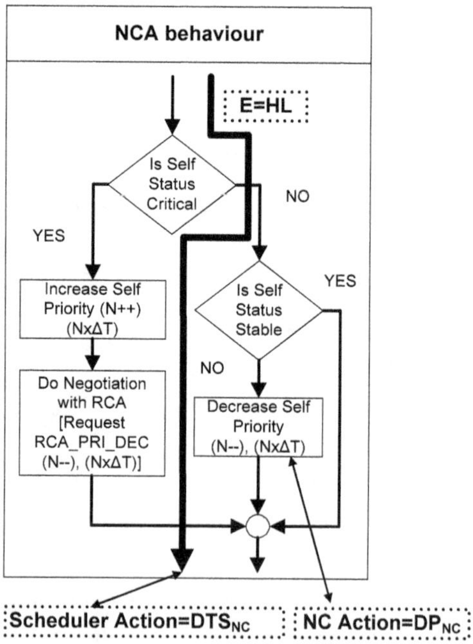


**Fig. 16** NCA behaviour algorithm

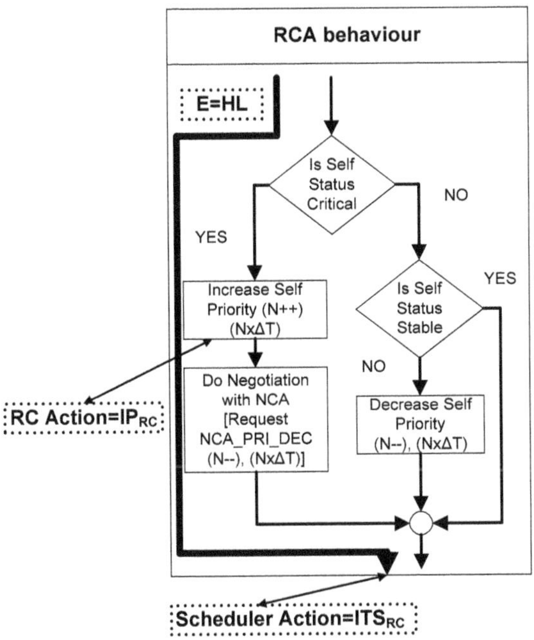


**Fig. 17** RCA behaviour algorithm

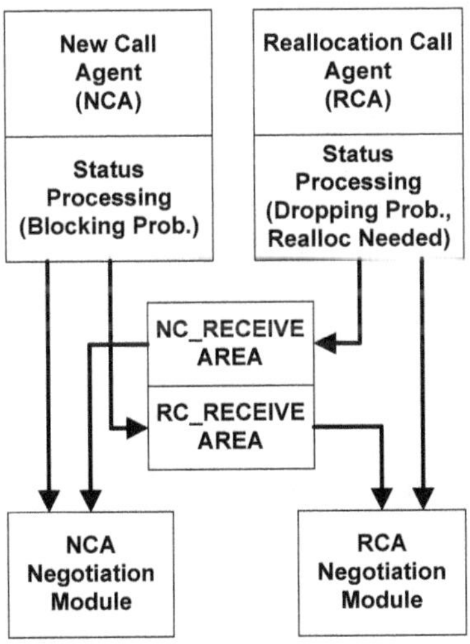


**Fig. 18** Agent communication

Figure 19 shows a sample of the message exchange dialog.

```
70
-RC- send message to NC for priority decrease
-NC- message received from RC: request NC priority decrement
-NC- request from RC:Rejected (NC-status=-1)
NC-priority=2, RC-priority=6
80
NC- priority =2, RC- priority =6
90
NC- priority =2, RC- priority =6
100
-RC- send message to NC for priority decrease
-NC- message received from RC: request NC priority decrement
-NC- request from RC:Accepted
NC- priority =1, RC- priority =7
110
-RC- self priority dec -1
NC- priority =1, RC- priority =6
120
-RC- send message to NC for priority decrease
-NC- message received from RC: request NC priority decrement
-NC- request from RC:Rejected (priority minimum)
NC- priority =1, RC- priority =7
```

**Fig. 19** Sample of message exchange dialog

## 3 Implementation Issues

*A. The Multi-Threading (MT) platform*

In the above approach each agent has been implemented as a thread. All the active threads are controlled by Java Virtual Machine (JVM) which is a multi-threading platform. Java is a very attractive programming language especially for building flexible, portable and high performance network applications and is the most popular language among programming community [32]. Several simulation tools and libraries have been developed for both general purpose discrete event simulation and simulation of wireless networks based on Java. On the other hand, the Java implementation can be viewed as an example; the developer may use another technology for the implementation of the MT concept.

The MT operation in the case of Java is exclusive responsibility of the Java Virtual Machine (JVM). Thus, any Java code can be executed directly in any platform. Threading methodology can be applied when concurrency is needed. In the context of the conceptual design of a program, this approach can be quite useful even if it runs on a single-processor machine. Thus, more reliable modelling and simulation of a real system can be achieved. More information about the threading capabilities of Java can be found in [33-36].

*Operating System Applications, JVM and JVM threads*
Modern OSs support the execution of multiple applications. Each application reserves a memory area for storing local data and uses Central Processing Unit (CPU) time when active (fig. 20). In single processor machines (conventional PCs) the CPU time is distributed among the applications (tasks).

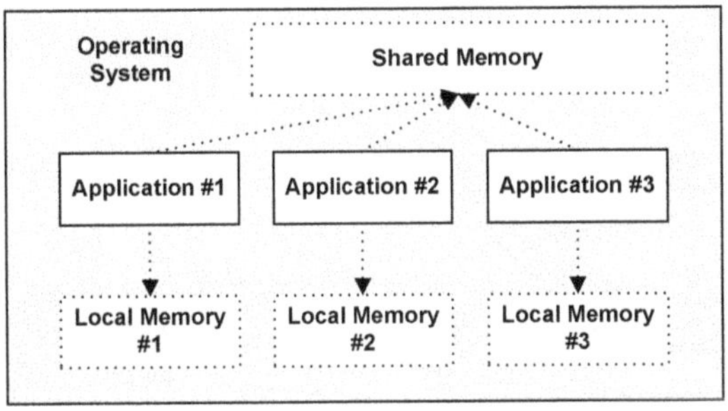


**Fig. 20** The Operating System environment

Every Java application runs within the JVM environment. In JVM multiple applications and/or multiple parts of a single application can be executed (fig. 21). In a multiprocessor system, threads can be executed in different CPUs. The high speed switching between threads, in the case of a single processor, gives the impression of parallel execution.

When the thread execution starts, the JVM gives equal priorities to each one of them. Within the running application, the default thread priorities can be changed according to current needs. The given CPU time slice for each thread is based on the defined priorities (possible values, 1 to 10). The OS gives a time slice to JVM and this time is distributed to java applications/threads.

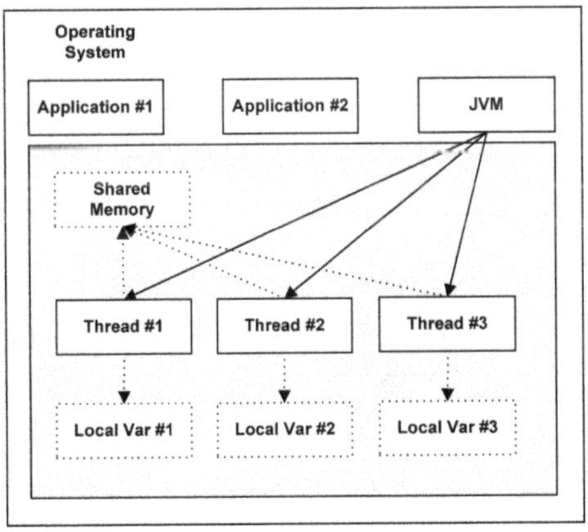


**Fig. 21** The Java Virtual Machine (JVM) Environment

*Creating Threads*

Multiple threads can be created within a Java application (fig. 2.13). Every application constitutes at least one thread. The main method of a thread is the run() method. When the method start() is activated for a thread, then the corresponding run() method is executed at the thread body. Figure 22 illustrates three concurrent threads created within a Java application.

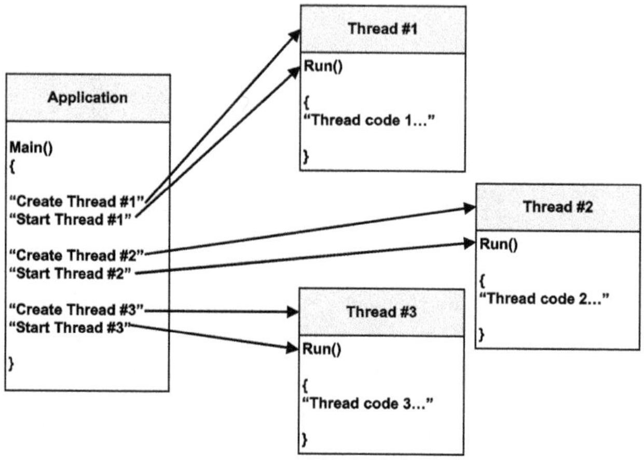


**Fig. 22** Creating threads

A variety of methods are available for changing the thread status within an application. Using these methods, efficient synchronization control between threads can be achieved. The start() method activates the selected thread that is running concurrently with other threads. A thread is terminated when the run() method returns or when the stop() method is activated. Figure 23 shows all the possible thread status and the corresponding status transitions.

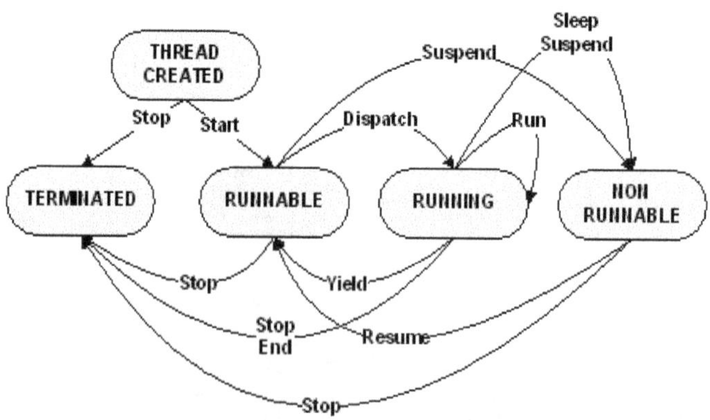


**Fig. 23** Thread status transitions

An internal mechanism of JVM, called scheduler, defines the real-time order of thread execution. Scheduling can be controlled by the programmer and is categorized as follows:

- Non pre-emptive
- Pre-emptive

In non pre-emptive, the scheduler runs the current thread forever and requires from this thread to tell explicitly if it is safe to start another thread. In pre-emptive, the scheduler runs a thread for a specific time-slice (usually a tiny period within a second) and then "pre-empts" it, (calling suspend()), and resumes another thread for the next time-slice. The non-pre-emptive scheduling can be very useful especially in time critical (e.g. real time) applications when the interruption of thread execution can happen in the wrong moment. Modern schedulers are usually pre-emptive, therefore the development of MT applications is easier. JVM uses priorities for scheduling threads. Initially, it gives equal priorities to all threads. ***A major drawback of the JVM is that the behavior of the scheduler (e.g. thread execution sequence) can not be predicted [37]. For that reason, a controlled scheduling mechanism is used in this study.*** Figure 24 illustrates three threads sharing the same processor. Note that there is no real parallel execution between threads but only high speed switching between them. The sleep() method deactivates temporarily the current thread in order to give time for the next thread execution. If the OS uses pre-emptive MTAS the sleep() method is not needed for switching between threads.

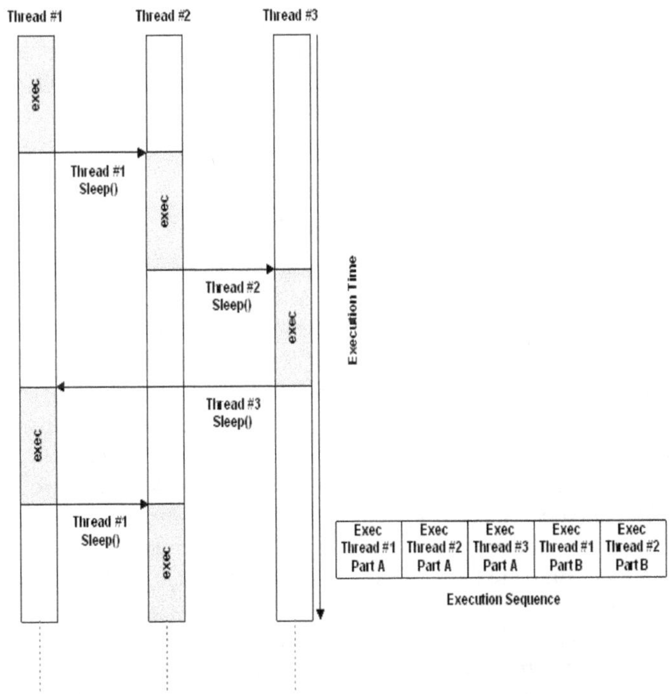


**Fig. 24** Three threads share the same processor (sample scenario)

*B. Controlling the thread execution sequence*

In a wireless cellular environment, where events happen concurrently, the performance of the network depends not only on the logical results of computation, but also on the time sequence at which the results are produced. In other words, it depends on the logical sequence of the various network procedures regarding the efficient bandwidth management. Network events such as new call admission, reallocation, call termination, etc, can be faced as tasks that have to be served by the network. The scheduling mechanism that manages task servicing plays a major role in the resulting network performance. In other words, this mechanism defines how the network will serve the concurrent tasks (events) in the most efficient way in terms of network performance regarding channel allocation.

When the network events are faced as concurrent, the Multi Tasking conceptual approach extended properly can be applied. In a cellular (concurrent) model, when a MU is under processing from the network, another MU is moving or trying for reallocation. Thus, the task (event) execution has to be partial, in order to handle the other concurrent MUs.

Assume that four events-threads have been scheduled to execute at the same time (simultaneous activities in the cellular network). Due to event code segment interleaving none of the events will be executed completely until the final event segments are reached. Figure 4.15, illustrates a sample of a controlled switching between the basic threads that may represent the network events. The fast switching between the four threads create a concurrent environment that represents the physical activities (events) of the cellular network. Thus, the execution sequence inside the Central Processing Unit (CPU) consists of different code segments of the active threads. In a conventional concurrent environment, the active threads share the same processor and so the processing time is divided among these threads. Figure 25 shows how the code interleaving is achieved.

The basic goals of the proposed event scheduling mechanism are:

- The control of thread execution sequence.
- The control of thread core code execution time.

The thread execution sequence and activation is controlled by a clock inside a thread (super thread) with maximum priority instead of minimum priority which is applied in the rest of the threads [38-41]. Figure 26 shows the pseudo code of clock implementation.

Using the *Thread.start()* method for each thread, all the threads become active but the core code of the threads (except clock) is disabled while *Thread#n_active=0*. The thread clock is under execution in most of the times because its priority has the maximum value as compared to the rest of the threads. Thus, the clock (super thread) synchronizes the thread core code activation and execution. Initially, the core code of *thread#1* is activated. The execution time inside thread is controlled by the corresponding sleep method. The containing sleep periods inside threads are adjusted with the corresponding sleep periods in the thread clock. Thus, the execution time of each thread is fully controlled. After *Thread#1_limit* time the core code is deactivated. Thus, the rest of the threads (core codes) are executed in a controlled order and for a specific time period.

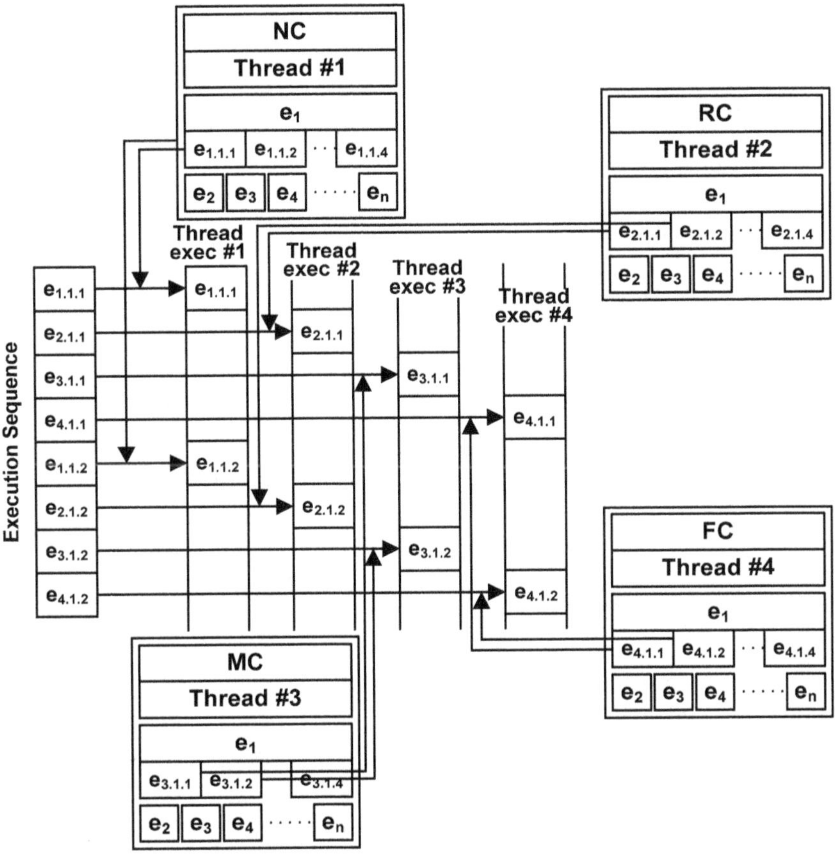


**Fig. 25** Sample of possible thread code interleaving (4 threads scenario)

```
Thread Clock
{
 While (active simulation time)
   {
       Thread#1_active=1;
       Sleep(Thread#1_limit);
       Thread#1_active=0;
       .
       .
       .
       Thread#n_active=1;
       Sleep(Thread#n_limit);
       Thread#n_active=0;
   }
}
```

**Fig. 26** Clock implementation

Figure 27 shows the thread implementation for a network agent.

```
Thread#1
{
  While (active simulation time)
    {
      While (Thread#1_active==0) {;}
      //PART_1A
      Sleep(Thread#1_Sleep)
      While (Thread#1_active==0) {;}
      //PART_1B
    }
}

Thread#n
{
  While (active simulation time)
    {
      While (Thread#n_active==0) {;}
      //PART_N1
      Sleep(Thread#n_Sleep)
      While (Thread#n_active==0) {;}
      //PART_N2
    }
}
```

**Fig. 27** Thread implementation

| **Scheduler** | **Threads** | | | | | | | | | | | |
|---|---|---|---|---|---|---|---|---|---|---|---|---|
| **JVM** (random sequence) | $T_2$ | $T_1$ | $T_1$ | $T_2$ | $T_1$ | $T_3$ | $T_2$ | $T_1$ | $T_3$ | $T_2$ | $T_1$ | $T_2$ |
| **Time Clock** (controlled sequence) | X | $T_1$ | $T_1$ | X | $T_1$ | $T_3$ | X | X | $T_3$ | $T_2$ | X | $T_2$ |

**Fig. 28** Real thread execution sequence under the JVM platform
Execution sequence: T1 (3ΔT), T3(2ΔT), T2(2ΔT),
X=Inactive thread code

Despite the random thread activation by the JVM, the proposed method defines and controls the activation period and execution sequence of the core code of the threads. Figure 28 shows that the core code of a thread is executed only when this thread is under processing by the JVM scheduler and at the same time is activated from the corresponding thread clock. The same figure also shows the wasted execution time periods that caused by the JVM unpredictability.

*Synchronization Issues*

As mentioned before, the simulation system uses a User Registry (UR) for keeping a detailed record of each connected MU. Due to that fact, the UR constitutes a shared resource area. When concurrent events take place, two or more threads try to access the UR at the "same time". An active thread can be pre-empted by the scheduler, when an access activity in a shared resource is not completed. While this thread is now pre-empted, another thread tries also to access the shared resource. Due to the given time slice to that thread by the scheduler, the access activity is completed. After the re-activation of the first thread, the semi-completed access activity of the first phase is now complete. If the above two threads try to access the same record in the UR, then the resulting data are incorrect depending on the thread switching. For the above reasons, the UR of the simulation system must be accessed through the synchronized method. This synchronization will prevent the shared resource area from simultaneous access by two or more threads. When a thread has locked an object (e.g. access method for UR), and is waiting for another thread to finish, while that other thread is waiting for the first thread to release that same object before it can finish then a deadlock occur.

When two or more threads try to access the shared resource area, in the most of the cases, they not refer to the same MU. Thus, the synchronization mechanism is necessary only in some cases. In order to avoid deadlocks, the following features must be designed and developed:

- Controlled thread switching (execution/interleaving)
- Control points where the UR will be accessed
- Conditional synchronization (synchronization only if needed)

Every time a thread tries to access UR for a specific MU, flag values are written in a local table that belongs to that thread. After the update of the local table, the thread scans all the local tables and decides to access the main UR through non-synchronized methods; otherwise, the access is completed through synchronized methods. Using this mechanism, high speed and data manipulation is achieved with no deadlocks.

*Limitations of the Multi-Threading Platform*

As mentioned previously, the JVM scheduler behavior cannot be predicted [37] and so a fully controlled scheduling mechanism must be applied (the proposed PQ-TDM). In other words, the execution sequence of the threads can not be guaranteed across all Java virtual machines [33].

The event execution sequence plays a major role for the network performance due to the fact that the events represent the modelled network services. An obvious approach for controlling threads is the exploitation of the build-in instructions of the JVM, *Thread.stop*, *Thread.suspend* and *Thread.resume*. Using *suspend* and *resume*, each thread can be activated or deactivated according to the scheduling needs (execution time slice). For example, a thread can be executed only for a specific time period between two successive suspend and resume instructions. When a thread accesses a critical section (shared resources via mutual exclusion) it holds the lock on the monitor for protecting this section. If this thread is suspended, no other thread can access the critical section (until the execution of the resume instruction for the suspended thread) and the system is driven to deadlock. For the above reason, suspend and resume instructions are deadlock-prone and are depreciated. On the other hand, the *Thread.stop* is inherently unsafe. When a thread is stopped (*ThreadDeath* exception), the corresponding monitor locks become available to other threads. A possible arbitrary behaviour may result if other threads operate on damaged objects (these objects maybe damaged due to inconsistent state when the protections of the initial monitors are released).

According to [42], the *thread.stop* instruction maybe replaced by simple code which modifies a variable to show if the selected thread must be stopped.

## 4 Experimental Model and Wireless Environment Validation

*A. Basic statistical metrics*

Blocking and dropping probability constitutes the most popular and applicable performance metrics for network behavior, especially for channel allocation and bandwidth management [43-51].

According to [52], blocking and dropping probabilities can be defined as follows:

- Call Blocking probability, $P_B$ = probability that a new call is denied access to the network.
- Call Dropping or Forced Termination probability, $P_D$ = probability that a call in progress is forced to terminate earlier.

The blocking probability $\mathbf{P_{blocking}}$ is calculated from the ratio

$$P_{blocking} = \frac{number\ of\ blocked\ calls}{number\ of\ calls} \tag{20}$$

The dropping probability $\mathbf{P_{fc}}$ is calculated from the ratio

$$P_{fc} = \frac{number\ of\ forced\ calls}{number\ of\ calls - number\ of\ blocked\ calls} \tag{21}$$

*B. Wireless platform validation*

The proposed models are evaluated through an implemented wireless network environment. This environment has been built on the known theoretical components for radio propagation, signal measurements and cellular network operation. The presented simulation results in this study derive from the proposed models that are implemented in the above wireless network environment. Thus, the validation of this environment is necessary in order to prove the correctness of the results. The validation procedure can be found also in [45] and consists of two phases which are:

- Monte Carlo simulations
- pdf (probability density function) evaluation based on theoretical solutions

The transmitted signal suffers from the two most important factors within the wireless environment which are path loss (distance attenuation) and shadowing (obstacles in the signal path). On the other hand, the CNR between BS and MU and the total interference from other co-channel MUs affect the received signal quality (fig. 29). for this example, $n=3$.

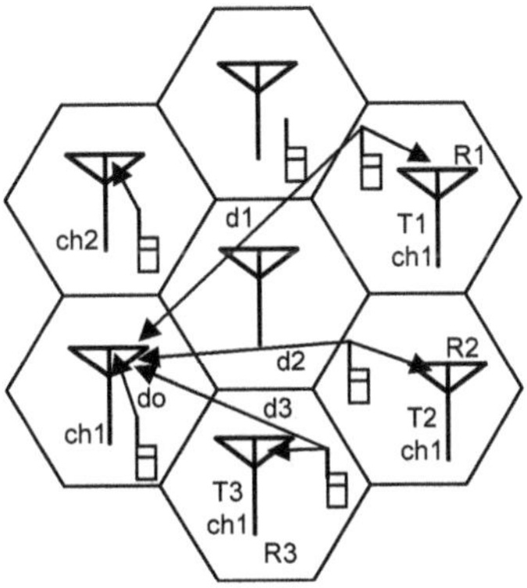


**Fig. 29** Co-Channel Interference

The CNIR for the MU $T_0$ can be derived from the following formula:

$$R_{cni} = \frac{AP_0 d_0^{-\alpha} 10^{\frac{\xi_0}{10}}}{N + \sum_{i-1}^{n} AP_i d_i^{-\alpha} 10^{\frac{\xi_i}{10}}} \tag{22}$$

where $A$ is a proportional coefficient, $P_i$ is the transmitted power of the MU $T_i$, $d_i$ is the distance between MU $T_i$ and BS $R_0$, $\xi_i$ is the shadowing distortion between $T_i$ and $R_0$ and $\alpha$ is a path loss factor. The average received signal strength decays as a power low of the distance transmitter-receiver. Path loss decays the transmitted signal with a factor α (eq. 22). For an urban area $\alpha=3.5$ [27].

Initially the distance

$$h = \sqrt{(x_2 - x_1)^2 + (y_2 - y_1)^2} \tag{23}$$

is calculated (fig. 30). The path loss factor between these points is $h^{-\alpha}$. The shadowing is subject to log-normal distribution with σ as standard deviation. Shadowing corresponds to $\xi_i$ of equation 22. The shadow attenuation [27] is obtained as follows:

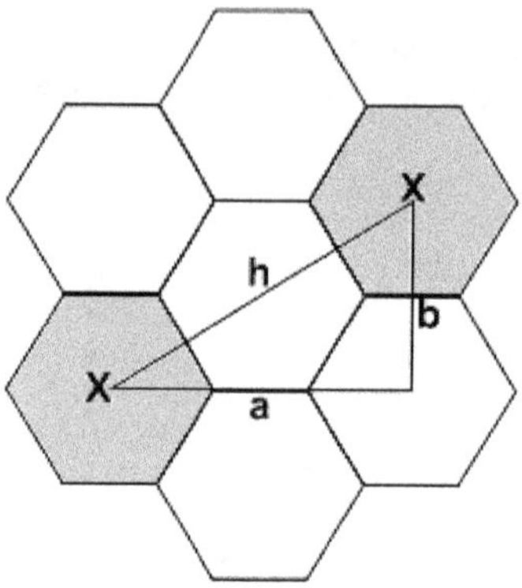


**Fig. 30** Distance calculation

$$sh = 10^{\frac{\sigma \cdot n}{10}} \tag{24}$$

where σ is the standard deviation of shadowing and n is a number from the normal distribution. Using the shadow attenuation and distance between MU and BS, the distance attenuation *dw* can be derived. The CNR is calculated between MU and BS [27]

$$cn = 10^{\frac{cnedge}{10}} \cdot dw \tag{25}$$

Now, the CNIR can be calculated as:

$$CNIR = \frac{C}{N+I} = \frac{1}{\frac{N}{C} + \frac{I}{C}} = \frac{1}{\left(\frac{C}{N}\right)^{-1} + \left(\frac{C}{I}\right)^{-1}} \tag{26}$$

The CNIR can be calculated by knowing the ratios C/N and C/I. The ratio C/N is already known from (25) and the C/I is determined from the ratio *dw/uw*, (*dw* for signal between MU and BS, *uw* for the total signal attenuation caused by other co-channel MUs). Appendix E contains information regarding the total CNIR calculation.

*Network Performance Validation*

The most known statistical metrics for the cellular network performance evaluation are blocking and dropping probabilities. Blocking probability represents the blocked calls, while dropping represents the unsuccessful channel reallocation for an ongoing call. The dropping probability is strongly connected to $R_{cni}$ (eq. 22), because when this ratio is not above the accepted threshold and the network can not allocate an appropriate channel, the call is dropped.

On the other hand, blocking probability can be theoretically calculated. If the received power of each MU is high enough, it is assumed that the interference from other MUs can be ignored. The theoretical formula [28] is as follows:

$$P_{blocking\_theoretical} = \frac{\binom{n-1}{s}(vh)^s}{\sum_{i=0}^{s}\binom{n-1}{i}(vh)^i} \tag{27}$$

where $n$ is the number of users, $s$ is the number of channels, $v$ is the average call arrival rate (for no connected MU) and $h$ is the average call holding time.

Equation 27 shows only the basic relation between channels and users and does not take in to consideration critical factors that affect the blocking probability such as traffic conditions, service type, channel allocation strategy, etc. Figure 31 shows the theoretical blocking probability that derives from eq. 27 as compared to simulated blocking probability. The simulated probability has been generated from the large scale network based on the UDCA and network services.

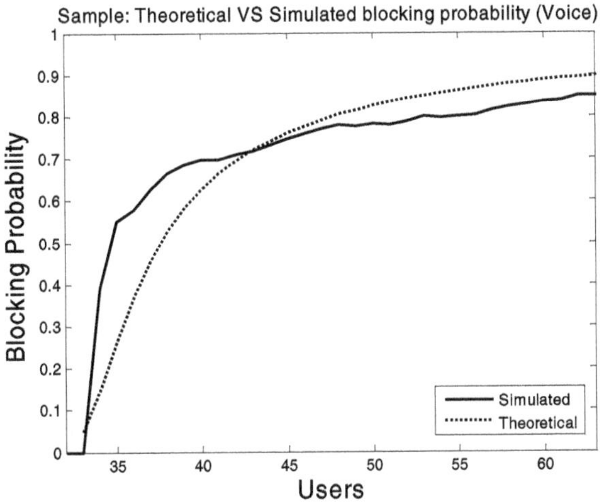


**Fig. 31** Theoretical blocking probability versus simulated

## 5 Simulation Results

*Large Scale Model*

The simulated cellular network consists of 490 cells distributed in ten super areas. Each super area contains seven clusters of 7 cells. Based on the used propagation model the radius of a single cell is about 1km. Based on this reference distance, the real size of the simulated cellular network is 46km (x-width) x 33km (y-width). Within the modelled network, 28 network agents have been totally used inside super area clusters (four network agents-thread per internal cluster, 4 agents x 7 clusters = 28 agents). During simulation process the model execution is sequential between the ten super areas. Thus, the sequential behaviour is restricted only in passing from one super area to another. Figure 32 shows the large scale cellular network environment.

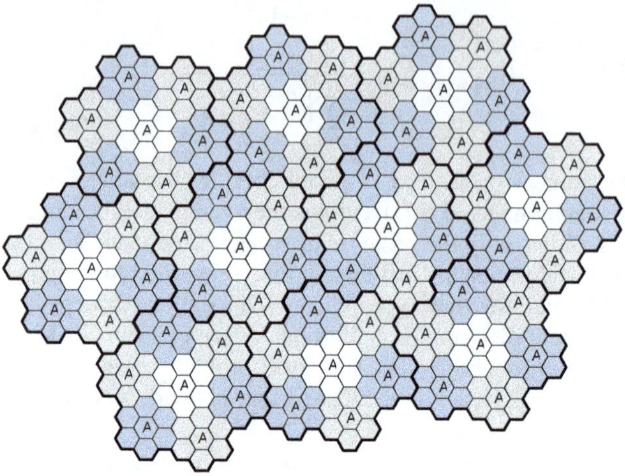

**Fig. 32** Large scale cellular network environment

Figures 33 and 34 show the network model behavior regarding the application of the proposed approach based on multi agent technology. In the first model (MT PQ-TDM) only a multi threading based approach that uses the PQ-TDM scheduler is used. This model is compared with the corresponding application of the multi agent approach in a second model which is also based on the multi threading and the PQ-TDM scheduler.

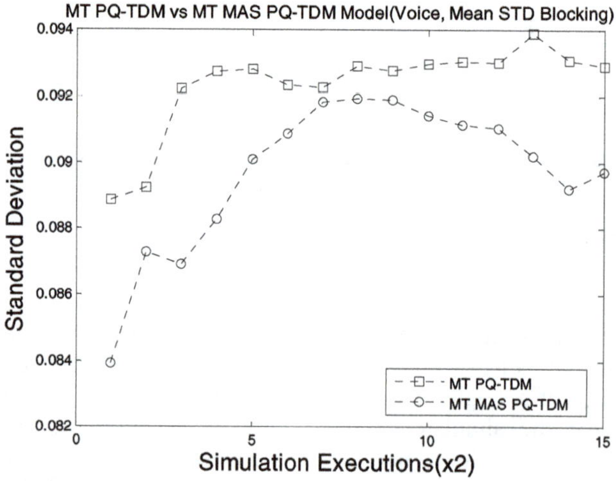


**Fig. 33** Multi Threaded (MT) PQ-TDM Versus Multi Threaded Multi Agent (MT MAS) PQ-TDM model (Voice, Mean STD Blocking Probability, Classical DCA)

Due to the dynamic adaptability of networks agents in the current network performance the resulted improvement in terms of stability is expected (fig. 5.4 and 5.5).

Simulation results (fig. 33 and 34) show how the multi agent based simulation model (Multi Threading Multi Agent System - MT MAS PQ-TDM) which uses an efficient negotiation strategy between network agents and is based on the

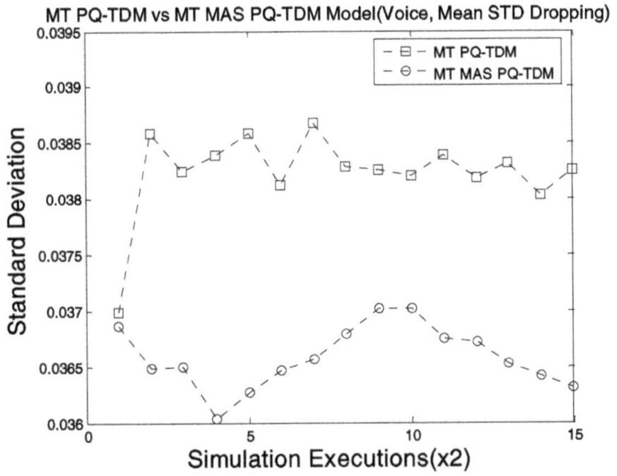


**Fig. 34** Multi Threaded (MT) PQ-TDM Versus Multi Threaded Multi Agent (MT MAS) PQ-TDM model (Voice, Mean STD Dropping Probability, Classical DCA)

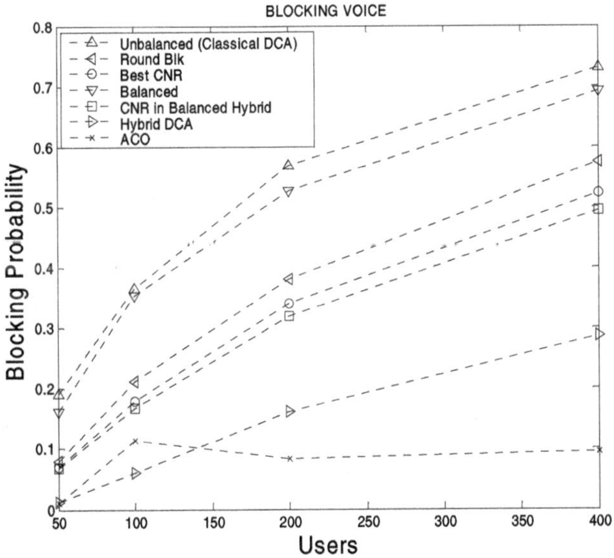


**Fig. 35** ACO algorithm blocking probability compared to other tested DCA variations

PQ-TDM scheduling mechanism performs better as compared to the single multi threading model (MT PQ-TDM) which does not use any negotiation. These figures also show that the multi agent based model has higher adaptability to the dynamically changing network conditions.

When negotiation between NCA and RCA exists, more balanced network performance is achieved in terms of blocking and dropping. When the current blocking or dropping exceeds a predefined ratio the result of the negotiation is the redefinition of the agents' priority in order to keep balanced performance conditions within the network. When a cooperative negotiation takes place between the network agents (one agent helps the other agent), the resulted network model behavior can be controlled more easily.

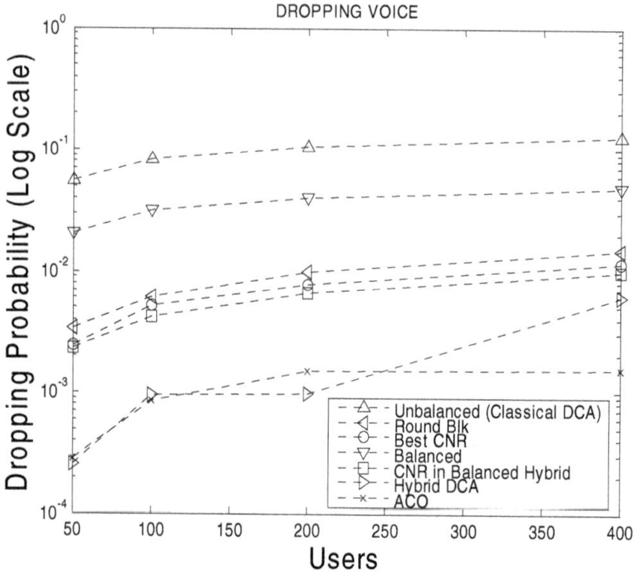


**Fig. 36** ACO algorithm dropping probability compared to other tested DCA variations

MRMRMRMRMRMRMRMRMRMRMRMFNRNRNR
NRNRNRNRNRNRNRNRNRNRNRNRNRFMRMRM
FNMNMNMNMNMNMNMNMNMNMNMNMNFRN
RNRNRNRNRFMRMRMRMRMRMRMRMRMRMR
MRMRMFNMNMNMNMNMNFRNRNRNRNRNRNR
NRNRNRNRNRNRFMRMRMRMRMRMFNMNMNM
NMNMNMNMNMNMNMNMNMNMNFRNRNRNRNRNR
FMRMRMRMRMRMRMRMRMRMRMRMRMFNMNM

**Fig. 37** Thread execution sequence, Default JVM, equal priorities(N=NC, R=RC, M=MC, F=FC)

In a large scale environment and when the traffic and user load is rising, the conventional channel allocation algorithms can not improve significantly the corresponding network performance. When wireless and traffic conditions are changing rapidly, then the adaptation of the network to the current conditions becomes a difficult goal. This adaptation can be achieved through more intelligent channel allocation approaches where new channels are given only in neighboring areas with better conditions (channel availability, wireless conditions –CNIR-).

```
FFFNRMFFFNRMFFFNRMFFFNRM
FFFRMNFFFRMNFFFRMNFFFRMN
FFFMRNFFFMRNFFFMRNFFFMRN
FFFRNMFFFRNMFFFRNMFFFRNM
FFFMNRFFFMNRFFFMNRFFFMNR
...
```

**Fig. 38** Network event execution inside cluster

```
111111......111111{NC:1031,332}
{RC:1031,0}
{MC:1031,0}
{FC:1031,0}
--------------------------------
111111......111111{NC:1235,14575}
{RC:1031,0}
{MC:1031,0}
{FC:1031,0}
--------------------------------
111111......111111{NC:1219,21591}
222222......222222{RC:1093,17704}
333333......333333{MC:1000,21929}
444444......444444{FC:1188,15734}
```

**Fig. 39** MODEL 2 (see table 6.4),Thread execution, Symmetrical Sleep, Clock priority=1, Network thread priorities=10

Figure 39 shows that some threads are not executed due to the fact that the clock is rarely activated by the default scheduler when minimum priority is applied and so the required thread control is not possible. Moreover, the thread activation time lead to different number of core code execution times for each thread (e.g. NC:1219=real thread core code active time in ms, 21591 executions, RC:1093=real thread core code active time in ms and 17704 executions)

```
111111......111111{NC:1000,70}
222222......222222{RC:1000,72}
333333......333333{MC:1000,72}
444444......444444{FC:1000,72}
--------------------------------
111111......111111{NC:1000,72}
222222......222222{RC:1000,72}
333333......333333{MC:1000,72}
444444......444444{FC:1000,72}
--------------------------------
111111......111111{NC:1000,72}
222222......222222{RC:1000,72}
333333......333333{MC:1000,72}
444444......444444{FC:1000,72}
```

**Fig. 40** MODEL 3 (see table 6.4),Thread execution, Symmetrical Sleep, Clock priority=10 (super thread), Network thread priorities=1

Figure 40 shows that the thread core code active time can be controlled and also the number of executions based on the proposed approach.

## 6 Conclusions

The efficiency of resource management schemes is a major issue in wireless communication networks. Channel allocation schemes should not be static due to the dynamically changing traffic conditions and network performance requirements. Thus, more sophisticated models need to be designed. This paper has investigated a novel strategy for implementing the Ant Colony Optimization (ACO) algorithm in the multi agent system framework for efficiently modelling resource management in cellular communication systems as well as for improving channel allocation results enriching decision making strategies for adaptable user needs for network services, supporting concurrency. Simulation results for generic large scale cellular systems presented in the experimental study show quite promising and illustrate the effectiveness of the proposed approach. However, more extensive evaluations and comparisons are needed in real world cases to clearly support the proposed methodology in order to incorporate the suggested schemes in real world simulators of cellular communications.

The Multi Agent concept gives the opportunity to adapt the network services to current user needs and network behavior. The sequence of channel allocation and call servicing based on defined priorities affect the network performance. Applying the multi agent concept combined with several herein proposed novel cooperative / competitive negotiation schemes to channel allocation, simulation model performance can be significantly improved. Further such schemes could be devised and explored.

On the other hand, in a large scale wireless network, even small differences in the channel allocation procedure may largely affect network performance. Thus, more efficient channel allocation schemes based on computational intelligence techniques (especially ant colony optimization and swarm intelligence) must be further investigated. Finally, although multi threading technology constitutes a valuable tool for alternative implementation of simulation models which support concurrent events, however, on the other hand, the existence of serious drawbacks, especially in large scale cases, (e.g. deadlocks, synchronization, etc) based on specific Java technology and platform constraints must be taken into consideration. New multithreading implementation strategies should be defined to facilitate large scale wireless network simulation and this would be an important and open research issue for the simulation community, not only in communications

## References

[1] Zhang, M., Yum, T.S.: Comparisons of Channel Assignment Strategies in Cellular Mobile Telephone Systems. IEEE Transactions on Vehicular Technology 38(4), 211–215 (1989)

[2] Lai, W.K., Coghill, G.C.: Channel Assignment through Evolutionary Optimization. IEEE Transactions on Vehicular Technology 45(1), 91–96 (1996)

[3] MacDonald, V.H.: The cellular Concepts. The Bell System Technical Journal 58, 15–42 (1979)

[4] Elnoubi, S.M., Singh, R., Gupta, S.C.: A New Frequency Channel Assignment Algorithm in High Capacity Mobile Communication Systems. IEEE Transactions on Vehicular Technology 21(3), 125–131 (1982)

[5] Xu, Z., Mirchandani, P.B.: Virtually Fixed Channel Assignment for Cellular Radio-Telephone Systems: A Model and Evaluation. In: IEEE International Conference on Communications, ICC 1992, Chicago, vol. 2, pp. 1037–1041 (1982)

[6] Cimini, L.J., Foschini, G.J.: Distributed Algorithms for Dynamic Channel Allocation in Microcellular Systems. In: IEEE Vehicular Technology Conference, pp. 641–644 (1992)

[7] Cox, D.C., Reudink, D.O.: Increasing Channel Occupancy in Large Scale Mobile Radio Systems: Dynamic Channel Reassignment. IEEE Transanctions on Vehicular Technology 22, 218–222 (1973)

[8] Del Re, E., Fantacci, R., Giambene, G.: A Dynamic Channel Allocation Technique based on Hopfield Neural Networks. IEEE Transanctions on Vehicular Technology 45(1), 26–32 (1996)

[9] Sivarajan, K.N., McEliece, R.J., Ketchum, J.W.: Dynamic Channel Assignment in Cellular Radio. In: IEEE 40th Vehicular Technology Conference, pp. 631–637 (1990)

[10] Kahwa, T.J., Georgans, N.D.: A Hybrid Channel Assignment Schemes in Large-Scale, Cellular Structured Mobile Communication Systems. IEEE Transactions on Communications 26, 432–438 (1978)

[11] Hayzelden, A., Bigham, J.: Software Agents for Future Communications Systems. Springer, Berlin (1999)

[12] Iraqi, Y., Boutaba, R.: A Multi-agent System for Resource Management in Wireless Mobile Multimedia Networks. In: Ambler, A.P., Calo, S.B., Kar, G. (eds.) DSOM 2000. LNCS, vol. 1960, pp. 218–229. Springer, Heidelberg (2000)

[13] Bigham, J., Du, L.: Cooperative Negotiation in a MultiAgent System for RealTime Load Balancing of a Mobile Cellular Network. In: AAMAS 2003, pp. 14–18 (July 2003)

[14] Beasley, D., Bull, D.R., Martin, R.R.: An overview of Genetic Algorithms: Part I, Fundamentals. University Computing 15(2), 58–69 (1993)

[15] Holland, J.H.: Adaptation in Natural and Artificial Systems. MIT Press, Cambridge (1975)

[16] Goldberg, D.E.: Genetic Algorithms in Search, Optimization and Machine Learning. Addison Wesley, New York (1989)

[17] Lima, M.A.C., Araujo, A.F.R., Cesar, A.C.: Dynamic channel assignment in mobile communications based on genetic algorithms. Personal, Indoor and Mobile Radio Communications (2002)

[18] Yener, A., Rose, C.: Genetic Algorithms Applied to Cellular Call Admission Problem: Local Policies. IEEE Vehicular Technology 46(1), 72–79 (1997)

[19] Kendall, G., Mohamad, M.: Solving the Fixed Channel Assignment Problem in Cellular Communications Using An Adaptive Local Search. In: Burke, E.K., Trick, M.A. (eds.) PATAT 2004. LNCS, vol. 3616, pp. 219–231. Springer, Heidelberg (2004)

[20] Kim, J.S., Park, S., Dowd, P., Nasrabadi, N.: Channel Assignment in Cellullar Radio using Genetic Algorithms. Wireless Personal Communications 3(3), (14), 273–286 (1996)

[21] Wang, L., Li, S., Lay, S.C., Yu, W.H., Wan, C.: Genetic Algorithms for Optimal Channel Assignments in Mobile Communications. Neural Network World 12(6), 599–619 (2002)

[22] Bonabeau, E., Dorigo, M., Theraulaz, G.: Swarm Intelligence: From Natural to Artificial Systems. Artificial Life 7, 315–319 (2001)

[23] Eberhart, R.C., Kennedy, J.: A New Optimizer Using Particle Swarm Theory. Presented at the Sixth International Symposium on Micro Machine and Human Science, Nagoya, Japan, pp. 39–43 (1995)

[24] Colorni, A., Dorigo, M., Maniezzo, V.: Distributed optimization by ant colonies. Presented at the First European Conference on Artificial Life (ECAL 1991), Paris, France, pp. 134–142 (1991)

[25] Bonabeau, E., Henaux, F., Guerin, S., Snyers, D., Kuntz, P., Theraulaz, G.: Routing in Telecommunications Networks with 'Smart' Ant-like Agents, Intelligent Agents for Telecommunications Applications. Frontiers in Artificial Intelligence & Applications, vol. 36 (1998)

[26] Bundgaard, M., Damgaard, T.C., Jacob, F.D., Winther, W.: Ant Routing System, IT University of Copenhagen (2002)

[27] Rappaport, T.S.: Wireless Communications Principles and Practice. Prentice-Hall, Englewood Cliffs (2002)

[28] Harada, H., Prasad, R.: Simulation and Software radio for mobile communications. Artech House (2002)

[29] Molisch, A.F.: Wireless Communications. IEEE Press/Wiley (2005)

[30] Jeffrey, G.A., Arunabha, G., Rias, M.: Fundamentals of WiMAX. Prentice-Hall, Englewood Cliffs (2007)

[31] Papazoglou, P.M., Karras, D.A., Papademetriou, R.C.: High Performance Novel Hybrid DCA algorithms for efficient Channel Allocation in Cellular Communications modelled and evaluated through a Java Simulation System. WSEAS Transactions on Communications 5(11), 2078–2085 (2006)
[32] http://www.tiobe.com/tpci.htm
[33] Oaks, S., Wong, H.: Java Threads, 3rd edn. O'Reilly, Sebastopol (2004)
[34] Kramer, J.M.: Concurrency: State Models & Java Programs, 2nd edn. John Wiley & Sons, Chichester (2006)
[35] Lindsey, C.S., Tolliver, J.S., Lindblad, T.: An Introduction to Scientific and Technical Computing with java. Cambridge University Press, Cambridge (2005)
[36] Pidd, M., Cassel, R.A.: Three phase simulation in java. In: Proceedings of the 1998 Winter Simulation Conference (1998)
[37] Mengistu, D., Lundberg, L., Davidsson, P.: Performance Prediction of Multi-Agent Based Simulation Applications on the Grid. International journal of intelligent technology 2(3) (2007) ISSN 1305-6417
[38] Papazoglou, P.M., Karras, D.A., Papademetriou, R.C.: On Integrated Ant Colony Optimization Strategies for Improved Channel Allocation in Large Scale Wireless Communications. In: Proceedings of the 10th WSEAS International Conference on Mathematical Methods, Computational Techniques and Intelligent Systems (MAMECTIS) (March 2008)
[39] Papazoglou, P.M., Karras, D.A., Papademetriou, R.C.: A Multi-agent Simulation Model for Wireless Communications involving an improved agent negotiation scheme based on real time event scheduling mechanisms. In: IEEE European Modelling Symposium EMS (June 2008)
[40] Papazoglou, P.M., Karras, D.A., Papademetriou, R.C.: On the Implementation of Ant Colony Optimization Scheme for Improved Channel Allocation in Wireless Communications. In: IEEE International Conference on Intelligent Systems, IS 2008 (2008)
[41] Papazoglou, P.M., Karras, D.A., Papademetriou, R.C.: An Efficient Scheduling Mechanism for Simulating Concurrent Events in Wireless Communications Based on an Improved Priority Queue (PQ) TDM Layered Multi-Threading Approach. WSEAS Transactions on Communications 7(3) (September 2008)
[42] http://java.sun.com/j2se/1.4.2/docs/guide/misc/threadPrimitiveDeprecation.html
[43] Katzela, I., Naghshineh, M.: Channel assignment schemes for cellular mobile telecommunication systems: A comprehensive survey. IEEE Personal Communications, 10–31 (1996)
[44] Wong, S.H.: Channel Allocation for Broadband Fixed Wireless Access Networks. Unpublished doctorate dissertation, University of Cambridge (2003)
[45] Haas, H.: Interference analysis of and dynamic channel assignment algorithms in TD–CDMA/TDD systems. Unpublished Doctoral dissertation, University of Edinburg (2000)
[46] Salgado, H., Sirbu, M., Peha, J.: Spectrum Sharing Through Dynamic Channel Assignment For Open Access To Personal Communications Services. In: Proc. of IEEE Intl. Communications Conference (ICC), pp. 417–422 (1995)
[47] Godara, L.C.: Applications of Antenna Arrays to Mobile Communications, Part I: Performance Improvement, Feasibility, and System Considerations. Proceedings of the IEEE 85(7) (1997)

[48] Tripathi, N.D., Jeffrey, N., Reed, H., VanLandingham, H.F.: Handoff in Cellular Systems. IEEE Personal Communications (1998)

[49] Hollos, D., Karl, H., Wolisz, A.: Regionalizing Global Optimization Algorithms to Improve the Operation of Large Ad Hoc Networks. In: Proceedings of the IEEE Wireless Communications and Networking Conference, Atlanta, Georgia, USA (2004)

[50] Cheng, M., Li, Y., Du, D.Z.: Combinatorial Optimization in Communication Networks. Kluwer Academic Publishers, Dordrecht (2005)

[51] Lee, W.C., Lee, J., Huff, K.: On Simulation Modeling of Information Dissemination Systems in Mobile Environments. In: Leong, H.V., Li, B., Lee, W.-C., Yin, L. (eds.) MDA 1999. LNCS, vol. 1748, pp. 45–57. Springer, Heidelberg (1999)

[52] Cherriman, P., Romiti, F., Hanzo, L.: Channel Allocation for Third-generation Mobile Radio Systems. In: ACTS 1998, vol. 1, pp. 255–261 (1998)

# Author Index

www.ingramcontent.com/pod-product-compliance
Lightning Source LLC
Chambersburg PA
CBHW071351220926
51613CB00008B/44
* 9 7 8 3 6 4 2 1 3 4 2 7 2 *